Dallas, Texas

# GOODNESS-OF-FIT TECHNIQUES

# STATISTICS: Textbooks and Monographs

*A SERIES EDITED BY*

### D. B. OWEN, Coordinating Editor

*Department of Statistics*
*Southern Methodist University*
*Dallas, Texas*

## OTHER VOLUMES IN PREPARATION

# GOODNESS-OF-FIT TECHNIQUES

*edited by*

## Ralph B. D'Agostino

Department of Mathematics
Boston University
Boston, Massachusetts

## Michael A. Stephens

Department of Mathematics and Statistics
Simon Fraser University
Burnaby, British Columbia, Canada

CRC Press
Taylor & Francis Group
Boca Raton  London  New York

CRC Press is an imprint of the
Taylor & Francis Group, an **informa** business

Published in 1986 by CRC Press
Taylor & Francis Group
6000 Broken Sound Parkway NW, Suite 300
Boca Raton, FL 33487-2742

First issued in paperback 2020

© 1986 Taylor & Francis Group, LLC
CRC Press is an imprint of Taylor & Francis Group, an Informa business

No claim to original U.S. Government works

ISBN-13: 978-0-367-58034-6 (pbk)
ISBN 13: 978-0-8247-7487-5 (hbk)

**Visit the Taylor & Francis Web site at**
**http://www.taylorandfrancis.com**

**and the CRC Press Web site at**
**http://www.crcpress.com**

Library of Congress catalog number: 86-4571

---

**Library of Congress Cataloging-in-Publication Data**

---

Catalog record is available from the Library of Congress

---

To Egon S. Pearson
who helped us both

# Preface

From the earliest days of statistics, statisticians have begun their analysis by proposing a distribution for their observations and then, perhaps with somewhat less enthusiasm, have checked on whether this distribution is true. Thus over the years a vast number of test procedures have appeared, and the study of these procedures has come to be known as goodness-of-fit. When several of the present authors met at the Annual Meeting of the American Statistical Association in Boston in 1976 and proposed writing a book on goodness-of-fit techniques, we certainly did not foresee the magnitude of the task ahead. Quite early on we asked Professor E. S. Pearson if he would join us. He declined and stated his view that the time was not yet ripe for a book on the subject. As we, nevertheless, have slowly written it, it has often appeared that his assessment was correct. As fast as we have tried to survey what we know, with every issue the journals produce new papers with new techniques and new information.

However, many colleagues have told us that the time is ready for a major summary of the literature, and for some sorting and sifting to take place. This we have tried to do. The emphasis of this book was determined by the writers to be mostly on the practical side. The intent is to give a survey of the leading methods of testing fit, to provide tables where necessary to make the tests available, to make (where possible) some assessment of the comparative merits of different test procedures, and finally to supply numerical examples to aid in understanding the techniques.

This applied emphasis has led to some difficult decisions. Many goodness-of-fit techniques are supported by elegant mathematics involving combinatorics, analysis, and geometric probability, mostly arising in the distribution theory, both small-sample and asymptotic, or in examining power and efficiency. Furthermore, there are many unsolved problems, especially in

v

discovering the relationships between different approaches, which would require sophisticated mathematics to resolve. However, for the book to be of manageable size, mathematical details have had to be held to the minimum necessary to describe clearly the various techniques. References to fuller mathematical treatments are given throughout the book. We also leave out tests comparing several samples. Although these are often closely related to the one-sample tests of this volume, they are not usually classified as goodness-of-fit tests. Including them here would have made the book too large.

In arranging the book it was necessary to decide whether to collect together all methods of testing for specific famous distributions, such as the normal or the exponential, or whether to group tests according to techniques such as chi-squared tests, empirical distribution function tests, or tests based on probability plotting. In the end, and perhaps because of the fact that many authors were involved, we reached the inevitable compromise to try to do both. In order to make chapters as complete as possible, there is some necessary overlap.

There is also some imbalance with respect to tables. Some major, well-established techniques require quite small tables—surely an attractive feature—while many new and unproven techniques need fairly extensive tables, often based on Monte Carlo studies. Where we have judged the techniques important, either as new methods or to complete a group of existing methods, we have included the necessary tables and, in fact, have considerably extended some of those in the literature. By doing so we hope not only to make the newer techniques available for practical use but also to make the book useful for further research in making the comparisons between methods which we feel are still necessary. On the other hand, we at times only refer to tables for some techniques which have never appeared to win much favor.

As we have surveyed the tests available, it has become clear that much work remains to be done. It sometimes seems that new test statistics, even for standard problems, are invented every day. In goodness-of-fit, where there is a wide range of problems and almost never a best solution, this appears to be easy to do. However, the simple invention of a test statistic is surely not enough. We suggest that, to gain acceptance, new methods should have a clear motivation, be easily understood by the practical statistician, and be well documented. Where new tables are necessary, they should be comprehensive. The day may well come when computer algorithms will replace tables, but for most statisticians this day has not yet arrived. Also new methods should be compared with the array of procedures which often already exist.

Finally, of course, this book is inevitably a reflection of the interests of the editors and the contributors. Although we have tried to cast our net wide, some special techniques for testing fit may seem to have received too much attention, while others have been neglected. For the latter cases it is, we believe, mostly because the practical aspects are not yet sufficiently developed. We have attempted throughout the book to at least summarize the present state of knowledge for these. We hope that by drawing attention to them, we can again encourage further research.

We also acknowledge gratefully Ms. Sylvia Holmes and Mr. Thomas Orowan of Simon Fraser University and Boston University, respectively, for much help with the typing of the manuscript, and the staff of Marcel Dekker, Inc., for their patient editorial work with this volume.

RALPH B. D'AGOSTINO
MICHAEL A. STEPHENS

# Acknowledgments

We acknowledge with thanks the permission of numerous publishers, editors, and authors to reprint their tables and graphs. We thank the Addison-Wesley publishers of Reading, Mass., for permission to publish the normal distribution table (from Mosteller, Rourke, and Thomas, Probability with Statistical Applications [1982; Table 1 (adapted)] and the chi-square table [from D. B. Owen, Handbook of Statistical Tables (1962)]. These tables are in the Appendix. We thank the editor of the Annals of Statistic and the Institute of Mathematical Statistics for permission to reprint an adaptation of a table from Lewis (1961). This table is in Chapter 4.

We thank the editor of Biometrika and the Biometrika Trustees for permission to reprint tables and graphs from a number of articles for various chapters in this book. The chapters, authors, and publication years are as follows: Chapter 2, Harter (1961); Chapter 4, Lockhart and Stephens (1985), Pettitt and Stephens (1976), Pettitt (1976, 1977), and Stephens (1977, 1979); Chapter 5, Shapiro and Wilk (1965); Chapter 7, Bowman and Shenton (1975); Chapter 8, Stephens (1966); and Chapter 9, D'Agostino (1971, 1972), D'Agostino and Tietjen (1971, 1973), and D'Agostino and Pearson (1973). Also in Chapter 4 is an adaptation of Table 54 from Biometrika Tables for Statisticians, Volume 2 (1972).

We thank the editor of Technometrics and the American Statistical Association for permission to reprint tables and data: Chapter 4, Proschen (1963), Koziol and Byar (1975), and Pettitt and Stephens (1977); Chapter 5, Shapiro and Wilk (1972), and Chapter 12, Grubbs (1965), Tietjen and Moore (1972), Stefansky (1972), Lund (1975), Galpin and Hawkin (1981), and Jain (1981). We thank the editor of the Journal of the American Statistical Association and the American Statistical Association for permission to reprint tables from Stephens (1974), and Chandra, Singpurwalla, and Stephens (1981) for Chapter 4.

We thank the editors of the <u>Journal of the Royal Statistical Society</u> and the Royal Statistical Society for permission to reprint tables from Stephens (1970), Pettitt (1977), Burrows (1979), and Stephens (1981) in Chapters 4 and 8. We thank the editor of <u>Communications in Statistics</u> and Marcel Dekker, Inc., for permission to reprint tables from Miller and Quesenberry (1979), and Solomon and Stephens (1983) in Chapter 8. Finally we thank the editors of the <u>Journal of Statistical Computation and Simulation</u> and Gordon and Breach Science Publishers, Inc., for permission to reprint a table from Quesenberry and Miller (1977).

# Contents

# Contributors

K. O. BOWMAN Engineering Physics and Mathematics Division, Oak Ridge National Laboratory, Oak Ridge, Tennessee

RALPH B. D'AGOSTINO Department of Mathematics, Boston University, Boston, Massachusetts

JOHN R. MICHAEL* Quality Assurance Center, Bell Telephone Laboratories, Holmdel, New Jersey

DAVID S. MOORE Department of Statistics, Purdue University, West Lafayette, Indiana

C. P. QUESENBERRY Department of Statistics, North Carolina State University, Raleigh, North Carolina

WILLIAM R. SCHUCANY Department of Statistics, Southern Methodist University, Dallas, Texas

LEONARD R. SHENTON† Computer Center, University of Georgia, Athens, Georgia

MICHAEL A. STEPHENS Department of Mathematics and Statistics, Simon Fraser University, Burnaby, British Columbia, Canada

---

*Current affiliation: Westat, Inc., Rockville, Maryland
†Current affiliation: Advanced Computational Methods Center, University of Georgia, Athens, Georgia

GARY L. TIETJEN  Analysis and Assessment Division, Los Alamos
    National Laboratory, Los Alamos, New Mexico

# GOODNESS-OF-FIT TECHNIQUES

# 1

# Overview

Ralph B. D'Agostino  Boston University, Boston, Massachusetts

Michael A. Stephens  Simon Fraser University, Burnaby, B.C., Canada

## 1.1  GOODNESS-OF-FIT TECHNIQUES

This book is devoted to the presentation and discussion of goodness-of-fit techniques. By these we mean methods of examining how well a sample of data agrees with a given distribution as its population. The techniques discussed are almost entirely for univariate data, for which there is a vast literature; methods for multivariate data are much less well developed.

In the formal framework of hypothesis testing the null hypothesis $H_0$ is that a given random variable x follows a stated probability law F(x) (for example, the normal distribution or the Weibull distribution); the random variable may come from a process which is under investigation. The goodness-of-fit techniques applied to test $H_0$ are based on measuring in some way the conformity of the sample data (a set of x-values) to the hypothesized distribution, or, equivalently, its discrepancy from it. The techniques usually give formal statistical tests and the measures of consistency or of discrepancy are test statistics.

The null hypothesis $H_0$ can be a simple hypothesis, when F(x) is specified completely, for example, normal with mean $\mu = 100$ and standard deviation $\sigma = 10$; or $H_0$ can give an incomplete specification and will then be a composite hypothesis, for example, when it states only that F(x) is normal with unspecified $\mu$ and $\sigma$.

In most applications of goodness-of-fit techniques, the alternative hypothesis $H_1$ is composite—it gives little or no information on the distribution of the data, and simply states that $H_0$ is false. The major focus is on the measure of agreement of the data with the null hypothesis; in fact, it is usually hoped to accept that $H_0$ is true.

1

There are several reasons for this. First, the distribution of sample data may throw light on the process that generated the data; if a suggested model for the process is correct, the sample data follow a specific distribution, which can be tested. Also, parameters of the distribution may be connected with important parameters in describing the basic model. Secondly, knowledge of the distribution of data allows for application of standard statistical testing and estimation procedures. For example, if the data follow a normal distribution, inferences concerning the means and variances can be made using t tests, analyses of variances, and F tests; similarly, if the residuals after fitting a regression model are normal, tests may be made on the model parameters. Estimation procedures such as the calculation of confidence intervals, tolerance intervals, and prediction intervals, often depend strongly on the underlying distribution. Finally, when a distribution can be assumed, extreme tail percentiles, which are needed, for example, in environmental work, can be computed.

The fact that it is usually hoped to accept the null hypothesis and proceed with other analyses as if it were true, sets goodness-of-fit testing apart from most statistical testing procedures. In many testing situations it is rejection of the null hypothesis which appears to prove a point. This might be so, for example, in a test for no treatment effects in a factorial analysis—rejection of $H_0$ indicates one or more treatments to be better than others. Even when one would like to accept a null hypothesis—for example, in a test for no interaction in the above factorial analysis—the statistical test is usually clear and the only problem is with the level of significance. In a test of fit, where the alternative is very vague, the appropriate statistical test will often be by no means clear and no general theory of Neyman-Pearson type appears applicable in these situations. Thus many different, sometimes elaborate, procedures have been generated to test the same null hypothesis, and the ideas and motivations behind these are diverse. Even when concepts such as statistical power of the procedures are considered it rarely happens that one testing procedure emerges as superior.

It may happen that the alternative hypothesis has some specification, although it could be incomplete; for example, an alternative to the null hypothesis of normality may be that the random variable has positive skewness. When the alternative distribution contains some such specification, tests of fit should be designed to be sensitive to it. Even in these situations uniquely best tests are rarities.

In addition to formal hypothesis testing procedures, goodness-of-fit techniques also include less formal methods, in particular, graphical techniques. These have a long history in statistical analysis. Graphs are drawn so that adherence to or deviation from the hypothesized distribution results in certain features of the graph. For example, in the probability plot the ordered observations are plotted against functions of the ranks. In such plots a straight line indicates that the hypothesized distribution is a reasonable model for the data and deviations from the straight line indicate inappropri-

ateness of the model. The type of departure from the straight line may indicate the nature of the true distribution. Historically the straight line has
been judged by eye, and it is only recently that more formal techniques have
been given.

## 1.2 OBJECTIVES OF THE BOOK

There are five major objectives of this book. They are:

1. To identify the major theories behind goodness-of-fit techniques;
2. To present an up-to-date picture of the status of these techniques;
3. To give references to the relevant literature;
4. To illustrate with numerical examples, and
5. To make some recommendations on the use of different techniques.

There are several features that bear mention. First, a substantial
number of numerical examples are included. These are for the most part
easy to find. In many chapters subsections containing numerical examples
are identified by the letter E before the section number. For example, in
Chapter 9, Section E 9.3.4.1.1 contains a numerical example of the Shapiro-
Wilk test for normality.

Second, a set of data sets is used throughout the book. These allow for
comparisons of some of the techniques on the same data sets. Some of these
data sets are real data and others are simulated. The data sets are given in
full in the appendix.

Third, the chapters contain specific recommendations for use of the
test methods. Nevertheless, we have avoided the attempt to present final
definitive recommendations. The authors for the chapters of this book each
have significant expertise, but there is not always complete agreement among
them on what is best. As we stated previously, theory does not exist which
can identify the uniquely best procedure for most goodness-of-fit situations,
and personal opinion and judgment will often enter any consideration. Each
author has made recommendations based on his or her understanding and
view of the problem.

Fourth, many references are given. There is an enormous literature
and we have made no attempt to survey all of it. We have especially
avoided heavy mathematical treatment and the details of theorems. A substantial list of references is given with each chapter, they include references
to earlier source material and to the theoretical background of the test procedures; it is hoped they will aid the development of further research.

Finally we recognize that it is impossible to include all goodness-of-fit
topics in this survey; our emphasis is largely on the practical aspects of
testing. Some techniques are still underdeveloped, and, for example, suggested tests may lack tables for practical application, or enough comparisons

have not been made to assess their merits; for these and similar reasons,
some subjects have been lightly treated, if at all.

In goodness-of-fit there are many areas with unsolved problems, or
unanswered questions. Some of the subjects on which there will surely be
much work in the future include tests for censored data, especially for randomly censored data, tests based on the empirical characteristic function,
tests based on spacings, and tests for multivariate distributions, especially
for multivariate normality. Many comparisons between techniques are still
needed, and also the exploration of wider questions such as the relationship
of formal goodness-of-fit testing (as, indeed, in other forms of testing) to
modern, more informal, approaches to statistical analysis where distributional models are not so rigidly specified. We hope this book sets forth the
major topics of its subject, and will act as a base from which these and many
other questions can be explored.

## 1.3 THE TOPICS OF THE BOOK

In addition to this chapter the book consists of eleven other chapters. These
are divided into three groups. The first consists of Chapters 2 to 7, containing general concepts applicable to testing for a variety of distributions.
Chapter 2 describes graphical procedures for evaluating goodness-of-fit.
These are informal procedures based mainly on the probability plot, useful
for exploring data and for supplementing the formal testing procedures of the
other chapters.

Chapter 3 reviews chi-square-type tests. The classical chi-square
goodness-of-fit tests are reviewed first and then recent developments involving general quadratic forms and nonstandard chi-squared statistics are
also discussed.

Chapter 4 presents tests based on the empirical distribution function
(edf). These tests include the classical Kolmogorov-Smirnov test and other
tests such as the Cramér-von Mises and Anderson-Darling tests. Consideration is given to simple and composite null hypotheses. The normal, exponential, extreme-value, Weibull, and gamma distributions among other
distributions are given individual discussion.

Chapter 5 deals with tests based on regression and correlation. Some
of these procedures can be viewed as arising from computing a correlation
coefficient from a probability plot and testing if it differs significantly from
unity. Also involved are tests based on comparisons of linear regression
estimates of the scale parameter of the hypothesized distribution to the estimate coming from the sample standard deviation. The Shapiro-Wilk test for
normality is one such test.

In Chapter 6 transformation techniques are reviewed. Here the data are
first transformed to uniformity and goodness-of-fit tests for uniformity are

applied to these transformed data. These techniques can deal with simple
and composite hypotheses.

Tests based on the third and fourth sample moments are presented in
Chapter 7. These techniques were first developed to test for normality. In
Chapter 7 they are extended to nonnormal distributions.

The second group of chapters consists of Chapters 8, 9, and 10. These
deal with tests for three distributions—the uniform, the normal, and the
exponential—which have played prominent roles in statistical methodology.
Many tests for these distributions have been devised, often based on the
methods of previous chapters, and they are brought together, for each distribution, in these three chapters.

Chapters 11 and 12 form the last group; they cover extra materials.
The problem of analyzing censored data is of great importance and Chapter 11
is devoted to this. Many of the previous chapters have sections on censored
data. Chapter 11 collects these together, fills in some omissions, and gives
examples; there is also a discussion on probability plotting of censored data.

The final chapter 12 is on the analysis and detection of outliers. This
material might be considered outside the direct scope of goodness-of-fit
techniques; however, it is closely related to them since they are often applied
with this problem in mind, so we felt it would be useful to close the book with
a chapter on outliers.

# 2
# Graphical Analysis

Ralph B. D'Agostino  Boston University, Boston, Massachusetts

## 2.1 INTRODUCTION

The purpose of this chapter is to illustrate the use of graphical techniques as they relate to goodness-of-fit problems. Graphical techniques as presented here are simple tools which can be implemented easily with the use of graph paper or simple computer programs. They are less formal than the numerical techniques that are presented in the following chapters and are great aids in understanding the numerous relationships present in data. For goodness-of-fit problems they can be used in at least two ways:

1. As an exploratory technique. Here the objective is to uncover characteristics of the data that are suggestive of mathematical properties of the underlying phenomena ranging from incomplete specifications such as symmetry or thick tailness to complete specification such as normality with specific mean and standard deviation.
2. In conjunction with formal numerical techniques. Here the objective is to test formally a preconceived hypothesis or one suggested by the graphs. The graphs can help reveal departures from the assumed models and statistical distributions. Often they uncover features of the data that were totally unanticipated prior to the analysis. The numerical techniques quantify the information and evidence in the data or graphs and act as a verification of inferences suggested from these. The use of graphs alone may lead to spurious conclusions and the use of numerical techniques is often essential in order to avoid this.

    In general, with goodness-of-fit problems, it is useful for numerical testing to be preceded and supplemented by graphical analysis. In

the following we will point out the specific relations between some
graphical procedures and those formal numerical tests that quantify the
information revealed in the graphs.

This chapter deliberately concerns itself with simple to use graphical
procedures involving arithmetic or log graph papers in conjunction possibly
with simple arithmetic and table look-ups, or else with procedures involving
readily available special probability plotting papers. Further most of the
procedures are or can be easily computerized. The view underlying this
approach is that graphical techniques are useful because of their ease and
informality. Involved, complicated procedures detract from this usefulness.
This chapter borrows heavily from the works, concepts, and spirit of
Wilk and Gnanadesikan (1968), Feder (1974), Daniel (1959), Bliss (1967),
W. Nelson and Thompson (1971), W. Nelson (1972), Tukey (1977), and
Chambers, Cleveland, Kleiner, and Tukey (1983).

## 2.2  EMPIRICAL CUMULATIVE DISTRIBUTION FUNCTION

### 2.2.1  Definition

Say we have a random sample $X_1, \ldots, X_n$ drawn from a distribution with
cumulative distribution function (cdf) F, then the empirical cumulative distribution function (ecdf) is defined as

$$F_n(x) = \frac{\#(X_j \le x)}{n}, \quad -\infty < x < \infty \tag{2.1}$$

where $\#(X_j \le x)$ is read, the number of $X_j$'s less than or equal to x. The
ecdf is also often called the edf, empirical distribution function. The plot of
the ecdf is done on arithmetic graph paper plotting i/n as ordinate against
the i'th ordered value of the sample, $X_{(i)}$, as abscissa. Figure 2.1a is an
ecdf plot of the data set NOR given in the appendix which is a random sample
of size 100 from the normal distribution with mean 100 and standard deviation 10.
The ecdf plot provides an exhaustive representation of the data. For all
x values $F_n(x)$ converges for large samples to F(x), the value of the underlying distribution's cdf at x. This convergence is actually strong convergence
uniformly for all x (Rényi, 1970, p. 400).
The use of the ecdf plot does not depend upon any assumptions concerning
the underlying parametric distribution and it has some definite advantages
over other statistical devices, viz.,
    1. It is invariant under monotone transformations with regard to quan-
tiles. However, its appearance may change.
    2. Its complexity is independent of the number of observations.

3. It supplies immediate and direct information regarding the shape of the underlying distribution (e.g., on skewness and bimodality).

4. It is an effective indicator of peculiarities (e.g., outliers).

5. It supplies robust information on location and dispersion.

6. It does not involve grouping difficulties that arise in using for example, a histogram.

7. It can be used effectively in censored samples.

There is, however, one serious potential drawback with the use of ecdf plots and other graphical techniques which was already mentioned in the last section. They can be sensitive to random occurrences in the data and sole reliance on them can lead to spurious conclusions. This is especially true if the sample size is small. This warning always should be kept in mind. In the following we will illustrate uses of the ecdf and related graphs. We will also indicate situations where the user may be misled by them and where further clarification or confirmation via other graphical analyses (e.g., probability plotting) or numerical techniques may be needed.

The ecdf is a standard item in a number of computer packages such as the Statistical Package for the Social Sciences (SPSS), the Statistical Analysis System (SAS), and Biomedical Computer Programs (BMDP).

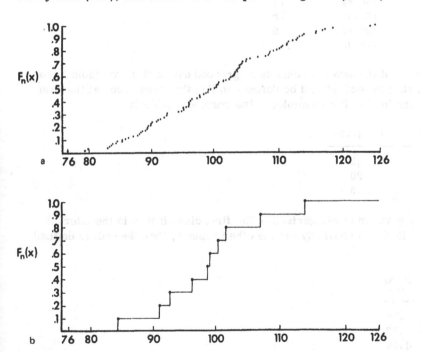

FIGURE 2.1 Empirical distribution function of NOR data set. (a) Ecdf of full data set (n = 100). (b) Ecdf of first ten observations.

Two other technical points are worth mentioning here. First, as defined by formula (2.1) the ecdf is actually a step function with steps or jumps at the values of the variable that occur in the data. Figure 2.1a does not display the ecdf as a step function. Very often it is not displayed as such, especially when the sample size is large and the underlying variable is continuous as is the case with the NOR data. Figure 2.1b displays the ecdf as a step function for the first ten observations of the NOR data set. The ordered values of these first ten observations along with their ecdf values are:

| Ordered observations | | $F_n(x) = i/n$ |
| Number (i) | Value | |
|---|---|---|
| 1 | 84.27 | .1 |
| 2 | 90.87 | .2 |
| 3 | 92.55 | .3 |
| 4 | 96.20 | .4 |
| 5 | 98.70 | .5 |
| 6 | 98.98 | .6 |
| 7 | 100.42 | .7 |
| 8 | 101.58 | .8 |
| 9 | 106.82 | .9 |
| 10 | 113.75 | 1.0 |

Second, if the data set consists of grouped data and the variable is continuous, then the ecdf should be defined so that the steps occur at the true upper class limits. For example, if the frequency table is

| Classes | Frequency |
|---|---|
| 10-13 | 15 |
| 14-17 | 20 |
| 18-21 | 15 |

and an observation is categorized in the first class if it is in the interval $9.5 < x \leq 13.5$ and similarly for the other classes, then the ecdf is defined as

| X | $F_n(x)$ |
|---|---|
| 13.5 | .30 |
| 17.5 | .70 |
| 21.5 | 1.00 |

## 2.2.2 Investigation of Symmetry

Figure 2.2 contains plots of three distributions to illustrate different situations one can encounter in attempting to determine if a distribution is symmetric or skewed. The three distributions are the normal (which is symmetric), the negative exponential (which is positively skewed—i.e., "its upper tail is longer than its lower tail" or "its upper percentage points are farther from the median than are the lower") and the Johnson unbounded $S_U$ (1,2) curve (which is negatively skewed—i.e., "its lower tail is longer than its upper tail"). The density functions for these three distributions are, respectively,

$$f(x) = \frac{1}{\sigma\sqrt{2\pi}}\, \exp\left[-\frac{1}{2}[(x-\mu)/\sigma]^2\right]$$

$$f(x) = \frac{1}{\theta}e^{-x/\theta}$$

and

$$f(x) = \frac{1}{\sigma\sqrt{2\pi}}\,\frac{\delta}{\sqrt{1+((x-\mu)/\sigma)^2}}\,\exp\left[-\frac{1}{2}\{\gamma + \delta\,\sinh^{-1}[(x-\mu)/\sigma]\}^2\right]$$

Here $\mu$, $\sigma$, $\theta$, $\gamma$, and $\delta$ are parameters of the distributions.

If a distribution is symmetric, then in the plot of the population cdf $F(x)$ the distance on the horizontal axis between the median (50-th percentile) and any percentile P below the median (0 < P < 50) is equal to the distance from the median to the (100 - P)th percentile. Figure 2.3a represents this relation

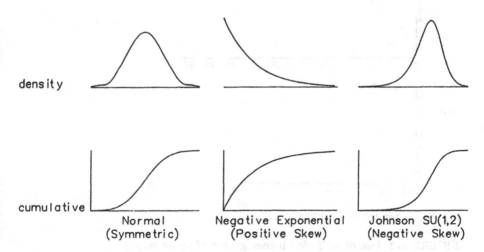

FIGURE 2.2 Differentiation of symmetric and skewed distributions.

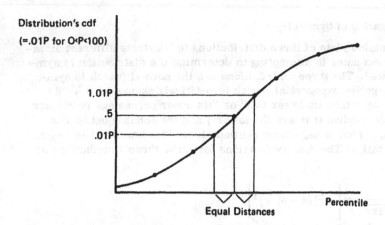

**(a) Relation of percentile about median for symmetric distributions.**

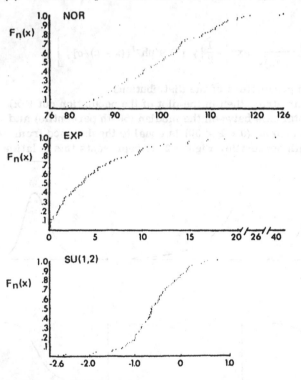

**(b) Ecdfs for three distributions.**

**FIGURE 2.3 Use of ecdfs for investigation of symmetry.**

in diagram form. This relation should be reflected in the ecdf. An examination of the ecdfs given in Figure 2.3b shows it clearly is in the NOR data set and clearly is not for the other two data sets (EXP for the negative exponential distribution and SU(1,2) for the Johnson unbounded distribution). Some rough numerical values from the ecdfs are:

Absolute Values of Distances from Sample Median to Percentiles

| Sample percentiles P 100 - P | NOR | EXP | SU(1,2) |
|---|---|---|---|
| 10 | 21 | 3.5 | 1.75 |
| 90 | 25 | 15.0 | 1.55 |
| 20 | 11 | 2.3 | .50 |
| 80 | 9 | 6.0 | .85 |
| 25 | 10 | 2.0 | .15 |
| 75 | 7 | 3.5 | .55 |
| 40 | 4 | .7 | .05 |
| 60 | 3 | 1.1 | .20 |

If the distribution has positive skewness the portion of the ecdf for $i/n$ values close to 1 (e.g., greater than .9) will usually be longer and flatter (almost parallel to the horizontal axis) than the rest of the ecdf. Similarly, if the distribution has negative skewness the long flat portion will lie in the lower end of its ecdf (e.g., $i/n$ values less than .1). The ecdfs from both the EXP and SU(1,2) data sets behave as expected.

Another, more sensitive and informative graph for studying asymmetry is a simple scatter diagram plotting the upper half of the ordered observations against the lower. That is, letting $X_{(1)}$, $X_{(2)}$, $\ldots$, $X_{(n)}$ represent the ordered observations, plot $X_{(n)}$ versus $X_{(1)}$, $X_{(n-1)}$ versus $X_{(2)}$, and in general, $X_{(n+1-i)}$ versus $X_{(i)}$ for $i \leq n/2$. Figure 2.4 contains these plots for the NOR, EXP, and SU(1,2) data sets. A negative unit slope indicates symmetry, a negative slope exceeding unity in absolute value indicates positive skewness, and a negative slope less than unity in absolute value indicates negative skewness. Notice how well this technique identifies the behavior of the distribution with respect to symmetry. Note also that not all of the observations are plotted. They are not needed usually for a correct visual identification.

Another useful plotting technique involves plotting the sums $X_{(n+1-i)} + X_{(i)}$ against the differences $X_{(n+1-i)} - X_{(i)}$, which would produce a horizontal configuration for a symmetric distribution (Wilk and Gnanadesikan, 1968). A plot of the (100 - P)th sample percentile versus the $P^{th}$ sample percentile

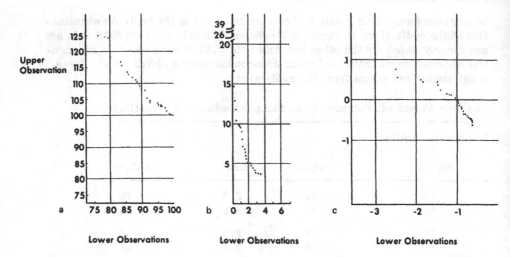

FIGURE 2.4 Plot of upper versus lower observations for investigation of
symmetry, slope computed on all data. (a) NOR data, slope = -1.06.
(b) EXP data, slope = -4.62. (c) SU(1,2) data, slope = -.75.

for $0 < p < 50$ is called a symmetry plot and is also useful (Chambers et al.,
1983).

Formal numerical techniques for investigating and testing for symmetry
are often based on the sample $\sqrt{b_1}$ statistic. A full treatment of this procedure is given in Chapter 7.

### 2.2.3 Detection of Outliers

Outliers, observations that appear to deviate markedly from other members
of the sample (Grubbs, 1969), often can be detected by the use of ecdf plots.
They usually appear as one or a cluster of observations separated from the
rest of the sample and are identifiable in the ecdf if, in addition to the plot,
some knowledge is available concerning the features of the underlying distribution which should be reflected in the data (e.g., the maximum permis-
sible range of the observations or the largest or smallest possible correct
values of the observations may be known or it may be known that the underlying distribution is symmetric.

Figure 2.5 illustrates the use of the ecdf for detecting an outlier. The
figure contains two ecdfs. The first (Figure 2.5a) is a plot of the first ten
observations of the NOR data set. These observations are: 92.55, 96.20,
84.27, 90.87, 101.58, 106.82, 98.70, 113.75, 98.98, and 100.42. The
second ecdf (Figure 2.5b) is a plot of the same data with the last observation,
100.42, replaced by an outlier equal to 140. This example is an exaggeration

of what usually happens in practice, but it illustrates well the type of configuration that results in an ecdf plot of a symmetric distribution such as
the normal distribution when an outlier or outliers are present. Note if it
were not known that the underlying distribution is symmetric or nearly symmetric, it would be impossible to judge if the ecdf of Figure 2.5b represents
data with an outlier present or data from a skewed distribution (see, for example, the ecdfs of the SU(1, 2) and EXP data sets given in Figure 2.3b).

We will illustrate later in this chapter the use of the probability plotting
technique for detecting outliers. Further, Chapter 12 is devoted solely to
the problems of detecting and testing for outliers. The formal techniques of
that chapter should be used in conjunction with informal graphical techniques.

### 2.2.4  Mixtures of Distributions—
Presence of Contamination

At times we may be dealing with samples that arise as mixtures of two or
more distributions. For example, the author once was involved in a study
dealing with taking measurements on parasite transmitting snails obtained
from field sampling. There was no nonstatistical way to separate the different generations (i.e., age groups) of snails in the sample. The parameters
that were desired were related to age. The author was also involved in

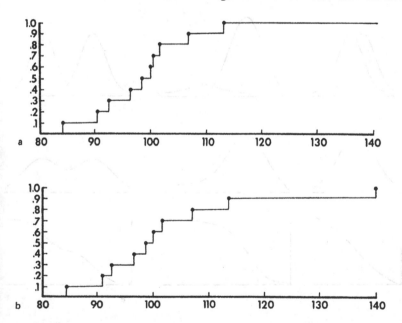

FIGURE 2.5  Plots of ecdfs from NOR data set illustrating effect of an outlier.
(a)  Ecdf of first ten observations. (b)  Ecdf of first nine plus one outlier.

another study dealing with oral glucose tolerance test data. In this study it was suggested that there might exist two subpopulations—normals and diabetics. The data set consisted mainly of normals. Again there was no simple nonstatistical way of removing the small "contaminating" subsample of diabetics. In both of these situations the graphical techniques of this chapter proved to be extremely useful.

Unless the component distributions of the mixture are very distinct (e.g., the difference between the means is much larger than the individual distributions' standard deviation), the ecdf of the combined sample may not supply much information to aid in determining if a mixture exists. Figure 2.6 illustrates the problem. It contains separate and combined densities of mixtures of normal distributions. If the component distributions are "close" as in (a) and (b) of Figure 2.6, the combined distribution may very well be unimodal.

Figure 2.7 further illustrates the problem. These are ecdfs from mixtures of two normal distributions. The main underlying distribution is the normal distribution with mean zero and standard deviation unity. However, the sampling was done in such a way that for each observation drawn there was a probability $\pi$ that the observation would come from the normal distribution with mean 3 and standard deviation unity. The data set for (a) of Figure 2.7 had $\pi = .1$ (data set LCN (.10, 3) of the appendix) and the set

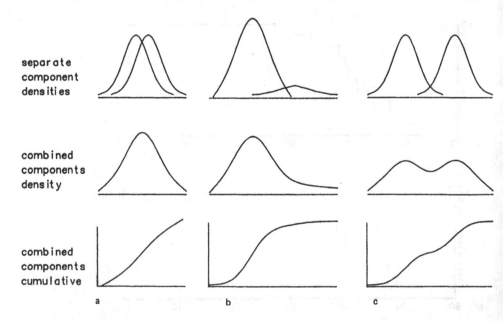

FIGURE 2.6 Mixtures of normal distributions. (a) Two equal close components. (b) Two close components. (c) Well separated components.

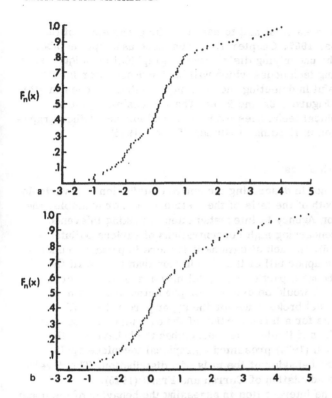

FIGURE 2.7  Ecdfs of contaminated distributions (main component is standard
normal: mean zero and standard deviation unity). (a) Standard normal with
10 percent contamination from normal with mean 3 and standard deviation 1.
(b) Standard normal with 20 percent contamination from normal with mean 3
and standard deviation 1.

for (b) of Figure 2.7 had $\pi = .2$ (data set LCN (.20,3)). The ecdfs in Figure 2.7 look very much like those that are produced by positively skewed
distributions. In fact the populations cdfs are positively skewed. The contamination "caused" the skewness.

   If the component distributions are "well separated" as in (c) of Figure
2.6, the resulting mixture will be bimodal and with sufficient data available,
the ecdf will show the changes from concavity to convexity to concavity as
does the cdf of (c) in Figure 2.6. In general only under the condition of substantial separation of the components will the ecdf reveal bimodality.

   There is an extensive literature on mixtures (see, for example, Johnson
and Kotz, 1970, Section 7.2) and the usual procedure is to assume some
functional form for the components or for the major distribution or distributions of the components. Specific parametric techniques are then employed

to establish if a mixture does exist and to estimate the parameters of the components (e.g., Bliss, 1967, Chapter 7). Given these assumptions about the functional form of the underlying distributions, graphical techniques such as the probability plotting techniques which will be discussed later in this chapter can be very useful in detecting the presence of mixtures even in situations such as those in Figure 2.6a and 2.6b. These probability plotting techniques are the graphical techniques we recommend for use. Other graphical procedures are given in Harding (1949) and Taylor (1965).

### 2.2.5 Assessing Tail Thickness

At times the interest is not in describing the entire distribution of a variable but rather only one or both of the tails of the distribution. For example, the Environmental Protection Agency is interested often in making inferences and issuing standards concerning high concentrations of various pollutants (Curran and Frank, 1975). In such situations it is more important to understand the behavior of the upper tail of the distribution than it is to fit the entire distribution. Although a particular model may adequately describe most of the distribution, it would be useless for predicting maximum or extreme values if the model broke down for the upper percentiles. Also, a model that is not accurate for a large portion of the data may still be useful for predicting upper values if it adequately describes the behavior of the upper percentiles. Bryson (1974) presented a graphical technique applicable to deal with assessing the behavior of the tails of a distribution. The development below is due to the adaptation of Curran and Frank (1975).

To be specific say the interest lies in assessing the behavior of the upper tail. Mathematically, this is equivalent to assessing the thickness of the upper tail or finding a mathematical model which "fits" the upper tail. The most convenient mathematical model is the negative exponential distribution. Here, the probability density function is

$$f(x) = \frac{1}{\theta}e^{-x/\theta}, \qquad \theta > 0, \ x > 0 \tag{2.2a}$$

and the cumulative distribution function is

$$F(x) = 1 - e^{-x/\theta} \tag{2.2b}$$

From this we have

$$1 - F(x) = e^{-x/\theta}$$

and

$$\ln(1 - F(x)) = -x/\theta \tag{2.2c}$$

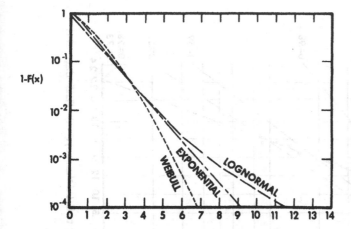

FIGURE 2.8 Relation of lognormal and Weibull distributions to negative exponential on semi-log graph paper (for investigation of tail thickness).

The implication is that if $1 - F(x)$ is plotted against x on semi-log graph paper the plot will be a straight line (see Figure 2.8). Because of this, it is convenient to use the negative exponential distribution as the reference distribution and compare other distributions to it. The Weibull and lognormal distributions are often the two major distributions of potential interest for this type of problem. The two-parameter Weibull has as its probability density and as its cdf

$$f(x) = \frac{k}{\theta}\left(\frac{x}{\theta}\right)^{k-1} e^{-(x/\theta)^k}, \qquad \theta > 0, \ x > 0, \ k > 0 \tag{2.3a}$$

$$F(x) = 1 - e^{-(x/\theta)^k} \tag{2.3b}$$

The lognormal distribution has as its probability density

$$f(x) = \frac{1}{x\sigma\sqrt{2\pi}} e^{-(\ln x - \mu)^2/2\sigma^2} \tag{2.4}$$

Its cdf, $F(x)$, does not have a closed form representation. Figure 2.8 contains plots of $1 - F(x)$ versus x on semi-log paper for a negative exponential, a Weibull with $k > 1$ and a lognormal distribution. Notice the negative exponential produces a straight line, the lognormal distribution curves upward and the Weibull with $k > 1$ curves downward. A distribution that curves downward is termed "light tailed." A heavy tailed distribution has a probability density function whose upper tail approaches zero less rapidly than the

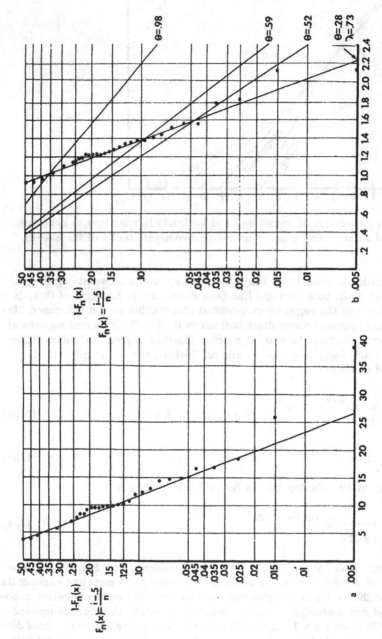

FIGURE 2.9  Plots of data sets on semi-log graph paper (for determining tail thickness).  (a)  EXP data set.  (b) WE2 data set.

exponential or, in other words, a heavy tailed distribution has a greater probability of yielding high values. On the other hand, a light tailed distribution has a probability density function whose upper tail approaches zero more rapidly than the exponential and, therefore, is less likely to yield high values. In particular it should be mentioned that all lognormal distributions are heavy tailed and all Weibull distributions with $k > 1$ are light tailed (when $k = 1$, the Weibull is the negative exponential distribution). So if these are the two models of interest for the upper tail, an examination of a plot of $1 - F_n(x)$ (i.e., one minus the ecdf) versus $X_{(i)}$, the ordered observations will often indicate which is the appropriate model.

The semi-log graph paper used in Figure 2.8 is four cycle paper. Three quarters of the vertical axis concerns only the upper $10^{-1}$ to $10^{-4}$ points of the distribution ($.90 \leq F(x) \leq .9990$). So this graph does focus almost exclusively on the upper 10 percent of the distribution. However, it does contain on the vertical axis the rest of the distribution and plotting the full distribution can cause confusion in attempting to judge the fit of the upper tail. In general, points for which $F_n(x) \leq .5$ should not be plotted. For samples as small as 100 the author has found it convenient to use two cycle semi-log paper, define $F_n(x)$ as

$$F_n(x) = \frac{\#(X_j \leq x) - .5}{n} = \frac{i - .5}{n} \tag{2.5}$$

and plot the data only for $F_n(x) \geq .50$. Note in (2.5) $i = \#(X_j \leq x)$.

### E 2.2.5.1 Example

Figure 2.9 contains the above described plots for the EXP and WE2 (Weibull with $k = 2$) data sets. Consider first the EXP data set plotted in Figure 2.9a. The dots represent the observed values of $1 - F_n(x)$ for $F_n(x) \geq .50$. These appear to lie roughly on a straight line. If the negative exponential distribution is an adequate model for these data, then a straight line for the theoretical exponential as in Figure 2.8 should fit the observed points. To obtain the theoretical line we need $\theta$. The parameter $\theta$ in (2.2) can be estimated by

$$\hat{\theta} = \frac{x}{\ln(1 - F_n(x))} \tag{2.6}$$

where $x$ represents any value for which the model is supposed to hold. In particular the $x$ for which $F_n(x) = .6321$ or $1 - F_n(x) = .3699$ yields a direct estimate of $\theta$. For the EXP data set the estimate of $\theta$ using this or almost any choice of $x$ is approximately 5 (i.e., $\hat{\theta} = 5$). The line drawn in Figure 2.9a is the line $1 - F(x)$ for the negative exponential with $\theta = 5$. Except for the last two data points it fits well to the data. The inference to be drawn from this exercise is that the negative exponential model is an appropriate model which accounts well for all the data points except possibly the last two. Note that in judging the goodness-of-fit of these points it is the horizontal distance from the points to the line that are important, not the vertical distances. The last data point, in particular, may appear to be further away from the line than might be expected. Such variability in the extreme observation is, however, often observed.

Consider next the WE2 data plotted in Figure 2.9b. Using the x for which $F_n(x) = .6321$ to obtain $\hat{\theta}$ in (2.6) we obtain $\hat{\theta} = .98$. The line $1 - F(x)$ for the negative exponential with $\theta = .98$ is drawn in Figure 2.9b. Notice how it lies above most of the data. Using Figure 2.8 as a guide this suggests (correctly) that the data is from a distribution with a thinner upper tail than the negative exponential. Also on Figure 2.9b are plotted two other lines representing $1 - F(x)$ for the negative exponential of (2.2). These arose from solving (2.6) for $\theta$ using $F_n(x) = .90$ and $F_n(x) = .95$. The estimators of $\theta$ are, respectively, .59 and .52. Again the inference is the same, viz., the negative exponential model of (2.2) is not appropriate and the distribution under consideration has a thinner tail than the negative exponential. Note, this inference is correct.

A further examination of the WE2 data plot in Figure 2.9b does reveal that the points do appear to lie on a straight line. The above analysis establishes that the data cannot be explained by a model such as (2.2). They can, however, be explained by a negative exponential model which incorporates a displacement value, viz.,

$$f(x) = \frac{1}{\theta} e^{-[(x-\lambda)/\theta]}, \qquad \theta > 0, \quad x > \lambda \tag{2.7}$$

where $\lambda$ is the displacement value. The cdf for this distribution is

$$F(x) = 1 - e^{-[(x-\lambda)/\theta]}$$

If we start with this model then any two distinct x values (or two $F_n(x)$ values) can be used to produce linear equations for $\hat{\lambda}$ and $\hat{\theta}$. The equations are

$$- \ln(1 - F_n(x_1))\theta + \lambda = x_1$$
$$- \ln(1 - F_n(x_2))\theta + \lambda = x_2 \tag{2.8}$$

Using $F_n(x_1) = .50$ and $F_n(x_2) = .90$ in the WE2 data set produces $\hat{\theta} = .28$, $\hat{\lambda} = .73$. The line of $1 - F(x)$ for model (2.7) with these parameters is also plotted in Figure 2.9b. This provides an excellent fit. So if we restrict our attention solely to the upper tail the WE2 data can be well explained by a negative exponential of the form (2.7). Of course, the correct model is the

Weibull of (2.3) with k = 2. Completion of the first part of the above analysis would have led correctly to the Weibull model.

### 2.2.5.2 Extensions

The above material can easily be modified to examine the lower tail of the distribution (viz., by plotting $F_n(x)$ of (2.5) versus the observations on semi-log paper).

Often the normal distribution is used as the reference distribution in discussing tail thickness and the standardized central fourth moment $\beta_2$ (the kurtosis measure) is used as the appropriate measure. For these problems, one is usually interested in fitting the complete distribution and not just the tail. We will discuss this tail thickness concept in Section 2.4 below. Also, Chapter 7 will discuss in detail the formal computational procedures associated with this concept.

### 2.2.6 Assessing the Fit of the Full Distribution

The ecdf can be used also for assessing how well a particular statistical distribution fits the entire data set. The procedure starts by plotting on the

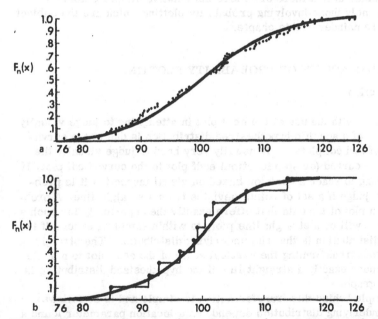

FIGURE 2.10 Comparison of population and empirical cumulative distribution functions for NOR data set. (a) Full data set (n = 100): population mean 100, standard deviation 10. (b) First ten observations: sample mean 98.41, standard deviation 8.28.

same grid of a piece of graph paper the ecdf of the sample and the cdf of the hypothetical distribution. For example, Figure 2.10a contains the ecdf of the NOR data along with the cdf for the normal distribution with mean 100 and standard deviation 10 (i.e., the true underlying distribution). If values of the parameters of the hypothetical distribution are unspecified, these must be estimated for the data set under investigation by means of some procedure such as the method of moments or the method of maximum likelihood and then the cdf of the hypothetical distribution using these estimates as parameter values are plotted. For an example, Figure 2.10b contains the plot of the ecdf of the first ten observations of the NOR data set along with the cdf of the normal distribution with mean and standard deviation equal to the sample mean and standard deviation, viz., $\bar{x} = 98.41$ and $s = 8.28$.

The next step in the informal graphical analysis involves comparing the two plots (ecdf and cdf) and deciding if they are "close." Usually this informal procedure is the first step in a more elaborate analysis which includes formal numerical techniques referred to as empirical cumulative distribution function techniques or more simply empirical distribution function (EDF) techniques. Chapter 4 contains a detailed account of these techniques.

While the above described graphical procedure has merit, especially when used with the formal numerical EDF techniques, it is deficient as an informal technique in that there are more informative simple graphical techniques—namely those involving probability plotting which are the subject matter of the remainder of this chapter.

## 2.3 GENERAL CONCEPTS OF PROBABILITY PLOTTING

### 2.3.1 Introduction

A major problem with the use of the ecdf plot in attempting to judge visually the correctness of a specific hypothesized distribution is due to the curvature of the ecdf and cdf plots. It is usually very hard to judge visually the closeness of the curved (or step function) ecdf plot to the curved cdf plot. If one is attempting to reach a decision based on visual inspection it is probably easiest to judge if a set of points deviates from a straight line. A probability plot is a plot of the data that offers exactly the opportunity for such a judgment, for it will be a straight line plot, to within sampling error, if the hypothesized distribution is the true underlying distribution. The straight line results from transforming the vertical scale of the ecdf plot to a scale which will produce exactly a straight line if the hypothesized distribution is plotted on the graph.

The principle behind this transformation is simple and is as follows. Say the true underlying distribution depends on a location parameter $\mu$ and a scale parameter $\sigma$. ($\mu$ and $\sigma$ need not be the mean and standard deviation, respectively). The cdf of such a distribution can be written as

$$F(x) = G\left(\frac{x - \mu}{\sigma}\right) = G(z) \qquad (2.9)$$

where

$$z = \frac{x - \mu}{\sigma}$$

is referred to as the standardized variable and $G(\cdot)$ is the cdf of the standardized random variable $Z$. The ecdf plot is based on plotting $F(x)$ on $x$. For sample data $F(x)$ is replaced by $F_n(x)$ and the plotted values of $x$ are the observed values of the random variable $X$. Now if the plot were one of <u>z on x</u> (or equivalently $G^{-1}(F(x))$ on $x$ where $G^{-1}(\cdot)$ is the inverse transformation which here transforms $F(x)$ into the corresponding standardized value $z$), the resulting plot would be the straight line

$$z = G^{-1}(F(x)) = \frac{x - \mu}{\sigma} = -\frac{\mu}{\sigma} + \frac{1}{\sigma}x \qquad (2.10a)$$

or in terms of <u>x on z</u>

$$x = \mu + z\sigma \qquad (2.10b)$$

A <u>probability plot</u> is a plot of

$$z = G^{-1}(F_n(x)) \quad \text{on } x \qquad (2.11a)$$

where $x$ represents the observed values of the random variable $X$. Notice $F(x)$ in (2.10a) is replaced by $F_n(x)$ in (2.11a). With observed ordered observations $x_{(1)} \le \cdots \le x_{(n)}$, a probability plot can also be described as a plot of

$$z_i = G^{-1}(F_n(x_{(i)})) \quad \text{on } x_{(i)} \qquad (2.11b)$$

For probability plotting the ecdf $F_n(x)$ of (2.11a) or $F_n(x_{(i)})$ of (2.11b) are usually not defined as in (2.1) but rather as either

$$F_n(x_{(i)}) = p_i = \frac{i - 0.5}{n} \quad \text{for } i = 1, \ldots, n \qquad (2.12)$$

or more generally as

$$\frac{i - c}{n - 2c + 1} \quad \text{for } 0 \le c \le 1 \qquad (2.13)$$

In (2.12) the $c$ of (2.13) is equal to $0.5$. See Barnett (1975) and Chapter 11 for further discussion of the selection of $c$. <u>In the following we will always</u>

use the $F_n(x)$ given by $p_i$ of (2.12). Given that F is the true cdf, the probability plot of (2.11) should be approximately a straight line. In fact there is strong convergence to a straight line for large samples.

### E 2.3.2  An Example—Logistic Distribution

As an example of the above consider the problem of investigating the appropriateness of the logistic distribution as the underlying distribution from which the LOG data set was obtained. (The LOG data set was drawn from a logistic distribution and is given in the Appendix.) The cdf of the logistic distribution is

$$F(x) = [1 + \exp\{-\pi(x - \mu)/(\sigma\sqrt{3})\}]^{-1} \qquad (2.14)$$

Here $\mu$ and $\sigma$ are the mean and standard deviation, respectively. The cdf of the standardized logistic distribution (i.e., of $Z = (X - \mu)/\sigma$) is

$$G(z) = [1 + \exp(-\pi z/\sqrt{3})]^{-1}$$

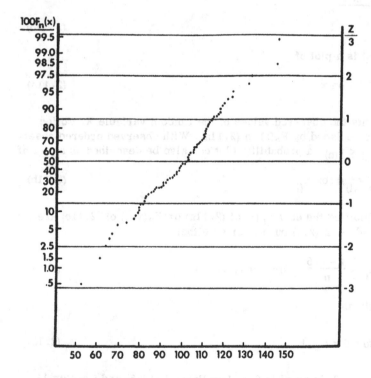

FIGURE 2.11  Logistic probability plot of LOG data set.

**TABLE 2.1** Partial Data for Logistic Probability Plot of LOG Data (Plotted in Figures 2.11 and 2.12)

| Ordered Observation Number (i) | $F_n(x) = p_i = \dfrac{i - .5}{n}$ | z | Ordered Observation $X_{(i)}$ |
|---|---|---|---|
| 1 | .005 | -2.90 | 51.90 |
| 2 | .015 | -2.31 | 60.57 |
| 3 | .025 | -2.02 | 63.35 |
| 4 | .035 | -1.83 | 65.87 |
| 5 | .045 | -1.68 | 66.35 |
| 6 | .055 | -1.56 | 68.44 |
| 7 | .065 | -1.47 | 74.29 |
| 8 | .075 | -1.39 | 76.52 |
| 9 | .085 | -1.31 | 78.32 |
| 10 | .095 | -1.24 | 78.48 |
| 11 | .105 | -1.18 | 79.07 |
| 12 | .115 | -1.12 | 79.32 |
| 13 | .125 | -1.07 | 81.17 |
| 14 | .135 | -1.02 | 81.61 |
| 15 | .145 | - .98 | 82.45 |
| 86 | .855 | .98 | 113.79 |
| 87 | .865 | 1.02 | 114.97 |
| 88 | .875 | 1.07 | 116.01 |
| 89 | .885 | 1.12 | 116.58 |
| 90 | .895 | 1.18 | 116.99 |
| 91 | .905 | 1.24 | 117.01 |
| 92 | .915 | 1.31 | 118.54 |
| 93 | .925 | 1.39 | 118.92 |
| 94 | .935 | 1.47 | 121.83 |
| 95 | .945 | 1.56 | 123.39 |
| 96 | .955 | 1.68 | 123.58 |
| 97 | .965 | 1.83 | 131.24 |
| 98 | .975 | 2.02 | 132.40 |
| 99 | .985 | 2.31 | 144.28 |
| 100 | .995 | 2.90 | 145.33 |

Recalling from (2.9) that

$$F(x) = G\left(\frac{x - \mu}{\sigma}\right) = G(z)$$

where $z = (x - \mu)/\sigma$ and solving for $z$ in terms of $F(x)$ we obtain from the above,

$$z = G^{-1}(F(x)) = \frac{\sqrt{3}}{\pi} \ln\left(\frac{F(x)}{1 - F(x)}\right)$$

Now, according to (2.11) a _logistic probability plot_ consists of plotting on arithmetic graph paper

$$z = \frac{\sqrt{3}}{\pi} \ln\left(\frac{F_n(x)}{1 - F_n(x)}\right) \tag{2.15}$$

where $F_n(x)$ is the ecdf defined by (2.12) on one axis (e.g., the vertical axis) versus $x$ on the other (horizontal) axis. Here $x$ represents the observed values in the sample. Figure 2.11 contains an appropriate graph set up for this problem.

Notice in Figure 2.11 there are two alternative ways of labeling the vertical axis. The first way, which is probably the most informative, is to label the axis in terms of $F_n(x)$ (or $100F_n(x)$ which is the more conventional way). The second way is in terms of the values of the standardized variable $z$. In Figure 2.11 we have labelled the left vertical axis as $100F_n(x)$ and the right vertical axis as $z$. Notice the vertical axis is linear in $z$. It is _not_ linear in $F_n(x)$.

Figure 2.11 contains the data plotted on it. To make more explicit the actual points plotted on this graph we list in Table 2.1 the values of $F_n(x)$ of (2.12) and $z$ obtained for (2.15) for the first and last fifteen ordered observations. Also listed are the corresponding ordered observations.

### 2.3.3. Informal Goodness-of-Fit and Estimation of Parameters

Once the data are plotted the next step is to determine the goodness-of-fit of the data. For a probability plot this means determining if a straight line "fits well" the data. This problem can be approached in a very formal manner and Chapter 5 (Regression Techniques) discusses this approach in detail. For the purposes of this chapter it means drawing a straight line through the points and deciding in an informal manner if the fit is good.

#### 2.3.3.1 First Procedure

The simplest procedure is to draw a line "by eye" through the points. One convenient way to do this is to locate a point on the plot corresponding to around the 10th percentile $(F_n(x) = .10)$ and another around the 90th percentile $(F_n(x) = .90)$ and connect these two. Figure 2.12a contains such a line for the logistic probability plot of the complete LOG data set. (Notice this is the same plot as Figure 2.11. Here we have the straight line imposed on the graph.) This line fits the data extremely well, accommodating even the extreme points. There are two comments which need mentioning here. The first concerns the non-random pattern of the points about the line. The ordered observations are not independent and the type of pattern shown in Figure 2.12a is to be expected. Second, in judging deviations from the line remember it is the <u>horizontal distances</u> from the points to the line that are important.

After the "by eye" line is drawn it can be used to supply quick estimates of the parameters of the distribution. For example, with the LOG data of Figure 2.12a we can obtain estimates of the mean $\mu$ and standard deviation $\sigma$ by recalling that $z = 0$ corresponds to the mean $\mu$ and $z = 1$ (86th percentile) corresponds to $\mu + \sigma$. In Figure 2.12a we have lines extending from $z = 0$ and 1 to the straight line and down to the x axis. From these we estimate $\hat{\mu} = 99$ and $\hat{\sigma} = 17$.

### 2.3.3.2 Other Procedures

A second procedure for obtaining the line and estimates is to recognize that from (2.10) we have that the desired line can be represented by

$$x = \mu + z\sigma \tag{2.16}$$

and estimates of $\mu$ and $\sigma$ can be obtained by using unweighted least squares (simple linear regression). The general solution for these are

$$\hat{\sigma} = \frac{\Sigma (z - \bar{z})x}{\Sigma (z - \bar{z})^2} \text{ and } \hat{\mu} = \bar{x} - \hat{\sigma}\bar{z} \tag{2.17}$$

If $\Sigma z = 0$, then

$$\hat{\mu} = \bar{x} \text{ and } \hat{\sigma} = \frac{\Sigma zx}{\Sigma z^2} \tag{2.18}$$

for the LOG data set, $\hat{\mu} = 99.78$ and $\hat{\sigma} = 16.70$.

Still a third procedure applicable if $\mu$ and $\sigma$ are the mean and standard deviation, as they are in the logistic distribution of (2.14), is to use $\bar{x}$ and $s$, the sample mean and standard deviation, as estimates. For the LOG data $\bar{x} = 99.78$ and $s = 16.67$. Notice for the LOG data there are very little differences among the results of these different procedures. The true parameter values are $\mu = 100$ and $\sigma = 18.14$.

More elaborate procedures involve finding the best linear unbiased estimators of $\mu$ and $\sigma$ (see D'Agostino and Lee, 1976, for the logistic distribution). These procedures lead to the regression techniques of Chapter 5.

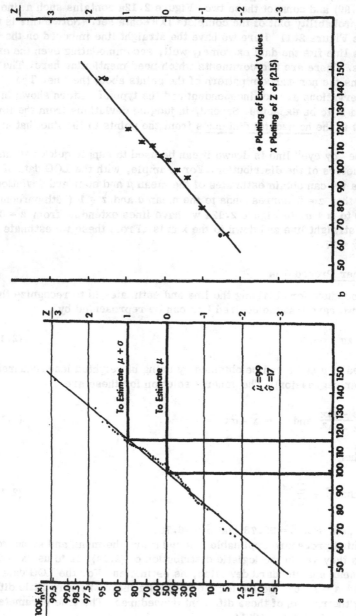

FIGURE 2.12 Logistic analysis of LOG data set. (a) Full data set (n = 100) line fit "by eye." (b) First ten observations, small sample analysis.

## 2.3.4  Small Samples

When the size of the sample is small (say 50 or less) the probability plots of z on x as given by (2.11) may display curvature in the tails even if the hypothesized distribution is correct. For these cases the usual recommendation is to use the <u>expected values of the ordered statistics</u> from the standardized distribution of the hypothesized distribution for the plotting positions of the vertical axis. These are used in place of the z of (2.11) which are the percentile points of the standardized distribution. The expected values are defined as follows. Say $Z_{(1)} \leq \cdots \leq Z_{(n)}$ represent the ordered observations for a sample of size n from a standardized distribution. Then the expected values are defined as $EZ_{(i)}$ for $i = 1, \ldots, n$ where E represents the expected value operator.

## E 2.3.4.1  Example

For the logistic distribution the expected values are readily available (Gupta and Shah, 1965, and Gupta, Qureshi and Shah, 1967). However, for this particular distribution there appears to be no reason to use them in plotting. Figure 2.12b contains a logistic probability plot (i.e., a logistic analysis) of the first ten unordered observations of the LOG data. The data along with the expected values of the ordered observations, $F_n(x)$ and z of (2.11) and (2.15) are as follows:

| Ordered observations | Expected values of standardized logistic ordered observations | $F_n(x) = p_i$ | z of (2.11) and (2.15) |
|---|---|---|---|
| 63.35 | -1.56 | .05 | -1.62 |
| 78.32 | -.95 | .15 | -.96 |
| 94.63 | -.60 | .25 | -.61 |
| 96.91 | -.34 | .35 | -.34 |
| 102.97 | -.11 | .45 | -.11 |
| 104.47 | .11 | .55 | .11 |
| 109.99 | .34 | .65 | .34 |
| 111.81 | .60 | .75 | .61 |
| 118.54 | .95 | .85 | .96 |
| 144.28 | 1.56 | .95 | 1.62 |

The differences between the expected values and the z's of (2.15) are not large enough to influence the plots. This is seen clearly in Figure 2.12b. Remember in judging the fit it is the <u>horizontal distance</u> from a point to the line that is important.

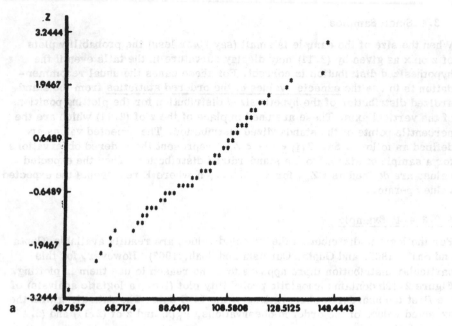

**FIGURE 2.13 Use of computer scatter diagrams for probability plotting.
(a) Logistic analysis of LOG data.**

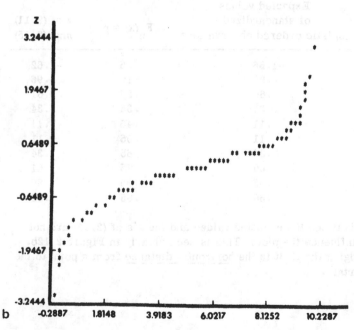

FIGURE 2.13 (b) Logistic analysis of UNI data.

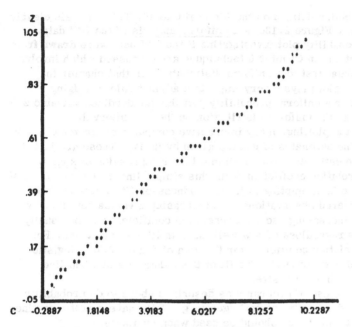

FIGURE 2.13 (c) Uniform analysis of UNI data.

## 2.3.5 Grouped Data/Ties in Data

For grouped data such as discussed in Section 2.2.1 the simplest procedure
for probability plotting is to plot only the data at the true upper class limit
for each interval. Of course, use (2.12) for the $F_n(x)$. This is equivalent
to representing all the observations in the interval at the upper end points
of the intervals. For ungrouped data with ties the simplest procedure is to
average the z values for the observations in the ties (see E 2.4.1.1 for a
numerical example).

## 2.3.6 Use of Simple Computer Graphics

Elaborate sophisticated computer graphics are not needed to produce probability plots. Many interactive systems have the capability of ordering the
observations of a data set and defining new variables. In such systems a
probability plot is simply a <u>scatter diagram</u> with z of (2.11) on the vertical
axis and the sample observations on the horizontal axis. Figure 2.13 displays such scatter diagram-probability plots. Figure 2.13a is the logistic
plot of the LOG data (already plotted in Figures 2.11 and 2.12a). Figure
2.13b is a logistic analysis of the UNI data. The UNI data were drawn from

the uniform distribution defined on the interval 0 to 10. This analysis clearly
indicates lack of fit. Figure 2.13c is a <u>uniform analysis</u> of the UNI data.
That is, it is a probability plot investigating if the UNI data were drawn from
a uniform distribution. In Chapter 6 techniques are discussed which involve
transforming the data first to a uniform distribution. In that chapter the
uniform probability plot plays a very important adjunct role in judging
goodness-of-fit. For a uniform probability plot the standardized variable z
usually is defined as the uniform distribution on the unit interval.

In addition to the plotting, many interactive computer programs can also
be used to obtain the estimates of $\mu$ and $\sigma$ given by (2.17). These are just the
intercept and slope estimates from a simple linear regression of <u>x on z</u>.
However, the correlation coefficient from this simple linear regression must
be viewed with care in attempting to judge goodness-of-fit. Because of the
matching of the ordered observations with increasing z values both x and z
are monotonically increasing, so the correlation coefficient will be usually
large in magnitude regardless of how well the data fit a straight line. For
example, the correlation coefficient for the data of Figure 2.13b (logistic
analysis of the UNI data) is .947. The fit of these data to a straight line
obviously leaves much to be desired.

In addition to the use of programs as described above to do probability
plotting, many standard software packages (e.g., SAS) have specific routines
for probability plotting. These should be used when available.

### 2.3.7 Summary Comments

As given above a <u>probability plot</u> is a plot of

$$z_i = G^{-1}(F_n(x_{(i)})) = G^{-1}(p_i) \text{ on } x_{(i)} \tag{2.19}$$

where $G^{-1}(\cdot)$ is the inverse transformation of the standardized distribution
of the population (hypothesized distribution) under consideration. We recommend for $F_n(x_{(i)})$

$$F_n(x_{(i)}) = p_i = \frac{i - .5}{n} \tag{2.20}$$

In the examples above we have used arithmetic graph paper placing z on the
vertical axis and x on the horizontal axis. Of course, it is not incorrect to
place x on the vertical axis and z on the horizontal axis. (In Chapter 11 probability plotting is done that way.) Nor is it essential to use arithmetic paper.
Many probability plotting papers, which have the axes appropriately labelled,
are available commercially. Logistic paper and many other probability
papers are available from the Codex Book Company, 74 Broadway in Norwood,
Massachusetts.

TABLE 2.2 Plotting Formulas for Some Familiar Distributions
$$\left(p_i = \frac{i - 0.5}{n}\right)$$

| Distribution | cdf F(x) | Horizontal Axis | Vertical Axis $z_i$ |
|---|---|---|---|
| Uniform | $\frac{x-\mu}{\sigma}$ for $\mu < x < \mu+\sigma$ | $x_{(i)}$ | $p_i = \frac{i - .5}{n}$ |
| Normal | $\Phi\left(\frac{x-\mu}{\sigma}\right)$ | $x_{(i)}$ | See (2.22) to (2.24) |
| Lognormal | $\Phi\left(\frac{\ln x - \mu}{\sigma}\right)$ | $\ln(x_{(i)})$ | See (2.22) to (2.24) |
| Weibull | $1 - \exp\left(-\left(\frac{t}{\theta}\right)^k\right)$ | $\ln(t_{(i)})$ | $\ln(-\ln(1 - p_i))$ |
| Extreme Value | $1 - \exp\left(-\exp\left(\frac{x-\mu}{\sigma}\right)\right)$ | $x_{(i)}$ | $\ln(-\ln(1 - p_i))$ |
| Logistic | $[1 + \exp\{-\pi(x-\mu)/\sigma\sqrt{3}\}]^{-1}$ | $x_{(i)}$ | $\frac{\sqrt{3}}{\pi}\ln(p_i/(1 - p_i))$ |
| Exponential | $1 - \exp(-(x/\theta))$ | $x_{(i)}$ | $-\ln(1 - p_i)$ |

    Once the points are plotted the major task is to judge if the plotted data form a straight line. If they do not, the task is then to decide what are the properties of the underlying distribution or data which cause this nonlinearity. We will now illustrate this probability procedure with the normal, lognormal and Weibull plotting. Table 2.2 contains the appropriate formulas for probability plotting for those and other familiar distributions.

## 2.4 NORMAL PROBABILITY PLOTTING

### 2.4.1 Probability Plotting

Normal probability plotting, normal plotting or normal analysis is the plotting of data in order to investigate the goodness-of-fit of the data to the normal distribution with density given by

$$f(x) = \frac{1}{\sqrt{2\pi}\sigma} e^{-\frac{1}{2}[(x-\mu)/\sigma]^2} \tag{2.21}$$

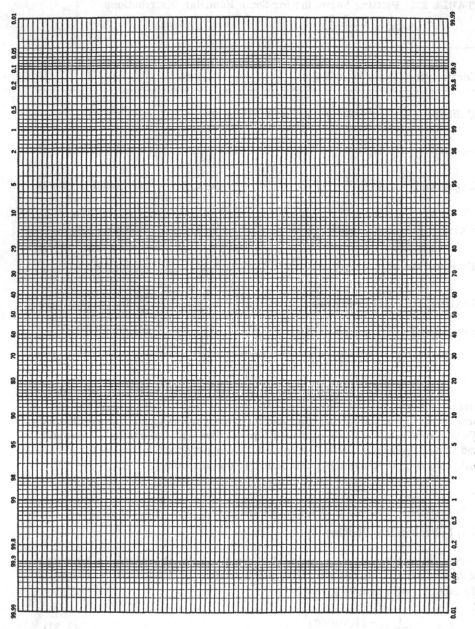

FIGURE 2.14  Normal probability paper.

TABLE 2.3  Plotting Positions z for Normal Probability Plotting (n ≤ 50)
(Expected Values of Standard Normal Order Statistics*)

(n = Sample Size,  i = Observation Number)

| i\n | 3 | 4 | 5 | 6 | 7 | 8 | 9 | 10 |
|---|---|---|---|---|---|---|---|---|
| 1 | -0.85 | -1.03 | -1.16 | -1.27 | -1.35 | -1.42 | -1.49 | -1.54 |
| 2 | 0.00 | -0.30 | -0.50 | -0.64 | -0.76 | -0.85 | -0.93 | -1.00 |
| 3 |  |  | 0.00 | -0.20 | -0.35 | -0.47 | -0.57 | -0.66 |
| 4 |  |  |  |  | 0.00 | -0.15 | -0.27 | -0.38 |
| 5 |  |  |  |  |  |  | 0.00 | -0.12 |

|  | 11 | 12 | 13 | 14 | 15 | 16 | 17 | 18 |
|---|---|---|---|---|---|---|---|---|
| 1 | -1.59 | -1.63 | -1.67 | -1.70 | -1.74 | -1.77 | -1.79 | -1.82 |
| 2 | -1.06 | -1.12 | -1.16 | -1.21 | -1.25 | -1.28 | -1.32 | -1.35 |
| 3 | -0.73 | -0.79 | -0.85 | -0.90 | -0.95 | -0.99 | -1.03 | -1.07 |
| 4 | -0.46 | -0.54 | -0.60 | -0.66 | -0.71 | -0.76 | -0.81 | -0.85 |
| 5 | -0.22 | -0.31 | -0.39 | -0.46 | -0.52 | -0.57 | -0.62 | -0.66 |
| 6 | 0.00 | -0.10 | -0.19 | -0.27 | -0.34 | -0.40 | -0.45 | -0.50 |
| 7 |  |  | 0.00 | -0.09 | -0.17 | -0.23 | -0.30 | -0.35 |
| 8 |  |  |  |  | 0.00 | -0.08 | -0.15 | -0.21 |
| 9 |  |  |  |  |  |  | 0.00 | -0.07 |

|  | 19 | 20 | 21 | 22 | 23 | 24 | 25 | 26 |
|---|---|---|---|---|---|---|---|---|
| 1 | -1.84 | -1.87 | -1.89 | -1.91 | -1.93 | -1.95 | -1.97 | -1.98 |
| 2 | -1.38 | -1.41 | -1.43 | -1.46 | -1.48 | -1.50 | -1.52 | -1.54 |
| 3 | -1.10 | -1.13 | -1.16 | -1.19 | -1.21 | -1.24 | -1.26 | -1.29 |
| 4 | -0.89 | -0.92 | -0.95 | -0.98 | -1.01 | -1.04 | -1.07 | -1.09 |
| 5 | -0.71 | -0.75 | -0.78 | -0.82 | -0.85 | -0.88 | -0.91 | -0.93 |
| 6 | -0.55 | -0.59 | -0.63 | -0.67 | -0.70 | -0.73 | -0.76 | -0.79 |
| 7 | -0.40 | -0.45 | -0.49 | -0.53 | -0.57 | -0.60 | -0.64 | -0.67 |
| 8 | -0.26 | -0.31 | -0.36 | -0.41 | -0.45 | -0.48 | -0.52 | -0.55 |
| 9 | -0.13 | -0.19 | -0.24 | -0.29 | -0.33 | -0.37 | -0.41 | -0.44 |
| 10 | 0.00 | -0.06 | -0.12 | -0.17 | -0.22 | -0.26 | -0.30 | -0.34 |
| 11 |  |  | 0.00 | -0.06 | -0.11 | -0.16 | -0.20 | -0.24 |
| 12 |  |  |  |  | 0.00 | -0.05 | -0.10 | -0.14 |
| 13 |  |  |  |  |  |  | 0.00 | -0.05 |

*z for order statistic $X_{(i)}$ where $i > n/2$ is -z of order statistic $X_{(j)}$ where
$j = n + 1 - i$.

(continued)

**TABLE 2.3 (continued)**

| i\n | 27 | 28 | 29 | 30 | 31 | 32 | 33 | 34 |
|---|---|---|---|---|---|---|---|---|
| 1 | -2.00 | -2.01 | -2.03 | -2.04 | -2.06 | -2.07 | -2.08 | -2.09 |
| 2 | -1.56 | -1.58 | -1.60 | -1.62 | -1.63 | -1.65 | -1.66 | -1.68 |
| 3 | -1.31 | -1.33 | -1.35 | -1.36 | -1.38 | -1.40 | -1.42 | -1.43 |
| 4 | -1.11 | -1.14 | -1.16 | -1.18 | -1.20 | -1.22 | -1.23 | -1.25 |
| 5 | -0.96 | -0.98 | -1.00 | -1.03 | -1.05 | -1.07 | -1.09 | -1.11 |
| 6 | -0.82 | -0.85 | -0.87 | -0.89 | -0.92 | -0.94 | -0.96 | -0.98 |
| 7 | -0.70 | -0.73 | -0.75 | -0.78 | -0.80 | -0.82 | -0.85 | -0.87 |
| 8 | -0.58 | -0.61 | -0.64 | -0.67 | -0.69 | -0.72 | -0.74 | -0.76 |
| 9 | -0.48 | -0.51 | -0.54 | -0.57 | -0.60 | -0.62 | -0.65 | -0.67 |
| 10 | -0.38 | -0.41 | -0.44 | -0.47 | -0.50 | -0.53 | -0.56 | -0.58 |
| 11 | -0.28 | -0.32 | -0.35 | -0.38 | -0.41 | -0.44 | -0.47 | -0.50 |
| 12 | -0.19 | -0.22 | -0.26 | -0.29 | -0.33 | -0.36 | -0.39 | -0.41 |
| 13 | 0.09 | -0.13 | -0.17 | -0.21 | -0.24 | -0.28 | -0.31 | -0.34 |
| 14 | 0.00 | -0.04 | -0.09 | -0.12 | -0.16 | -0.20 | -0.23 | -0.26 |
| 15 | | | 0.00 | -0.04 | -0.08 | -0.12 | -0.15 | -0.18 |
| 16 | | | | 0.00 | -0.04 | -0.08 | -0.11 |
| 17 | | | | | | | 0.00 | -0.04 |

| i\n | 35 | 36 | 37 | 38 | 39 | 40 | 41 | 42 |
|---|---|---|---|---|---|---|---|---|
| 1 | -2.11 | -2.12 | -2.13 | -2.14 | -2.15 | -2.16 | -2.17 | -2.18 |
| 2 | -1.69 | -1.70 | -1.72 | -1.73 | -1.74 | -1.75 | -1.76 | -1.78 |
| 3 | -1.45 | -1.46 | -1.48 | -1.49 | -1.50 | -1.52 | -1.53 | -1.54 |
| 4 | -1.27 | -1.28 | -1.30 | -1.32 | -1.33 | -1.34 | -1.36 | -1.37 |
| 5 | -1.13 | -1.14 | -1.16 | -1.17 | -1.19 | -1.20 | -1.22 | -1.23 |
| 6 | -1.00 | -1.02 | -1.03 | -1.05 | -1.07 | -1.08 | -1.10 | -1.11 |
| 7 | -0.87 | -0.91 | -0.92 | -0.94 | -0.96 | -0.98 | -0.99 | -1.01 |
| 8 | -0.79 | -0.81 | -0.83 | -0.85 | -0.86 | -0.88 | -0.90 | -0.91 |
| 9 | -0.69 | -0.71 | -0.73 | -0.75 | -0.77 | -0.79 | -0.81 | -0.83 |
| 10 | -0.60 | -0.63 | -0.65 | -0.67 | -0.69 | -0.71 | -0.73 | -0.75 |
| 11 | -0.52 | -0.54 | -0.57 | -0.59 | -0.61 | -0.63 | -0.65 | -0.67 |
| 12 | -0.44 | -0.47 | -0.49 | -0.51 | -0.54 | -0.56 | -0.58 | -0.60 |
| 13 | -0.36 | -0.39 | -0.42 | -0.44 | -0.46 | -0.49 | -0.51 | -0.53 |
| 14 | -0.29 | -0.32 | -0.34 | -0.37 | -0.39 | -0.42 | -0.44 | -0.46 |
| 15 | -0.22 | -0.24 | -0.27 | -0.30 | -0.33 | -0.35 | -0.37 | -0.40 |
| 16 | -0.14 | -0.17 | -0.20 | -0.23 | -0.26 | -0.28 | -0.31 | -0.33 |
| 17 | -0.07 | -0.10 | -0.14 | -0.16 | -0.19 | -0.22 | -0.25 | -0.27 |
| 18 | 0.00 | -0.03 | -0.07 | -0.10 | -0.13 | -0.16 | -0.18 | -0.21 |
| 19 | | | 0.00 | -0.03 | -0.06 | -0.09 | -0.12 | -0.14 |
| 20 | | | | | 0.00 | -0.03 | -0.06 | -0.09 |
| 21 | | | | | | | 0.00 | -0.03 |

*z for order statistic $X_{(i)}$ where $i > n/2$ is -z of order statistic $X_{(j)}$ where $j = n + 1 - i$.

TABLE 2.3 (continued)

| i\n | 43 | 44 | 45 | 46 | 47 | 48 | 49 | 50 |
|---|---|---|---|---|---|---|---|---|
| 1 | -2.19 | -2.20 | -2.21 | -2.22 | -2.22 | -2.23 | -2.24 | -2.25 |
| 2 | -1.79 | -1.80 | -1.81 | -1.82 | -1.83 | -1.84 | -1.85 | -1.85 |
| 3 | -1.55 | -1.57 | -1.58 | -1.59 | -1.60 | -1.61 | -1.62 | -1.63 |
| 4 | -1.38 | -1.40 | -1.41 | -1.42 | -1.43 | -1.44 | -1.45 | -1.46 |
| 5 | -1.25 | -1.26 | -1.27 | -1.28 | -1.30 | -1.31 | -1.32 | -1.33 |
| 6 | -1.13 | -1.14 | -1.16 | -1.17 | -1.18 | -1.19 | -1.21 | -1.22 |
| 7 | -1.02 | -1.04 | -1.05 | -1.07 | -1.08 | -1.09 | -1.11 | -1.12 |
| 8 | -0.93 | -0.95 | -0.96 | -0.98 | -0.99 | -1.00 | -1.02 | -1.03 |
| 9 | -0.84 | -0.86 | -0.88 | -0.89 | -0.91 | -0.92 | -0.94 | -0.95 |
| 10 | -0.76 | -0.78 | -0.80 | -0.81 | -0.83 | -0.84 | -0.86 | -0.87 |
| 11 | -0.69 | -0.71 | -0.72 | -0.74 | -0.76 | -0.77 | -0.79 | -0.80 |
| 12 | -0.62 | -0.64 | -0.65 | -0.67 | -0.69 | -0.70 | -0.72 | -0.74 |
| 13 | -0.55 | -0.57 | -0.59 | -0.60 | -0.62 | -0.64 | -0.66 | -0.67 |
| 14 | -0.48 | -0.50 | -0.52 | -0.54 | -0.56 | -0.58 | -0.59 | -0.61 |
| 15 | -0.42 | -0.44 | -0.46 | -0.48 | -0.50 | -0.52 | -0.53 | -0.55 |
| 16 | -0.36 | -0.38 | -0.40 | -0.42 | -0.44 | -0.46 | -0.48 | -0.49 |
| 17 | -0.29 | -0.32 | -0.34 | -0.36 | -0.38 | -0.40 | -0.42 | -0.44 |
| 18 | -0.23 | -0.26 | -0.28 | -0.30 | -0.32 | -0.34 | -0.36 | -0.38 |
| 19 | -0.17 | -0.20 | -0.22 | -0.25 | -0.27 | -0.29 | -0.31 | -0.33 |
| 20 | -0.12 | -0.14 | -0.17 | -0.19 | -0.21 | -0.24 | -0.26 | -0.28 |
| 21 | -0.06 | -0.09 | -0.11 | -0.14 | -0.16 | -0.18 | -0.21 | -0.23 |
| 22 | 0.00 | -0.03 | -0.06 | -0.08 | -0.11 | -0.13 | -0.15 | -0.18 |
| 23 | | | 0.00 | -0.03 | -0.05 | -0.08 | -0.10 | -0.13 |
| 24 | | | | | 0.00 | -0.03 | -0.05 | -0.07 |
| 25 | | | | | | | 0.00 | -0.02 |

*z for order statistic $X_{(i)}$ where $i > n/2$ is -z of order statistic $X_{(j)}$ where $j = n + 1 - i$.

This plotting can be achieved by using already prepared normal probability paper such as shown in Figure 2.14 or by using arithmetic paper where the z of (2.19) is approximated by

$$z = \text{sign}(F_n(x) - .5)(1.238t(1 + 0.0262t)) \qquad (2.22)$$

Here

$$t = \{-\ln[4F_n(x)(1 - F_n(x))]\}^{\frac{1}{2}} \qquad (2.23)$$

and

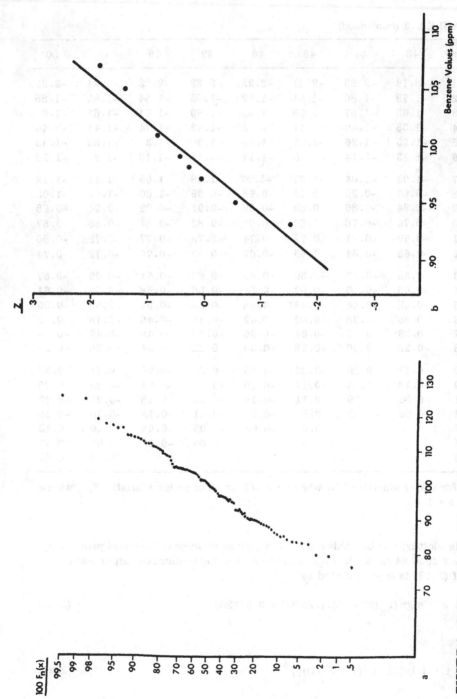

FIGURE 2.15 Normal probability plots. (a) NOR data set. (b) Dosimeter data set.

$$\text{sign}(F_n(x) - .5) = \begin{cases} +1 \text{ if } F_n(x) - .5 > 0 \\ -1 \text{ if } F_n(x) - .5 < 0 \end{cases} \qquad (2.24)$$

This approximation to z is given in Hamaker (1978) and appears to be of sufficient accuracy for plotting. Notice this z function defined by (2.22) and (2.24) can be programmed easily and so permits the use of simple computer graphics for performing normal probability plotting (see Section 2.3.6 for further details).

For small samples (say less than 50 observations) the z of (2.22) should be replaced with the expected values of the order statistics from the standard normal distribution—i.e., the distribution with $\mu = 0$ and $\sigma = 1$ (Harter, 1961). Table 2.3 contains these for sample sizes up to 50. The normal probability plots for samples smaller than 25 can show substantial variation and nonlinearity even if the underlying distribution is normal (see, for example, Daniel and Wood, 1971, and Hahn and Shapiro, 1967). We caution the reader against placing too much reliance upon a plot in these situations. Remember, in general and especially for these situations, graphs should be used for informal preliminary judgments and/or as adjuncts to formal numerical techniques. Chapter 9 contains the formal techniques for testing for normality.

## E 2.4.1.1 Examples

Figure 2.15 contains two normal probability plots. Figure 2.15a is a plot of the NOR data set already extensively discussed in Section 2.2. Figure 2.15b is a plot from a sample of 20 dosimeter readings of benzene (D'Agostino and Gillespie, 1978). A dosimeter is a portable device for measuring a person's exposure to various gases. The dosimeter data are in parts per million (ppm). The frequency distribution and plotting points z are:

Dosimeter Data for Measuring Benzene

| Data values (ppm) | Frequencies | Expected values order statistics (z) |
|---|---|---|
| .93 | 3 | -1.47 |
| .95 | 6 | -.53 |
| .97 | 3 | .06 |
| .98 | 1 | .32 |
| .99 | 1 | .45 |
| 1.01 | 4 | .85 |
| 1.05 | 1 | 1.41 |
| 1.07 | 1 | 1.87 |
| | 20 | |

Notice in Figure 2.15b we plotted only 8 points since only eight different values appeared in the sample. The z values are averaged in the case of the ties. The line drawn in Figure 2.15b is the line $\bar{x} + zs$, where $\bar{x} = .98$ and $s = .04$.

For grouped data (i.e., data grouped into frequency classes) only one value per class should be plotted. This plotted value should be the true upper limit of the class (see Section 2.2.1 for an illustration of true upper limits and Section 2.3.5 for more details).

### 2.4.2 Deviations from Normality

#### 2.4.2.1 Unimodal Distributions

A useful way to distinguish unimodal non-normal distributions from the normal is in terms of the skewness and kurtosis measures defined as

$$\text{Skewness: } \sqrt{\beta_1} = \frac{\mu_3}{\mu_2^{3/2}} = \frac{E(X - \mu)^3}{\{E(X - \mu)^2\}^{3/2}} \tag{2.25}$$

and

$$\text{Kurtosis: } \beta_2 = \frac{\mu_4}{\mu_2^2} = \frac{E(X - \mu)^4}{\{E(X - \mu)^2\}^2} \tag{2.26}$$

For the normal distribution $\sqrt{\beta_1} = 0$ and $\beta_2 = 3$. The sample estimators of these and the tests of fit based on them are discussed in Chapters 7 and 9. Figure 2.16 contains normal probability plots of four data sets of the appendix which represent various combinations of $\sqrt{\beta_1}$ and $\beta_2$. Figure 2.16a and 2.16b are plots of symmetric distributions. Notice for $\beta_2 < 3$ (UNI data, $\beta_2 = 1.80$) the plot is, within sampling error, antisymmetric about the median, being concave for $x <$ median and convex for $x >$ median. For $\beta_2 > 3$ (SU(0,2) data, $\beta_2 = 4.51$) the plot is again, within sampling error, antisymmetric about the median. Now, however, it is convex for $x <$ median and concave for $x >$ median. (See Figure 2.18 for further illustrations.) Notice also for skewed distributions (Figures 2.16c and 2.16d) the plots are either convex or concave throughout.

#### 2.4.2.2 Outliers, Mixtures and Contamination

Figure 2.17 illustrates the use of normal probability plotting for the detection of outliers and the presence of mixtures (or contamination). For previous discussions see Sections 2.2.3 and 2.2.4, respectively. Figure 2.17a is a plot of the data whose ecdf is given in Figure 2.5b (i.e., first nine observations of the NOR data set plus one outlier equal to 140). Notice how the point

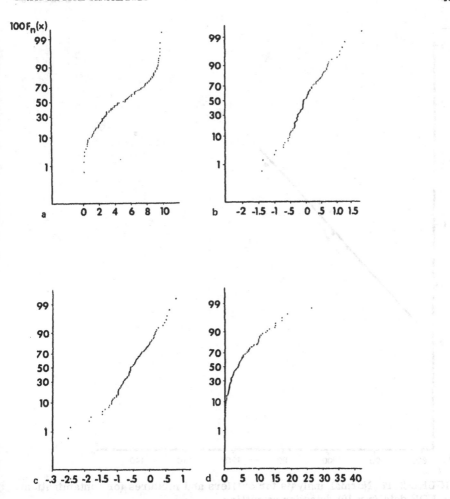

FIGURE 2.16  Normal probability plots for nonnormal unimodal distributions. Symmetric distributions ($\sqrt{\beta_1} = 0$). (a)  UNI data $\beta_2 = 1.80$. (b)  SU(0,2), $\beta_2 = 4.51$. Skewed distributions ($\sqrt{\beta_1} \neq 0$). (c) SU(1,2) data $\sqrt{\beta_1} = -.87$, $\beta_2 = 5.59$. (d) EXP data $\sqrt{\beta_1} = 2$, $\beta_2 = 9$.

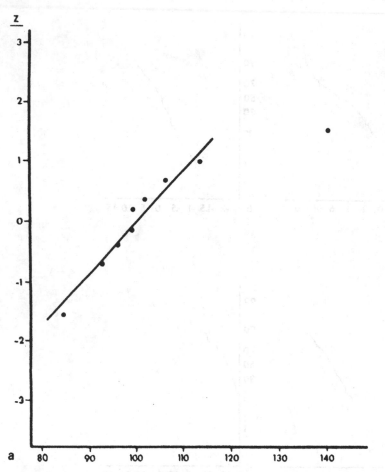

FIGURE 2.17 Normal analysis for outliers and mixtures (or contamination).
(a) NOR data (n = 10) detection of outlier.

corresponding to the observation 140 is clearly out of line with the rest of
the data. In practice the techniques of Chapter 12 should now be used to confirm that this point is an outlier.

Figures 2.17b and 2.17c are normal probability plots of the contaminated normal data sets LCN(.10,3) and LCN(.20,3) whose ecdfs are given,
respectively, in Figures 2.7a and 2.7b. The reader should note two important related points concerning both Figures 2.17b and 2.17c. First, both
reveal the presence of two straight lines. This is seen, for example, in
Figure 2.17b [LCN(.1,3) data set] where one straight line can be fit nicely
through the data below the 80th percentile of the sample and a second straight
line can be fit through the data from about the 92nd percentile up. The points

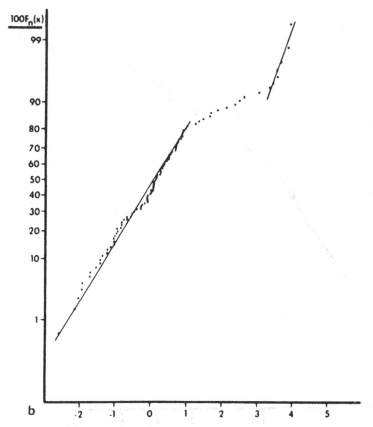

(b)  LCN(.10,3) data, contaminated normal.

from the 80th to the 92nd percentiles represent a contaminated or transition zone where the two distributions cannot be clearly separated. Second, neither plot displays a convex nor a concave pattern throughout. One of these patterns would be the case if we had simply skewed distributions under analysis here as we did in Figures 2.16c and 2.16d. Recall the ecdfs of the LCN(.10, 3 and LCN(.20,3) data sets could not be distinguished from those of skewed distributions. With probability plotting the underlying components can surface as straight line segments in the plot, and so do produce a completely different effect than what is produced by a unimodal skewed distribution (see Figure 2.18 for further illustration). Once it is established that there are two or more components in the data, the next step is to estimate the param-

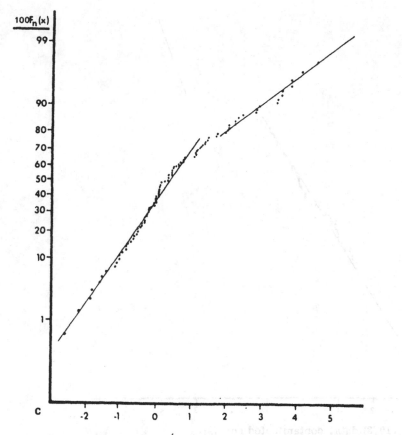

FIGURE 2.17 (continued)  (c)  LCN(.20,3) data, contaminated normal.

eters of the components. The reader is referred to Bliss (1967) and Johnson and Kotz (1970) for further details.

## 2.4.2.3 Recognizing and Responding to Nonnormality

Figure 2.18 provides guidelines to aid the user in interpreting normal probability plots. Notice in the drawings of Figure 2.18 the empirical cumulative distribution function and/or z scale is on the <u>vertical axis</u>. Some graphs have these on the horizontal axis. The resulting configurations will be different if this is done.

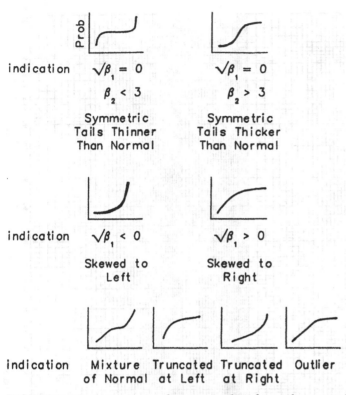

indication   $\sqrt{\beta}_1 = 0$        $\sqrt{\beta}_1 = 0$

             $\beta_2 < 3$               $\beta_2 > 3$

             Symmetric               Symmetric
             Tails Thinner           Tails Thicker
             Than Normal             Than Normal


indication   $\sqrt{\beta}_1 < 0$        $\sqrt{\beta}_1 > 0$

             Skewed to               Skewed to
             Left                    Right


indication   Mixture  Truncated  Truncated  Outlier
             of Normal  at Left   at Right

FIGURE 2.18 Indications of nonnormality from the normal probability plots.

## 2.5 LOGNORMAL PROBABILITY PLOTTING

### 2.5.1 Probability Plotting for Two Parameter Lognormal

The two parameter lognormal distribution has density

$$f(x) = \frac{1}{x\sigma\sqrt{2\pi}} e^{-(\ln x - \mu)^2/2\sigma^2}, \quad x > 0 \tag{2.27}$$

The random variable $Y = \ln X$ has a normal distribution with mean $\mu$ and standard deviation $\sigma$. Probability plotting for this distribution can be achieved in a number of ways: (1) on already prepared lognormal probability paper (Figure 2.19), (2) on normal probability paper such as shown in Figure 2.14 where x of the horizontal axis is replaced by log x, (3) on arithmetic graph paper where z of (2.22) to (2.24) is the variable of the vertical axis and log x is the variable of the horizontal axis, or (4) on semi-log graph paper

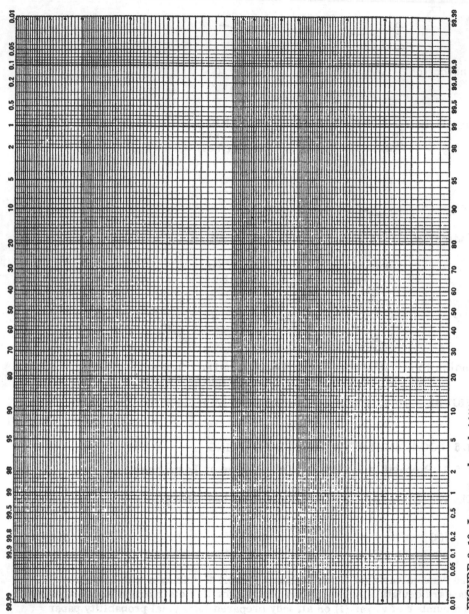

FIGURE 2.19 Lognormal probability paper.

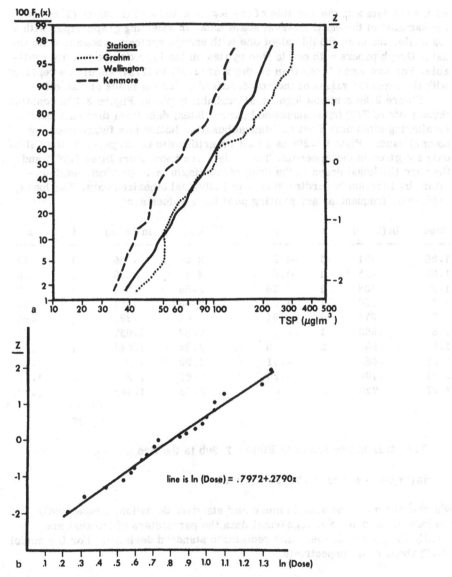

FIGURE 2.20  Lognormal probability plots.  (a)  Total suspended particulates.  (b)  CHEN data set.

where the data x is the variable of the log axis and z of (2.22) to (2.24) is the variable of the equal interval scale axis. In selecting graph paper with a log scale, the user should select one with enough cycles to accommodate the data. Graph papers with one to five cycles on the log axis are readily available. For samples of less than 50 the z of (2.22) to (2.24) should be replaced with the expected values of the standardized order statistics of Table 2.3.

Figure 2.20 contains lognormal probability plots. Figure 2.20a contains three plots of TSP (total suspended particulates) data from three air quality monitoring sites near Boston, Massachusetts. Notice this figure uses log normal paper. Figure 2.20b is a plot on arithmetic graph paper of the CHEN data set given in the Appendix. These data are taken from Bliss (1967) and they are the lethal doses of the drug cinobrifagin in 10 (mg/kg), as determined by titration to cardiac arrest in individual etherized cats. The loses, ln (doses), frequencies and plotting position z values are:

| Dose | ln (Dose) | f | z | Dose | ln (Dose) | f | z |
|------|-----------|---|------|------|-----------|---|------|
| 1.26 | .231 | 1 | −1.97 | 2.34 | .850 | 1 | .10 |
| 1.37 | .315 | 1 | −1.52 | 2.41 | .880 | 1 | .20 |
| 1.55 | .438 | 1 | −1.26 | 2.56 | .940 | 1 | .30 |
| 1.71 | .536 | 1 | −1.07 | 2.63 | .967 | 2 | .46 |
| 1.77 | .571 | 1 | −.91 | 2.67 | .982 | 1 | .64 |
| 1.81 | .593 | 1 | −.76 | 2.82 | 1.037 | 2 | .84 |
| 1.89 | .637 | 2 | −.58 | 2.84 | 1.044 | 1 | 1.07 |
| 1.98 | .683 | 1 | −.41 | 2.99 | 1.095 | 1 | 1.26 |
| 2.03 | .708 | 3 | −.20 | 3.65 | 1.295 | 1 | 1.52 |
| 2.07 | .728 | 1 | .0 | 3.83 | 1.343 | 1 | 1.97 |

25

The straight line drawn in Figure 2.20b is the line

$$\ln (\text{Dose}) = \hat{\mu} + z\hat{\sigma} = .7972 + z(.2790)$$

where $\hat{\mu}$ and $\hat{\sigma}$ are the sample mean and standard deviation, respectively, of the logs of the data. For lognormal data the parameters of interest are usually the geometric mean and geometric standard deviation. For the model (2.27) these are, respectively,

$$e^{\mu} \text{ and } e^{\sigma}$$

Estimates of these based on the data in Figure 2.20b are exp (.7972) = 2.2193 and exp (.2790) = 1.3218. For data plotted directly on log normal paper exp ($\mu$) is estimated as the 50th percentile (i.e., x value corresponding to z = 0) and exp ($\sigma$) is estimated as the ratio of the 84th percentile to the 50th percentile (i.e., the ratio of the x value corresponding to z = 1 to the x value corresponding to z = 0).

### 2.5.1.1 Zero Data Values

At times, when dealing with a set of data that appears to be lognormally distributed, there may be a subset of these observations that are all equal to zero. Before the data can be plotted these zeros must be "adjusted." First, it is possible that they represent a contamination and simply should be removed. Second, they may reflect a measurement limitation of the measurement instrument. In this case it may be justified to replace them with the "least detectable level" of the instrument. If this is not known then it may be possible to adjust the zeros by adding a small arbitrary constant to them or to all the data values before they are plotted. Careful consideration should be given before any of these suggestions are employed.

### 2.5.2 Three Parameter Lognormal

The three parameter lognormal distribution has density

$$f(x) = \frac{1}{x\sigma \sqrt{2\pi}} e^{-(\ln(x-\lambda)-\mu)^2/2\sigma^2}, \quad x > \lambda \qquad (2.28)$$

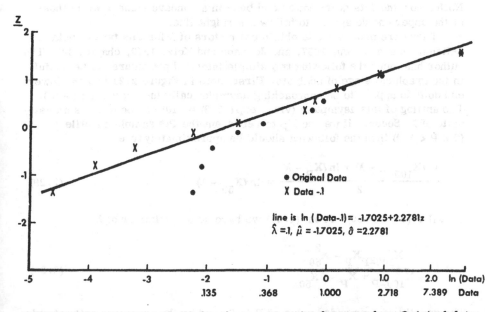

FIGURE 2.21 Lognormal plot for three parameter lognormal. ● Original data; X Data −.1. Line is ln(Data −.1) = −1.7025 + 2.2781z; $\hat{\lambda}$ = .1, $\hat{\mu}$ = −1.7025, $\hat{\sigma}$ = 2.2781.

A plot of data from this distribution for a lognormal analysis will not produce a straight line unless the $\lambda$ value or an estimate of it is subtracted from all the data. The data in Figure 2.21 illustrate the situation. These data come from Leidel, Busch and Lynch (1977) and represent readings of hydrogen fluoride. The dots represent the unadjusted data. These data and the plotting positions for z are given in the first four columns below:

| Data (ppm) | f | z | ln (Data) | ln (Data - .1) |
|---|---|---|---|---|
| .11 | 2 | -1.38 | -2.21 | -4.61 |
| .12 | 1 | -.79 | -2.12 | -3.91 |
| .14 | 2 | -.42 | -1.97 | -3.22 |
| .21 | 1 | -.10 | -1.56 | -2.21 |
| .33 | 1 | .10 | -1.11 | -1.47 |
| .80 | 1 | .31 | -.22 | -.36 |
| .91 | 1 | .54 | -.09 | -.21 |
| 1.30 | 1 | .79 | .26 | .18 |
| 2.60 | 1 | 1.12 | .96 | .92 |
| 10.00 | 1 | 1.63 | 2.30 | 2.29 |
| | 12 | | | |

Notice how the dots at the lower end bend in a concave manner while those at the upper end do appear to follow a straight line.

There are many ways to obtain estimators of $\lambda$ for this type of data (Aitchinson and Brown, 1957, and Johnson and Kotz, 1970, chapter 14). The author has found the following two simple informal procedures to be useful in the graphical stage of analysis. First, note in Figure 2.21 that the lower end dots do appear to be approaching asymptotically the log value of -2.3. The antilog of this asymptote (viz., $\exp(-2.3) = .10$) can be used as an estimate of $\lambda$. Second, if we use $X_P$ to represent the Pth sample quantile $(0 < P < 100)$ then the following should be approximately true

$$\frac{\ln (X_{100-P} - \lambda) + \ln (X_P - \lambda)}{2} = \ln (X_{50} - \lambda) \qquad (2.29)$$

for all P $(0 < P < 50)$. From (2.29) we have as an estimator of $\lambda$

$$\hat{\lambda} = \frac{X_{100-P} X_P - X_{50}^2}{X_{100-P} + X_P - 2X_{50}} \qquad (2.30)$$

The usually recommended value of P is 5, and so the suggested estimator is

$$\hat{\lambda} = \frac{X_{95}X_5 - X_{50}^2}{X_{95} + X_5 - 2X_{50}} \tag{2.31}$$

As a value for the Pth sample quantile, the user can use either the ith order statistics $X_{(i)}$ where

$$i = [n(.01P)] + 1 \tag{2.32}$$

(here [y] is the largest integer in y) or else obtain it directly from a graph. That is, draw a curve by hand through the data and use of the Pth quantile the x value on the horizontal axis corresponding to $100F_n(x) = $ Pth value on the vertical axis. Applying (2.31) to the data of Figure 2.21 we again obtain .1 as a good approximation to λ. Figure 2.21 also contains a plot of the data minus this .1 value (i.e., ln(Data - .1)). This plot is given as x values on the graph. A straight line now does fit reasonably well these values indicating the appropriateness of the lognormal distribution.

### 2.5.3 Responding to Lognormal Plots

Figure 2.22 provides guidelines to aid the user in interpreting lognormal probability plots. As with Figure 2.18 the drawings in Figure 2.22 have the

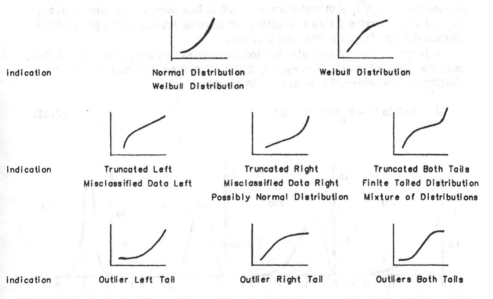

FIGURE 2.22 Indications of non-lognormality from nonlinearity of lognormal probability plots.

empirical cumulative distribution function and/or z scale on the <u>vertical</u> <u>axis</u>.

## 2.6 WEIBULL PROBABILITY PLOTTING

### 2.6.1 Probability Plotting for Two Parameter Weibull

The two parameter Weibull distribution has density

$$f(t) = \frac{k}{\theta}\left(\frac{t}{\theta}\right)^{k-1} e^{-(t/\theta)^k} \tag{2.33}$$

and cdf

$$F(t) = 1 - e^{-(t/\theta)^k} \tag{2.34}$$

where $\theta$, $k$, $t > 0$. This is a very versatile distribution and by varying the parameter k it can assume a large number of different shapes. For example, when $k = 1$ the Weibull distribution is the negative exponential distribution, when k is in the neighborhood of 3.6 the Weibull distribution is similar in shape to the normal distribution, when $k < 3.6$ the Weibull has positive skewness (i.e., $\sqrt{\beta_1} > 0$) and when $k > 3.6$ it has negative skewness (i.e., $\sqrt{\beta_1} < 0$). To illustrate this versatility we have in Figure 2.23a plots of the Weibull density for four different k values.

A large number of probability plotting papers are available for a Weibull analysis (see Nelson and Thompson, 1971). However, Weibull probability plotting can be achieved also simply by plotting

$$z = \ln\left(-\ln\left(1 - F_n(t)\right)\right) \text{ on } \ln t \tag{2.35}$$

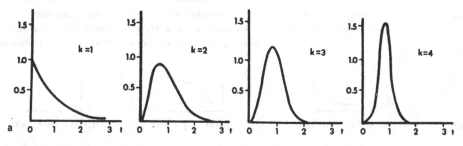

FIGURE 2.23 Weibull analysis. (a) Weibull densities for different k values ($\theta = 1$).

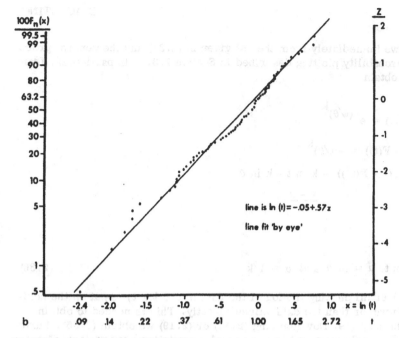

line is ln (t) = −.05+.57z

line fit 'by eye'

(b)  Weibull analysis of WE2 data.

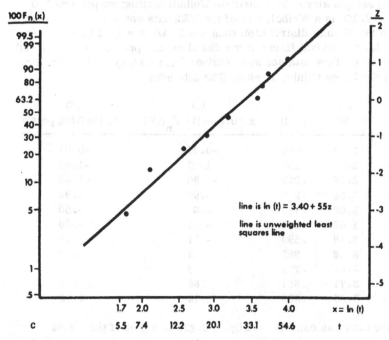

line is ln (t) = 3.40 + 55z

line is unweighted least
squares line

(c)  Weibull analysis of cancer survival data.

This follows immediately from the cdf given in (2.34) and the general procedure for probability plotting described in Section 2.3.1. In particular, from
(2.34) we obtain

$$1 - F(t) = e^{-(t/\theta)^k}$$

$$\ln(1 - F(t)) = -(t/\theta)^k$$

$$\ln(-\ln(1 - F(t))) = k \ln t - k \ln \theta$$

$$= \frac{x - \mu}{\sigma}$$

where

$$x = \ln t, \ \mu = \ln \theta \text{ and } \sigma = 1/k \tag{2.36}$$

Note with Weibull plotting the log of the data (i.e., log t) is put on the horizontal axis rather than the data values directly. This is needed to obtain
linearity in the plots. Now applying (2.11) or (2.19) we obtain (2.35). Further
while small sample expected values of order statistics are available (Mann,
1968) Weibull probability plotting usually works well by simply employing
(2.35) for all sample sizes. To illustrate Weibull plotting we present two
plots. Figure 2.23b is a Weibull plot of the WE2 data set (i.e., 100 observations from the Weibull distribution with k = 2 and $\theta$ = 1). Figure 2.23c
is a plot of eleven survival times in months of cancer patients who have had
an adrenalectomy. This data set was obtained from a study by Dr. Richard
Oberfield of the Lahey Clinic, Boston. The data are:

| (1) Rank | (2) t | (3) $x = \ln t$ | (4) $F_n(t)$ | (5) $z = \ln(-\ln(1 - F_n(t)))$ | (6) $z_i$ Plotting points |
|---|---|---|---|---|---|
| 1 | 6 | 1.79 | .045 | -3.07 | -3.07 |
| 2 | 9 | 2.20 | .136 | -1.92 | -1.92 |
| 3 | 13 | 2.56 | .227 | -1.36 | -1.36 |
| 4 | 18 | 2.89 | .318 | -.96 | -.96 |
| 5 | 22 | 3.09 | .409 | -.64 | -.50 |
| 6 | 22 | 3.09 | .500 | -.37 | -.50 |
| 7 | 36 | 3.58 | .590 | -.11 | .02 |
| 8 | 36 | 3.58 | .682 | .14 | .02 |
| 9 | 37 | 3.61 | .773 | .39 | .39 |
| 10 | 41 | 3.71 | .864 | .69 | .69 |
| 11 | 52 | 3.95 | .955 | 1.13 | 1.13 |

Following our previous convention Figure 2.23c is a plot of the z's of
column (6) on the x's of column (3).

The straight line drawn in Figure 2.23b was drawn using the "by eye" technique described in Section 2.3.3.1. From the line we obtain

$$x = \ln t = \hat{\mu} + z\hat{\sigma} = -.05 + z(.47)$$

Notice $\mu$ and $\sigma$ are location and scale parameters of the distribution of ln T. However, they are not the mean and standard deviation. Using (2.36) we next obtain

$$\hat{\theta} = \exp(\hat{\mu}) = .95 \quad \text{and} \quad \hat{k} = 1/\hat{\sigma} = 2.13$$

Recall the true parameter values are $\theta = 1$ and $k = 2$. The informal "by eye" technique did well in this case.

Because the sample size is small for the data in Figure 2.23c we use the unweighted least squares estimates of (2.17) to obtain the estimates of $\mu$ and $\sigma$ from this data (columns (3) and (5) of the data above were used for the x's and z's, respectively). The estimates are $\hat{\mu} = 3.40$ and $\hat{\sigma} = .55$. From these we obtain $\hat{\theta} = \exp(3.40) = 29.96$ and $\hat{\sigma} = 1/.55 = 1.82$.

### 2.6.1.1 Zero Data

As with the lognormal distribution there may be a subset of zero values in data that otherwise appears to have a Weibull distribution. For an example dealing with wind speed data see Takle and Brown (1978). The recommended procedures for dealing with these zero values are exactly those given for the lognormal in Section 2.5.1.1.

### 2.6.2 Three Parameter Weibull

The three parameter Weibull has density

$$f(t) = \frac{k}{\theta}\left(\frac{t-\lambda}{\theta}\right)^{k-1} e^{-[(t-\lambda)/\theta]^k}, \quad t > \lambda \tag{2.37}$$

Similar to the three parameter lognormal distribution the $\lambda$ value must be subtracted from all the data before a Weibull plot will produce a straight line. In many cases a value close to the minimum t is adequate as an estimate of $\lambda$. Other more precise techniques can be employed (Johnson and Kotz, 1970).

## 2.7 OTHER TOPICS

There are a number of other topics, for which due to space limitation, we cannot give detail treatments. Some of these are:

1. <u>Analysis of Residuals from a Model</u>. All of the above material
applies directly if we are dealing with residuals from a model. That is,
say we have a mathematical model

$$Y = h(X, \beta) + \eta \qquad (2.38)$$

where Y represents a random variable, X a vector of random variables or
known constants, $\beta$ a vector of unknown parameters and $\eta$ an error term.
For example, (2.38) can represent a multiple regression model. If the $\beta$
vector is estimated from data, say estimator is $\hat{\beta}$, then the estimator of Y
is

$$\hat{Y} = h(X, \hat{\beta}) \qquad (2.39)$$

and

$$\hat{\eta} = Y - \hat{Y} \qquad (2.40)$$

is the residual. While the residuals comprise a dependent sample the graphical techniques described above can be used to analyze them. Chapter 12 on
outliers discusses further the analysis of residuals.

2. <u>Analysis of Censored Samples</u>. There are no restrictions in the
above which make it necessary for all the data to be available for plotting.
The above techniques can be used on censored data. There are, however,
special concerns which arise with censored data (e.g., see Nelson, 1972)
that require careful and complete discussion. Chapter 11 is devoted solely
to this problem.

3. <u>Q-Q Plots and P-P Plots</u>. Wilk and Gnanadesikan (1968) discuss in
detail the quantile-quantile probability plots (Q-Q plots) and the percentage-
percentage probability plots (P-P plots). A Q-Q plot is the plot of the quantiles (or, as we call them above, the percentiles) of one distribution on the
quantiles of a second distribution. If one of these distributions is a hypothesized theoretical distribution a Q-Q plot is just a probability plot as developed in Section 2.3. A P-P plot is the plot of the percentages of one distribution on the percentages of a second. Wilk and Gnanadesikan (1968) state
that while the P-P plots are limited in their usefulness, they are useful for
detecting discrepancies in the middle of a distribution (i.e., about the median)
and also may be useful for multivariate analysis.

4. <u>Transformation to Normality</u>. In some settings (e.g., analysis of
variance) it is suggested first to transform to normality and then analyze
the transformed data. Box and Cox (1964) suggest a power transformation for
this. Their transformation is as follows:

$$y = \begin{cases} x^{\theta} & \text{if } \theta > 0 \\ \log x & \text{if } \theta = 0 \\ -x^{\theta} & \text{for } \theta < 0 \end{cases} \qquad (2.41)$$

Here x refers to the original data and $\theta$ is the power exponent. Box and Cox develop a maximum likelihood estimator for $\theta$. Once it is computed normal probability plotting as developed in Section 2.4 can be applied directly to the transformed data y of (2.41). The techniques of Chapter 9 can be used to test formally the normality of the transformed data.

5. Probability Plotting for the Gamma Distribution. One important distribution that does not lend itself immediately to the probability plotting techniques described above is the Gamma distribution. Even with the aid of transformations it cannot be put in the simple form of a distribution dependent upon a location and scale parameter. Wilk, Gnanadesikan and Huyett (1962) present a technique and accompanying tables to handle this situation. See Chapter 11 for further comments on this.

6. Multivariate Normality. There has been much attention paid to the problem of probability plotting for the multivariate normal distribution and a number of techniques have been suggested (Gnanadesikan, 1973). The author has found the following technique to be very informative. First transform the data to principal components and then do univariate normal probability plotting (Section 2.4) for each component. Each component can be considered an independent variable. If the original data set is from a multivariate normal distribution then each component should produce a straight line in the univariate plots. More will be said about multivariate normality in Chapter 9.

## 2.8 CONCLUDING COMMENT

The aim of this chapter has been to present to the reader simple informal graphical techniques which can be used in conjunction with the formal techniques to be discussed in the following chapters. In performing an analysis we suggest that the reader should draw a graph, examine it and judge if other graphs are needed. As the formal numerical techniques are being applied use the graphs to interpret them and to gain insight into the phenomenon under investigation.

## REFERENCES

Aitchinson, J. and Brown, J. A. C. (1957). The Lognormal Distribution. Cambridge University Press, London.

Andrews, D. F. (1972). Plots of high dimensional data. Biometrika 28, 125-136.

Anscombe, F. J. (1973). Graphs in statistical analysis. The Amer. Statistician 27(1), 17-21.

Barnett, V. (1975). Probability plotting methods and order statistics. J. R. Statist. Soc. C 24, 95-108.

Bliss, C. I. (1967). Statistics in Biology, Vol. 1. McGraw-Hill, New York.

Box, G. E. P. and Cox, D. R. (1964). An analysis of transformations. J. R. Statist. Soc. B 26, 211-252.

Bryson, M. C. (1974). Heavy-tailed distributions: properties and test. Technometrics 16, 61-68.

Chambers, J. M., Cleveland, W. S., Kleiner, B. and Tukey, P. A. (1983). Graphical Methods for Data Analysis. Duxbury Press, Boston.

Chernoff, H. and Lieberman, G. (1954). Use of normal probability paper. Jour. Amer. Stat. Assoc. 49, 778-785.

Chernoff, H. and Lieberman, G. (1956). The use of generalized probability paper for continuous distributions. Annals of Math. Stat. 27, 806-818.

Curran, T. C. and Frank, N. H. (1975). Assessing the validity of the log-normal and model when predicting maximum air pollution concentrations. Presented at the 68th Annual Meeting of the Air Pollution Control Association, Boston, Mass.

D'Agostino, R. B. and Lee, A. F. S. (1976). Linear estimation of the logistic parameters for complete or tail-censored samples. Jour. Amer. Stat. Assoc. 71, 462-464.

D'Agostino, R. B. and Gillespie, J. C. (1978). Comments on the OSHA accuracy of measurement requirement for monitoring employee exposure to benzene. Amer. Industrial Hygiene Assoc. Journ. 39, 510-513.

Daniel, C. (1959). Use of Rulf-normal plots in interpreting factorial two-level experiments. Technometrics 1, 311-411.

Daniel, D. and Wood, F. S. (1971). Fitting Equations to Data. Wiley, New York.

Feder, P. I. (1974). Graphical techniques in statistical data analysis. Technometrics 16, 287-299.

Gentleman, J. F. (1977). It's all a plot (using interactive computer graphics in teaching statistics). The Amer. Statistician 31(4), 166-175.

Gnanadesikan, R. (1973). Graphical methods for informal inference in multivariate data analysis. Bull. of the International Stat. Institute. Proceedings of the 39th Session.

Gnanadesikan, R. and Lee, E. T. (1970). Graphical techniques for internal comparisons amongst equal degree of freedom groupings in multiresponse experiments. Biometrika 57, 229-237.

Grubbs, F. E. (1969). Procedures for detecting outlying observations in samples. Technometrics 11, 1-21.

Gupta, S. S. and Shah, R. K. (1965). Exact moments and percentage points of the order statistics and the distribution of the range from the logistic distribution. Ann. Math. Statist. 36, 907-920.

Gupta, S. S., Qureishi, A. S., and Shah, B. K. (1967). Best linear unbiased estimators of the parameters of the logistic distribution using order statistics. Technometrics 9, 43-56.

Hahn, G. J. and Shapiro, S. S. (1967). Statistical Models in Engineering. Wiley, New York.

Hamaker, H. C. (1978). Approximating the cumulative normal distribution and its inverse. Applied Statist. 27, 76-79.

Harding, J. P. (1949). The use of probability paper for the graphical analysis of polynomial frequency distributions. Jour. of Marine Biology Assoc., United Kingdom 28, 141-153.

Harter, H. L. (1961). Expected values of normal order statistics. Biometrika 48, 151-165.

Johnson, N. L. and Kotz, S. (1970). Continuous Univariate Distributions, Vol. 1. Wiley, New York.

Kimball, B. Z. (1960). On the choice of plotting positions on probability paper. Jour. Amer. Stat. Assoc. 55, 546-560.

Leidel, N. A., Busch, K. A., and Lynch, J. R. (1977). Occupational Exposure Sampling Strategy Manual. U.S. Dept. of H. E. W., Cincinnati, Ohio.

Mann, N. R. (1968). Results on statistical estimation and hypothesis testing with application to the Weibull and extreme-value distributions. Aerospace Research Laboratories Report ARL 68-0068, U.S. Air Force, Wright-Patterson Air Force Base, Ohio.

Mosteller, F. and Tukey, J. W. (1977). Data Analysis and Regression. Addison-Wesley, Reading, Mass.

Nelson, L. S. (1977). Graphical aid for drawing normal distributions. Jour. Quality Technology 9, 42-43.

Nelson, W. (1972). Theory and application of hazard plotting for censored failure data. Technometrics 14, 945-966.

Nelson, W. and Thompson, V. C. (1971). Weibull probability papers. Jour. of Quality Technology 3, 45-50.

Rényi, A. (1970). Probability Theory. North Holland Publishing, Amsterdam.

Takle, E. S. and Brown, J. M. (1978). Note on the use of Weibull statistics to characterize wind-speed data. Journal of Applied Meteorology 17, 556-559.

Taylor, B. J. R. (1965). The analysis of polymodal frequency distributions. Jour. of Animal Ecology 34, 445–452.

Tukey, J. W. (1971). Dynamic and static display of multidimensional data: a number of suggestions. Unpublished mimeo report.

Tukey, J. W. (1977). Exploratory Data Analysis. Addison-Wesley, Reading, Mass.

Wilk, M. B., Gnanadesikan, R. and Huyett, M. J. (1962). Probability plots for the gamma distribution. Technometrics 4, 1–15.

Wilk, M. B. and Gnanadesikan, R. (1964). Graphical methods for internal comparisons in multiresponse experiments. Ann. Math. Statist. 35, 613–631.

Wilk, M. B. and Gnanadesikan, R. (1968). Probability plotting methods for the analysis of data. Biometrika 55, 1–19.

# 3

# Tests of Chi-Squared Type

David S. Moore  Purdue University, West Lafayette, Indiana

## 3.1 INTRODUCTION

In the course of his <u>Mathematical Contributions to the Theory of Evolution</u>, Karl Pearson abandoned the assumption that biological populations are normally distributed, introducing the Pearson system of distributions to provide other models. The need to test fit arose naturally in this context, and in 1900 Pearson invented his chi-squared test. This statistic and others related to it remain among the most used statistical procedures.

Pearson's idea was to reduce the general problem of testing fit to a multinomial setting by basing a test on a comparison of observed cell counts with their expected values under the hypothesis to be tested. This reduction in general discards some information, so that tests of chi-squared type are often less powerful than other classes of tests of fit. But chi-squared tests apply to discrete or continuous, univariate or multivariate data. They are therefore the most generally applicable tests of fit.

Modern developments have increased the flexibility of chi-squared tests, especially when unknown parameters must be estimated in the hypothesized family. This chapter considers two classes of chi-squared procedures. One, called "classical" because it contains such familiar statistics as the log likelihood ratio, Neyman modified chi-squared, and Freeman-Tukey, is discussed in Section 3.2. The second, consisting of nonnegative definite quadratic forms in the standardized cell frequencies, is the main subject of Section 3.3. Other newer developments relevant to both classes of statistics, especially the use of data-dependent cells, are also treated primarily in 3.3, while such practical considerations as choice of cells and accuracy of asymptotic approximate distributions appear in 3.2. Both sections contain a number of examples.

Tests of the types considered here are also used in assessing the fit of models for categorical data. The scope of this volume forbids venturing into this closely related territory. Bishop, Fienberg, and Holland (1975) discuss the methods of categorical data analysis most closely related to the contents of this chapter.

## 3.2  CLASSICAL CHI-SQUARED STATISTICS

To test the simple hypothesis that a random sample $X_1, \ldots, X_n$ has the distribution function $F(x)$, Pearson partitioned the range of $X_j$ into M cells, say $E_1, \ldots, E_M$. If $N_1, \ldots, N_M$ are the observed number of $X_j$'s in these cells, then $N_i$ has the binomial distribution with parameters n and

$$p_i = P(X_j \text{ falls in } E_i) = \int_{E_i} dF(x) \tag{3.1}$$

when the null hypothesis is true. Pearson reasoned that the differences $N_i - np_i$ between observed and expected cell frequencies express lack of fit of the data to F, and he sought an appropriate function of these differences for use as a measure of fit.

Pearson's argument here was in three stages: (i) The quantities $N_i - np_i$ have in large samples approximately a multivariate normal distribution, and this distribution is nonsingular if only $M - 1$ of the cells are considered. (ii) If $Y = (Y_1, \ldots, Y_p)'$ has a nonsingular p-variate normal distribution $N_p(\mu, \Sigma)$, then the quadratic form $(Y - \mu)'\Sigma^{-1}(Y - \mu)$ appearing in the exponent of the density function has the $\chi^2(p)$ distribution as a function of Y. Here of course $\mu$ is the p-vector of means, and $\Sigma$ is the $p \times p$ covariance matrix of Y. (iii) Computation shows that if $Y = (N_1 - np_1, \ldots, N_{M-1} - np_{M-1})'$, this quadratic form is

$$X^2 = \sum_{i=1}^{M} \frac{(N_i - np_i)^2}{np_i}$$

which therefore has approximately the $\chi^2(M - 1)$ null distribution in large samples. This is the Pearson chi-squared statistic.

This elegant argument will reappear in our survey of recent advances in chi-squared tests. Pearson reduced the problem of testing fit to the problem of testing whether a multinomial distribution has cell probabilities $p_i$ given by (3.1). This problem, and the statistic $X^2$, do not depend on whether F is univariate or multivariate, discrete or continuous. But if F is continuous, consideration of only the cell frequencies $N_i$ does not fully use the information available in the observations $X_j$. Thus the flexibility and relative lack of power of $X^2$ stem from the same source.

### 3.2.2 Composite Hypothesis

It is common to wish to test the composite hypothesis that the distribution function of the observations $X_j$ is a member of a parametric family $\{F(\cdot|\theta) : \theta \text{ in } \Omega\}$, where $\Omega$ is a p-dimensional parameter space. Pearson recommended estimating $\theta$ by an estimator $\tilde{\theta}_n$ (a function of $X_1, \ldots, X_n$), and testing fit to the distribution $F(\cdot|\tilde{\theta}_n)$. Thus the estimated cell probabilities become

$$p_i(\tilde{\theta}_n) = \int_{E_i} dF(x|\tilde{\theta}_n)$$

and the Pearson statistic is

$$X^2(\tilde{\theta}_n) = \sum_{i=1}^{M} \frac{[N_i - np_i(\tilde{\theta}_n)]^2}{np_i(\tilde{\theta}_n)}$$

Pearson did not think that estimating $\theta$ changes the large sample distribution of $X^2$, at least when $\tilde{\theta}_n$ is consistent. In this he was wrong. It was not until 1924 that Fisher showed that the limiting null distribution of $X^2(\tilde{\theta}_n)$ is not $\chi^2(M - 1)$, and that this distribution depends on the method of estimation used.

Fisher argued that the appropriate method of estimation is maximum likelihood estimation based on the cell frequencies $N_i$. This grouped data MLE is the solution of the equations

$$\sum_{i=1}^{M} \frac{N_i}{p_i(\theta)} \frac{\partial p_i(\theta)}{\partial \theta_k} = 0, \quad k = 1, \ldots, p \tag{3.2}$$

obtained by differentiating the logarithm of the multinomial likelihood function. Fisher noted that the log likelihood ratio statistic

$$G^2 = 2 \sum_{i=1}^{M} N_i \log \frac{N_i}{np_i}$$

is asymptotically equivalent to $X^2$. He further observed that an estimator asymptotically equivalent to the grouped data MLE can be obtained by choosing $\theta$ to minimize $X^2(\theta)$ for the observed $N_i$. This minimum chi-squared estimator is the solution of

$$\sum_{i=1}^{M} \left\{ \frac{N_i}{p_i(\theta)} \right\}^2 \frac{\partial p_i(\theta)}{\partial \theta_k} = 0, \quad k = 1, \ldots, p \tag{3.3}$$

Let us denote either estimator by $\bar{\theta}_n$. Then $X^2(\bar{\theta}_n)$ is conceptually the Pearson statistic for testing fit to $F(\cdot\,|\,\bar{\theta}_n)$, the member of the family $\{F(x|\theta)\}$ which is closest to the data if the Pearson statistic is used as a measure of distance. Fisher showed that the Pearson-Fisher statistic $X^2(\bar{\theta}_n)$ has the $\chi^2(M - p - 1)$ distribution under the null hypothesis, no matter what $\theta$ in $\Omega$ is the true value. This is the famous "lose one degree of freedom for each parameter estimated" result.

Neyman (1949) noted that another estimator asymptotically equivalent to $\bar{\theta}_n$ can be obtained by minimizing the modified chi-squared statistic

$$X^2_m = \sum_{i=1}^{M} \frac{[N_i - np_i(\theta)]^2}{N_i}$$

This minimum modified chi-squared estimator is the solution of

$$\sum_{i=1}^{M} \frac{p_i(\theta)}{N_i} \frac{\partial p_i(\theta)}{\partial \theta_k} = 0, \quad k = 1, \ldots, p \tag{3.4}$$

Since for the purposes of large sample theory under the null hypothesis this estimator is interchangeable with the previous two, call it also $\bar{\theta}_n$ to minimize notation. Neyman's remark is important because equations (3.4) are more often solvable in closed form than are (3.3) and (3.2).

### 3.2.2.1 Example

Consider a chi-squared test of fit to the family of density functions

$$f(x|\theta) = \frac{1}{2}(1 + \theta x), \quad -1 \le x \le 1 \tag{3.5}$$

with $\Omega = (-1, 1)$. This family has been used as a model for the distribution of the cosine of the scattering angle in some beam-scattering experiments in physics. For cells $E_i = (a_{i-1}, a_i]$ with

$$-1 = a_0 < a_1 < \cdots < a_M = 1$$

we have

$$p_i(\theta) = \int_{a_{i-1}}^{a_i} f(x|\theta)\, dx$$

$$= \frac{\theta}{4}(a_i^2 - a_{i-1}^2) + \frac{1}{2}(a_i - a_{i-1})$$

It is easily seen that neither (3.2) nor (3.3) has a closed solution, while ((3.4) has solution

$$\bar{\theta}_n = -2 \frac{\sum\limits_{i=1}^{M} (a_i - a_{i-1})(a_i^2 - a_{i-1}^2)/N_i}{\sum\limits_{i=1}^{M} (a_i^2 - a_{i-1}^2)^2/N_i}$$

Substituting this value in the Pearson statistic produces an easily computed test of fit for the family (3.5) using $\chi^2(M - 2)$ critical points.

But even the minimum modified chi-squared estimator must often be obtained by numerical solution of its defining equations. If cells $E_i = (a_{i-1}, a_i]$ are used in a chi-squared test of fit to the normal family

$$F(x|\mu, \sigma) = \Phi\left(\frac{x - \mu}{\sigma}\right), \quad -\infty < x < \infty$$

($\Phi$ is the standard normal distribution function), then

$$p_i(\mu, \sigma) = \Phi\left(\frac{a_i - \mu}{\sigma}\right) - \Phi\left(\frac{a_{i-1} - \mu}{\sigma}\right)$$

It takes only a moment to see that none of the three versions of $\bar{\theta}_n$ can be obtained algebraically, so that recourse to numerical solution is required. Most computer libraries contain efficient routines using (for example) Newton's method to accomplish the solution.

This circumstance calls to mind Fisher's warning that his "lose one degree of freedom for each parameter estimated" result is <u>not</u> true when estimators not asymptotically the same as $\bar{\theta}_n$ are used. For example, in testing univariate normality we may <u>not</u> simply use the raw data MLE's

$$\bar{X} = \frac{1}{n} \sum_{j=1}^{n} X_j$$

$$\hat{\sigma} = \left\{ \frac{1}{n} \sum_{j=1}^{n} (X_j - \bar{X})^2 \right\}^{\frac{1}{2}}$$

in the Pearson statistic. Chernoff and Lehmann (1954) studied the conse-quences of using the raw data MLE $\hat{\theta}_n$ in the Pearson statistic. They found that $X^2(\hat{\theta}_n)$ has as its limiting distribution under $F(\cdot|\theta)$ the distribution of

$$\chi^2(M - p - 1) + \sum_{k=1}^{p} \lambda_k(\theta)\chi_k^2(1) \tag{3.6}$$

Here $\chi^2(M - p - 1)$ and $\chi_k^2(1)$ are independent chi-squared random variables with the indicated numbers of degrees of freedom. The numbers $\lambda_k(\theta)$ satisfy $0 \le \lambda_k(\theta) < 1$. So the large sample distribution of $X^2(\hat{\theta})$ is not $\chi^2$ and depends on the true value of $\theta$. All that can be said in general is that the correct critical points fall between those of $\chi^2(M - p - 1)$ and those of $\chi^2(M - 1)$. These bounds often make $X^2(\hat{\theta}_n)$ usable in practice, especially when the number of cells M is large and the number of parameters p is small.

### 3.2.3 A Family of Statistics

We have already mentioned the Pearson chi-squared, modified chi-squared, and log likelihood ratio statistics. Another statistic recommended by some statisticians is the <u>Freeman-Tukey statistic</u>

$$FT^2 = 4 \sum_{i=1}^{M} \{N_i^{\frac{1}{2}} - (np_i)^{\frac{1}{2}}\}^2$$

Cressie and Read (1984) have systematized the theory of classical chi-squared procedures by introducing a class of test statistics based on measures of divergence between discrete distributions on M points. If $q = (q_1, \ldots, q_M)$ and $p = (p_1, \ldots, p_M)$ are such probability distributions, the directed divergence of order $\lambda$ of q from p is

$$I^\lambda(q:p) = \frac{1}{\lambda(\lambda + 1)} \sum_{i=1}^{M} q_i [(q_i/p_i)^\lambda - 1]$$

$I^\lambda$ is a metric only for $\lambda = -1/2$, but is a useful generalized information measure of "distance" for all real $\lambda$. If N is the vector of cell frequencies $N_i$, and $p(\theta)$ the vector of probabilities $p_i(\theta)$, the Cressie-Read statistics are the divergences of the empiric distribution N/n from the estimated hypothesized distribution $p(\tilde{\theta}_n)$,

$$R^\lambda(\tilde{\theta}_n) = 2n I^\lambda(N/n : p(\tilde{\theta}_n))$$

If $I^\lambda$ is defined by continuity at $\lambda = -1$, 0, this class includes $X^2(\lambda = 1)$, $G^2(\lambda = 0)$, $FT^2(\lambda = -1/2)$ and $X_m^2(\lambda = -1)$.

These statistics are all asymptotically equivalent to $X^2(\tilde{\theta}_n)$ under $F(\cdot|\theta_0)$ for any estimator $\tilde{\theta}_n$ such that $n^{1/2}(\tilde{\theta}_n - \theta_0)$ is bounded in probability. Moreover, the "minimum distance" estimators of $\theta$ derived from the statistics $R^\lambda$ are all asymptotically equivalent under the null hypothesis to the grouped data MLE and minimum chi-squared estimators. So if $\bar\theta_n$ is any of these estimators and $\lambda$ is any real number, $R^\lambda(\bar\theta_n)$ has the $\chi^2(M - p - 1)$ limiting null distribution. The Cressie-Read statistics remain asymptotically equivalent under contiguous alternative distributions, but not under alternatives distant from the hypothesized family.

If the Cressie-Read family is taken as a completion of the class of statistics equivalent to $X^2$ in large samples, there remain the practical problems of use for finite n. How large must n be before the asymptotic distribution theory is trustworthy? How many cells should be used, and how should they be chosen? What of these statistics should be used? We now turn to these questions.

### 3.2.4 Choosing Cells

An objection to the use of chi-squared tests has been the arbitrariness introduced by the necessity to choose cells. This choice is guided by two considerations: the power of the resulting test, and the desire to use the asymptotic distribution of the statistic as an approximation to the exact distribution for sample size n. These issues have been studied in detail for the case of a simple hypothesis, i.e., the case of testing fit to a completely specified distribution F. Recommendations can be made in this case which may reasonably be extended to the case of testing fit to a parametric family $\{F(\cdot|\theta)\}$.

Mann and Wald (1942) initiated the study of the choice of cells in the Pearson test of fit to a continuous distribution F. They recommended, first, that the cells be chosen to have equal probabilities under the hypothesized distribution F. The advantages of such a choice are: (1) The Pearson test is unbiased. (Mann and Wald proved only local unbiasedness, but Cohen and Sackrowitz (1975) establish unbiasedness of both $X^2$ and $G^2$. This is not true when the cells have unequal probabilities under F.) (2) The distance $\sup|F_1(x) - F(x)|$ to the nearest alternative $F_1$ indistinguishable from F by $X^2$ is maximized (Mann-Wald), and $X^2$ maximizes the determinant of the matrix of second partial derivatives of the power function among all locally unbiased tests of the same size (Cohen-Sackrowitz). (3) Empirical studies have shown that the $\chi^2$ distribution is a more accurate approximation to the exact null distribution of $X^2$, $G^2$ and $FT^2$ when equiprobable cells are employed (see Section 3.2.5 for references).

Mann and Wald then made recommendations on the number M of equiprobable cells to be used. Their work rests on large-sample approximations and on a somewhat complex minimax criterion, so that it is at best a rough guide in practice. Mann and Wald found that for a sample of size n (large) and significance level $\alpha$, one should use approximately

$$M = 4\left\{\frac{2n^2}{c(\alpha)^2}\right\}^{1/5} \tag{3.7}$$

where $c(\alpha)$ is the upper $\alpha$-point of the standard normal distribution. The optimum is quite broad. In particular, the M of (3.7) can be halved with little effect on power. Retracing the Mann-Wald calculations using better approximations, as in Schorr (1974), confirms that the "optimum" M is smaller than the value given by (3.7). Since the exact optimum depends on the criterion, a choice of error probabilities, and of course on the assumption that the hypothesized F contains no unknown parameters, the practitioner need not go beyond the following recommendation: <u>Choose a number M of equiprobable cells falling between the value (3.7) <u>for</u> $\alpha = 0.05$ <u>and half that value</u>. Since half the value (3.7) is $1.88n^{2/5}$, the choice $M \doteq 2n^{2/5}$ is convenient. This recommendation is not an endorsement of the use of $\alpha = 0.05$ (or any fixed $\alpha$) in tests of fit. Because (3.7) increases slowly with $\alpha$, but overstates the number of cells required, the value for $\alpha = 0.05$ can also be used when larger significance levels are in mind.

For small n, accuracy of the $\chi^2$ approximation to the exact null distribution becomes of paramount concern. We shall see (Section 3.2.5) that the recommendations above, especially that of equiprobable cells, are sustained by this concern. When parameters must be estimated, cells equiprobable under the estimated parameter value can be employed. This requires data-dependent cells, a major modern innovation to be discussed in Section 3.3.1 below. Since an "objective" procedure for choosing cells is desirable, all examples in this chapter will use equiprobable cells with (3.7) for $\alpha = 0.05$ as a guide to choosing M.

## 3.2.5 Small-Sample Distributions

The distribution theory of chi-squared statistics (and most other formal tests of fit) is a large-sample theory. Indeed, Pearson's discovery of $X^2$ rested on the normal limiting distribution of the cell frequencies. How usable in practice are critical points or P-values for $X^2$ or $R^\lambda$ obtained from the chi-squared distribution? Cochran (1954) gave a commonly accepted rule of thumb: all expected cell frequencies $np_i$ should be at least 1, with at least 80 percent being at least 5. The availability of inexpensive computing has led to extensive study of this issue in recent years. Several recommended papers summarizing this work are Roscoe and Byars (1971), Larntz (1978) and Koehler and Larntz (1980), and Read (1984).

Each of these papers has a different emphasis. Roscoe and Byars present a simulation study of the Pearson test of fit to a simple hypothesis and summarize much earlier work. Larntz (1978) compares the Pearson, log likelihood ratio and Freeman-Tukey statistics with regard to the accuracy of the chi-squared approximation. He includes the simple hypothesis case and four cases in which parameters must be estimated. Koehler and Larntz (1980) study $X^2$ and $G^2$ when the number of cells M increases with n rather than remaining fixed. In this case the limiting distribution is normal rather than chi-squared when a simple hypothesis is being tested (see Section 3.3.4

below. Read (1984) investigates the family of $R^\lambda$ statistics for testing fit to the simple hypothesis of equiprobable cells, and considers the usefulness of two improved approximations to the exact distribution.

The consensus of these and other studies is that the traditional rule of thumb is very conservative, especially when the estimated cell probabilities are not too unequal. Here are the recommendations of Roscoe and Byars for the Pearson $X^2$, which may serve as a guide for practitioners.

1. With equiprobable cells, the average expected cell frequency should be at least 1 (that is, $n \geq M$) when testing fit at the $\alpha = 0.05$ level; for $\alpha = 0.01$, the average expected frequency should be at least 2 (that is, $n \geq 2M$).

2. When cells are not approximately equiprobable, the average expected frequencies in (1) should be doubled.

3. These recommendations apply when $M \geq 3$. For $M = 2$ (1 degree of freedom), the chi-squared test should be replaced by the test based on the exact binomial distribution.

Note that the Roscoe-Byars recommendations are based on the average rather than the minimum cell expectation. Any such rule may be defeated, as Koehler and Larntz (1980) remark, by a sufficiently skewed assignment of cell probabilities. They suggest the guidelines $M \geq 3$, $n \geq 10$, $n^2/M \geq 10$ as adequate for use of the $\chi^2$ approximation to the Pearson statistic. These are somewhat conservative when, as we recommend, cell probabilities are approximately equal. The Mann-Wald suggestion (3.7) meets both the Roscoe-Byars and Koehler-Larntz guidelines. Simulations suggest that when these guidelines are met, the true $\alpha$ for $X^2$ is usually slightly less than the nominal $\alpha$ given by $\chi^2$. But the true $\alpha$ generally exceeds the nominal $\alpha$ for $R^\lambda$ with $\lambda$ not close to 1, often substantially, when approximately equiprobable cells are employed.

Though these recommendations rest on study of the simple $H_0$ case, Larntz (1978) gives some grounds for adopting them when parameters must be estimated.

The comparative studies of Larntz (1978) and Read (1984) establish clearly that the $\chi^2$ approximation is notably more accurate for $X^2$ than for such common competitors as $G^2$ and $FT^2$. Read, for equiprobable cells, finds close agreement between the exact and approximate critical levels of $R^\lambda$ for $1/3 \leq \lambda \leq 1.5$ when $n \leq 20$ and $2 < M \leq 6$. Only $X^2$ ($\lambda = 1$) among the more common members of the $R^\lambda$ family falls in this class. Moreover, although increasing n for fixed M enlarges the class of $\lambda$ for which the $\chi^2$ approximation is reasonable, Read finds that as M increases for fixed n, the error in this approximation "increases dramatically" for values of $\lambda$ outside the recommended interval.

Statisticians, including the authors of the papers we have cited, differ on criteria for an "adequate" large sample approximation. Readers may therefore want to examine these papers in detail for additional information, particularly if the use of $R^\lambda$ statistics other than $X^2$ is contemplated.

### 3.2.6 Choosing a Statistic

Since both hypotheses and alternatives of interest for an omnibus test of fit are very general, it is difficult to give comprehensive recommendations based on power for choosing among a class of such tests. Asymptotic results (for the simple $H_0$ case) are ambiguous. When M is held fixed as n increases, all $R^\lambda$ are equivalent against local alternatives, and $G^2$ is favored against distant alternatives (Hoeffding, 1965). But if M increases with n, the limiting distributions of $R^\lambda$ vary with $\lambda$ under both hypothesis and local alternatives, and $X^2$ appears to be favored (Holst 1972, Morris 1975, Cressie and Read 1984).

In many practical situations, power considerations are secondary to the accuracy of the $\chi^2$ approximation to the exact null distribution. In such cases, the Pearson $X^2$ is the statistic of choice. Some quite limited computations of exact power by Koehler and Larntz (1980) and Read (1984) shed some light on the dependence of power on the alternative hypothesis and on the choice of $\lambda$. Read suggests $1/3 \leq \lambda \leq 2/3$ as a compromise with reasonable power against the alternatives he considers. Again $X^2$ fares better than its common competitors $G^2$, $X_m^2$ and $FT^2$.

A different approach that may aid the choosing of a statistic is to examine the type of lack of fit measured by each statistic. The sample measure of the <u>degree</u> of lack of fit accompanying $R^\lambda(\tilde{\theta}_n)$ (which measures the <u>significance</u> of lack of fit) is $R^\lambda(\tilde{\theta}_n)/n$. If G is the true distribution of the observations $X_j$, all common estimators $\tilde{\theta}_n$ converge under G to a $\theta_0$ such that $F(\cdot | \theta_0)$ is "closest" to G in some sense. When G is a member of the hypothesized family $\{F(\cdot | \theta): \theta \text{ in } \Omega\}$, this is just consistency of $\tilde{\theta}_n$. When G is not in this family and $\tilde{\theta}_n$ is the minimum-$R^\lambda$ estimator, $\theta_0$ is the point such that $p(\theta_0)$ is closest to the vector $\pi_G$ of cell probabilities under G by the discrepancy measure $I^\lambda(\pi_G : p(\theta))$. Moreover, $R^\lambda(\tilde{\theta}_n)/n$ converges w.p. 1 to $2I^\lambda(\pi_G : p(\theta_0))$. For example, $X^2(\tilde{\theta}_n)/n$ converges to

$$2I^1(\pi_G : p(\theta_0)) = \sum_{i=1}^{M} \frac{(\pi_i - p_i)^2}{p_i}$$

where $\bar{\theta}$ is the minimum chi-squared estimator, $\theta_0$ is the point closest to G by the $I^1$ measure, and $p_i = p_i(\theta_0)$. See Moore (1984) for details of these results.

A choice of $\lambda$ can be based on a choice of distance measure, and power against an alternative of interest will depend on the distance of that alternative from the hypothesis under the given measure. For a specific alternative, $\lambda$ can be chosen to maximize the distance of this alternative from $\{F(\cdot | \theta)\}$. This generalizes the conclusions of Read (1984). <u>For general alternatives, we recommend (pending further study) that the Pearson $X^2$ statistic be employed in practice</u> when a choice is made among the statistics $R^\lambda$. We will see below that consideration of a broader class of chi-squared-like statistics will modify this recommendation. But $X^2$ will remain the statistic of choice when the null hypothesis is simple or when minimum chi-squared estimation is used.

### E3.2.7 Examples of the Pearson Test

Because of its relative lack of power, $X^2$ cannot be recommended for testing fit to standard distributions for which special-purpose tests are available, or for which the special tables of critical points needed to apply tests based on the empirical distribution function (EDF) when parameters are estimated have been computed. Testing fit to the family (3.5) is, on the other hand, a realistic application of the Pearson-Fisher statistic $X^2(\hat{\theta}_n)$. The examples below of $X^2$ applied to the NOR data set are intended only as illustrations of the mechanics of applying the test.

#### 3.2.7.1 Example

Since NOR purports to be data simulating a normal sample with $\mu = 100$ and $\sigma = 10$, let us first assess the simulation by testing fit to this specific distribution. The Mann-Wald recipe (3.7) with $\alpha = 0.05$ and n = 100 gives M = 24. For computational convenience, we use M = 25 cells chosen to be equiprobable under N(100, 100). The cell boundaries are $100 + 10z_i$, where $z_i$ is the 0.04i point from the standard normal table, i = 1, 2, ..., 24. For example, the 0.04 point is -1.75, so the upper boundary of the leftmost cell is $100 + (10)(-1.75) = 82.5$. Table 3.1 shows the cells and their observed frequencies. The expected frequencies are all $(100)(0.04) = 4$. When $p_i = 1/M$ for all i, we have

$$X^2 = \frac{M}{n} \sum_{i=1}^{M} \left( N_i - \frac{n}{M} \right)^2$$

So in this example,

$$X^2 = \frac{1}{4} \sum_{i=1}^{25} (N_i - 4)^2$$

$$= \frac{112}{4} = 28$$

The appropriate distribution is $\chi^2(24)$, and the P-value (attained significance level) of $X^2 = 28$ is 0.260.

To test the NOR data for fit to the family of univariate normal distributions, an intuitively reasonable procedure is to estimate $\mu$, $\sigma$ by $\bar{X}$, $\hat{\sigma}$ and use cells with boundaries $\bar{X} + z_i\hat{\sigma}$, where $z_i$ are as before. These cells are

TABLE 3.1 Chi-squared Tests for Normality of the NOR Data

| Cell | Fit to N(100,100) Upper boundary | Frequency | Fit to normal family Upper boundary | Frequency |
|------|----------|-----------|----------|-----------|
| 1 | 82.5 | 3 | 81.2 | 3 |
| 2 | 85.9 | 8 | 84.8 | 5 |
| 3 | 88.3 | 5 | 87.3 | 5 |
| 4 | 90.1 | 8 | 89.2 | 5 |
| 5 | 91.6 | 4 | 90.7 | 6 |
| 6 | 92.9 | 2 | 92.1 | 4 |
| 7 | 94.2 | 1 | 93.5 | 3 |
| 8 | 95.3 | 5 | 94.6 | 1 |
| 9 | 96.4 | 6 | 95.8 | 4 |
| 10 | 97.5 | 1 | 96.9 | 6 |
| 11 | 98.5 | 3 | 98.0 | 3 |
| 12 | 99.5 | 3 | 99.0 | 3 |
| 13 | 100.5 | 4 | 100.1 | 2 |
| 14 | 101.5 | 2 | 101.1 | 5 |
| 15 | 102.5 | 2 | 102.2 | 2 |
| 16 | 103.6 | 7 | 103.3 | 5 |
| 17 | 104.7 | 7 | 104.5 | 9 |
| 18 | 105.8 | 3 | 105.6 | 3 |
| 19 | 107.1 | 1 | 107.0 | 1 |
| 20 | 108.4 | 2 | 108.3 | 1 |
| 21 | 109.9 | 4 | 109.9 | 5 |
| 22 | 111.7 | 6 | 111.8 | 6 |
| 23 | 114.1 | 6 | 114.3 | 6 |
| 24 | 117.5 | 4 | 117.8 | 4 |
| 25 | $\infty$ | 3 | $\infty$ | 3 |

equiprobable under the normal distribution with $\mu = \bar{X}$ and $\sigma = \hat{\sigma}$. It will be remarked in Section 3.3.1 that the Pearson statistic with these data-dependent cells has the same large sample distribution as if the fixed cell boundaries $100 + 10z_k$ to which the random boundaries converge were used. This distribution is <u>not</u> $\chi^2(24)$, since $\mu$ and $\sigma$ were estimated by their raw data MLE's $\bar{X}$ and $\hat{\sigma}$ in computing the cell probabilities $p_i(\bar{X}, \hat{\sigma}) = 0.04$. The appropriate distribution has the form (3.6), so that its critical points fall between those of $\chi^2(24)$ and $\chi^2(22)$. Calculation shows that $\bar{X} = 99.54$ and $\hat{\sigma} = 10.46$. The cell boundaries $\bar{X} + \hat{\sigma} z_k$ and the observed cell frequencies are given at the right of Table 3.1. The observed chi-squared value is $X^2 = 22$, reflecting the somewhat better fit when parameters are estimated from the data. The P-value falls between $0.460$ (from $\chi^2(22)$) and $0.579$ (from $\chi^2(24)$).

For comparison, the same procedure was applied to test the LOG data set for normality. In this case, $\bar{X} = 99.84$ and $\hat{\sigma} = 16.51$, and the observed chi-squared value using cell boundaries $\bar{X} + \hat{\sigma} z_k$ is $X^2 = 31.5$. The corresponding P-value lies between $0.086$ (from $\chi^2(22)$) and $0.140$ (from $\chi^2(24)$). Thus this test has correctly concluded that NOR fits the normal family well, while the fit of LOG is marginal. Since the logistic distributions are difficult to distinguish from the normal family, this is a pleasing performance. In contrast, the same procedure with $M = 10$ has $X^2 = 9.4$ for the LOG data, so that the P-value lies between $0.225$ (from $\chi^2(7)$) and $0.402$ (from $\chi^2(9)$). Using three cells gives $X^2 = 0.98$ and again fails to suggest that the LOG data set is not normally distributed. Thus for these particular data, the larger M suggested by (3.7) produces a more sensitive test.

### 3.2.7.2 Example

The same procedure can be applied to the EMEA data, but a glance shows that these data as given are discrete and therefore not normal. Indeed, with 15 cells equiprobable under the $N(\bar{X}, \hat{\sigma})$ distribution for these data, $X^2 = 554$. Since the data are grouped in classes centered at integers, a more intelligent procedure is to use fixed cells of unit width centered at the integers, with cell probabilities computed from $N(\bar{X}, \hat{\sigma})$. Of course, $\bar{X}$ and $\hat{\sigma}$ from the grouped data are only approximate. Sheppard's correction for $\hat{\sigma}$ improves the approximation, and gives $\bar{X} = 14.540$ and $\hat{\sigma} = 2.216$. Calculating the cell probabilities and computing the Pearson statistic, we obtain $X^2 = 7.56$. The P-value lies between $0.819$ (from $\chi^2(12)$) and $0.911$ (from $\chi^2(14)$), so that the EMEA data fit the normal family very well indeed. The applicability of $X^2$ to grouped data such as these is an advantage of chi-squared methods.

## 3.3 GENERAL CHI-SQUARED STATISTICS

### 3.3.1 Data-Dependent Cells

As already noted in Section 3.2.7, the use of data-dependent cells increases the flexibility of chi-squared tests, fortunately without increasing their complexity in practice. The essential requirement is that as the sample size increases, the random cell boundaries must converge in probability to a set of fixed boundaries. The limiting cells will usually be unknown, since they depend on the true parameter value $\theta_0$. Random cells are used in chi-squared tests by "forgetting" that the cells are data-dependent and proceeding as if fixed cells had been chosen. Since the cell frequencies are no longer multinomial, the theory of such tests is mathematically difficult. But in practice, the limiting distribution of $R^\lambda$ with random cells is exactly the same as if the limiting fixed cells had been used. This is true even when parameters are estimated. Details and regularity conditions appear in Section 4 of Moore and Spruill (1975) for k-dimensional rectangular cells. Pollard (1979) has extended the theory to cells of very general shape. Therefore, <u>any statistic, such as the Pearson-Fisher $X^2(\bar{\theta}_n)$, that has a $\theta_0$-free limiting null distribution with fixed cells has that same distribution for any choice of converging random cells.</u>

A statistic such as the Chernoff-Lehmann $X^2(\hat{\theta}_n)$ which has a $\theta_0$-dependent limiting null distribution for fixed cells, has in general this same deficiency with random cells. But if the hypothesized family $\{F(\cdot|\theta)\}$ is a location-scale family, a proper choice of random cells eliminates this $\theta_0$-dependency and also allows cells to be chosen equiprobable under the estimated $\theta$, thus matching the recommended practice in the simple hypothesis case. Such cell choices should be made whenever possible. Theorem 4.3 of Moore and Spruill (1975) is a general account of this. Let us here illustrate it by returning to the $X^2$ statistic for testing univariate normality.

When the parameter $\theta = (\mu, \sigma)$ is estimated by $\hat{\theta}_n = (\bar{X}, \hat{\sigma})$ and cell boundaries $\bar{X} + z_i\hat{\sigma}$ are used, the estimated cell probabilities are

$$p_i(\bar{X}, \hat{\sigma}) = \int_{\bar{X}+z_{i-1}\hat{\sigma}}^{\bar{X}+z_i\hat{\sigma}} (2\pi\hat{\sigma}^2)^{-\frac{1}{2}} e^{-(t-\bar{X})^2/2\hat{\sigma}^2} \, dt$$

$$= \int_{z_{i-1}}^{z_i} (2\pi)^{-\frac{1}{2}} e^{-u^2/2} \, du$$

These are not dependent on $(\bar{X}, \hat{\sigma})$, and are equiprobable if $z_i$ are the successive $i/M$ points of the standard normal distribution. Since this choice of cells leaves both $N_i$ and $p_i$ unchanged when any location-scale transformation is applied to all observations $X_j$, the Pearson statistic (and indeed, any $R^\lambda$) has the same distribution for all $(\mu, \sigma)$. The limiting null distribution has the form (3.6) but the $\lambda_k$ are now free of any unknown parameter. Critical points may therefore be computed. Two methods for doing so, and tables for testing normality, appear in Dahiya and Gurland (1972) and Moore (1971). Dahiya and Gurland (1973) study the power of this test. The idea of using random cells in this fashion is due to A. R. Roy (1956) and G. S. Watson (1957, 1958, 1959). We will refer to the Pearson statistic using the raw data MLE and random cells as the <u>Watson-Roy statistic</u>. Section 3.2.7.1, an example in Section 3.2.7, illustrated its use.

Note that the Watson-Roy statistic has $\theta$-free limiting null distribution only for location-scale families, that this distribution is not a standard tabled distribution, and that a separate calculation of critical points is required for testing fit to each location-scale family. These statements are also true for EDF tests of fit. Since the latter are more powerful, the Watson-Roy statistic has few advantages when $F(\cdot|\theta)$ is univariate and continuous.

Nonetheless, data-dependent cells move the cells to the data without essentially changing the asymptotic distribution theory of the chi-squared statistic. They should be routinely employed in practice, and this is done in most of the examples in this chapter.

### 3.3.2 General Quadratic Forms

Some of the most useful recent work on chi-squared tests involves the study of quadratic forms in the standardized cell frequencies other than the sum of squares used by Pearson. Random cells are commonly recommended in these statistics, for the reasons outlined in Section 3.3.1, and do not affect the theory. A statement of the nature and behavior of these general statistics of chi-squared type is necessarily somewhat complex. Practitioners may find it helpful to study the examples computed in Section 3.3.3 and in Rao and Robson (1974) before approaching the summary treatment below.

Random cells should be denoted by $E_{in}(X_1, \ldots, X_n)$ in a precise notation, but here the notation $E_i$ for cells and $N_i$ for cell frequencies will be continued. The "cell probabilities" under $F(\cdot | \theta)$ are

$$p_i(\theta) = \int_{E_i} dF(x|\theta), \quad i = 1, \ldots, M$$

Denote by $V_n(\theta)$ the M-vector of standardized cell frequencies having ith component

$$[N_i - np_i(\theta)]/(np_i(\theta))^{\frac{1}{2}}$$

If $Q_n = Q_n(X_1, \ldots, X_n)$ is a possibly data-dependent $M \times M$ symmetric nonnegative definite matrix, the general form of statistic to be considered is

$$V_n(\tilde{\theta}_n)' Q_n V_n(\tilde{\theta}_n) \tag{3.8}$$

when $\theta$ is estimated by $\tilde{\theta}_n$. The Pearson statistic is the special case for which $Q_n \equiv I_M$, the $M \times M$ identity matrix. The large-sample theory of these statistics is given in Moore and Spruill (1975). The basic idea is that of Pearson's proof: Show that $V_n(\tilde{\theta}_n)$ is asymptotically multivariate normal (even with random cells) and then apply the distribution theory of quadratic forms in multivariate normal random variables. All statistics of form (3.8) have as their limiting null distribution that of a linear combination of independent chi-squared random variables. References on the calculation of such distributions may be found in Davis (1977).

To avoid the necessity to compute special critical points, it is advantageous to seek statistics (3.8) which have a chi-squared limiting null distribution. This idea is due to D. S. Robson. Rao and Robson (1974) treat the important case of raw data MLEs. They give the quadratic form in $V_n(\hat{\theta}_n)$ having the $\chi^2(M-1)$ limiting null distribution. The appropriate matrix is $Q(\hat{\theta}_n)$, where

$$Q(\theta) = I_M + B(\theta)[J(\theta) - B(\theta)'B(\theta)]^{-1}B(\theta)'$$

$J(\theta)$ is the $p \times p$ Fisher information matrix for $F(\cdot | \theta)$, and $B(\theta)$ is the $M \times p$ matrix with $(i,j)$th entry

$$p_i(\theta)^{-\frac{1}{2}} \frac{\partial p_i(\theta)}{\partial \theta_j}$$

The Rao-Robson statistic is

$$R_n = V_n(\hat{\theta}_n)'Q(\hat{\theta}_n)V_n(\hat{\theta}_n)$$

This test can be used whenever $J - B'B$ is positive definite. Since $nJ$ is the information matrix from the raw data and $nB'B$ the information matrix from the cell frequencies, $J - B'B$ is always nonnegative definite. Notice that $R_n$ is just the Pearson statistic $X^2(\hat{\theta}_n)$ plus a term that conceptually builds up the distribution (3.6) to $\chi^2(M-1)$. This term simplifies considerably, since $\sum_1^M \partial p_i / \partial \theta_j = 0$ implies that

$$V_n'B = n^{-\frac{1}{2}} \left( \sum_{i=1}^{M} \frac{N_i}{p_i} \frac{\partial p_i}{\partial \theta_1}, \ldots, \sum_{i=1}^{M} \frac{N_i}{p_i} \frac{\partial p_i}{\partial \theta_p} \right) \tag{3.9}$$

and

$$R_n = X^2(\hat{\theta}_n) + (V_n'B)(J - B'B)^{-1}(V_n'B)' \tag{3.10}$$

all terms being evaluated at $\theta = \hat{\theta}_n$. Further simplification can be achieved in location-scale cases by the use of random cells for which $p_i(\hat{\theta}_n) \equiv 1/M$. Rao and Robson (1974) give several examples of the use of this statistic, using random cells in some cases.

Simulations by Rao and Robson show that $R_n$ has generally greater power than either the Pearson-Fisher or Watson-Roy statistics. Spruill (1976) gives a theoretical treatment showing that $R_n$ dominates the Watson-Roy statistic for any location-scale family $\{F(\cdot | \theta)\}$. Since $R_n$ is powerful, has tabled critical points, and is easy to compute whenever the MLE $\hat{\theta}_n$ can be obtained, it is recommended as a standard chi-squared test of fit.

Moore (1977) gives a general recipe for the quadratic form having the chi-squared limiting null distribution with maximum degrees of freedom when nearly arbitrary estimators $\tilde{\theta}_n$ are used. First compute the limiting multi-variate normal law of $V_n(\tilde{\theta}_n)$, which under $F(\cdot|\theta_0)$ has covariance matrix $\Sigma(\theta_0)$ whose form depends on the large-sample properties of the estimators $\tilde{\theta}_n$. If $\Sigma_n^-$ is a consistent estimator of the generalized inverse $\Sigma(\theta_0)^-$, the desired statistic is $V_n(\tilde{\theta}_n)'\Sigma_n^-V_n(\tilde{\theta}_n)$. The derivation of this <u>Wald's method</u> <u>statistic</u> clearly follows the lines of Pearson's original proof. The statistic can be computed in closed form more often than might be expected. It is the Pearson statistic when $\tilde{\theta}_n = \bar{\theta}_n$, the Rao-Robson statistic when $\tilde{\theta}_n = \hat{\theta}_n$, and can even in some cases be used when the $X_j$ are dependent (Moore, 1982). LeCam, Mahan and Singh (1983) have studied these statistics in depth, and show that they have certain asymptotic optimality properties given the choice of estimator $\tilde{\theta}_n$. This strengthens the case for use of the Rao-Robson statistic when raw data MLEs are chosen.

If (3.6) can be built up to $\chi^2(M-1)$, it can also be chopped down to $\chi^2(M-p-1)$. Dzhaparidze and Nikulin (1974) point out that the appropriate statistic is

$$Z_n(\tilde{\theta}_n) = V_n'(I_M - B(B'B)^{-1}B')V_n$$

where $V_n$ and B are evaluated at $\theta = \tilde{\theta}_n$. $Z_n$ has the $\chi^2(M-p-1)$ limiting distribution whenever $\tilde{\theta}_n$ approaches $\theta_0$ at the usual $n^{1/2}$ rate, and can there-fore be used with any reasonable estimator of $\theta$. Computation of $Z_n$ is again simplified by (3.9). As might be expected, simulations suggest that $Z_n(\tilde{\theta}_n)$ is inferior in power to both the Watson-Roy and Rao-Robson statistics.

### E 3.3.3 Examples of General Chi-Squared Tests

#### 3.3.3.1 Example

It is desired to test fit to the negative exponential family

$$f(x|\theta) = \theta^{-1}e^{-x/\theta}, \quad 0 < x < \infty$$

where $\Omega = \{\theta : 0 < \theta < \infty\}$. Since the MLE of $\theta$, $\hat{\theta}_n = \bar{X}$, is available, the Rao-Robson statistic is the recommended chi-squared test. When $p = 1$, (3.9) and (3.10) reduce to

$$R_n = \sum_{i=1}^{M} \frac{(N_i - np_i)^2}{np_i} + \frac{1}{nD}\Big(\sum_{i=1}^{M} \frac{N_i}{p_i} \frac{dp_i}{d\theta}\Big)^2$$

where

$$D = J - \sum_{i=1}^{M} \frac{1}{p_i}\Big(\frac{dp_i}{d\theta}\Big)^2$$

and J, $p_i$, $dp_i/d\theta$ are all evaluated at $\theta = \hat{\theta}_n$. For a sample of size n = 100, we will once more use M = 25 equiprobable cells. In this scale-parameter family, equiprobable cells are achieved by the use of random cell boundaries of the form $z_i \bar{X}$. From

$$p_i(\theta) = \int_{z_{i-1}\bar{X}}^{z_i\bar{X}} \theta^{-1} e^{-x/\theta} \, dx \qquad (3.11)$$

the condition $p_i(\bar{X}) \equiv 1/25$ gives $z_0 = 0$, $z_{25} = \infty$ and

$$z_i = -\log\left(1 - \frac{i}{25}\right), \qquad i = 1, \ldots, 24$$

Differentiating (3.11) under the integral sign, then substituting $\theta = \bar{X}$, gives

$$\frac{dp_i}{d\theta} = \bar{X}^{-1}\left[\left(1 - \frac{i}{25}\right)\log\left(1 - \frac{i}{25}\right) - \left(1 - \frac{i-1}{25}\right)\log\left(1 - \frac{i-1}{25}\right)\right] = \frac{v_i}{\bar{X}}$$

Because of their iterative nature, the quantities $v_i$ are easily computed on a programmable calculator. The Fisher information is $J(\theta) = \theta^{-2}$ so that

$$D = \bar{X}^{-2}\left[1 - 25\sum_{i=1}^{25} v_i^2\right]$$

Finally

$$R_{100} = \frac{1}{4}\sum_{i=1}^{25}(N_i - 4)^2 + \frac{(25)^2}{100}\frac{(\Sigma_1^{25} N_i v_i)^2}{1 - 25\Sigma_1^{25} v_i^2}$$

Table 3.2 records $z_i$ and $v_i$, from which

$$1 - 25\sum_1^{25} v_i^2 = 0.04255$$

For the WE2 data set, $\bar{X} = 0.878$. The resulting cell boundaries and cell frequencies appear in Table 3.2, and

$$R_{100} = \frac{1}{4}(351) + \frac{(25)^2}{100}\frac{(-0.0519)^2}{0.04255}$$

$$= 87.75 + 0.40 = 88.15$$

TABLE 3.2 The Rao-Robson Test for the Negative Exponential Family, with 25 Equiprobable Cells

| $i$ | $z_i$ | $v_i$ | WE2 | | EXP | |
|---|---|---|---|---|---|---|
| | | | $z_i \bar{X}$ | $N_i$ | $z_i \bar{X}$ | $N_i$ |
| 1 | .0408 | −.0392 | 0.036 | 1 | 0.221 | 6 |
| 2 | .0834 | −.0375 | 0.073 | 0 | 0.451 | 5 |
| 3 | .1278 | −.0358 | 0.112 | 1 | 0.692 | 3 |
| 4 | .1743 | −.0340 | 0.153 | 1 | 0.944 | 2 |
| 5 | .2231 | −.0321 | 0.196 | 3 | 1.208 | 5 |
| 6 | .2744 | −.0301 | 0.241 | 1 | 1.486 | 5 |
| 7 | .3285 | −.0279 | 0.288 | 2 | 1.779 | 7 |
| 8 | .3857 | −.0257 | 0.338 | 3 | 2.088 | 2 |
| 9 | .4463 | −.0234 | 0.392 | 5 | 2.416 | 4 |
| 10 | .5108 | −.0209 | 0.448 | 5 | 2.766 | 3 |
| 11 | .5798 | −.0182 | 0.509 | 1 | 3.140 | 3 |
| 12 | .6539 | −.0153 | 0.574 | 5 | 3.541 | 4 |
| 13 | .7340 | −.0123 | 0.644 | 3 | 3.974 | 6 |
| 14 | .8210 | −.0089 | 0.721 | 5 | 4.445 | 3 |
| 15 | .9163 | −.0053 | 0.804 | 8 | 4.962 | 4 |
| 16 | 1.0216 | −.0013 | 0.897 | 4 | 5.532 | 4 |
| 17 | 1.1394 | .0032 | 1.000 | 16 | 6.170 | 3 |
| 18 | 1.2730 | .0082 | 1.118 | 9 | 6.893 | 3 |
| 19 | 1.4271 | .0139 | 1.253 | 11 | 7.728 | 4 |
| 20 | 1.6094 | .0206 | 1.413 | 7 | 8.715 | 2 |
| 21 | 1.8326 | .0287 | 1.609 | 5 | 9.923 | 7 |
| 22 | 2.1203 | .0388 | 1.861 | 1 | 11.481 | 3 |
| 23 | 2.5257 | .0524 | 2.217 | 3 | 13.676 | 3 |
| 24 | 3.2189 | .0733 | 2.826 | 0 | 17.430 | 6 |
| 25 | $\infty$ | .1288 | $\infty$ | 0 | $\infty$ | 3 |

This gives a P-value of $3 \times 10^{-9}$ using the $\chi^2(24)$ distribution. In contrast, the EXP data set has $\bar{X} = 5.415$, cell boundaries and frequencies given at the right of Table 3.2, and

$$R_{100} = \frac{1}{4}(54) + \frac{(25)^2}{100} \frac{(-0.1231)^2}{0.04255}$$

$$= 13.5 + 2.23 = 15.73$$

The P-value from $\chi^2(24)$ is 0.898.

Table 3.2 reveals an important practical advantage of chi-squared tests, especially when equiprobable cells are employed: examination of the deviations of the cell frequencies $N_i$ from their common expected value (here 4) shows clearly the nature of the lack of fit detected by the test. In this case, the Weibull with power parameter $k = 2$ has far too few observations in the lower tail, too many in the middle slope of the density function, and too few in the extreme upper tail. A glance at graphs of the Weibull and exponential density functions (e.g., on pp. 379-380 of Derman, Gleser, and Olkin 1973) shows how accurately the $N_i$ mirror the differences between the two distributions.

As these examples suggest, the Pearson statistic $X^2(\hat{\theta}_n)$, which is the first component of $R_n$, is usually adequate for drawing conclusions when M is large and p is small. In this example, the critical points of $X^2(\hat{\theta}_n)$ fall between those of $\chi^2(22)$ and those of $\chi^2(24)$. A reasonable strategy is to compute $X^2(\hat{\theta}_n)$ first, completing the computation of $R_n$ only if the results after the first stage are ambiguous.

### 3.3.3.2 Example

The BAEN data are to be tested for fit to the double-exponential family

$$f(x|\theta) = \frac{1}{2\theta_2} e^{-|x-\theta_1|/\theta_2}, \quad -\infty < x < \infty$$

$$\Omega = \left\{ (\theta_1, \theta_2) : -\infty < \theta_1 < \infty, \ 0 < \theta_2 < \infty \right\}$$

The MLE $\hat{\theta}_n = (\hat{\theta}_{1n}, \hat{\theta}_{2n})$ from a random sample $X_1, \ldots, X_n$ is

$$\hat{\theta}_{1n} = \text{median} (X_1, \ldots, X_n)$$

$$\hat{\theta}_{2n} = \frac{1}{n} \sum_{j=1}^{n} |X_j - \hat{\theta}_{1n}|$$

In this location-scale setting, equiprobable cells with boundaries $\hat{\theta}_{1n} + a_i \hat{\theta}_{2n}$ will again be employed. Using an even number of cells, say $M = 2\nu$, and choosing the $a_i$ symmetrically as $a_{\nu+i} = -a_{\nu-i} = c_i$, where

$$c_i = -\log\left(1 - \frac{i}{\nu}\right), \quad i = 0, \ldots, \nu$$

(in particular, $a_0 = -\infty$, $a_\nu = 0$, $a_M = \infty$) gives $p_i(\hat{\theta}_n) \equiv 1/M$.
Computations similar to those shown in Section 3.3.3.1 yield

$$\frac{\partial p_i}{\partial \theta_1}(\hat{\theta}_n) = -1/M\hat{\theta}_{2n} \qquad i = 1, \ldots, \nu$$

$$= 1/M\hat{\theta}_{2n} \qquad i = \nu+1, \ldots, M \qquad\qquad (3.12)$$

$$\frac{\partial p_i}{\partial \theta_2}(\hat{\theta}_n) = \frac{1}{2\hat{\theta}_{2n}}(c_{k-1}e^{-c_{k-1}} - c_k e^{-c_k}) \qquad \begin{array}{l} i = \nu+k, \ \nu-k+1 \\[4pt] k = 1, \ldots, \nu \end{array}$$

If $d_k = c_{k-1}e^{-c_{k-1}} - c_k e^{-c_k}$, then

$$B(\hat{\theta}_n)'B(\hat{\theta}_n) = \hat{\theta}_{2n}^{-1}\begin{pmatrix} 1 & 0 \\ 0 & \nu\Sigma_1^{\nu}d_i^2 \end{pmatrix}$$

Since the information matrix is $\theta_2^{-1}I_2$, the matrix $J(\hat{\theta}_n) - B(\hat{\theta}_n)'B(\hat{\theta}_n)$ has rank 1 and the Rao-Robson statistic is not defined. (The reason for this unusual situation is that for this choice of cells, the median is both the raw data MLE and the grouped data MLE for $\theta_1$.) The Dzhaparidze-Nikulin statistic is

$$Z_n(\hat{\theta}_n) = \frac{M}{n}\sum_{i=1}^{M}\left(N_i - \frac{n}{M}\right)^2 - \frac{M}{n}\frac{1}{2\Sigma_1^{\nu}d_i^2}\left[\sum_{i=1}^{\nu}d_i(N_{\nu+1} + N_{\nu-i+1})\right]^2$$

This computation was simplified by the fact that $B'B$ is diagonal and the first term of (3.9) is 0 by (3.12) and the definition of the median.

The BAEN data contain $n = 33$ observations, for which $\hat{\theta}_{1n} = 10.13$ and $\hat{\theta}_{2n} = 3.36$. Table 3.3 contains $c_i$, upper cell boundaries $\hat{\theta}_{1n} + c_i\hat{\theta}_{2n}$, and cell frequencies for these data. The statistic $Z_n$ is, after some arithmetic,

$$Z_n = \frac{10}{33}\sum_{i=1}^{10}(N_i - 3.3)^2 - \frac{10}{33}\frac{1}{(2)(.1574)}[-1.2828]^2$$

$$= 7.30 - 1.59 = 5.71$$

The P-value from $\chi^2(7)$ is $0.426$. The Pearson statistic $X^2 = 7.30$ has critical points falling between those of $\chi^2(7)$ and $\chi^2(8)$, taking advantage of the fact that the grouped data MLE was used to estimate one of the two unknown parameters. The corresponding bounds on the P-value are $0.398$ and $0.505$. The double exponential model clearly fits the BAEN data very well. Even

TABLE 3.3  Testing the Fit of the BAEN Data
to the Double Exponential Family

| Cell | $c_i$ | $\hat{\theta}_{1n} + c_i \hat{\theta}_{2n}$ | $N_i$ |
|------|-------|---------------------------------------------|-------|
| 1 | -1.609 | 4.722 | 4 |
| 2 | -0.916 | 7.051 | 7 |
| 3 | -0.511 | 8.414 | 3 |
| 4 | -0.223 | 9.380 | 2 |
| 5 | 0 | 10.130 | 1 |
| 6 | 0.223 | 10.880 | 3 |
| 7 | 0.511 | 11.846 | 4 |
| 8 | 0.916 | 13.209 | 3 |
| 9 | 1.609 | 15.538 | 4 |
| 10 | $\infty$ | $\infty$ | 2 |

though an anomaly reduced from 2 to 1 the difference in the degrees of freedom of the $\chi^2$ distributions bounding $X^2$, there is a considerable spread in
the corresponding P-values. This is typical when n (and therefore M) is
small. In examples where the goodness of fit is less clear than here, use of
$R_n$ or $Z_n$ can be essential to a clear conclusion.

### 3.3.3.3 Example

In testing for multivariate normality, a natural choice of cell boundaries are
the concentric hyperellipses centered at the sample mean and with shape
determined by the inverse of the sample covariance matrix. These are level
surfaces of the multivariate normal density function with parameters estimated. Equiprobable cells of this form have the advantage of revealing by
the observed cell counts the presence of such common types of departure
from normality as peakedness or heavy tails. Chi-squared statistics in this
setting are computed and applied by Moore and Stubblebine (1981). Here we
consider the special case of testing fit to the circular bivariate normal family, a common model for "targeting" problems. It represents the effect of
independent normal horizontal and vertical components with equal variances.
The density function is

$$f(x,y|\theta) = \frac{1}{2\pi\sigma^2} e^{-\frac{1}{2\sigma^2}\{(x-\mu_1)^2+(y-\mu_2)^2\}}, \quad -\infty < x,y < \infty$$

$$\Omega = \{\theta = (\mu_1,\mu_2,\sigma): -\infty < \mu_1,\mu_2 < \infty, 0 < \sigma < \infty\}$$

The MLE of $\theta$ from a random sample $(X_1,Y_1), \ldots, (X_n,Y_n)$ is $\hat{\theta}_n = (\hat{\mu}_1, \hat{\mu}_2, \hat{\sigma})$, where

$$\hat{\mu}_1 = \bar{X} \qquad \hat{\mu}_2 = \bar{Y}$$

$$\hat{\sigma}^2 = \frac{1}{2n}\left\{\sum_{j=1}^{n}(X_j - \bar{X})^2 + \sum_{j=1}^{n}(Y_j - \bar{Y})^2\right\}$$

In constructing a test of fit to this family, it is natural to use as cells annuli centered at $(\bar{X}, \bar{Y})$ with successive radii $c_i \hat{\sigma}$ for

$$0 = c_0 < c_1 < \cdots < c_{M-1} < c_M = \infty$$

Thus

$$E_i = \{(x,y) : c_{i-1}^2 \hat{\sigma}^2 \le (x - \bar{X})^2 + (y - \bar{Y})^2 < c_i^2 \hat{\sigma}^2\}$$

The cell probabilities are

$$p_i(\theta) = \iint_{E_i} f(x,y|\theta)\,dx\,dy$$

and calculation shows that $p_i(\hat{\theta}_n) \equiv 1/M$ when

$$c_i = \left\{-2\log\left(1 - \frac{i}{M}\right)\right\}^{\frac{1}{2}}, \qquad i = 1, \ldots, M-1$$

The recommended test is based on the Rao-Robson statistic. Differentiating $p_i(\theta)$ under the integral sign, then substituting $\theta = \hat{\theta}_n$ gives

$$\left.\frac{\partial p_i}{\partial \mu_1}\right|_{\hat{\theta}} = \left.\frac{\partial p_i}{\partial \mu_2}\right|_{\hat{\theta}} = 0$$

$$\left.\frac{\partial p_i}{\partial \sigma}\right|_{\hat{\theta}} = \hat{\sigma}^{-1}(c_{i-1}^2 e^{-\frac{1}{2}c_{i-1}^2} - c_i^2 e^{-\frac{1}{2}c_i^2})$$

$$= v_i/\hat{\sigma}$$

Hence

$$B'B\big|_{\hat{\theta}} = \frac{M}{\hat{\sigma}^2}\begin{bmatrix} 0 & 0 & 0 \\ 0 & 0 & 0 \\ 0 & 0 & \sum_1^M v_i^2 \end{bmatrix}$$

The Fisher information matrix for the circular bivariate normal family is also diagonal,

$$J(\theta) = \frac{1}{\sigma^2} \begin{bmatrix} 1 & 0 & 0 \\ 0 & 1 & 0 \\ 0 & 0 & 4 \end{bmatrix}$$

so that $(J - B'B)^{-1}$ is trivially obtained. Moreover, from (3.9) it follows that

$$V_n'B = n^{-\frac{1}{2}}(0, 0, \Sigma_1^M N_i v_i / \hat{\sigma})$$

The Rao-Robson statistic is therefore

$$\begin{aligned} R_n &= X^2(\hat{\theta}_n) + (V_n'B)(J - B'B)^{-1}(V_n'B)' \\ &= \frac{M}{n} \sum_{i=1}^{M} \left(N_i - \frac{n}{M}\right)^2 + \frac{M^2}{n} \frac{\left(\Sigma_1^M N_i d_i\right)^2}{1 - M \Sigma_1^M d_i^2} \end{aligned}$$

where

$$d_i = v_i/2 = \left(1 - \frac{i}{M}\right) \log\left(1 - \frac{i}{M}\right) - \left(1 - \frac{i-1}{M}\right) \log\left(1 - \frac{i-1}{M}\right)$$

The limiting null distribution is $\chi^2(M - 1)$, while that of the Pearson statistic $X^2(\hat{\theta}_n)$ has critical points falling between those of $\chi^2(M - 1)$ and $\chi^2(M - 4)$. The Rao-Robson correction term will often be necessary for a clear picture of the fit of this three-parameter family.

### 3.3.3.4 Example

The negative exponential distribution with density function

$$f(x|\theta) = \theta^{-1} e^{-x/\theta}, \quad 0 < x < \infty$$
$$\Omega = \{\theta : 0 < \theta < \infty\}$$

is often assumed in life testing situations. Such studies may involve not a full sample, but rather Type II censored data. That is, order statistics are observed up to the sample $\alpha$-quantile,

$$X_{(1)} < X_{(2)} < \cdots < X_{([n\alpha])}$$

where $[n\alpha]$ is the greatest integer in $n\alpha$ and $0 < \alpha < 1$. It is natural to make use of random cells with sample quantiles $\xi_i = X_{([n\delta_i])}$ as cell boundaries. Here $\xi_0 = 0$, $\xi_M = \infty$ and

$$0 = \delta_0 < \delta_1 < \cdots < \delta_{M-1} = \alpha < \delta_M = 1$$

so that the $n - [n\alpha]$ unobserved $X_i$ fall in the rightmost cell. Although the cell frequencies $N_i$ are now **fixed**, the general theory of Moore and Spruill (1975) applies to this choice of cells. A full treatment of this type of problem is given in Mihalko and Moore (1980). Chi-squared tests are immediately applicable to data censored at fixed points. We now see that allowing random cells allows Type II censored data to be handled as well.

The Pearson-Fisher Statistic. Estimate $\theta$ by the grouped data MLE found as the solution of (3.2). That equation becomes in this case

$$\sum_{i=1}^{M} N_i \frac{\xi_{i-1} e^{-\xi_{i-1}/\theta} - \xi_i e^{-\xi_i/\theta}}{e^{-\xi_{i-1}/\theta} - e^{-\xi_i/\theta}} = 0$$

which is easily solved iteratively to obtain $\bar{\theta}_n = \bar{\theta}_n(\xi_1, \cdots, \xi_{M-1})$. The test statistic is

$$X^2(\bar{\theta}_n) = \sum_{i=1}^{M} \frac{[N_i - np_i(\bar{\theta}_n)]^2}{np_i(\bar{\theta}_n)}$$

where

$$N_i = [n\delta_i] - [n\delta_{i-1}] \qquad \text{(nonrandom)}$$

$$p_i(\theta) = e^{-\xi_{i-1}/\theta} - e^{-\xi_i/\theta} \qquad \text{(random)}$$

The limiting null distribution is $\chi^2(M - 2)$.

The Wald's Method Statistic. A more powerful chi-squared test can be obtained by use of the raw data MLE of $\theta$ from the censored sample, namely (Epstein and Sobel, 1953),

$$\tilde{\theta}_n = \frac{1}{[n\alpha]} \left( \sum_{i=1}^{[n\alpha]} X_{(i)} + (n - [n\alpha]) X_{([n\alpha])} \right)$$

By obtaining the limiting distribution of $V_n(\tilde{\theta}_n)$ and then finding the appropriate quadratic form, a generalization of the Rao-Robson statistic to censored samples can be obtained. This is done in Mihalko and Moore (1980).

The resulting statistic for the present example is

$$R_n = X^2(\tilde{\theta}_n) + (nD)^{-1}\left(\sum_{i=1}^{M} N_i v_i / p_i(\tilde{\theta}_n)\right)^2$$

where $N_i$ and $p_i(\theta)$ are as above, and

$$v_i = \tilde{\theta}_n^{-1}(\xi_{i-1} e^{-\xi_{i-1}/\tilde{\theta}_n} - \xi_i e^{-\xi_i/\tilde{\theta}_n})$$

$$D = 1 - e^{-\xi_{M-1}/\tilde{\theta}_n} - \sum_{i=1}^{M} v_i^2 / p_i(\tilde{\theta}_n)$$

In the full sample case, $\alpha = 1$, $\xi_{M-1} = \infty$, $N_M = 0$, $\tilde{\theta}_n = \bar{X}$ and the statistic $R_n$ reduces to the Rao-Robson statistic of Section 3.3.3.1 (with M - 1 cells bounded by the $\xi_i$).

The motivation for using censored data when lifetimes or survival times are being measured is apparent from the EXP data set. The sample 80th percentile is 9.46, while the maximum of the 100 observations is 39.12. The MLE of $\theta$ from the data censored at $\alpha = 0.8$ is $\tilde{\theta}_n = 5.471$, compared with the full sample MLE, $\bar{X} = 5.415$. Experience shows that the Roscoe-Byars guidelines are not adequate to ensure accurate critical points from the $\chi^2$ distribution in the present situation, where the $np_i$ are random and unequal. Tests of the EXP data will therefore be made with (a) the full sample using 10 cells having the sample deciles as boundaries; and (b) the data censored at $\alpha = 0.8$ using 9 cells with the first 8 sample deciles as boundaries. All cells except the rightmost in case (b) contain 10 observations. The results are, for the full sample,

$$R_n = 6.132 + 0.0220 = 6.352$$

with a P-value of 0.704 from $\chi^2(9)$. For the censored sample,

$$R_n = 5.153 + 0.065 = 5.218$$

with a P-value of 0.734 from $\chi^2(8)$. These results are comparable to those obtained for the same data in Section 3.3.3.1.

## 3.3.4 Nonstandard Chi-Squared Statistics

We have considered two classes of "standard" chi-squared statistics, the Cressie-Read class based on measures of divergence and the Moore-Spruill class of nonnegative definite quadratic forms. The Pearson $X^2$ is the only common member of these classes. All of the Cressie-Read statistics are asymptotically equivalent to $X^2$ under the null hypothesis when the same (possibly random) cells and the same estimators are used. But different divergence measures may be sensitive to different types of divergence of $N_i$ from $np_i$, and this fact can be used to choose a statistic when a specific type of alternative is to be guarded against. The Moore-Spruill statistics differ in asymptotic behavior under the null hypothesis. The choice of statistics within this class is most often made to obtain a $\chi^2$ limiting null distribution for given estimator $\hat{\theta}_n$. (The Cressie-Read statistics have a $\chi^2$ limiting null distribution only for estimators equivalent to the grouped-data MLE, a class that includes all minimum-$R^\lambda$ estimators.)

The theory of these standard chi-squared statistics assumes independent observations and a fixed number of cells M. Relaxing these assumptions leads to situations that are incompletely explored, and some other statistics have also been suggested. In this section we mention a few of these nonstandard cases.

1. Increasing M with n. Usual practice is to increase the number of cells M as the sample size n increases (recall the Mann-Wald recommendation (3.7)). This practice is not explicitly recognized in the standard theory. The large-sample theory of the usual chi-squared statistics for increasing M is available in the case of a simple null hypothesis (Holst 1972, Morris 1975, Cressie and Read, 1984). The limiting null distributions of the $R^\lambda$ are normal, with mean and variance depending on $\lambda$. The statistics are therefore no longer asymptotically equivalent, and $X^2$ is the optimal member of the class in terms of Pitman efficiency. The behavior of these statistics when parameters are estimated has not been explored.

Two possible variations in practice suggest themselves. (1) Allow M to increase with n at a rate faster than the Mann-Wald suggestion $n^{2/5}$. Kempthorne (1968) proposed the use of the Pearson statistic with M = n equiprobable cells. Simulation studies suggest that standard statistics with fewer cells have superior power except against very short-tailed alternatives. (2) Use a normal rather than a $\chi^2$ approximation for the distribution of standard statistics. For $X^2$, the $\chi^2$ approximation is generally both adequate in practice and superior to the normal. The $\chi^2$ is also easier to use, since it does not require computing the asymptotic mean and variance. For other $R^\lambda$ (such as $G^2$), the $\chi^2$ approximation is much less good, and the normal approximation may be superior. See Koehler and Larntz (1980). But Read (1984) gives an adjustment of the $\chi^2$ approximation that is easier to use than the normal and should also be considered.

2. Dependent observations. Since many data are collected as time series, tests of fit that assume independence may often be applied to data that are in fact dependent. Positive dependence among the observations will cause omnibus tests of fit to reject a true hypothesis about the distribution of

the individual observations too often. That is, positive dependence is confounded with lack of fit. This is shown in considerable generality for both chi-squared and EDF tests by Gleser and Moore (1983). If a model for the dependence is assumed, it may be possible to compute the effect of dependence or even to construct a valid chi-squared test using the distributional results in Moore (1982). But in general, data should be checked for serial dependence before testing fit, as the tests are sensitive to dependence as well as to lack of fit.

3. Sequentially adjusted cells. By use of the conditional probability integral transformation (see Chapter 6), O'Reilly and Quesenberry (1973) obtain particular members of the following class of nonstandard chi-squared tests. Rather than base cell frequencies on cells $E_i$ (fixed) or $E_{in}(X_1, \ldots, X_n)$ (data-dependent) into which all of $X_1, \ldots, X_n$ are classified, the cells used to classify each successive $X_j$ are functions $E_{ij}$ of $X_1, \ldots, X_j$ only. Thus additional observations do not require reclassification of earlier observations, as in the usual random cell case. No general theory of chi-squared statistics based on such sequentially adjusted cells is known. O'Reilly and Quesenberry obtain by their transformation approach specific functions $E_{ij}$ such that the cell frequencies are multinomially distributed and the Pearson statistic has the $\chi^2(M - 1)$ limiting null distribution. The transformation approach requires the computation of the minimum variance unbiased estimator of $F(\cdot|\theta)$. Testing fit to an uncommon family thus requires the practitioner to do a hard calculation. Moreover, any test using sequentially adjusted cells has the disadvantage that the value of the statistic depends on the order in which the observations were obtained. These are serious barriers to use.

4. Easterling's approach. Easterling (1976) provides an interesting approach to parameter estimation based on tests of fit. Roughly speaking, he advocates replacing the usual confidence intervals for $\theta$ in $F(\cdot|\theta)$ based on the acceptance regions of a test of

$$H_0: \theta = \theta_0$$
$$H_1: \theta \neq \theta_0$$

with intervals based on the acceptance regions of tests of fit to completely specified distributions,

$$H_0^*: G(\cdot) = F(\cdot|\theta_0)$$
$$H_1^*: G(\cdot) \neq F(\cdot|\theta_0)$$

In the course of his discussion, Easterling suggests rejecting the family $\{F(x|\theta): \theta \text{ in } \Omega\}$ as a model for the data if the (say) 50% confidence interval for $\theta$ based on acceptance regions for $H_0^*$ is empty. This "implicit test of fit"

deserves comment, using the chi-squared case to make some observations
that apply as well when other tests of $H_0^*$ are employed.

Taking then the standard chi-squared statistic for $H_0^*$,

$$X^2(\theta_0) = \sum_{i=1}^{M} \frac{[N_i - np_i(\theta_0)]^2}{np_i(\theta_0)}$$

and denoting by $\chi_\alpha^2(M - 1)$ the upper $\alpha$-point of the $\chi^2(M - 1)$ distribution,
the $(1 - \alpha)$-confidence interval is empty if and only if

$$X^2(\theta) > \chi_\alpha^2(M - 1) \quad \text{for all } \theta \text{ in } \Omega \tag{3.13}$$

But if $\bar{\theta}_n$ is the minimum chi-squared estimator, (3.13) holds if and only if

$$X^2(\bar{\theta}_n) > \chi_\alpha^2(M - 1) \tag{3.14}$$

When any $F(x|\theta)$ is true, $X^2(\bar{\theta}_n)$ has the $\chi^2(M - p - 1)$ distribution, and the
probability of the event (3.14) can be explicitly computed. It is less than $\alpha$,
but close to $\alpha$ when M is large. Thus Easterling's suggestion essentially
reduces to the use of standard tests of fit with parameters estimated by the
minimum distance method corresponding to the test statistic employed.
Moreover, his method by-passes a proper consideration of the distributional
effects of estimating unknown parameters.

## 3.4 RECOMMENDATIONS ON USE OF
    CHI-SQUARED TESTS

Chi-squared tests are generally less powerful than EDF tests and special-
purpose tests of fit. It is difficult to assess the seriousness of this lack of
power from published sources. Comparative studies have generally used the
Pearson statistic rather than the more powerful Watson-Roy and Rao-Robson
statistics. Moreover, such studies have often dealt with problems of parameter estimation in ways which tend to understate the power of general purpose
tests such as chi-squared and Kolmogorov-Smirnov tests. This is true of the
study by Shapiro, Wilk and Chen (1968), for example. Reliable information
about the power of chi-squared tests for normality can be gained from Table
IV of Rao and Robson (1974) and from Tables 1 and 2 of Dahiya and Gurland
(1973). The former demonstrates strikingly the gain in power (always at
least 40% in the cases considered, and usually much greater) obtained by
abandoning the Pearson-Fisher statistic for more modern chi-squared statistics. Nonetheless, chi-squared tests cannot in general match EDF and special purpose tests of fit in power.

This relative lack of power implies three theses on the practical use of chi-squared techniques. First, chi-squared tests of fit must compete for use primarily on the basis of flexibility and ease of use. Discrete and/or multivariate data do not discomfit chi-squared methods, and the necessity to estimate unknown parameters is more easily dealt with by chi-squared tests than by other tests of fit.

Second, chi-squared statistics actually having a (limiting) chi-squared null distribution have a much stronger claim to practical usefulness. Ease of use requires the ability to obtain (1) the observed value of the test statistic, and (2) critical points for the test statistic. The calculations required for (1) in chi-squared statistics are at most iterative solutions of nonlinear equations and evaluation of quadratic forms, perhaps with matrix expressed as the inverse of a given symmetric pd matrix. These are not serious barriers to practical use, given the current availability of computer library routines. Computation of critical points of an untabled distribution is a much harder task for a user of statistical methods. Chi-squared and EDF statistics both have as their limiting null distributions the distributions of linear combinations of central chi-squared random variables. General statistics of both classes require a separate table of critical points for each hypothesized family. The effort needed is justified when the hypothesized family is common, but should be expended on a test more powerful than chi-squared tests. In less common cases, or when no more powerful test with $\theta$-free null distribution is available, there are several chi-squared tests requiring only tables of the $\chi^2$ distribution. These include the Pearson-Fisher, Rao-Robson, and Dzhaparidze-Nikulin tests, and others which can be constructed by Wald's method. Among the chi-squared statistics proposed and studied to date, the Rao-Robson statistic $R_n$ of (3.10) appears to have generally superior power and is therefore the statistic of choice for protection against general alternatives. Computation of $R_n$ in the nonstandard cases most appropriate for chi-squared tests of fit does require some mathematical work. However, the Pearson statistic $X^2(\hat{\theta}_n)$ with raw-data MLEs is the first and usually dominant component of $R_n$. If $X^2(\hat{\theta}_n)$ itself lies in the upper tail of the $\chi^2(M-1)$ distribution, the fit can be rejected without computing $R_n$.

The third thesis rests on the exposition and examples in this chapter. Chi-squared tests are the most practical tests of fit in many situations. When parameters must be estimated in non-location-scale families or in uncommon distributions, when the data are discrete, multivariate, or even censored, chi-squared tests remain easily applicable.


ACKNOWLEDGMENT

This work was supported in part by National Science Foundation Grant DMS-8501966.

REFERENCES

Bishop, Y. M. M., Fienberg, S. E. and Holland, P. W. (1975). Discrete Multivariate Analysis. Cambridge: The MIT Press.

Chernoff, H. and Lehmann, E. L. (1954). The use of maximum-likelihood estimates in $\chi^2$ test for goodness of fit. Ann. Math. Statist. 25, 579-586.

Cochran, W. G. (1954). Some methods of strengthening the common $\chi^2$ tests. Biometrics 10, 417-451.

Cohen, A. and Sackrowitz, H. B. (1975). Unbiasedness of the chi-square, likelihood ratio, and other goodness of fit tests for the equal cell case. Ann. Statist. 4, 959-964.

Cressie, N. and Read, T. R. C. (1984). Multinomial goodness-of-fit tests. J. Roy. Statist. Soc. B46, 440-464.

Dahiya, R. C. and Gurland, J. (1972). Pearson chi-square test of fit with random intervals. Biometrika 59, 147-153.

Dahiya, R. C. and Gurland, J. (1973). How many classes in the Pearson chi-square test? J. Amer. Statist. Assoc. 68, 707-712.

Davis, A. W. (1977). A differential equation approach to linear combinations of independent chi-squares. J. Amer. Statist. Assoc. 72, 212-214.

Derman, C., Gleser, L. J. and Olkin, I. (1973). A Guide to Probability Theory and Application. Holt, Rinehart and Winston, New York.

Dzhaparidze, K. O. and Nikulin, M. S. (1974). On a modification of the standard statistics of Pearson. Theor. Probability Appl. 19, 851-853.

Easterling, R. G. (1976). Goodness-of-fit and parameter estimation. Technometrics 18, 1-9.

Epstein, B. and Sobel, M. (1953). Life testing. J. Amer. Statist. Assoc. 48, 486-502.

Fisher, R. A. (1924). The conditions under which $\chi^2$ measures the discrepancy between observation and hypothesis. J. Roy. Statist. Soc. 87, 442-450.

Gleser, L. J. and Moore, D. S. (1983). The effect of dependence on chi-squared and empiric distribution tests of fit. Ann. Statist. 11, 1100-1108.

Hoeffding, W. (1965). Asymptotically optimal tests for multinomial distributions. Ann. Math. Statist. 36, 369-408.

Holst, L. (1972). Asymptotic normality and efficiency for certain goodness-of-fit tests. Biometrika 59, 127-145.

Kempthorne, O. (1968). The classical problem of inference—goodness of fit. Proc. Fifth Berkeley Symp. Math. Statist. Prob. 1, 235-249.

Koehler, K. J. and Larntz, K. (1980). An empirical investigation of goodness-of-fit statistics for sparse multinomials. J. Amer. Statist. Assoc. 75, 336-344.

Larntz, K. (1978). Small-sample comparisons of exact levels for chi-squared goodness-of-fit statistics. J. Amer. Statist. Assoc. 73, 253-263.

LeCam, L., Mahan, C. and Singh, A. (1983). An extension of a theorem of H. Chernoff and E. L. Lehmann. In: Rizvi, M. H., Rustagi, J. S. and Siegmund, D. (Eds.). Recent Advances in Statistics: Papers in Honor of Herman Chernoff, 303-337. Academic Press, New York.

Mann, H. B. and Wald, A. (1942). On the choice of the number of class intervals in the application of the chi-square test. Ann. Math. Statist. 13, 306-317.

Mihalko, D. and Moore, D. S. (1980). Chi-square tests of fit for type II censored samples. Ann. Statist. 8, 625-644.

Moore, D. S. (1971). A chi-square statistic with random cell boundaries. Ann. Math. Statist. 42, 147-156.

Moore, D. S. (1977). Generalized inverses, Wald's method and the construction of chi-squared tests of fit. J. Amer. Statist. Assoc. 7, 131-137.

Moore, D. S. (1982). The effect of dependence on chi-squared tests of fit. Ann. Statist. 10, 1163-1171.

Moore, D. S. (1984). Measures of lack of fit from tests of chi-squared type. J. Statist. Planning Inf. 10, 151-166.

Moore, D. S. and Spruill, M. C. (1975). Unified large-sample theory of general chi-squared statistics for tests of fit. Ann. Statist. 3, 599-616.

Moore, D. S. and Stubblebine, J. B. (1981). Chi-square tests for multivariate normality, with application to common stock prices. Comm. Statist. A10, 713-733.

Morris, C. (1975). Central limit theorems for multinomial sums. Ann. Statist. 3, 165-188.

Neyman, J. (1949). Contribution to the theory of the $\chi^2$ test. Proc. Berkeley Symp. Math. Statist. and Prob., 239-273.

O'Reilly, F. J. and Quesenberry, C. P. (1973). The conditional probability integral transformation and applications to obtain composite chi-square goodness-of-fit tests. Ann. Statist. 1, 74-83.

Pollard, D. (1979). General chi-square goodness-of-fit tests with data-dependent cells. Z. Wahrscheinlichkeitstheorie verw. Geb. 50, 317-331.

Rao, K. C. and Robson, D. S. (1974). A chi-square statistic for goodness-of-fit within the exponential family. Comm. Statist. 3, 1139-1153.

Read, T. R. C. (1984). Small sample comparisons for the power divergence
goodness-of-fit statistics. J. Amer. Statist. Assoc. 79, 929-935.

Roscoe, J. T. and Byars, J. A. (1971). An investigation of the restraints
with respect to sample size commonly imposed on the use of the chi-
square statistic. J. Amer. Statist. Assoc. 66, 755-759.

Roy, A. R. (1956). On $\chi^2$ statistics with variable intervals. Technical Report
No. 1, Stanford Univ., Department of Statistics.

Schorr, B. (1974). On the choice of the class intervals in the application of
the chi-square test. Math. Operations Forsch. u. Statist. 5, 357-377.

Shapiro, S. S., Wilk, M. B. and Chen, H. J. (1968). A comparative study
of various tests for normality. J. Amer. Statist. Assoc. 63, 1343-1372.

Spruill, M. C. (1976). A comparison of chi-square goodness-of-fit tests
based on approximate Bahadur slope. Ann. Statist. 4, 409-412.

Watson, G. S. (1957). The chi-squared goodness-of-fit test for normal distributions. Biometrika 44, 336-348.

Watson, G. S. (1958). On chi-square goodness-of-fit tests for continuous
distributions. J. Roy. Statist. Soc., Ser. B 20, 44-61.

Watson, G. S. (1959). Some recent results in chi-square goodness-of-fit
tests. Biometrics 15, 440-468.

# 4

# Tests Based on EDF Statistics

Michael A. Stephens  Simon Fraser University, Burnaby, B.C., Canada

## 4.1 INTRODUCTION

Graphical methods have a wide appeal in deciding if a random sample appears to come from a given distributional form. Some of these methods have already been considered in Chapter 2. In this chapter we consider tests of fit based on the empirical distribution function (EDF). The EDF is a step function, calculated from the sample, which estimates the population distribution function. EDF statistics are measures of the discrepancy between the EDF and a given distribution function, and are used for testing the fit of the sample to the distribution; this may be completely specified (Case 0 below) or may contain parameters which must be estimated from the sample.

## 4.2 EMPIRICAL DISTRIBUTION FUNCTION STATISTICS

### 4.2.1 The Empirical Distribution Function (EDF)

Suppose a given random sample of size n is $X_1, \ldots, X_n$ and let $X_{(1)} < X_{(2)} < \cdots < X_{(n)}$ be the order statistics; suppose further that the distribution of X is F(x). For the present and in most of this chapter we assume this distribution to be continuous. The empirical distribution function (EDF) is $F_n(x)$ defined by

$$F_n(x) = \frac{\text{number of observations} \leq x}{n} \; ; \; -\infty < x < \infty$$

TABLE 4.1  Leghorn Chick Data

| $X^a$ | $Z_1^b$ | $Z_2^c$ |
|-------|---------|---------|
| 156   | 0.104   | 0.040   |
| 162   | 0.139   | 0.060   |
| 168   | 0.180   | 0.087   |
| 182   | 0.304   | 0.184   |
| 186   | 0.345   | 0.221   |
| 190   | 0.388   | 0.261   |
| 190   | 0.388   | 0.261   |
| 196   | 0.455   | 0.329   |
| 202   | 0.523   | 0.402   |
| 210   | 0.612   | 0.505   |
| 214   | 0.655   | 0.557   |
| 220   | 0.716   | 0.633   |
| 226   | 0.771   | 0.704   |
| 230   | 0.804   | 0.747   |
| 230   | 0.804   | 0.747   |
| 236   | 0.848   | 0.805   |
| 236   | 0.848   | 0.805   |
| 242   | 0.885   | 0.855   |
| 246   | 0.906   | 0.883   |
| 270   | 0.977   | 0.976   |

[a]Original values X of weights of
20 chicks; in grams.
[b]Values $Z_1$ are given by the Probability Integral Transformation for a
test for normality, with given mean
200 and given standard deviation 35
for use in a Case 0 test.
[c]Values $Z_2$ are given by the Probability Integral Transformation using
sample mean 209.6 and sample
standard deviation 30.65 for use in
a Case 3 test.

More precisely, the definition is

$$F_n(x) = 0, \qquad x < X_{(1)}$$

$$F_n(x) = \frac{i}{n}, \qquad X_{(i)} \le x < X_{(i+1)}, \qquad i = 1, \ldots, n-1 \qquad (4.1)$$

$$F_n(x) = 1, \qquad X_{(n)} \le x$$

Thus $F_n(x)$ is a step function, calculated from the data; as x increases it takes a step up of height $1/n$ as each sample observation is reached. For any x, $F_n(x)$ records the proportion of observations less than or equal to x, while $F(x)$ is the probability of an observation less than or equal to x. We can expect $F_n(x)$ to estimate $F(x)$, and it is in fact a consistent estimator of $F(x)$; as $n \to \infty$, $|F_n(x) - F(x)|$ decreases to zero with probability one.

### E 4.2.1 Example

Table 4.1 gives the weight X in grams of twenty 21-day-old leghorn chicks, given by Bliss (1967) and taken from Appendix A. Figure 4.1 gives the EDF

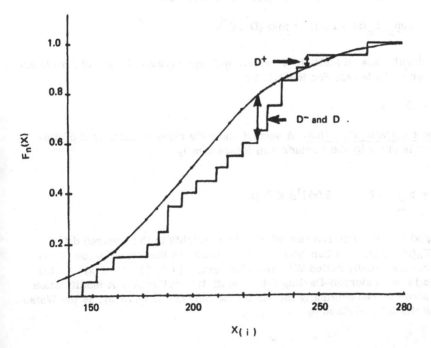

FIGURE 4.1 EDF of X. Also drawn is the normal distribution, mean 200, variance 1225.

of these sample data. It is clear that the weights have been rounded to the nearest gram, so that strictly the parent population is discrete, but with these large numbers this approximation will make negligible difference. The $F_n(x)$ suggests that $F(x)$ will have the characteristic S-shape of many distributions, including the normal distribution. A typical normal distribution, with mean $\mu = 200$ and variance $\sigma^2 = 1225$, has also been drawn in Figure 4.1; this will be used in later work to illustrate tests that the sample of weights comes from this distribution.

### 4.2.2 EDF Statistics

A statistic measuring the difference between $F_n(x)$ and $F(x)$ will be called an EDF statistic; we shall concentrate on seven which have attracted most attention. They are based on the vertical differences between $F_n(x)$ and $F(x)$, and are conveniently divided into two classes, the supremum class and the quadratic class.

The supremum statistics. The first two EDF statistics, $D^+$ and $D^-$, are, respectively, the largest vertical difference when $F_n(x)$ is greater than $F(x)$, and the largest vertical difference when $F_n(x)$ is smaller than $F(x)$; formally, $D^+ = \sup_x \{ F_n(x) - F(x) \}$ and $D^- = \sup_x \{ F(x) - F_n(x) \}$. The most well-known EDF statistic is D, introduced by Kolmogorov (1933):

$$D = \sup_x | F_n(x) - F(x) | = \max(D^+, D^-)$$

A closely related statistic V, given by Kuiper (1960), is useful for observations on a circle (see Section 4.5.3):

$$V = D^+ + D^-$$

The quadratic statistics. A second and wide class of measures of discrepancy is given by the Cramér-von Mises family

$$Q = n \int_{-\infty}^{\infty} \{ F_n(x) - F(x) \}^2 \psi(x) dF(x)$$

where $\psi(x)$ is a suitable function which gives weights to the squared difference $\{ F_n(x) - F(x) \}^2$. When $\psi(x) = 1$ the statistic is the Cramér-von Mises statistic, now usually called $W^2$, and when $\psi(x) = [ \{ F(x) \} \{ (1 - F(x) \} ]^{-1}$ the statistic is the Anderson-Darling (1954) statistic, called $A^2$. A modification of $W^2$, also devised originally for the circle (see Section 4.5.3), is the Watson (1961) statistic $U^2$ defined by

$$U^2 = n \int_{-\infty}^{\infty} \left\{ F_n(x) - F(x) - \int_{-\infty}^{\infty} [F_n(x) - F(x)] dF(x) \right\}^2 dF(x)$$

### 4.2.3 Computing Formulas

From the basic definitions of the supremum statistics and the quadratic statistics given above, suitable computing formulas must be found. This is done by using the Probability Integral Transformation (PIT), $Z = F(X)$; when $F(x)$ is the true distribution of X, the new random variable Z is uniformly distributed between 0 and 1. Then Z has distribution function $F^*(z) = z$, $0 \leq z \leq 1$. Suppose that a sample $X_1, \ldots, X_n$ gives values $Z_i = F(X_i)$, $i = 1, \ldots, n$, and let $F_n^*(z)$ be the EDF of the values $Z_i$.

EDF statistics can now be calculated from a comparison of $F_n^*(z)$ with the uniform distribution for Z. It is easily shown that, for values z and x related by $z = F(x)$, the corresponding vertical differences in the EDF diagrams for X and for Z are equal; that is,

$$F_n(x) - F(x) = F_n^*(z) - F^*(z) = F_n^*(z) - z \, ;$$

consequently EDF statistics calculated from the EDF of the $Z_i$ compared with the uniform distribution will take the same values as if they were calculated from the EDF of the $X_i$, compared with $F(x)$. This leads to the following formulas for calculating EDF statistics from the Z-values.

The formulas involve the Z-values arranged in ascending order, $Z_{(1)} < Z_{(2)} < \cdots < Z_{(n)}$. Then, with $\bar{Z} = \Sigma_i Z_i / n$,

$$D^+ = \max_i \{i/n - Z_{(i)}\} \, ; \, D^- = \max_i \{Z_{(i)} - (i-1)/n\} \, ; \, D = \max(D^+, D^-)$$

$$V = D^+ + D^-$$

$$W^2 = \Sigma_i \{Z_{(i)} - (2i-1)/(2n)\}^2 + 1/(12n) \, ; \, U^2 = W^2 - n(\bar{Z} - 0.5)^2$$

$$A^2 = -n - (1/n) \Sigma_i (2i-1)[\log Z_{(i)} + \log\{1 - Z_{(n+1-i)}\}]$$

$$\text{(4.2)}$$

Another formula for $A^2$ is

$$A^2 = -n - (1/n) \Sigma_i [(2i-1) \log Z_{(i)} + (2n+1-2i) \log\{1 - Z_{(i)}\}]$$

In these expressions, log x means $\log_e x$, and the sums and maxima are over $1 \leq i \leq n$. All these formulas are very straightforward to calculate, particularly with a modern computer, or programmable desk calculator. Note that statistic $D^-$ can be easily miscalculated, using $i/n$ instead of $(i-1)/n$.

## 4.3 GOODNESS-OF-FIT TESTS BASED ON THE EDF (EDF TESTS)

### 4.3.1 General Comments

The general test of fit is a test of

$H_0$: a random sample of n X-values comes from $F(x; \theta)$

where $F(x; \theta)$ is a (continuous) distribution and $\theta$ is a vector of parameters. For example, for a normal distribution under test, $\theta = (\mu, \sigma^2)$, and for a Gamma distribution defined as in Section 4.12, $\theta = (\alpha, \beta, m)$. When $\theta$ is fully specified, we call the situation Case 0. Then $Z_{(i)} = F(X_{(i)}; \theta)$ gives a set $Z_{(i)}$ which, on $H_0$, are ordered uniforms and equations (4.2) are used to give EDF statistics. On the other hand, $F(x; \theta)$ may be defined only as a member of a family of distributions such as the normal or Gamma, but all or part of the vector $\theta$ may be unknown. As an example of Case 0, the data in Table 4.1 might be tested to be normal with mean $\mu = 200$ and variance $\sigma^2 = 1225$, and as an example of the second situation it would be tested only to be normal, with unknown mean and variance. For the second case it would be natural to use the sample mean and variance as estimates of the components of $\theta = (\mu, \sigma^2)$.

For Case 0, distribution theory of EDF statistics is well-developed, even for finite samples, and tables have been available for some time. When $\theta$ contains one or more unknown parameters, these parameters may be replaced by estimates, to give $\hat{\theta}$ as the estimate of $\theta$. Then formulas (4.2) may still be used to calculate EDF statistics, with $Z_{(i)} = F(X_{(i)}; \hat{\theta})$. However, even when $H_0$ is true, the $Z_{(i)}$ will now not be an ordered uniform sample, and the distributions of EDF statistics will be very different from those for Case 0; they will depend on the distribution tested, the parameters estimated, and the method of estimation, as well as on the sample size. New points should then be used for the appropriate test, even for large samples, otherwise a serious error in significance level will result.

### 4.3.2 Unknown Location and Scale Parameters

When the unknown components of $\theta$ are location or scale parameters, and if these are estimated by appropriate methods, the distributions of EDF statistics will not depend on the true values of the unknown parameters. Thus percentage points for EDF tests for such distributions, for example, the normal, exponential, extreme-value, and logistic distributions, depend only on the family tested and on the sample size n. Nevertheless, the exact distributions of EDF statistics are very difficult to find and except for the exponential distribution, Monte Carlo studies have been extensively used to find points for finite n. Fortunately, for the quadratic statistics $W^2$, $U^2$, and $A^2$, asymptotic theory is available; furthermore, the percentage points of these statistics for finite n converge rapidly to the asymptotic points. For the statistics $D^+$, $D^-$, D, and V, there is no general asymptotic theory (except for Case 0), and even asymptotic points must be estimated. This may be done by plotting, for a fixed $\alpha$, the Monte Carlo points for samples of size n against m = 1/n, and then extrapolating to m = 0; alternatively, since the statistics are functions of a process which is asymptotically Gaussian, points may be found by simulating the Gaussian process. Serfling and Wood (1975) and Wood (1978a) have obtained asymptotic points by this method. Both techniques are of course subject to sampling variation and errors due to extrapolation; Chandra, Singpurwalla, and Stephens (1981) have given some comparisons of the two methods in obtaining points for tests for the extreme value distribution.

For the tests corresponding to many distributional families, Stephens (1970, 1974b, 1977, 1979) has given modifications of the test statistics; if the statistic is, say, T, the modification is a function of n and T which is then referred to the asymptotic points of T or of $T\sqrt{n}$. Asymptotic theory depends on using asymptotically efficient estimators for the estimates of unknown components of $\theta$; the asymptotic points given will then be valid for any such estimators. Points for finite n will depend on which estimators are used; usually these are maximum likelihood estimators although an exception is the Cauchy distribution below. In the test situations in following sections, percentage points for finite n, using the estimators given, were found from extensive Monte Carlo studies, often done by the author, although other studies referenced have also been used. The modifications are then derived from an examination of how these points, for $\alpha = 0.05$ say, converge to the asymptotic point. A feature of the modifications is that, at least in the appropriate tail, they do not depend on $\alpha$; thus when such modifications have been found, the usual tables of percentage points, with entries for n and $\alpha$, can be reduced to one line for each test situation. The modifications hold only if the estimators given are used. They have been calculated to be most accurate at about $\alpha = 0.05$, but usually give good results, for practical purposes, for $\alpha$ less than about 0.2.

### 4.3.3 Unknown Shape Parameters

When unknown parameters are not location or scale parameters, for example when the shape parameter of a Gamma or a Weibull distribution is unknown, null distribution theory, even asymptotic, when the parameters are estimated, will depend on the true values of these parameters. However, if this dependence is very slight, a set of tables, to be used with the estimated value of the shape parameter, can still be valuable (see, for example, Section 4.12, concerning tests for the Gamma distribution). Other methods of dealing with unknown parameters are discussed in Section 4.16.3.

There is now a vast literature on EDF statistics and tests, and only the principal references related directly to the tests and tables are included here. Surveys have been given by Sahler (1968) and by Neuhaus (1979); a comprehensive review of the theory, and many references, is given by Durbin (1973).

We next give tests and applications of EDF statistics for Case 0, and in subsequent sections give tests for the major distributions.

## 4.4 EDF TESTS FOR A FULLY SPECIFIED DISTRIBUTION (CASE 0)

The following procedure can now be set out for EDF tests for Case 0, that is, for the null hypothesis

$H_0$: a random sample $X_1, \ldots, X_n$ comes from $F(x;\theta)$, a completely specified continuous distribution

(a) Put the $X_i$ in ascending order, $X_{(1)} < X_{(2)} < \cdots < X_{(n)}$.
(b) Calculate $Z_{(i)} = F(X_{(i)};\theta)$, $i = 1, \ldots, n$.
(c) Calculate the appropriate test statistic using (4.2).
(d) Modify the test statistic as in Table 4.2 using the modifications for the upper tail, and compare with the appropriate line of percentage points. If the statistic exceeds the value in the upper tail given at level $\alpha$, $H_0$ is rejected at significance level $\alpha$.

### E 4.4.1 Example

Suppose the data in Table 4.1 are to be tested to come from a normal distribution with mean $\mu = 200$ and standard deviation $\sigma = 35$. This is the distribution drawn in Figure 4.1. The Probability Integral Transformation gives the values in column $Z_1$ of Table 4.1. These have then been used to draw the EDF in Figure 4.2. The calculation of $D^+$ and $D^-$ is also illustrated in the figure. Formulas (4.2) give, to 3 d.p.: $D^+ = 0.044$, $D^- = 0.171$, $D = 0.171$, $V = 0.216$, for the supremum statistics, and $W^2 = 0.187$, $U^2 = 0.051$, and $A^2 = 1.019$ for the quadratic class. From Table 4.2 the modified value $D^*$ is found from $D^* = D(\sqrt{n} + 0.12 + 0.11/\sqrt{n})$ and the value is 0.790. Reference to the percentage points on the same line as the modification (the asymptotic percentage points of $D\sqrt{n}$) shows D to be not significant at the 15% level. The modified values of the other statistics (using, for example, $W^*$ for modified $W^2$) are: $D^{+*} = 0.203$, $D^{-*} = 0.790$, $V^* = 1.011$, $W^* = .177$, $U^* = .048$, $A^* = 1.019$. These give levels of significance $\alpha$ (or p-levels) well below the 25% point for all the statistics, so that the hypothesis $H_0$ will not be rejected.

It will be observed from the modifications in Table 4.2 that the percentage points of $W^2$ and $U^2$ for finite n converge rapidly in the upper tail to the asymptotic points, and even if the modifications were not included in the table, the use of the asymptotic percentage points for $n \geq 20$ would give negligible error in $\alpha$. Even more striking is the fact that, for $n > 3$, the distribution of $A^2$ is accurately given by the asymptotic distribution. This

TABLE 4.2 Modifications and Percentage Points for EDF Statistics for Testing a Completely Specified Distribution (Case 0; Section 4.4)

| Statistic T | Modified form T* | Significance level $\alpha$ | | | | | | | |
|---|---|---|---|---|---|---|---|---|---|
| | | .25 | .15 | .10 | .05 | .025 | .01 | .005 | .001 |
| | | Upper tail percentage points | | | | | | | |
| $D^+(D^-)$ | $D^+(\sqrt{n} + 0.12 + 0.11/\sqrt{n})$ | 0.828 | 0.973 | 1.073 | 1.224 | 1.358 | 1.518 | 1.628 | 1.859 |
| D | $D(\sqrt{n} + 0.12 + 0.11/\sqrt{n})$ | 1.019 | 1.138 | 1.224 | 1.358 | 1.480 | 1.628 | 1.731 | 1.950 |
| V | $V(\sqrt{n} + 0.155 + 0.24/\sqrt{n})$ | 1.420 | 1.537 | 1.620 | 1.747 | 1.862 | 2.001 | 2.098 | 2.303 |
| $W^2$ | $(W^2 - 0.4/n + 0.6/n^2)(1.0 + 1.0/n)$ | 0.209 | 0.284 | 0.347 | 0.461 | 0.581 | 0.743 | 0.869 | 1.167 |
| $U^2$ | $(U^2 - 0.1/n + 0.1/n^2)(1.0 + 0.8/n)$ | 0.105 | 0.131 | 0.152 | 0.187 | 0.222 | 0.268 | 0.304 | 0.385 |
| $A^2$ | For all $n \geq 5$ | 1.248 | 1.610 | 1.933 | 2.492 | 3.070 | 3.880 | 4.500 | 6.000 |
| | | Lower tail percentage points | | | | | | | |
| D | $D(\sqrt{n} + 0.275 - 0.04/\sqrt{n})$ | – | 0.610 | 0.571 | 0.520 | 0.481 | 0.441 | – | – |
| V | $V(\sqrt{n} + 0.41 - 0.26/\sqrt{n})$ | – | 0.976 | 0.928 | 0.861 | 0.810 | 0.755 | – | – |
| $W^2$ | $(W^2 - 0.03/n)(1.0 + 0.05/n)$ | – | 0.054 | 0.046 | 0.037 | 0.030 | 0.025 | – | – |
| $U^2$ | $(U^2 - 0.02/n)(1 + 0.35/n)$ | – | 0.038 | 0.033 | 0.028 | 0.024 | 0.020 | – | – |
| $A^2$ | For all $n \geq 5$ | – | 0.399 | 0.346 | 0.283 | 0.240 | 0.201 | – | – |

Adapted from Stephens (1970), with permission of the Royal Statistical Society.

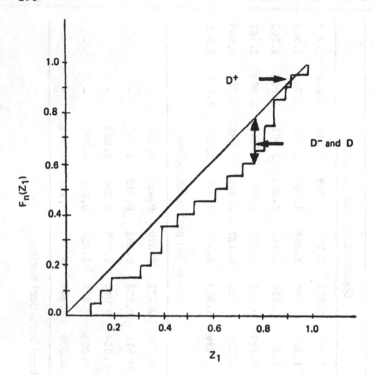

FIGURE 4.2 EDF of $Z_1$.

was demonstrated by Lewis (1961), and is valid all along the distribution. The modified forms are taken from Stephens (1970) where references to original sources of tables for EDF statistics are given.

## 4.5 COMMENTS ON EDF TESTS FOR CASE 0

### 4.5.1 Use of the Lower Tail

The test as described above is a one-tail test, using only the upper tail of the test statistics. This is because, in general, we should expect the difference between $F_n(x)$ and $F(x)$ or between $F_n(z)$ and $F(z)$ to be large when $H_0$ is not true. If a test statistic appears to be significant in the lower tail it suggests that the Z-sample is too regular to be a random uniform sample, and perhaps the original X-data have been tampered with. Such Z-values are called superuniform; tests for superuniformity can be made using the modifications and lower tail points also given in Table 4.2. Superuniform observations can arise in other ways also, particularly in connection with tests for the exponential distribution, or for randomness of points in time. An interesting data set which appears to be superuniform is the dates of the kings and queens of England (Pearson, 1963). Further comments on superuniformity are in Chapters 8 and 10.

### 4.5.2 Calculation of Significance Levels (p-Levels) for Given Statistics

Suppose a test statistic T takes the value t; the significance level, or p-value, of the statistic will then be the value $p = P(T > t)$. In some contexts the term is also applied to the lower tail probability $P(T < t)$ but here q, or q-level, will be used for this quantity; thus $q = 1 - p$. It is useful (especially in combining several independent tests, see Sections 4.18 and 8.15) to be able to calculate the significance level all along the distribution of T, and not merely in the tails. For $A^2$, Case 0, the table of q-values of the asymptotic distribution given by Lewis (1961) is reproduced in Table 4.3. Since $A^2$ needs no modification for sample size greater than three, Table 4.3 may be used to give q-values for all $n \geq 3$. Tables to find p or q in the tails were given for other EDF statistics by Stephens (1970).

### E 4.5.2 Example

In example E 4.4.1, $A^2 = 1.019$; from Table 4.3 the q-value is approximately 0.653, so that $p = 0.347$.

### 4.5.3 Observations on a Circle

A special problem arises if the observations X are measurements recording points on a circle. For example, suppose the circumference is of length 1, and let X be the arc length around the circumference, measured clockwise from an origin O. Clearly the value of $F(x;\theta)$ at a given point x varies with the choice of O, and the EDF statistics $D^+$, $D^-$, D, $W^2$, and $A^2$ take different values with different choices of origin. However, statistics V and $U^2$ do not; they were introduced by Kuiper and Watson to adapt the statistics D and $W^2$ for this problem, and V and $U^2$ should be used for observations recorded on a circle.

### E 4.5.3 Example

Consider a small sample of four values, which are to be tested for uniformity around a circle of unit circumference. With North as origin, and positive direction clockwise, suppose the X values are 0.3, 0.4, 0.5, 0.9. If East were regarded as origin, these values would change to 0.05, 0.15, 0.35, and 0.65. When the EDF are drawn for these two cases, the values of $D^+$, $D^-$, and D are, respectively, 0.15, 0.30, 0.30 and 0.40, 0.05, 0.40, but in both cases V is 0.45. The corresponding values of $W^2$, $A^2$, and $U^2$ are

TABLE 4.3  Distribution of $A^2$, Case 0:  The Table Gives $q = P (A^2 < z)$

| z | q | z | q | z | q |
|---|---|---|---|---|---|
| .025 | 0.0000 | 1.250 | 0.7503 | 3.100 | 0.9756 |
| .050 | 0.0000 | 1.300 | 0.7677 | 3.150 | 0.9770 |
| .075 | 0.0000 | 1.350 | 0.7833 | 3.200 | 0.9783 |
| .100 | 0.0000 | 1.400 | 0.7973 | 3.250 | 0.9795 |
| .125 | 0.0003 | 1.450 | 0.8111 | 3.300 | 0.9807 |
| .150 | 0.0014 | 1.500 | 0.8235 | 3.350 | 0.9818 |
| .175 | 0.0042 | 1.550 | 0.8350 | 3.400 | 0.9828 |
| .200 | 0.0096 | 1.600 | 0.8457 | 3.450 | 0.9837 |
| .225 | 0.0180 | 1.650 | 0.8556 | 3.500 | 0.9846 |
| .250 | 0.0296 | 1.700 | 0.8648 | 3.550 | 0.9855 |
| .275 | 0.0443 | 1.750 | 0.8734 | 3.600 | 0.9863 |
| .300 | 0.0618 | 1.800 | 0.8814 | 3.650 | 0.9870 |
| .325 | 0.0817 | 1.850 | 0.8888 | 3.700 | 0.9878 |
| .350 | 0.1036 | 1.900 | 0.8957 | 3.750 | 0.9884 |
| .375 | 0.1269 | 1.950 | 0.9021 | 3.800 | 0.9891 |
| .400 | 0.1513 | 2.000 | 0.9082 | 3.850 | 0.9897 |
| .425 | 0.1764 | 2.050 | 0.9138 | 3.900 | 0.9902 |
| .450 | 0.2019 | 2.100 | 0.9190 | 3.950 | 0.9908 |
| .475 | 0.2276 | 2.150 | 0.9239 | 4.000 | 0.9913 |
| .500 | 0.2532 | 2.200 | 0.9285 | 4.050 | 0.9917 |
| .525 | 0.2786 | 2.250 | 0.9328 | 4.100 | 0.9922 |
| .550 | 0.3036 | 2.300 | 0.9368 | 4.150 | 0.9926 |
| .575 | 0.3281 | 2.350 | 0.9405 | 4.200 | 0.9930 |
| .600 | 0.3520 | 2.400 | 0.9441 | 4.250 | 0.9934 |
| .625 | 0.3753 | 2.450 | 0.9474 | 4.300 | 0.9938 |
| .650 | 0.3930 | 2.500 | 0.9504 | 4.350 | 0.9941 |
| .675 | 0.4199 | 2.550 | 0.9534 | 4.400 | 0.9944 |
| .700 | 0.4412 | 2.600 | 0.9561 | 4.500 | 0.9950 |
| .750 | 0.4815 | 2.650 | 0.9586 | 4.600 | 0.9955 |
| .800 | 0.5190 | 2.700 | 0.9610 | 4.700 | 0.9960 |
| .850 | 0.5537 | 2.750 | 0.9633 | 4.800 | 0.9964 |
| .900 | 0.5858 | 2.800 | 0.9654 | 4.900 | 0.9968 |
| .950 | 0.6154 | 2.850 | 0.9674 | 5.000 | 0.9971 |
| 1.000 | 0.6427 | 2.900 | 0.9692 | 5.500 | 0.9983 |
| 1.050 | 0.6680 | 2.950 | 0.9710 | 6.000 | 0.9990 |
| 1.100 | 0.6912 | 3.000 | 0.9726 | 7.000 | 0.9997 |
| 1.150 | 0.7127 | 3.050 | 0.9742 | 8.000 | 0.9999 |
| 1.200 | 0.7324 | | | | |

Adapted from Lewis (1961), with permission of the author and of the Institute of Mathematical Statistics.

0.053, 0.337, 0.043 for North as origin, and 0.203, 1.116, 0.043 for East; $W^2$ and $A^2$ change in value but $U^2$, like V, remains constant.

### 4.5.4 Use of D to Give Confidence Intervals for a Distribution

The EDF, with statistic D, may also be used to provide confidence intervals for the true distribution function. This is done as follows. Let the critical value of D, for given n and $\alpha$, be $D_\alpha$, and draw a band of vertical height $D_\alpha$ on either side of $F_n(x)$; this gives a confidence band for the true distribution function, with confidence level $100(1 - \alpha)\%$. (Strictly speaking, there should be a slight modification of the band at the lower and upper tails, because otherwise the band contains negative values for the distribution, or values greater than 1, but the modifications are very small for a sample of reasonable size, say n = 20.)

### 4.5.5 Use of EDF Statistics to Give Confidence Sets for Parameters

When parameters are not known in $F(x;\theta)$, confidence sets may be provided for them by the following device. Suppose $\theta$ is a vector of unknown parameters, which need not be only location or scale parameters, and suppose values are given to unknown components of $\theta$ to give vector $\theta_0$; then $F(x;\theta_0)$ is completely specified and EDF statistics can be calculated. Suppose T is such a statistic. The confidence set for $\theta$, derived from T, and with level $100(1 - \alpha)\%$, includes all those values of $\theta_0$ which make T not significant at level $\alpha$. The confidence set is sometimes called a <u>consonance</u> set or region. Easterling (1976) and Littell and Rao (1978) have investigated the use of $A^2$ and D for finding consonance sets for parameters. The technique affords an interesting mixture of goodness-of-fit and parameter estimation methods.

### 4.5.6 Other EDF Statistics for Case 0

Many other statistics have been proposed to measure the discrepancy between $F_n(x)$ and $F(x;\theta)$; they are often closely related to the seven statistics discussed above, and have similar properties.

(a) Anderson and Darling (1952) suggested using a variance-weighted D, obtained by incorporating a weight function into the definition of D, much as it is included in $A^2$. Asymptotics for this statistic have been given by Doksum, Fenstad and Aaberge (1977), and tables for finite n by Niederhausen (1981).

(b) Suppose $s_i = F^{-1}(i/n)$, and let $b_i = i/n - F_n(s_i)$, where $F_n(x)$ is the EDF of the original sample X; Riedwyl (1967) suggested the test statistic $\sum_{i=1}^{n-1}|b_i|$. On the Z-diagram, this statistic is based on the discrepancy between $F_n^*(z)$ and $F(z)$ at equal intervals along the z-axis between 0 and 1; the statistic has a discrete distribution.

(c)  Let $\delta_i = \max_i \{ |Z_{(i)} - (i - 1)/n|, |Z_{(i)} - i/n| \}$; the Kolmogorov statistic D is then max $\delta_i$. Finkelstein and Schafer (1971) have proposed the statistic $S = \Sigma_i \delta_i$ and have given a table of percentage points for n up to 30.

(d)  Hegazy and Green (1975) and Green and Hegazy (1976) have discussed several statistics calculated from slight modifications of the computing formulas in Section 4.2. Berk and Jones (1979) gave other statistics based on $F_n(x)$ and similar to the Kolmogorov statistics. Hegazy and Green (1975) have demonstrated that their modified statistics can increase power against certain alternatives, and Berk and Jones showed certain optimal properties in the sense of Bahadur efficiency for their statistics.

(e)  A set of statistics closely related to EDF statistics, although not derived from the EDF, is the C and K set, described in Section 8.8.

## 4.6  POWER OF EDF STATISTICS FOR CASE 0

In Case 0, as we have seen, the final test is that a set of variables Z is uniformly distributed $U(0, 1)$, and a discussion of power properties of EDF statistics is therefore deferred to Chapter 8, where tests for uniformity are discussed in detail. However, for later comparisons in this chapter, we summarize certain properties of EDF statistics in Case 0 situations.

(a)  EDF statistics are usually much more powerful than the Pearson chi-square statistic; this might be explained by the fact that for the chi-square statistic the data must be grouped, with a resulting loss of information, especially for small samples.

(b)  The most well-known EDF statistic is D, but it is often much less powerful than the quadratic statistics $W^2$ and $A^2$.

(c)  Statistics $D^+$ and $D^-$ will be powerful in detecting whether or not the Z-set tends to be close to 0 or to 1, respectively; $A^2$, $W^2$, and D will detect either of these two alternatives, and $U^2$ and V are powerful in detecting a clustering of Z values at one point, or a division into two groups near 0 and 1. In terms of the original observations X, statistics $D^+$, $D^-$, $A^2$, $W^2$, and D will detect an error in mean in $F(x; \theta)$ as specified, and $U^2$ and V will detect an error in variance.

(d)  $A^2$ often behaves similarly to $W^2$, but is on the whole more powerful for tests when $F(x; \theta)$ departs from the true distribution in the tails, especially when there appears to be too many outlying X-values for the $F(x; \theta)$ as specified. In goodness-of-fit work, departure in the tails is often important to detect, and $A^2$ is the recommended statistic.

## 4.7  EDF TESTS FOR CENSORED DATA: CASE 0

### 4.7.1  Introduction

If some of the observations $X_1$, $X_2$, ..., $X_n$ of a random sample are missing, the sample is said to be censored. If all observations less than $X_{(s)}$ are missing the sample is left-censored, and if all observations greater than $X_{(r)}$ are missing, it is right-censored; in either case, the sample is said to be singly-censored. If observations are missing at both ends, the sample is doubly-censored.

Censoring may occur for random values of s or r (Type 1 censoring) or for fixed values (Type 2 censoring). These may be illustrated by lifetime measurements $X_i$ of equipment. If the experiment is continued for a fixed time t, the number of items which fail in that time would be a random variable and the censoring would be Type 1; if, on the other hand, it is decided to follow the experiment until 20 items have failed, then r is fixed at 20 and we have Type 2 censoring. Another form of censoring is random censoring, where, for example, observation $X_i$ may not be known, but it is known that $X_i > T_i$, where $T_i$ is another random variable. In the lifetesting experiment, this could occur if items were removed from the test for reasons other than because they failed in the manner investigated in the experiment.

EDF statistics have been adapted for all these forms of censoring. Consider Case 0, where the distribution under test, say F(x), is fully specified. Then the Probability Integral Transformation may be made for the observations available, giving a set $Z_i = F(X_i)$ which is itself censored. Suppose the X-set is right-censored, of Type 1; the values of X are known to be less than the fixed value $X^*$, and the available $Z_i$ are then $Z_{(1)} < Z_{(2)} < \cdots < Z_{(r)} < t$, where $t = F(X^*)$. If the censoring is Type 2, there are again r values $Z_{(i)}$, with $Z_{(r)}$ the largest and r fixed.

### 4.7.2  The Kolmogorov-Smirnov Statistics $D^+$, $D^-$, and $D$

The Kolmogorov-Smirnov statistic, modified for Type 1 censored data, is $_1D_{t,n}$, calculated from the EDF $F_n(z)$ of the r ordered Z-values:

$$_1D_{t,n} = \sup_{0 \le z \le t} |F_n(z) - z|$$

$$= \max_{1 \le i \le r} \left\{ \frac{i}{n} - Z_{(i)}, \; Z_{(i)} - \frac{i-1}{n}, \; t - \frac{r}{n} \right\} \qquad (4.3)$$

For Type 2 censored data, the Kolmogorov-Smirnov statistic is

TABLE 4.4 Upper Tail Asymptotic Percentage Points for $\sqrt{n}D$, $W^2$ and $A^2$, for Type 1 or Type 2 Censored Data from U(0, 1) (Section 4.7)

| p | Significance level $\alpha$ | | | | | | | |
|---|---|---|---|---|---|---|---|---|
| | 0.50 | 0.25 | 0.15 | 0.10 | 0.05 | 0.025 | 0.01 | 0.005 |
| **Statistic $\sqrt{n}D$** | | | | | | | | |
| .2 | .4923 | .6465 | .7443 | .8155 | .9268 | 1.0282 | 1.1505 | 1.2361 |
| .3 | .5889 | .7663 | .8784 | .9597 | 1.0868 | 1.2024 | 1.3419 | 1.4394 |
| .4 | .6627 | .8544 | .9746 | 1.0616 | 1.1975 | 1.3209 | 1.4696 | 1.5735 |
| .5 | .7204 | .9196 | 1.0438 | 1.1334 | 1.2731 | 1.3997 | 1.5520 | 1.6583 |
| .6 | .7649 | .9666 | 1.0914 | 1.1813 | 1.3211 | 1.4476 | 1.5996 | 1.7056 |
| .7 | .7975 | .9976 | 1.1208 | 1.2094 | 1.3471 | 1.4717 | 1.6214 | 1.7258 |
| .8 | .8183 | 1.0142 | 1.1348 | 1.2216 | 1.3568 | 1.4794 | 1.6272 | 1.7306 |
| .9 | .8270 | 1.0190 | 1.1379 | 1.2238 | 1.3581 | 1.4802 | 1.6276 | 1.7308 |
| 1.0 | .8276 | 1.0192 | 1.1379 | 1.2238 | 1.3581 | 1.4802 | 1.6276 | 1.7308 |
| **Statistic $W^2$** | | | | | | | | |
| .2 | .010 | .025 | .033 | .041 | .057 | .074 | .094 | .110 |
| .3 | .022 | .046 | .066 | .083 | .115 | .147 | .194 | .227 |
| .4 | .037 | .076 | .105 | .136 | .184 | .231 | .295 | .353 |
| .5 | .054 | .105 | .153 | .186 | .258 | .330 | .427 | .488 |
| .6 | .070 | .136 | .192 | .241 | .327 | .417 | .543 | .621 |
| .7 | .088 | .165 | .231 | .286 | .386 | .491 | .633 | .742 |
| .8 | .103 | .188 | .259 | .321 | .430 | .544 | .696 | .816 |
| .9 | .115 | .204 | .278 | .341 | .455 | .573 | .735 | .865 |
| 1.0 | .119 | .209 | .284 | .347 | .461 | .581 | .743 | .869 |
| **Statistic $A^2$** | | | | | | | | |
| .2 | .135 | .252 | .333 | .436 | .588 | .747 | .962 | 1.129 |
| .3 | .204 | .378 | .528 | .649 | .872 | 1.106 | 1.425 | 1.731 |
| .4 | .275 | .504 | .700 | .857 | 1.150 | 1.455 | 1.872 | 2.194 |
| .5 | .349 | .630 | .875 | 1.062 | 1.419 | 1.792 | 2.301 | – |
| .6 | .425 | .756 | 1.028 | 1.260 | 1.676 | 2.112 | 2.707 | – |
| .7 | .504 | .882 | 1.184 | 1.451 | 1.920 | 2.421 | 3.083 | – |
| .8 | .588 | 1.007 | 1.322 | 1.623 | 2.146 | 2.684 | 3.419 | – |
| .9 | .676 | 1.131 | 1.467 | 1.798 | 2.344 | 2.915 | 3.698 | – |
| 1.0 | .779 | 1.248 | 1.610 | 1.933 | 2.492 | 3.070 | 3.880 | 4.500 |

Table for $\sqrt{n}D$ adapted from Koziol and Byar (1975), with permission of the authors and of the American Statistical Association. Tables for $W^2$ and $A^2$ adapted from Pettitt and Stephens (1976), with permission of the author and of the Biometrika Trustees.

$$_2D_{r,n} = \sup_{0 \le z \le Z_{(r)}} |F_n(z) - z|$$

$$= \max_{1 \le i \le r} \left\{ \frac{i}{n} - Z_{(i)}, \; Z_{(i)} - \frac{i-1}{n} \right\} \qquad (4.4)$$

$$= \max_{1 \le i \le r} \left| \frac{i - .5}{n} - Z_{(i)} \right| + \frac{.5}{n}$$

The one-sided versions of these statistics are denoted by $_1D_{t,n}^+$, etc. Tables of percentage points of the null distribution of $_1D_{t,n}$ and $_2D_{r,n}$ were given by Barr and Davidson (1973). For both types of censoring, these converge to one asymptotic distribution, given by Koziol and Byar (1975); points from this distribution are in the first part of Table 4.4. Dufour and Maag (1978) gave useful formulas so that the asymptotic distributions could be used with finite samples. The technique is as follows.

Suppose the sample is right-censored, and $H_0$ is

$H_0$: the censored sample $X_{(1)} < X_{(2)} < \cdots < X_{(r)}$ comes from the fully specified continuous distribution $F(x)$

The values $Z_{(i)} = F(X_{(i)})$, $i = 1, \ldots, r$, are calculated. The steps in testing $H_0$ are then the following.

For Type 1 censoring:

(a)  Calculate $_1D_{t,n}$ from formula (4.3).
(b)  Modify $_1D_{t,n}$ to $D_t^*$ calculated from

$$D_t^* = \sqrt{n} \, _1D_{t,n} + 0.19/\sqrt{n}, \text{ for } n \ge 25 \text{ and } t \ge 0.25$$

(c)  Refer to Table 4.4 and reject $H_0$ at significance level $\alpha$ if $D_t^*$ exceeds the tabulated value for $\alpha$.

For Type 2 censoring:

(a)  Calculate $_2D_{r,n}$ from formula (4.4).
(b)  Modify $_2D_{r,n}$ to $D_r^*$ calculated from

$$D_r^* = \sqrt{n} \, _2D_{r,n} + 0.24/\sqrt{n}, \text{ for } n \ge 25 \text{ and } r/n \ge 0.4$$

(c)  Refer to Table 4.4 and reject $H_0$ at significance level $\alpha$ if $D_r^*$ exceeds the tabulated value at level $\alpha$.

For values of $n \le 25$ or for censoring more extreme than the ranges given above, refer to tables of Barr and Davidson (1973) or Dufour and Maag (1978).

In order to approximate the tail areas of the finite sample distribution, that is, to obtain p-values of test statistics, a relationship between the asymptotic distributions of one and two-sided statistics can be used.

The asymptotic distribution for the one-sided test statistic $\sqrt{n}(_1D^+_{t,n})$ has been given by Schey (1977) as

$$\lim_{n\to\infty} P\{\sqrt{n}(_1D^+_{t,n}) \leq y)\} = G_t(y)$$

$$= \phi(A_t y) - \phi(B_t y) e^{-2y^2}, \quad y > 0$$

where $A_t = (t - t^2)^{\frac{1}{2}}$ and $B_t = (2t - 1)A_t$. The tail area for the two-sided statistic is approximated well for significance levels less than 0.20 by doubling the one-sided value. Thus to obtain a p-value, the test statistic $_1D_{t,n}$ or $_2D_{r,n}$ is first adapted to obtain $D^*_t$ or $D^*_r$ as described above, and then the p-value, for a two-sided test, is well-approximated by $p_t = 2\{1 - G_t(D^*_t)\}$ or $p_r = 2\{1 - G_t(D^*_r)\}$. Examples of calculations of Kolmogorov-Smirnov statistics for censored data are given in Section 11.3.1.

### 4.7.3  Cramér-von Mises Statistics

A second group of statistics for censored samples is of the general Cramér-von Mises type. Pettitt and Stephens (1976) introduced versions of the Cramér-von Mises $W^2$, Watson $U^2$, and Anderson-Darling $A^2$ statistics, obtained (for right-censored data) by modifying the upper limit of integration in the definitions of these statistics; $W^2$ would then become $_1W^2_{t,n}$ for Type 1 censoring, and $_2W^2_{t,n}$ for Type 2 censoring. The computing formulas differ slightly for the two types of censoring. The formulas for Type 2 censoring, given $Z_{(1)} < \cdots < Z_{(r)}$, are

$$_2W^2_{r,n} = \sum_{i=1}^{r} \left(Z_{(i)} - \frac{2i-1}{2n}\right)^2 + \frac{r}{12n^2} + \frac{n}{3}\left(Z_{(r)} - \frac{r}{n}\right)^3$$

$$_2U^2_{r,n} = _2W^2_{r,n} - nZ_{(r)}\left[\frac{r}{n} - \frac{Z_{(r)}}{2} - \frac{r\bar{Z}}{nZ_{(r)}}\right]^2$$

where $\bar{Z} = \sum_{i=1}^{r} Z_{(i)}/r$ ;                                                                             (4.5)

$$_2A^2_{r,n} = -\frac{1}{n}\sum_{i=1}^{r}(2i-1)[\log Z_{(i)} - \log\{1 - Z_{(i)}\}] - 2\sum_{i=1}^{r}\log\{1 - Z_{(i)}\}$$

$$-\frac{1}{n}[(r-n)^2\log\{1 - Z_{(r)}\} - r^2\log Z_{(r)} + n^2 Z_{(r)}]$$

For Type 1 right-censored data, suppose t (t < 1) is the fixed censoring value of Z. This value is added to the sample set, and the statistics are now calculated by using the above formulas with r replaced by r + 1, and with $Z_{(r+1)} = t$; they will be called $_1W^2_{t,n}$, $_1U^2_{t,n}$, and $_1A^2_{t,n}$. Note that it is possible to have r = n observations less than t, so that when the value t is added, the new sample has size n + 1. Statistics $_1W^2_{t,n}$ and $_2W^2_{r,n}$ have the same asymptotic distributions for the two types of censoring; similarly for the other statistics. Asymptotic points for $W^2$ and $A^2$ are given in Table 4.4.

Thus the steps in making a test, for right-censored data, are:

(a) Calculate the statistic required as described above.
(b) Refer to Table 4.4 for Type 1 data, and Table 4.5 for Type 2 data.

For Type 1 censored data, Table 4.4 is entered at p = t, for all n. For Type 2 censoring, Table 4.5 is entered at p = r/n, with appropriate n. $H_0$ is rejected at significance level $\alpha$ if the statistic is greater than the point given for level $\alpha$.

The asymptotic points in Tables 4.4 and 4.5 are those given by Pettitt and Stephens (1976), with some additions. Points for finite n have been obtained by extensive Monte Carlo studies, which showed that points for Type 1 censoring converge so rapidly to the asymptotic distribution that a new table is not needed. Tables for $U^2$ are probably not so valuable for censored data and have been omitted. More extensive tables, including tables for $U^2$, are in Stephens (1986).

Smith and Bain (1976) have suggested another version of $W^2$ for use with Type 2 right-censored data from the uniform distribution; the statistic, say $_2^*W^2_{r,n}$, is $\Sigma_{i=1}^{r}\{Z_{(i)} - (2i - 1)/2n\}^2 + 1/(12n)$. $_2^*W^2_{r,n}$ will have the same asymptotic distribution as $_2W^2_{r,n}$, and Smith and Bain (1976) have given Monte Carlo points for finite n. Some comparisons of statistics for censored data, including $_2^*W^2_{r,n}$, $_2W^2_{r,n}$, and $_2A^2_{r,n}$ were made by Michael and Schucany (1979). For the alternatives and censoring factors which were studied, there were noticeable differences in the sensitivity of $_2^*W^2_{r,n}$ and $_2W^2_{r,n}$. In general, the statistic $_2A^2_{r,n}$ displayed the best power. Examples of calculations of Cramér-von Mises statistics for censored data are given in Section 11.3.1.

### 4.7.3.1 Left-Censored Data

For left-censored data, the values $Z^*_{(i)} = 1 - Z_{(n+1-i)}$, i = 1, ..., r may be calculated from the r largest observations and the set Z* used in tests for right-censored data. In Type 1 censoring, the left-censoring value t converts to t* = 1 - t, to be used as the right-censoring point with the Z* values.

TABLE 4.5 Upper Tail Percentage Points for $_2W^2_{r,n}$ and $_2A^2_{r,n}$ for Type 2 Right-Censored Data from the Uniform $U(0,1)$ Distribution (Section 4.7.3). The table should be entered at n and at p = r/n.

| Statistic | n | Significance level $\alpha$ | | | | | | |
|---|---|---|---|---|---|---|---|---|
| | | 0.50 | 0.25 | 0.15 | 0.10 | 0.05 | 0.025 | 0.01 |
| $_2W^2_{r,n}$ | 20 | 0.006 | 0.018 | 0.038 | 0.058 | 0.099 | 0.152 | 0.243 |
| | 40 | 0.008 | 0.018 | 0.032 | 0.046 | 0.084 | 0.128 | 0.198 |
| p = 0.2 | 60 | 0.009 | 0.020 | 0.031 | 0.044 | 0.074 | 0.107 | 0.154 |
| | 80 | 0.009 | 0.021 | 0.031 | 0.043 | 0.069 | 0.097 | 0.136 |
| | 100 | 0.009 | 0.022 | 0.031 | 0.043 | 0.066 | 0.092 | 0.127 |
| | ∞ | 0.010 | 0.025 | 0.031 | 0.041 | 0.057 | 0.074 | 0.094 |
| | 10 | 0.022 | 0.056 | 0.101 | 0.144 | 0.229 | 0.313 | 0.458 |
| | 20 | 0.029 | 0.062 | 0.095 | 0.132 | 0.209 | 0.297 | 0.419 |
| | 40 | 0.033 | 0.067 | 0.100 | 0.128 | 0.191 | 0.267 | 0.381 |
| p = 0.4 | 60 | 0.034 | 0.070 | 0.102 | 0.130 | 0.189 | 0.256 | 0.354 |
| | 80 | 0.035 | 0.071 | 0.103 | 0.132 | 0.187 | 0.251 | 0.342 |
| | 100 | 0.035 | 0.072 | 0.103 | 0.132 | 0.187 | 0.248 | 0.335 |
| | ∞ | 0.037 | 0.076 | 0.105 | 0.135 | 0.184 | 0.236 | 0.307 |
| | 10 | 0.053 | 0.107 | 0.159 | 0.205 | 0.297 | 0.408 | 0.547 |
| | 20 | 0.062 | 0.122 | 0.172 | 0.216 | 0.302 | 0.408 | 0.538 |
| | 40 | 0.067 | 0.128 | 0.180 | 0.226 | 0.306 | 0.398 | 0.522 |
| p = 0.6 | 60 | 0.068 | 0.131 | 0.184 | 0.231 | 0.313 | 0.404 | 0.528 |
| | 80 | 0.068 | 0.132 | 0.186 | 0.233 | 0.316 | 0.407 | 0.531 |
| | 100 | 0.069 | 0.133 | 0.187 | 0.235 | 0.318 | 0.409 | 0.532 |
| | ∞ | 0.070 | 0.136 | 0.192 | 0.241 | 0.327 | 0.417 | 0.539 |
| | 10 | 0.085 | 0.158 | 0.217 | 0.266 | 0.354 | 0.453 | 0.593 |
| | 20 | 0.094 | 0.172 | 0.235 | 0.289 | 0.389 | 0.489 | 0.623 |
| | 40 | 0.099 | 0.180 | 0.247 | 0.303 | 0.401 | 0.508 | 0.651 |
| p = 0.8 | 60 | 0.100 | 0.183 | 0.251 | 0.308 | 0.410 | 0.520 | 0.667 |
| | 80 | 0.101 | 0.184 | 0.253 | 0.311 | 0.415 | 0.526 | 0.675 |
| | 100 | 0.101 | 0.185 | 0.254 | 0.313 | 0.418 | 0.529 | 0.680 |
| | ∞ | 0.103 | 0.188 | 0.259 | 0.320 | 0.430 | 0.544 | 0.700 |
| | 10 | 0.094 | 0.183 | 0.246 | 0.301 | 0.410 | 0.502 | 0.645 |
| | 20 | 0.109 | 0.194 | 0.263 | 0.322 | 0.431 | 0.536 | 0.675 |
| | 40 | 0.112 | 0.199 | 0.271 | 0.330 | 0.437 | 0.546 | 0.701 |
| p = 0.9 | 60 | 0.113 | 0.201 | 0.273 | 0.333 | 0.442 | 0.553 | 0.713 |
| | 80 | 0.114 | 0.202 | 0.274 | 0.335 | 0.445 | 0.558 | 0.718 |
| | 100 | 0.114 | 0.202 | 0.275 | 0.336 | 0.447 | 0.561 | 0.722 |
| | ∞ | 0.115 | 0.204 | 0.278 | 0.341 | 0.455 | 0.573 | 0.735 |

(continued)

TABLE 4.5 (continued)

| Statistic | n | Significance level $\alpha$ | | | | | | |
|---|---|---|---|---|---|---|---|---|
| | | 0.50 | 0.25 | 0.15 | 0.10 | 0.05 | 0.025 | 0.01 |
| $_2W^2_{r,n}$ | 10 | 0.103 | 0.198 | 0.266 | 0.324 | 0.430 | 0.534 | 0.676 |
| | 20 | 0.115 | 0.201 | 0.275 | 0.322 | 0.444 | 0.551 | 0.692 |
| | 40 | 0.115 | 0.205 | 0.280 | 0.329 | 0.448 | 0.557 | 0.715 |
| p = 0.95 | 60 | 0.116 | 0.207 | 0.280 | 0.338 | 0.451 | 0.562 | 0.724 |
| | 80 | 0.117 | 0.208 | 0.281 | 0.340 | 0.453 | 0.566 | 0.729 |
| | 100 | 0.117 | 0.208 | 0.282 | 0.341 | 0.454 | 0.569 | 0.735 |
| | ∞ | 0.118 | 0.208 | 0.283 | 0.346 | 0.460 | 0.579 | 0.742 |
| | 10 | 0.117 | 0.212 | 0.288 | 0.349 | 0.456 | 0.564 | 0.709 |
| | 20 | 0.116 | 0.212 | 0.288 | 0.350 | 0.459 | 0.572 | 0.724 |
| p = 1.0 | 40 | 0.115 | 0.211 | 0.288 | 0.350 | 0.461 | 0.576 | 0.731 |
| | 100 | 0.115 | 0.211 | 0.288 | 0.351 | 0.462 | 0.578 | 0.736 |
| | ∞ | 0.119 | 0.209 | 0.284 | 0.347 | 0.461 | 0.581 | 0.743 |
| $_2A^2_{r,n}$ | 20 | 0.107 | 0.218 | 0.337 | 0.435 | 0.626 | 0.887 | 1.278 |
| | 40 | 0.119 | 0.235 | 0.337 | 0.430 | 0.607 | 0.804 | 1.111 |
| p = 0.2 | 60 | 0.124 | 0.241 | 0.341 | 0.432 | 0.601 | 0.785 | 1.059 |
| | 80 | 0.127 | 0.243 | 0.344 | 0.433 | 0.598 | 0.775 | 1.034 |
| | 100 | 0.128 | 0.245 | 0.345 | 0.434 | 0.596 | 0.769 | 1.019 |
| | ∞ | 0.135 | 0.252 | 0.351 | 0.436 | 0.588 | 0.747 | 0.962 |
| | 10 | 0.214 | 0.431 | 0.627 | 0.803 | 1.127 | 1.483 | 2.080 |
| | 20 | 0.241 | 0.462 | 0.653 | 0.824 | 1.133 | 1.513 | 2.011 |
| | 40 | 0.261 | 0.487 | 0.681 | 0.843 | 1.138 | 1.460 | 1.903 |
| p = 0.4 | 60 | 0.265 | 0.493 | 0.686 | 0.848 | 1.142 | 1.458 | 1.892 |
| | 80 | 0.268 | 0.496 | 0.688 | 0.850 | 1.144 | 1.457 | 1.887 |
| | 100 | 0.269 | 0.497 | 0.689 | 0.851 | 1.145 | 1.457 | 1.884 |
| | ∞ | 0.275 | 0.504 | 0.695 | 0.857 | 1.150 | 1.455 | 1.872 |
| | 10 | 0.354 | 0.673 | 0.944 | 1.174 | 1.577 | 2.055 | 2.774 |
| | 20 | 0.390 | 0.713 | 0.984 | 1.207 | 1.650 | 2.098 | 2.688 |
| | 40 | 0.408 | 0.730 | 1.001 | 1.229 | 1.635 | 2.071 | 2.671 |
| p = 0.6 | 60 | 0.413 | 0.739 | 1.011 | 1.239 | 1.649 | 2.084 | 2.683 |
| | 80 | 0.416 | 0.743 | 1.017 | 1.244 | 1.655 | 2.091 | 2.689 |
| | 100 | 0.418 | 0.746 | 1.020 | 1.248 | 1.659 | 2.095 | 2.693 |
| | ∞ | 0.425 | 0.756 | 1.033 | 1.260 | 1.676 | 2.112 | 2.707 |

(continued)

TABLE 4.5 (continued)

| Statistic | n | Significance level $\alpha$ | | | | | | |
|---|---|---|---|---|---|---|---|---|
| | | 0.50 | 0.25 | 0.15 | 0.10 | 0.05 | 0.025 | 0.01 |
| $_2A^2_{r,n}$ | 10 | 0.503 | 0.913 | 1.237 | 1.498 | 2.021 | 2.587 | 3.254 |
| | 20 | 0.547 | 0.952 | 1.280 | 1.558 | 2.068 | 2.570 | 3.420 |
| | 40 | 0.568 | 0.983 | 1.321 | 1.583 | 2.088 | 2.574 | 3.270 |
| p = 0.8 | 60 | 0.574 | 0.991 | 1.330 | 1.596 | 2.107 | 2.610 | 3.319 |
| | 80 | 0.578 | 0.995 | 1.335 | 1.603 | 2.117 | 2.629 | 3.344 |
| | 100 | 0.580 | 0.997 | 1.338 | 1.607 | 2.123 | 2.640 | 3.359 |
| | ∞ | 0.588 | 1.007 | 1.350 | 1.623 | 2.146 | 2.684 | 3.419 |
| | 10 | 0.639 | 1.089 | 1.435 | 1.721 | 2.281 | 2.867 | 3.614 |
| | 20 | 0.656 | 1.109 | 1.457 | 1.765 | 2.295 | 2.858 | 3.650 |
| | 40 | 0.666 | 1.124 | 1.478 | 1.778 | 2.315 | 2.860 | 3.628 |
| p = 0.9 | 60 | 0.670 | 1.128 | 1.482 | 1.784 | 2.325 | 2.878 | 3.648 |
| | 80 | 0.671 | 1.130 | 1.485 | 1.788 | 2.330 | 2.888 | 3.661 |
| | 100 | 0.673 | 1.131 | 1.486 | 1.790 | 2.332 | 2.893 | 3.668 |
| | ∞ | 0.676 | 1.136 | 1.492 | 1.798 | 2.344 | 2.915 | 3.698 |
| | 10 | 0.707 | 1.170 | 1.525 | 1.842 | 2.390 | 2.961 | 3.745 |
| | 20 | 0.710 | 1.177 | 1.533 | 1.853 | 2.406 | 2.965 | 3.750 |
| | 40 | 0.715 | 1.184 | 1.543 | 1.860 | 2.416 | 2.968 | 3.743 |
| p = 0.95 | 60 | 0.717 | 1.186 | 1.545 | 1.263 | 2.421 | 2.977 | 3.753 |
| | 80 | 0.718 | 1.187 | 1.546 | 1.865 | 2.423 | 2.982 | 3.760 |
| | 100 | 0.719 | 1.188 | 1.547 | 1.866 | 2.424 | 2.984 | 3.763 |
| | ∞ | 0.720 | 1.190 | 1.550 | 1.870 | 2.430 | 2.995 | 3.778 |
| p = 1.0 | all n | 0.775 | 1.248 | 1.610 | 1.933 | 2.492 | 3.070 | 3.880 |

#### 4.7.3.2  Doubly-Censored Data

For doubly-censored data, suppose the values $Z_{(s)}$ to $Z_{(r)}$ are available, $s < r$. Pettitt and Stephens (1976) defined Cramér-von Mises statistics for such data; in terms of the definitions above, the Cramér-von Mises statistic is, for Type 2 censoring,

$$_2W^2_{sr,n} = {_2W^2_{r,n}} - {_2W^2_{s,n}};$$

similar definitions hold for Type 1 censoring and for $U^2$ and $A^2$. Pettitt and

Stephens have given asymptotic percentage points for symmetric double-censoring, where limits of $r/n$ and $s/n$ are p and q, and $p = 1 - q$.

### 4.7.4 Random Censoring

An important type of censoring is <u>random censoring</u>, which can occur as follows. Suppose a full random sample consists of the values $X_1^0$, $X_2^0$, ..., $X_n^0$ from a distribution $F^0(x)$ and consider a set of <u>censoring variables</u> $T_1$, $T_2$, ..., $T_n$ drawn, independently of each other and of the $X^0$-set, from a censoring distribution $F_C(t)$. Whenever $X_i^0 > T_i$, $X_i^0$ is replaced by $T_i$, so that the available observations are the pairs $(X_i, \delta_i)$ defined as follows, for $i = 1$, ..., n:

$$X_i = \min(X_i^0, T_i) \text{ and } \delta_i = 1 \text{ if } X_i = X_i^0$$

$$= 0 \text{ if } X_i = T_i$$

Such data could occur when $X_i^0$ are lifetimes of patients who enter a study of a certain disease; then if the patient dies from the disease before the study ends, $X_i^0$ is recorded, but if the patient is still alive at the end of the study, or withdraws, or dies of another cause, the time $T_i$ is recorded for which he or she was observed. The distribution function $F(x)$ of X is then given by $1 - F(x) = \{1 - F^0(x)\}\{1 - F_C(x)\}$. There has been much recent interest in testing fit in the presence of random censoring, or in estimating and giving confidence intervals for $F^0(x)$ or the related <u>survival function</u> $S^0(x) = 1 - F^0(x)$.

### 4.7.4.1 Estimation of the Distribution Function

An estimate of $F^0(x)$ for randomly censored data, analogous to the EDF, is the Kaplan–Meier (1958) estimate. This is formed from the pairs $(X_i, \delta_i)$ as follows. First place the pairs in ascending order of the $X_i$; if $X_i = X_{(j)}$, define $R_i = j$ (i.e., the rank of $X_i$ in the ordering). The estimate of $F^0(x)$ is then $_cF_n^0(x)$ defined by

$$_cF_n^0(x) = 0, \qquad\qquad\qquad\qquad x < X_{(1)}$$

$$= 1 - \prod_{i:X_i \le x} \left\{ \frac{n - R_i}{n - R_i + 1} \right\}^{\delta_i}, \quad x < X_{(n)}$$

$$= 1 \qquad\qquad\qquad\qquad x > X_{(n)}$$

If no observation is censored, the estimate $_cF_n^0(x)$ becomes the EDF $F_n^0(x)$.

Clearly EDF statistics may be defined using $_cF_n^0(x)$ instead of $F_n^0(x)$, when random censoring is present. For Case 0, to test $H_0$: that $F^0(x)$ is a completely specified distribution, suppose $Z_{(j)} = F^0(X_{(j)})$, and let $_cF_n^Z(z)$ be the Kaplan-Meier estimate of the distribution of $Z$: suppose also that $w_j = _cF_n^Z\{Z_{(j)}\}$. The statistic corresponding to $W^2$ is then $_cW^2$ given by

$$_cW^2 = n \sum_{j=1}^{n+1} w_{j-1}\{Z_{(j)} - Z_{(j-1)}\}[w_{j-1} - \{Z_{(j)} - Z_{(j-1)}\}] + \frac{n}{3}$$

(Koziol and Green, 1976). If there is no censoring, $_cW^2$ becomes $W^2$. In general, the null distribution of $_cW^2$ will depend on the censoring distribution, although the tested distribution is completely specified. Koziol and Green (1976) have given asymptotic percentage points of $_cW^2$ for the specific censoring model with $1 - F_c(x) = \{1 - F^0(x)\}^\beta$, $\beta$ a positive constant. $\beta$ must be estimated from the proportion censored.

Koziol (1980) and Csörgő and Horvath (1981) have also given tests for Case 0 with random censoring. Gillespie and Fisher (1979) and Hall and Wellner (1980) have shown how confidence bands for $F^0(x)$ may be constructed from the Kaplan-Meier estimate; the Hall-Wellner bands reduce to those given by D (Section 4.5.4) when no censoring is present. The articles quoted give many references to related work.

#### 4.7.4.2 Replacement of Censored Values

Another possible technique for randomly censored data (Case 0) is to make the Probability Integral Transformation on the observations $X_i$ above and then to replace those values which come from the $T_i$ by new ones so that, on $H_0$, the final set of transformed values is $U(0,1)$. Let $H_0$ be, as before, that the $X_i^0$ come from $F^0(x)$, completely specified. The PIT is applied using $F^0(x)$, on the values $X_i^0$ and $T_i$; then let $U_i = F^0(X_i^0)$, and let $t_i = F^0(T_i)$.

Suppose $F_t(t)$ is the distribution of the $t_i$, and let $G_t(t) = \int_0^t F_t(s)\,ds$. Then replace $t_i$ by $U_i^*$ given by $G_t(U_i^*) = (1 - t_i) F_t(t_i) + G_t(t_i)$; it may be shown that the resulting combined set consisting of the values U and U* is distributed, on $H_0$, as $U(0,1)$. Then any of the many tests for Case 0 above may be applied to the combined set. Here the censoring distribution of t must be known to make the transformation; however, it may be possible to replace $F_t(t)$ and $G_t(t)$ by the EDF of the t-values and its integral (Stephens and Wagner, 1986). Other methods of analyzing randomly censored data, using, for example, probability plots, are given in Chapter 11.

### 4.7.5 Renyi Statistics

Renyi (1953) discussed a number of statistics based on the difference between $F_n(x)$ and $F(x)$, or on the ratio of $F_n(x)$ to $F(x)$ over a restricted range. These include

$$R_1 = \sup_{a \le F_n(x)} \{F_n(x) - F(x)\}$$

$$R_2 = \sup_{a \le F_n(x)} \left\{ \frac{F_n(x) - F(x)}{F(x)} \right\}$$

$$R_3 = \sup_{a \le F(x) \le b} \frac{|F_n(x) - F(x)|}{F(x)}$$

$$R_4 = \sup_{F(x) \le b} \{F_n(x) - F(x)\}$$

$$R_5 = \sup_{P} \frac{F_n(x)}{F(x)}$$

$$R_6 = \inf_{P} \frac{F_n(x)}{F(x)}$$

where P is the interval $0 < F_n(x) \le r/n$, with r an integer in the range $1 \le r \le n$. Birnbaum and Lientz (1969a, b) have given exact and asymptotic theory for some of these statistics for Case 0, and have produced tables of percentage points for $R_1$, $R_2$, and $R_4$; they also gave examples of the use of the statistics, particularly in giving confidence limits for $F(x)$ over a restricted range. Niederhausen (1981) has given tables of points for variance-weighted Kolmogorov-Smirnov D, that is, $R_3$ above but with denominator $[F(x)\{1 - F(x)\}]$ instead of $F(x)$ (see Section 4.5.6), and for the analogues of $D^+$ and $D^-$. Other statistics of Renyi type, or closely related, have been discussed by a number of authors but, despite the potential applications for censored data, they have not been much developed for practical use.

### 4.7.6 Transformation to Complete Samples

Before leaving the subject of censored data, we point out that, for Case 0 in particular, several techniques are available to transform a censored sample of uniforms to a complete sample of uniforms. Then Case 0 tests for uniformity, for complete samples, or any of the methods of Chapter 8 may be used to test $H_0$. These methods can even be applied when there are blocks of missing observations. This is essentially a procedure which does not employ statistics specially adapted for censored data, and it is discussed more fully in Section 11.3.3.

## 4.8 EDF TESTS FOR THE NORMAL DISTRIBUTION WITH UNKNOWN PARAMETERS

We now turn to EDF tests for distributions with one or more parameters unknown, beginning with the normal distribution.

### 4.8.1 Tests for Normality, Cases 1, 2, and 3

The null hypothesis is

$H_0$: a random sample $X_1, \ldots, X_n$ comes from $F(x, \theta)$, the normal distribution $N(\mu, \sigma^2)$, with one or both of $\theta \equiv (\mu, \sigma^2)$ unknown

Following Stephens (1974b, 1976a), three cases are distinguished according to which parameter or parameters are unknown.

Case 1: The variance $\sigma^2$ is known and $\mu$ is unknown, estimated by $\bar{X}$, the sample mean.

Case 2: The mean $\mu$ is known and $\sigma^2$ is unknown, estimated by $\Sigma_i(X_i - \mu)^2/n$ $(= s_1^2$, say).

Case 3: Both $\mu$ and $\sigma^2$ are unknown, and are estimated by $\bar{X}$ and $s^2 = \Sigma_i(X_i - \bar{X})^2/(n - 1)$.

**TABLE 4.6** Upper-Tail Asymptotic Percentage Points for Tests for Normality with $\mu$ Unknown (Section 4.8.1, Case 1) or $\sigma^2$ Unknown (Section 4.8.1, Case 2)

| Statistic | Significance level $\alpha$ | | | | | | | |
|---|---|---|---|---|---|---|---|---|
|  | .25 | .15 | .10 | .05 | .025 | .01 | .005 | .0025 |
| $W^2$ Case 1 | .094 | .117 | .134 | .165 | .197 | .238 | .270 | .302 |
| $W^2$ Case 2 | .190 | .263 | .327 | .442 | .562 | .725 | .851 | .978 |
| $U^2$ Case 1 | .088 | .110 | .127 | .157 | .187 | .228 | .259 | .291 |
| $U^2$ Case 2 | .085 | .105 | .122 | .151 | .180 | .221 | .252 | .284 |
| $A^2$ Case 1 | .644 | .782 | .894 | 1.087 | 1.285 | 1.551 | 1.756 | 1.964 |
| $A^2$ Case 2 | 1.072 | 1.430 | 1.743 | 2.308 | 2.898 | 3.702 | 4.324 | 4.954 |

Adapted from Stephens (1974b), with permission of the American Statistical Association.

TABLE 4.7 Modifications and Percentage Points for a Test for Normality with $\mu$ and $\sigma^2$ Unknown (Section 4.8.1, Case 3)

**Upper tail**

| Statistic | Modified statistic | .50 | .25 | .15 | .10 | .05 | .025 | .01 | .005 |
|---|---|---|---|---|---|---|---|---|---|
| $D$ | $D(\sqrt{n} - 0.01 + 0.85/\sqrt{n})$ | | — | 0.775 | 0.819 | 0.895 | 0.995 | 1.035 | |
| $V$ | $V(\sqrt{n} + 0.05 + 0.82/\sqrt{n})$ | | — | 1.320 | 1.386 | 1.489 | 1.585 | 1.693 | |
| $W^2$ | $W^2(1.0 + 0.5/n)$ | .051 | .074 | .091 | .104 | .126 | .148 | .179 | .201 |
| $U^2$ | $U^2(1.0 + 0.5/n)$ | .048 | .070 | .085 | .096 | .117 | .136 | .164 | .183 |
| $A^2$ | $A^2(1.0 + 0.75/n + 2.25/n^2)$ | .341 | .470 | .561 | .631 | .752 | .873 | 1.035 | 1.159 |

**Lower tail**

| Statistic | Modified statistic | .50 | .25 | .15 | .10 | .05 | .025 | .01 | .005 |
|---|---|---|---|---|---|---|---|---|---|
| $W^2$ | $W^2(1.0 + 0.5/n)$ | .051 | .036 | .029 | .026 | .022 | .019 | .017 | |
| $U^2$ | $U^2(1.0 + 0.5/n)$ | .048 | .033 | .027 | .025 | .021 | .018 | .016 | |
| $A^2$ | $A^2(1.0 + 0.75/n + 2.25/n^2)$ | .341 | .249 | .226 | .188 | .160 | .139 | .119 | |

Adapted, with additions, from Table 54 of Pearson and Hartley (1972) and from Stephens (1974b), with permission of the Biometrika Trustees and of the American Statistical Association.

Of these three cases, Case 3 is the most important in most practical situations. For the three cases, the steps in making the substitution $Z_{(i)} = F(X_{(i)}; \hat{\theta})$ are:

(a)  Calculate $w_i$, for $i = 1, \ldots, n$, from

$$w_i = (X_{(i)} - \bar{X})/\sigma \quad \text{(Case 1)}$$

$$w_i = (X_{(i)} - \mu)/s_1 \quad \text{(Case 2)}$$

$$w_i = (X_{(i)} - \bar{X})/s \quad \text{(Case 3)}$$

(b)  Calculate $Z_{(i)} = \Phi(w_i)$ $(i = 1, \ldots, n)$, where $\Phi(x)$ denotes the cumulative probability of a standard normal distribution $N(0, 1)$ to the value $x$, found from tables or computer routines.

(c)  Calculate the test statistics from the formulas (4.2).

(d)  For Cases 1 or 2, use Table 4.6. For Case 3, use Table 4.7 and calculate the modified statistic. If the value of the statistic used, or, in Case 3, its modified value, exceeds the appropriate percentage point at level $\alpha$, $H_0$ is rejected with significance level $\alpha$.

The percentage points given for statistics $W^2$, $U^2$, and $A^2$ are those of their asymptotic distributions, and can be found theoretically. The points for D and V (Case 3) are the asymptotic points of $\sqrt{n}$ times the statistic; these have so far not been found theoretically (but see Nesenko and Tjurin, 1978) and those given have been obtained by extrapolation of points for finite n obtained by Monte Carlo studies. The modifications for all the statistics were calculated from points for finite n obtained by Monte Carlo methods. The tables now given are extended and revised from previous tables, for example, those given in Stephens (1974b), quoted also in Pearson and Hartley (1972). Percentage points for $W^2$, $U^2$, and $A^2$ were calculated by Stephens (1971, 1974b, 1976a), by Durbin, Knott, and Taylor (1975), and by Martynov (1976); Monte Carlo studies for D, Case 3, were given by van Soest (1967), by Lilliefors (1967), and by Stephens (1974b); for D, Cases 1 and 2, similar studies have been made by van Tilmann-Deutler, Griesenbrock, and Schwensfeier (1975), for V, Case 3, by Louter and Koerts (1970), and for $W^2$, Case 3, by van Soest (1967). Asymptotic results were also given by Wood (1978) and by Nesenko and Tjurin (1978).

No modifications have been calculated for $W^2$, $U^2$, and $A^2$, Cases 1 and 2. The percentage points for finite n converge rapidly to the asymptotic points, so that the points given could be used with good accuracy for $n \geq 20$. For other references see Durbin (1973), Stephens (1974b), and Neuhaus (1979).

### E 4.8.1 Example

We return to the data on weights of chickens given in Table 4.1, and now suppose that the tests is for

$H_0$: the sample is from a normal distribution but with mean and variance unknown

The situation is therefore Case 3, and the appropriate estimates for $\mu$ and $\sigma^2$ are given by $\bar{x} = 209.6$, and $s^2 = 939.25$. The transformations give the values in column $Z_2$ of Table 4.1, and Figure 4.3 shows their EDF. Equations (4.2) give for the test statistics the values: $D^+ = 0.089$, $D^- = 0.104$, $D = 0.104$, $V = 0.192$, $W^2 = 0.034$, $U^2 = 0.034$, $A^2 = 0.214$. The modified values are $D^* = 0.483$, $V^* = 0.906$, $W^* = 0.035$, $U^* = 0.034$, $A^* = 0.223$. It can be seen from Table 4.7, using Case 3 percentage points, that these are not nearly significant at the 15% level, so that at this level the sample would not be rejected as coming from a normal population.

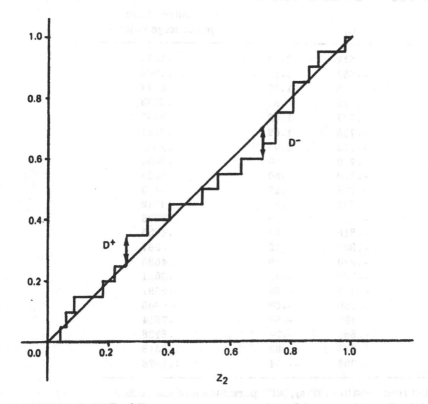

FIGURE 4.3 EDF of $Z_2$.

## 4.8.2 Significance Levels for Tests of Normality (Case 3)

Tables 4.8 and 4.9 can be used to find the p-level of a test statistic in an EDF test for normality (Case 3). Table 4.8 is adapted from Pettitt (1977a) and gives formulas to give the percentage point of $A^2$, for a sample of size n, corresponding to a given q-value. The value a for which $P(A^2 < a) = q$ is given by $a = a_\infty(1 + b_0/n + b_1/n^2)$, where $b_0$, $b_1$, and $a_\infty$ are given in the table against the value of q. Table 4.9 gives formulas for log p in the upper tail, or log q in the lower tail, for Case 3 tests and for modified values of $W^2$, $U^2$, $A^2$. They are more accurate in the upper tail (where the modifications of Table 4.7 are more accurate) but also give good approximations in the lower tail; these are useful for combining several test results (Section 4.18).

TABLE 4.8 Constants for Calculating the Significance Level of a Value of $A^2$ in a Test for Normality with Parameters Unknown (Case 3, Section 4.8.2)

| q | $b_0$ | $b_1$ | Asymptotic percentage point $a_\infty$ |
|---|---|---|---|
| .05 | -.512 | 2.10 | .1674 |
| .10 | -.552 | 1.25 | .1938 |
| .15 | -.608 | 1.07 | .2147 |
| .20 | -.643 | .93 | .2333 |
| .25 | -.707 | 1.03 | .2509 |
| .30 | -.735 | 1.02 | .2681 |
| .35 | -.772 | 1.04 | .2853 |
| .40 | -.770 | .90 | .3030 |
| .45 | -.778 | .80 | .3213 |
| .50 | -.779 | .67 | .3405 |
| .55 | -.803 | .70 | .3612 |
| .60 | -.818 | .58 | .3836 |
| .65 | -.818 | .42 | .4085 |
| .70 | -.801 | .12 | .4367 |
| .75 | -.800 | -.09 | .4695 |
| .80 | -.756 | -.39 | .5091 |
| .85 | -.749 | -.59 | .5597 |
| .90 | -.750 | -.80 | .6305 |
| .95 | -.795 | -.89 | .7514 |
| .975 | -.881 | -.94 | .8728 |
| .99 | -1.013 | -.93 | 1.0348 |
| .995 | -1.063 | -1.34 | 1.1578 |

Adapted from Pettitt (1977a), with permission of the author and of the Royal Statistical Society.

TABLE 4.9 Formulas for Significance Levels, Tests for Normality with Parameters Unknown (Case 3, Section 4.8.2)[a]

| | Statistic | | |
|---|---|---|---|
| | $W^2$, Case 3 | $U^2$, Case 3 | $A^2$, Case 3 |
| $z < z_1$ | $\log q = -13.953 + 775.5z - 12542.61z^2$ | $\log q = -13.642 + 766.31z - 12432.74z^2$ | $\log q = -13.436 + 101.14z - 223.73z^2$ |
| $z_1$ | 0.0275 | 0.0262 | 0.200 |
| $z_1 < z < z_2$ | $\log q = -5.903 + 179.546z - 1515.29z^2$ | $\log q = -6.3328 + 214.57z - 2022.28z^2$ | $\log q = -8.318 + 42.796z - 59.938z^2$ |
| $z_2$ | 0.051 | 0.048 | 0.340 |
| $z_2 < z < z_3$ | $\log p = 0.886 - 31.62z + 10.897z^2$ | $\log p = 0.8510 - 32.006z - 3.45z^2$ | $\log p = 0.9177 - 4.279z - 1.38z^2$ |
| $z_3$ | 0.092 | 0.094 | 0.600 |
| $z > z_3$ | $\log p = 1.111 - 34.242z + 12.832z^2$ | $\log p = 1.325 - 38.918z + 16.45z^2$ | $\log p = 1.2937 - 5.709z + 0.0186z^2$ |

[a]Suppose z is a modified value of $W^2$, $U^2$ or $A^2$ (see Table 4.7). For a given value z, find the interval in which z lies. The formula gives the value of $\log q$ (q = lower tail significance level) or $\log p$ (p = upper tail significance level).

E 4.8.2 Example

Suppose, for n = 20, $A^2$ were 0.435. Reference to Table 4.8 suggests a
q-level near 0.70. The percentage point for n = 20, q = 0.70 is given approximately by z = 0.4367(1 - .801/20 + .12/400) = .419. Similar calculations
give the percentage point for n = 20, q = 0.75 to be z ≈ 0.4506. Interpolation
between these values gives q for $A^2$ = 0.435 to be about 0.725. To use
Table 4.9 we first calculate modified $A^2$ (from Table 4.7) to be 0.454; then
Table 4.9 gives log p ≈ 0.9177 - 4.279(.454) - 1.38(.454)$^2$ = -1.309; then
p ≈ 0.270, and q = .730.

## 4.8.3 Related Tests for Normality

Green and Hegazy (1976) have shown that slight modifications of the basic
EDF statistics can improve power in tests for normality against selected
alternatives. Hegazy and Green (1975) have discussed tests based on values
$V_i = \{(X_{(i)} - \bar{X})/s\} - m_i$, where $m_i$ is the expected value of the i-th order
statistic of a sample of size n from $N(0, 1)$.

## 4.8.4 Tests for Normality with Censored Data

Pettitt (1976) has given percentage points for modified versions of $W^2$, $U^2$,
and $A^2$, for use in tests of normality with singly- or doubly-censored data.
The parameters $\mu$ and $\sigma$ can be estimated by maximum likelihood or by estimates given by Gupta (1952). Maximum likelihood estimates are complicated
to calculate and percentage points of the test statistics for finite n appear to
converge more slowly to the asymptotic points when these estimates are
used, so Gupta's estimates are suggested. These are linear combinations of
the available order statistics, for example, $\mu^* = \Sigma_{i=1}^{r} b_i X_{(i)}$ and
$\sigma^* = \Sigma_{i=1}^{r} c_i X_{(i)}$; for n ≤ 10 Gupta gives coefficients $c_i$ and $b_i$ (there called
$\beta_i$ and $\gamma_i$) for the most efficient estimates. For n ≥ 10, which would be the
situation most needed in practice, Gupta gives the easily calculated coefficients

$$b_i = \frac{1}{r} - \frac{\bar{m}(m_i - \bar{m})}{\Sigma_{i=1}^{r} (m_i - \bar{m})^2} \quad \text{and} \quad c_i = \frac{m_i - \bar{m}}{\Sigma_{i=1}^{r} (m_i - \bar{m})^2}$$

where $m_i$ is the expected value of the i-th order statistic of a sample of size
n from $N(0, 1)$ and where $\bar{m} = \Sigma_{i=1}^{r} m_i/r$. Values of $m_i$ are tabulated or can
be well approximated (see Section 5.7.2), and the estimates $\mu^*$ and $\sigma^*$ have
been shown to be asymptotically efficient (Ali and Chan, 1964). These estimates are the same as those obtained by least squares when $X_{(i)}$ is regressed

TABLE 4.10  Upper Tail Percentage Points for Statistics $W^2$ and $A^2$ for a Test for Normality (Parameters Unknown) with Complete or Type 2 Right-Censored Data (Section 4.8.4). p = r/n is the censoring ratio.

| Statistic | n | \multicolumn{7}{c}{Significance level $\alpha$} |
|---|---|---|---|---|---|---|---|---|

| Statistic | n | 0.5 | 0.25 | 0.15 | 0.10 | 0.05 | 0.025 | 0.01 |
|---|---|---|---|---|---|---|---|---|
| $W^2$ | 20 | 0.001 | 0.002 | 0.004 | 0.006 | 0.010 | 0.016 | 0.024 |
| | 40 | 0.002 | 0.004 | 0.006 | 0.008 | 0.013 | 0.021 | 0.041 |
| | 60 | 0.002 | 0.004 | 0.006 | 0.008 | 0.014 | 0.021 | 0.039 |
| p = 0.2 | 80 | 0.001 | 0.004 | 0.006 | 0.008 | 0.013 | 0.021 | 0.035 |
| | 100 | 0.001 | 0.004 | 0.006 | 0.008 | 0.013 | 0.020 | 0.032 |
| | ∞ | 0.000 | 0.004 | 0.006 | 0.008 | 0.009 | 0.017 | 0.020 |
| | 10 | 0.007 | 0.011 | 0.014 | 0.017 | 0.028 | 0.041 | 0.057 |
| | 20 | 0.009 | 0.014 | 0.019 | 0.024 | 0.037 | 0.055 | 0.090 |
| | 40 | 0.009 | 0.015 | 0.020 | 0.026 | 0.038 | 0.057 | 0.089 |
| p = 0.4 | 60 | 0.010 | 0.015 | 0.021 | 0.026 | 0.036 | 0.052 | 0.077 |
| | 80 | 0.010 | 0.016 | 0.021 | 0.026 | 0.036 | 0.049 | 0.073 |
| | 100 | 0.010 | 0.016 | 0.021 | 0.026 | 0.035 | 0.047 | 0.071 |
| | ∞ | 0.009 | 0.019 | 0.021 | 0.026 | 0.034 | 0.038 | 0.066 |
| | 10 | 0.017 | 0.026 | 0.033 | 0.040 | 0.054 | 0.075 | 0.109 |
| | 20 | 0.019 | 0.029 | 0.037 | 0.044 | 0.060 | 0.080 | 0.113 |
| | 40 | 0.020 | 0.031 | 0.040 | 0.047 | 0.061 | 0.077 | 0.106 |
| p = 0.6 | 60 | 0.020 | 0.031 | 0.040 | 0.047 | 0.060 | 0.077 | 0.105 |
| | 80 | 0.021 | 0.032 | 0.040 | 0.048 | 0.061 | 0.077 | 0.103 |
| | 100 | 0.021 | 0.032 | 0.040 | 0.048 | 0.061 | 0.076 | 0.101 |
| | ∞ | 0.025 | 0.032 | 0.044 | 0.052 | 0.064 | 0.074 | 0.092 |
| | 10 | 0.030 | 0.044 | 0.054 | 0.062 | 0.078 | 0.094 | 0.115 |
| | 20 | 0.032 | 0.046 | 0.057 | 0.067 | 0.083 | 0.100 | 0.122 |
| | 40 | 0.033 | 0.049 | 0.060 | 0.070 | 0.084 | 0.101 | 0.124 |
| p = 0.8 | 60 | 0.033 | 0.049 | 0.060 | 0.070 | 0.086 | 0.103 | 0.125 |
| | 80 | 0.034 | 0.050 | 0.061 | 0.071 | 0.087 | 0.105 | 0.127 |
| | 100 | 0.035 | 0.050 | 0.062 | 0.072 | 0.089 | 0.106 | 0.129 |
| | ∞ | 0.039 | 0.051 | 0.069 | 0.080 | 0.098 | 0.114 | 0.140 |
| | 10 | 0.037 | 0.054 | 0.066 | 0.076 | 0.093 | 0.110 | 0.137 |
| | 20 | 0.039 | 0.056 | 0.069 | 0.079 | 0.097 | 0.113 | 0.137 |
| | 40 | 0.040 | 0.058 | 0.072 | 0.082 | 0.099 | 0.116 | 0.142 |
| p = 0.9 | 60 | 0.040 | 0.058 | 0.072 | 0.082 | 0.100 | 0.118 | 0.141 |
| | 80 | 0.041 | 0.059 | 0.073 | 0.084 | 0.102 | 0.120 | 0.144 |
| | 100 | 0.042 | 0.060 | 0.074 | 0.085 | 0.103 | 0.122 | 0.146 |
| | ∞ | 0.045 | 0.067 | 0.082 | 0.094 | 0.114 | 0.135 | 0.163 |

(continued)

TABLE 4.10 (continued)

| Statistic | n | Significance level $\alpha$ | | | | | | |
|---|---|---|---|---|---|---|---|---|
| | | 0.5 | 0.25 | 0.15 | 0.10 | 0.05 | 0.025 | 0.01 |
| $W^2$ | 10 | 0.042 | 0.061 | 0.074 | 0.084 | 0.103 | 0.122 | 0.145 |
| | 20 | 0.043 | 0.062 | 0.076 | 0.087 | 0.106 | 0.124 | 0.147 |
| | 40 | 0.044 | 0.064 | 0.078 | 0.089 | 0.108 | 0.126 | 0.154 |
| p = 0.95 | 60 | 0.044 | 0.064 | 0.078 | 0.089 | 0.109 | 0.128 | 0.152 |
| | 80 | 0.045 | 0.065 | 0.079 | 0.090 | 0.110 | 0.130 | 0.154 |
| | 100 | 0.045 | 0.066 | 0.080 | 0.091 | 0.112 | 0.132 | 0.156 |
| | ∞ | 0.049 | 0.072 | 0.087 | 0.099 | 0.120 | 0.142 | 0.172 |
| | 10 | 0.049 | 0.070 | 0.086 | 0.098 | 0.119 | 0.141 | 0.167 |
| | 20 | 0.049 | 0.071 | 0.087 | 0.100 | 0.121 | 0.142 | 0.171 |
| | 40 | 0.050 | 0.073 | 0.088 | 0.101 | 0.122 | 0.141 | 0.169 |
| p = 1.0 | 60 | 0.050 | 0.073 | 0.088 | 0.101 | 0.123 | 0.144 | 0.171 |
| | 80 | 0.050 | 0.073 | 0.089 | 0.101 | 0.124 | 0.146 | 0.173 |
| | 100 | 0.050 | 0.073 | 0.089 | 0.102 | 0.125 | 0.146 | 0.174 |
| | ∞ | 0.051 | 0.074 | 0.091 | 0.104 | 0.126 | 0.148 | 0.179 |
| $A^2$ | 20 | 0.015 | 0.043 | 0.054 | 0.061 | 0.092 | 0.131 | 0.182 |
| | 40 | 0.028 | 0.053 | 0.067 | 0.079 | 0.112 | 0.158 | 0.253 |
| p = 0.2 | 60 | 0.035 | 0.053 | 0.069 | 0.084 | 0.114 | 0.160 | 0.246 |
| | 80 | 0.036 | 0.056 | 0.073 | 0.087 | 0.116 | 0.159 | 0.236 |
| | 100 | 0.036 | 0.059 | 0.075 | 0.089 | 0.119 | 0.158 | 0.228 |
| | ∞ | 0.030 | 0.077 | 0.094 | 0.099 | 0.133 | 0.149 | 0.185 |
| | 10 | 0.063 | 0.090 | 0.108 | 0.121 | 0.172 | 0.236 | 0.319 |
| | 20 | 0.072 | 0.107 | 0.135 | 0.162 | 0.220 | 0.297 | 0.439 |
| | 40 | 0.078 | 0.117 | 0.150 | 0.177 | 0.236 | 0.316 | 0.433 |
| p = 0.4 | 60 | 0.079 | 0.119 | 0.148 | 0.174 | 0.228 | 0.299 | 0.410 |
| | 80 | 0.082 | 0.121 | 0.153 | 0.178 | 0.229 | 0.292 | 0.395 |
| | 100 | 0.085 | 0.123 | 0.157 | 0.182 | 0.231 | 0.288 | 0.385 |
| | ∞ | 0.106 | 0.134 | 0.190 | 0.215 | 0.250 | 0.279 | 0.340 |
| | 10 | 0.111 | 0.158 | 0.198 | 0.233 | 0.304 | 0.405 | 0.592 |
| | 20 | 0.122 | 0.178 | 0.222 | 0.259 | 0.339 | 0.437 | 0.607 |
| | 40 | 0.130 | 0.191 | 0.238 | 0.278 | 0.348 | 0.430 | 0.570 |
| p = 0.6 | 60 | 0.132 | 0.193 | 0.239 | 0.275 | 0.348 | 0.430 | 0.557 |
| | 80 | 0.134 | 0.196 | 0.241 | 0.278 | 0.350 | 0.429 | 0.548 |
| | 100 | 0.136 | 0.198 | 0.244 | 0.280 | 0.351 | 0.429 | 0.541 |
| | ∞ | 0.151 | 0.212 | 0.261 | 0.300 | 0.359 | 0.426 | 0.512 |

(continued)

TABLE 4.10 (continued)

| Statistic | n | Significance level $\alpha$ | | | | | | |
|---|---|---|---|---|---|---|---|---|
| | | 0.50 | 0.25 | 0.15 | 0.10 | 0.05 | 0.025 | 0.01 |
| $A^2$ | 10 | 0.172 | 0.246 | 0.302 | 0.352 | 0.440 | 0.542 | 0.698 |
| | 20 | 0.185 | 0.267 | .330 | 0.380 | 0.473 | 0.574 | 0.743 |
| | 40 | 0.197 | 0.282 | 0.344 | 0.394 | 0.478 | 0.575 | 0.711 |
| p = 0.8 | 60 | 0.197 | 0.284 | 0.345 | 0.396 | 0.482 | 0.574 | 0.705 |
| | 80 | 0.200 | 0.288 | 0.349 | 0.401 | 0.489 | 0.580 | 0.707 |
| | 100 | 0.202 | 0.291 | 0.353 | 0.405 | 0.494 | 0.585 | 0.709 |
| | ∞ | 0.220 | 0.311 | 0.380 | 0.432 | 0.528 | 0.619 | 0.732 |
| | 10 | 0.214 | 0.303 | 0.368 | 0.425 | 0.530 | 0.642 | 0.825 |
| | 20 | 0.229 | 0.326 | 0.397 | 0.453 | 0.549 | 0.654 | 0.807 |
| | 40 | 0.242 | 0.342 | 0.414 | 0.473 | 0.566 | 0.669 | 0.814 |
| p = 0.9 | 60 | 0.243 | 0.343 | 0.415 | 0.472 | 0.571 | 0.675 | 0.805 |
| | 80 | 0.245 | 0.348 | 0.420 | 0.478 | 0.579 | 0.683 | 0.811 |
| | 100 | 0.248 | 0.352 | 0.425 | 0.483 | 0.585 | 0.689 | 0.818 |
| | ∞ | 0.265 | 0.380 | 0.456 | 0.517 | 0.623 | 0.729 | 0.871 |
| | 10 | 0.243 | 0.344 | 0.414 | 0.474 | 0.584 | 0.696 | 0.840 |
| | 20 | 0.257 | 0.366 | 0.440 | 0.500 | 0.600 | 0.708 | 0.853 |
| | 40 | 0.273 | 0.382 | 0.456 | 0.519 | 0.624 | 0.721 | 0.865 |
| p = 0.95 | 60 | 0.272 | 0.383 | 0.459 | 0.520 | 0.626 | 0.733 | 0.874 |
| | 80 | 0.276 | 0.388 | 0.465 | 0.528 | 0.633 | 0.744 | 0.885 |
| | 100 | 0.279 | 0.392 | 0.470 | 0.534 | 0.640 | 0.753 | 0.893 |
| | ∞ | 0.301 | 0.420 | 0.502 | 0.580 | 0.686 | 0.802 | 0.942 |
| | 10 | 0.309 | 0.425 | 0.511 | 0.578 | 0.700 | 0.818 | 0.964 |
| | 20 | 0.323 | 0.446 | 0.530 | 0.601 | 0.714 | 0.831 | 0.993 |
| | 40 | 0.330 | 0.456 | 0.541 | 0.611 | 0.723 | 0.833 | 0.981 |
| p = 1.0 | 60 | 0.331 | 0.458 | 0.545 | 0.614 | 0.734 | 0.847 | 0.993 |
| | 80 | 0.333 | 0.460 | 0.548 | 0.616 | 0.740 | 0.853 | 1.000 |
| | 100 | 0.334 | 0.461 | 0.550 | 0.618 | 0.742 | 0.857 | 1.005 |
| | ∞ | 0.341 | 0.470 | 0.561 | 0.631 | 0.752 | 0.873 | 1.035 |

Some asymptotic points taken from Pettitt (1976), with permission of the author and of the Biometrika Trustees.

against $m_i$, $i = 1, \ldots, r$, as, for example, in Section 5.7.3. The steps in making a test, with right-censored data, are then:

(a)  Calculate $\mu^*$ and $\sigma^*$.

(b)  Find $w_i = \{X_{(i)} - \mu^*\}/\sigma^*$, $i = 1, \ldots, r$.

(c)  Calculate $Z_{(i)} = \Phi(w_i)$ as in Section 4.8.1, step (b), above.

(d)  For Type 2 censored data use the $Z_i$, $i = 1, \ldots, r$, in the formulas of Section 4.7.3 to obtain statistics $_2W^2_{r,n}$, $_2A^2_{r,n}$, and $_2U^2_{r,n}$.

(e)  Refer $_2W^2_{r,n}$ or $_2A^2_{r,n}$ to the percentage points in Tables 4.10. The table is entered at $p = r/n$ and at n. The points for finite n were found by extensive Monte Carlo sampling, using 10,000 samples for each sample size n, and some asymptotic points were taken from Pettitt (1976). The test is easily adapted for left-censored data by changing the sign of all values given and observing that the sample is now right-censored. Tables for $_2U^2_{r,n}$ are in Stephens (1986).

An example of these tests is given in Example E 11.4.1.1.1; the same data set is used, with the correlation test described in Section 5.7, in E 11.4.1.2.1. For these data, the EDF statistics are the more sensitive.

It is possible to make a rough test for, say, large outliers, mixed with an otherwise normal sample, by first testing the whole sample and then the sample with suspect values removed. This procedure would be difficult to formalize since the two tests would be correlated, and also the censoring fraction will probably be chosen after observing the data.

For Type 1 right-censored data, test statistics $_1W^2_{t,n}$, $_1U^2_{t,n}$, and $_1A^2_{t,n}$ can be found as follows. Suppose the upper censoring value of X is t*, and let $\hat{p} = \Phi\{(t^* - \mu^*)/\sigma^*\}$. The procedure of Section 4.7.3 is followed, for Type 1 censored data; that is, set $Z_{(r+1)} = \hat{p}$, consider the sample to be now of size $r+1$, and use the formulas of that section to calculate the statistics. Tables can be constructed, by taking Monte Carlo samples from $N(0,1)$, censoring at $t = \Phi^{-1}(p)$, for given p, and calculating the test statistics just described; however, a correct test cannot be made, since for any given real-life sample, only $\hat{p}$ will be known, and not the correct p to enter the table. As the points vary considerably with p, entering the table at an estimate of p could produce substantial errors. However, Table 4.10 can be used to give an approximate test, especially for large samples; the asymptotic points are the same for both types of censoring, and tables for Type 1, produced as described above, have values close to those for Type 2 censored data. The same problem will arise for Type 1 censored data from other distributions.

### 4.8.5  Tests for Normality of Residuals in Regression

Mukantseva (1977), Pierce and Kopecky (1978), and Loynes (1980) have studied the asymptotic behavior of the EDF of the residuals when a regression model has been fitted, with the intention of testing these residuals for normality. If the model, of any order, is correct, the residuals will be normal, with known mean equal to zero, but with unknown variance. At first sight this situation would appear to be Case 2 of Section 4.8.1 above, but this is not so, because the residuals are not independent. However, the above authors have shown that if EDF statistics are calculated from the residuals their asymptotic distributions are the same as for Case 3 above.

As an example, consider simple linear regression using the model $y_i = \beta_0 + \beta_1 x_i + \epsilon_i$, $i = 1, \ldots, n$, with $\epsilon_i = N(0, \sigma^2)$. Let $\hat{\beta}_0$ and $\hat{\beta}_1$ be the usual least squares estimators, and let $\hat{\sigma}^2$ be the usual estimate of $\sigma^2$ obtained from the error sum of squares in the ANOVA table; if $\hat{y}_i = \hat{\beta}_0 + \hat{\beta}_1 x_i$, $\hat{\epsilon}_i = y_i - \hat{y}_i$, then $\hat{\sigma}^2 = \Sigma_i \hat{\epsilon}_i^2 / (n - 2)$. The studentized residuals are (Pierce and Gray, 1982, with slight change in notation)

$$w_i = \hat{\epsilon}_i / [\hat{\sigma} \{1 - 1/n - (x_i - \bar{x})^2 / S_{xx}\}^{\frac{1}{2}}]$$

where $S_{xx} = \Sigma_i (x_i - \bar{x})^2$. Let $Z_{(i)} = \Phi(w_{(i)})$; for an approximate test that the $\epsilon_i$ are normal, EDF statistics are then calculated from (4.2) and referred to the asymptotic points for Case 3 in Table 4.7. The modifications given in the table will not be valid for this problem.

From Monte Carlo studies, Pierce and Gray (1982) concluded the asymptotic points can be used with good accuracy for the simple linear regression model, for n as low as 20. White and MacDonald (1980) gave some results for multiple regression situations. It seems clear that the tests would be affected considerably if the experimental model were not correct—for example, if the correct regression model were a quadratic function, but only a linear fit was made; see also comments on the multiple regression situation by White and MacDonald (1980), and in the discussion to that paper, and by Pierce and Gray (1982). Wood (1978b) has discussed asymptotic theory for EDF statistics obtained from residuals in an analysis-of-variance model. In residual analysis, of course, other questions are also of great importance, for example, systematic variation of the residuals. For further discussion see Anscombe and Tukey (1963) or textbooks such as Draper and Smith (1966) or Kleinbaum and Kupper (1978).

## 4.9 EDF TESTS FOR THE EXPONENTIAL DISTRIBUTION

### 4.9.1 Tests for Exponentiality, Cases 1, 2, and 3

The exponential distribution, denoted by $\text{Exp}(\alpha, \beta)$, is the distribution

$$F(x; \alpha, \beta) = 1 - \exp\{-(x - \alpha)/\beta\}, \qquad x > \alpha; \beta > 0$$

In this section we consider tests of

$H_0$: a random sample $X_1, \ldots, X_n$ comes from distribution $\text{Exp}(\alpha, \beta)$

As for the test for normality, we can distinguish three cases:

Case 1: the origin or location parameter $\alpha$ is unknown, but $\beta$ is known;
Case 2: the scale parameter $\beta$ is unknown, but $\alpha$ is known;
Case 3: both parameters are unknown.

### 4.9.2 Tests for Case 1

The first method we shall describe for Case 1 uses a special property of the exponential distribution, as follows. Let $X'_{(i-1)} = X_{(i)} - X_{(1)}$, for $i = 2, \ldots, n$; on $H_0$, the $X'$ will be a random sample from $\text{Exp}(0, \beta)$ (see Section 10.3.1, Result 2) and, since $\beta$ is known, a Case 0 test can be made using the $n - 1$ values of $X'$.

Alternatively, $\alpha$ may be estimated unbiasedly by $\hat{\alpha} = (X_{(1)} - 1/n)$; this estimate is derived from the maximum likelihood estimate $X_{(1)}$, and has variance diminishing as $1/n^2$. Then $Z_{(i)}$ are found from $Z_{(i)} = 1 - \exp[-(X_{(i)} - \hat{\alpha})/\beta]$, $i = 1, \ldots, n$, and EDF statistics calculated from the $Z_{(i)}$ by formulas (4.2) will have their Case 0 distributions asymptotically so that the percentage points in Table 4.2 may be used for large samples. However, in contrast to the previous test procedure, the modifications given there will not apply, and since the two procedures are likely to have very similar power properties, the first procedure is more practical for relatively small samples.

### 4.9.3 Tests for Case 2

For this case (Case 2) suppose first that $\alpha$ is known to be zero. The maximum likelihood estimate of $\beta$ is given by $\hat{\beta} = \bar{X}$ where $\bar{X}$ is the sample mean.

The steps in testing $H_0$ are as follows:

(a)  Calculate $Z_{(i)} = 1 - \exp(-X_{(i)}/\bar{X})$, $i = 1, \ldots, n$.
(b)  Calculate the EDF statistics from (4.2).
(c)  Modify and compare with the percentage points given in Table 4.11, or alternatively obtain p-levels from Table 4.12.

If the known origin is $\alpha = \alpha_0$, not zero, the substitution $X'_i = X_i - \alpha$, $i = 1, \ldots, n$, can be made, and the $X'_i$ tested for $\text{Exp}(0, \beta)$ as just described. See Result 1 of Section 10.3.1.

The percentage points given are asymptotic points for the statistics $W^2$, $U^2$, and $A^2$; see Stephens (1974b, 1976a). The asymptotic distribution of $W^2$ was earlier tabulated by van Soest (1969), and points for $W^2$ and $U^2$ have also been calculated by Durbin, Knott, and Taylor (1975). The modifications were based on Monte Carlo points for finite $n$; for D and V, these were extrapolated

TABLE 4.11 Modifications and Percentage Points for EDF Tests for Exponentiality, Case 2: Origin Known, Scale Unknown (Section 4.9.3)

Upper tail

| Statistic T | Modified form T* | Significance level α | | | | | | | | |
|---|---|---|---|---|---|---|---|---|---|---|
| | | .25 | .20 | .15 | .10 | .05 | .025 | .01 | .005 | .0025 |
| D | $(D - 0.2/n)(\sqrt{n} + 0.26 + 0.5/\sqrt{n})$ | | | .926 | .995 | 1.094 | 1.184 | 1.298 | | |
| V | $(V - 0.2/n)(\sqrt{n} + 0.24 + 0.35/\sqrt{n})$ | | | 1.445 | 1.527 | 1.655 | 1.774 | 1.910 | | |
| W² | $W^2(1.0 + 0.16/n)$ | .116 | .130 | .148 | .175 | .222 | .271 | .338 | .390 | .442 |
| U² | $U^2(1.0 + 0.16/n)$ | .090 | .099 | .112 | .129 | .159 | .189 | .230 | .261 | .293 |
| A² | $A^2(1.0 + 0.6/n)$ | .736 | .816 | .916 | 1.062 | 1.321 | 1.591 | 1.959 | 2.244 | 2.534 |

Lower tail

| | | Significance level α | | | | | | | |
|---|---|---|---|---|---|---|---|---|---|
| | | .01 | .025 | .05 | .10 | .15 | .20 | .25 | .50 |
| W² | Asymptotic percentage points. | .0192 | .0233 | .0276 | .0338 | .039 | .044 | .048 | .074 |
| U² | | .0172 | .0207 | .0243 | .0293 | .0339 | .0373 | .0409 | .0601 |
| A² | | .150 | .178 | .208 | .249 | .280 | .312 | .342 | .502 |

Adapted from Table 54 of Pearson and Hartley (1972) and from Stephens (1974b), with permission of the Biometrika Trustees and of the American Statistical Association.

**TABLE 4.12** Formulas for Significance Levels, Tests for Exponentiality, Case 2: Origin Known, Scale Unknown (Section 4.9.3)[a]

| | Statistic | | |
|---|---|---|---|
| | $W^2$ | $U^2$ | $A^2$ |
| $z < z_1$ | $\log q = -11.334 + 459.098z - 5652.1z^2$ | $\log q = -11.703 + 542.5z - 7574.59z^2$ | $\log q = -12.2204 + 67.459z - 110.3z^2$ |
| $z_1$ | 0.035 | 0.029 | 0.260 |
| $z_1 < z < z_2$ | $\log q = -5.779 + 132.89z - 866.58z^2$ | $\log q = -6.3288 + 178.1z - 1399.49z^2$ | $\log q = -6.1327 + 20.218z - 18.663z^2$ |
| $z_2$ | 0.074 | 0.062 | 0.510 |
| $z_2 < z < z_3$ | $\log p = 0.586 - 17.87z + 7.417z^2$ | $\log p = 0.8071 - 25.166z + 8.44z^2$ | $\log p = 0.9209 - 3.353z + 0.300z^2$ |
| $z_3$ | 0.160 | 0.120 | 0.950 |
| $z > z_3$ | $\log p = 0.447 - 16.592z + 4.849z^2$ | $\log p = 0.7663 - 24.359z + 4.539z^2$ | $\log p = 0.731 - 3.009z + 0.15z^2$ |

[a]Suppose z is a modified value of $W^2$, $U^2$, or $A^2$ (See Table 4.11). For a given z, find the interval in which z lies. The formula gives the value of $\log q$ (q = lower tail significance level) or $\log p$ (p = upper tail significance level).

to obtain asymptotic values. Stephens (1974b) has given references to other Monte Carlo studies, the first of which, for D only, was given by Lilliefors (1969). Subsequently, Durbin (1975) has produced exact null distribution theory for $D^+$, $D^-$, and D, and has given percentage points for n up to 100; see also Margolin and Maurer (1976) for work on these statistics. Table 4.11 is an extended and revised version of that given in Stephens (1974b) making use of later results where possible. Table 4.12 gives formulas for obtaining p-values or q-values of a given value z of a modified statistic.

### E 4.9.3 Example

Proschan (1963, Table 1) has given a number of samples of data, consisting of intervals between failures of air conditioning equipment in aircraft. We

TABLE 4.13 Original Values X in a Test for Exponentiality

| X | $Z_1$[a] | $Z_2$[b] |
|---|---|---|
| 12 | 0.113 | 0.094 |
| 21 | 0.189 | 0.159 |
| 26 | 0.229 | 0.193 |
| 27 | 0.237 | 0.200 |
| 29 | 0.252 | 0.213 |
| 29 | 0.252 | 0.213 |
| 48 | 0.381 | 0.327 |
| 57 | 0.434 | 0.375 |
| 59 | 0.446 | 0.385 |
| 70 | 0.503 | 0.439 |
| 74 | 0.522 | 0.457 |
| 153 | 0.783 | 0.717 |
| 326 | 0.962 | 0.932 |
| 386 | 0.979 | 0.959 |
| 502 | 0.993 | 0.984 |

[a]Values $Z_1$ using the Probability Integral Transformation with $\theta$ given equal to 0.01.
[b]Values $Z_2$ using the transformation with $\theta$ estimated from the data, i.e., $\hat{\theta} = .0083$.
Taken from Proschan (1963), with permission of the author and of the American Statistical Association.

illustrate the test for exponentiality by taking his aircraft numbered 7910, for which the 15 intervals are as listed in Table 4.13. In order to emphasize the contrast between Case 0 of Section 4.6 and the test with unknown parameter $\beta$, we first test the null hypothesis of exponentiality with both parameters given: $\alpha = 0$, $\beta = 100$. Thus the null hypothesis is $H_0$: the data are a random sample from Exp$(0, 100)$.

The transformation $Z_i = 1 - \exp(-X_i/100)$ gives Z-values listed as $Z_1$ in Table 4.13; on $H_0$, $Z_1$ will be $U(0,1)$. The values give test statistics $D^+ = 0.210$, $D^- = 0.161$, $D = 0.210$, $V = 0.372$, $W^2 = 0.133$, $U^2 = 0.130$, $A^2 = 1.055$. The modified forms, using Table 4.2, are $D^{+*} = 0.973$, $D^{-*} = 0.650$, $D^* = 0.846$, $V^* = 1.522$, $W^* = 0.116$, $U^* = 0.130$, $A^* = 1.055$. Statistics $V^*$ and $U^*$ are almost significant at the 15% level; the others are far from significant even at this level.

If $\beta$ is not known in the test for exponentiality, the estimate is $\hat{\beta} = \bar{X} = 121.2$. With $\alpha = 0$ and $\hat{\beta} = 121.2$, the Probability Integral Transformation gives the values in column $Z_2$ of Table 4.13 and the test statistics modified as in Table 4.11 become: $D^* = 1.122$, $V^* = 1.661$, $W^* = 0.221$, $U^* = 0.172$, $A^* = 1.210$. All the statistics are now significant at the 10% level, with $D^*$, $V^*$, $U^*$, significant (and $W^*$ almost so) at the 5% level. Thus, given the freedom to choose the parameter, it appears that the assumption of an exponential parent population is suspect. The comparison with the previous test for Case 0 may appear paradoxical, since apparently in Case 0 one makes use of more information, and we shall return to this point in Section 4.16.

Many other tests for exponentiality with known origin are given in Chapter 10. In particular, two other test statistics based on the EDF ($\tilde{D}$ and $S^*$) are discussed in Section 10.8.1.

## 4.9.4 Tests for Case 3

Relatively few tests have been proposed to test for exponentiality with $\alpha$ and $\beta$ unknown, probably because Result 2 of Section 10.3.1 can be used to reduce the test to a test with $\alpha = 0$. However, this may not always be the best

TABLE 4.14 Modifications and Upper Tail Percentage Points for a Test for Exponentiality, Case 3: Origin and Scale Unknown (Section 4.9.4)

| Statistic | Modification | Significance level | | | | | |
|---|---|---|---|---|---|---|---|
| | | .25 | .15 | .10 | .05 | .025 | .01 |
| $W^2$ | $W^2(1 + 2.8/n - 3/n^2)$ | .116 | .148 | .175 | .222 | .271 | .338 |
| $U^2$ | $U^2(1 + 2.3/n - 3/n^2)$ | .090 | .112 | .129 | .159 | .189 | .230 |
| $A^2$ | $A^2(1 + 5.4/n - 11/n^2)$ | .736 | .916 | 1.062 | 1.321 | 1.591 | 1.959 |

TABLE 4.15  Upper Tail Percentage Points for $\sqrt{n}D^+$, $\sqrt{n}D^-$, $\sqrt{n}D$, $\sqrt{n}V$, $W^2$, $U^2$, and $A^2$, for a Test of Exponentiality, Case 3. Origin and Scale Unknown (Section 4.9.4)

| n | Upper tail significance level $\alpha$ | | | | | |
|---|---|---|---|---|---|---|
|   | .25 | .15 | .10 | .05 | .025 | .01 |
| Statistic $\sqrt{n}D^+$ | | | | | | |
| 5 | .491 | .569 | .639 | .743 | .825 | .917 |
| 10 | .580 | .674 | .745 | .851 | .952 | 1.038 |
| 15 | .610 | .700 | .768 | .872 | .978 | 1.077 |
| 20 | .624 | .716 | .785 | .894 | .995 | 1.108 |
| 25 | .635 | .725 | .799 | .909 | 1.010 | 1.125 |
| 50 | .660 | .758 | .832 | .943 | 1.051 | 1.163 |
| 100 | .682 | .778 | .853 | .967 | 1.074 | 1.189 |
| ∞ | .723 | .820 | .886 | .996 | 1.094 | 1.211 |
| Statistic $\sqrt{n}D^-$ | | | | | | |
| 5 | .627 | .705 | .753 | .821 | .891 | .955 |
| 10 | .671 | .761 | .825 | .916 | .993 | 1.089 |
| 15 | .688 | .783 | .842 | .933 | 1.022 | 1.111 |
| 20 | .696 | .791 | .855 | .949 | 1.041 | 1.132 |
| 25 | .702 | .795 | .860 | .958 | 1.052 | 1.149 |
| 50 | .710 | .807 | .874 | .976 | 1.072 | 1.178 |
| 100 | .717 | .814 | .879 | .984 | 1.089 | 1.192 |
| ∞ | .723 | .820 | .886 | .996 | 1.094 | 1.211 |
| Statistic $\sqrt{n}D$ | | | | | | |
| 5 | .683 | .749 | .793 | .865 | .921 | .992 |
| 10 | .753 | .833 | .889 | .977 | 1.048 | 1.119 |
| 15 | .771 | .865 | .912 | 1.002 | 1.079 | 1.163 |
| 20 | .786 | .872 | .927 | 1.021 | 1.099 | 1.198 |
| 25 | .792 | .878 | .936 | 1.033 | 1.115 | 1.215 |
| 50 | .813 | .879 | .960 | 1.061 | 1.149 | 1.257 |
| 100 | .824 | .911 | .972 | 1.072 | 1.171 | 1.278 |
| ∞ | .840 | .927 | .995 | 1.094 | 1.184 | 1.298 |
| Statistic $\sqrt{n}V$ | | | | | | |
| 5 | 1.098 | 1.186 | 1.234 | 1.314 | 1.400 | 1.494 |
| 10 | 1.194 | 1.294 | 1.363 | 1.461 | 1.556 | 1.662 |
| 15 | 1.225 | 1.325 | 1.392 | 1.504 | 1.596 | 1.701 |
| 20 | 1.245 | 1.346 | 1.419 | 1.536 | 1.635 | 1.769 |

(continued)

TABLE 4.15 (continued)

| n | .25 | .15 | .10 | .05 | .025 | .01 |
|---|-----|-----|-----|-----|------|-----|
| | | | Upper tail significance level $\alpha$ | | | |

Statistic $\sqrt{n}V$, continued

| n | .25 | .15 | .10 | .05 | .025 | .01 |
|---|-----|-----|-----|-----|------|-----|
| 25 | 1.260 | 1.366 | 1.438 | 1.559 | 1.658 | 1.796 |
| 50 | 1.292 | 1.400 | 1.481 | 1.600 | 1.701 | 1.847 |
| 100 | 1.310 | 1.419 | 1.502 | 1.647 | 1.740 | 1.897 |
| ∞ | 1.334 | 1.445 | 1.527 | 1.655 | 1.774 | 1.910 |

Statistic $W^2$

| 5 | .083 | .102 | .117 | .141 | .166 | .197 |
|---|------|------|------|------|------|------|
| 10 | .097 | .122 | 142 | .176 | .211 | .259 |
| 15 | .103 | .130 | .151 | .188 | .229 | .281 |
| 20 | .106 | .133 | .157 | .195 | .237 | .293 |
| 25 | .107 | .135 | .160 | .199 | .247 | .301 |
| 50 | .111 | .141 | .166 | .209 | .256 | .319 |
| 100 | .113 | .144 | .170 | .215 | .263 | .328 |
| ∞ | .116 | .148 | .175 | .222 | .271 | .338 |

Statistic $U^2$

| 5 | .068 | .083 | .093 | .113 | .131 | .153 |
|---|------|------|------|------|------|------|
| 10 | .075 | .094 | .108 | .131 | .155 | .187 |
| 15 | .080 | .099 | .114 | .139 | .165 | .200 |
| 20 | .082 | .102 | .117 | .143 | .170 | .207 |
| 25 | .083 | .104 | .119 | .146 | .173 | .212 |
| 50 | .087 | .108 | .124 | .152 | .180 | .223 |
| 100 | .089 | .110 | .126 | .155 | .184 | .229 |
| ∞ | .090 | .112 | .129 | .159 | .189 | .230 |

Statistic $A^2$

| 5 | .460 | .555 | .621 | .725 | .848 | .989 |
|---|------|------|------|------|------|------|
| 10 | .545 | .660 | .747 | .920 | 1.068 | 1.352 |
| 15 | .575 | .720 | .816 | 1.009 | 1.198 | 1.495 |
| 20 | .608 | .757 | .861 | 1.062 | 1.267 | 1.580 |
| 25 | .625 | .784 | .890 | 1.097 | 1.317 | 1.635 |
| 50 | .680 | .838 | .965 | 1.197 | 1.440 | 1.775 |
| 100 | .710 | .875 | 1.008 | 1.250 | 1.510 | 1.855 |
| ∞ | .736 | .916 | 1.062 | 1.321 | 1.591 | 1.959 |

(continued)

procedure. In this section we give EDF tests for Case 3, using estimates for both $\alpha$ and $\beta$, similar to other EDF tests. The null hypothesis is

$H_0$: the random sample $X_1, \ldots, X_n$ comes from the distribution
$\quad$ $Exp(\alpha,\beta)$, with $\alpha$, $\beta$ unknown

The test procedure is as follows:

(a) Calculate estimates $\hat{\beta} = n(\bar{X} - X_{(1)})/(n - 1)$ and $\hat{\alpha} = X_{(1)} - \hat{\beta}/n$.

(b) Calculate $W_i = (X_{(i)} - \hat{\alpha})/\hat{\beta}$, $i = 1, \ldots, n$.

(c) Calculate $Z_i = 1 - \exp(-W_i)$, $i = 1, \ldots, n$.

(d) Find the EDF statistics from (4.2), modify $W^2$, $U^2$, and $A^2$ using Table 4.14 and compare with the asymptotic percentage points given; for $D^+$, $D^-$, D, and V use Table 4.15 without modification.

The estimate of $\alpha$ is superefficient and so asymptotic theory is the same as for Case 2 in the previous section. For finite n, however, the distributions are different. The modifications for $W^2$, $U^2$, and $A^2$ and the points for $D^+$, $D^-$, D, and V were found from extensive Monte Carlo studies (Spinelli and Stephens, 1987). Van Soest (1969) has simulated the probability distribution for $W^2$, for n = 10 and 20. For comments on power, see Section 10.14.

### 4.9.5 Tests for Exponentiality with Censored Data

Suppose, for example, in a life-testing experiment, the observations are recorded only up to a fixed time t (Type 1 censoring) or until a fixed number r out of n are observed (Type 2 censoring). In either case let $X_{(r)}$ be the largest order statistic, so that the sample is right-censored. The parameter $\beta$ in $Exp(0,\beta)$ is estimated by

$$\hat{\beta} = \left\{ \sum_{i=1}^{r} X_{(i)} + (n - r)t \right\} \Big/ r \qquad \text{for Type 1 data}$$

and by

$$\hat{\beta} = \left\{ \sum_{i=1}^{r} X_{(i)} + (n - r)X_{(r)} \right\} \Big/ r \qquad \text{for Type 2 data}$$

A test for exponentiality $Exp(0,\beta)$ may then be made as follows:

(a) Calculate $Z_{(i)} = 1 - \exp(-X_{(i)}/\hat{\beta})$, $i = 1, \ldots, r$.

(b) For Type 2 censoring, use the $Z_{(i)}$ in the formulas of Section 4.7.3 to calculate statistics $_2W^2_{r,n}$, $_2U^2_{r,n}$, and $_2A^2_{r,n}$.

TABLE 4.16 Upper Tail Percentage Points for Statistics $W^2$ and $A^2$ for a Test for Exponentiality with Unknown Scale Parameter and Known Origin, for Complete or Right-Censored Data of Type 2 (Section 4.9.5) $p = r/n$ is the censoring ratio.

| Statistic | n | Significance level $\alpha$ | | | | | | |
|---|---|---|---|---|---|---|---|---|
|  |  | 0.50 | 0.25 | 0.15 | 0.10 | 0.05 | 0.025 | 0.01 |
| $W^2$ | 20 | 0.005 | 0.009 | 0.012 | 0.014 | 0.018 | 0.021 | 0.025 |
|  | 40 | 0.005 | 0.008 | 0.011 | 0.013 | 0.017 | 0.020 | 0.025 |
| $p = 0.2$ | 60 | 0.005 | 0.008 | 0.011 | 0.013 | 0.017 | 0.020 | 0.026 |
|  | 80 | 0.005 | 0.008 | 0.011 | 0.013 | 0.017 | 0.020 | 0.026 |
|  | 100 | 0.005 | 0.008 | 0.011 | 0.013 | 0.017 | 0.020 | 0.026 |
|  | $\infty$ | 0.005 | 0.008 | 0.011 | 0.013 | 0.016 | 0.021 | 0.026 |
|  | 10 | 0.019 | 0.030 | 0.038 | 0.045 | 0.055 | 0.066 | 0.079 |
|  | 20 | 0.017 | 0.028 | 0.037 | 0.044 | 0.056 | 0.068 | 0.083 |
|  | 40 | 0.017 | 0.028 | 0.036 | 0.044 | 0.056 | 0.068 | 0.084 |
| $p = 0.4$ | 60 | 0.017 | 0.028 | 0.036 | 0.044 | 0.056 | 0.068 | 0.085 |
|  | 80 | 0.017 | 0.028 | 0.036 | 0.044 | 0.056 | 0.069 | 0.086 |
|  | 100 | 0.017 | 0.027 | 0.036 | 0.043 | 0.056 | 0.069 | 0.086 |
|  | $\infty$ | 0.017 | 0.027 | 0.036 | 0.043 | 0.056 | 0.070 | 0.087 |
|  | 10 | 0.036 | 0.056 | 0.072 | 0.084 | 0.104 | 0.124 | 0.149 |
|  | 20 | 0.035 | 0.055 | 0.071 | 0.084 | 0.106 | 0.131 | 0.161 |
|  | 40 | 0.035 | 0.055 | 0.072 | 0.085 | 0.109 | 0.132 | 0.161 |
| $p = 0.6$ | 60 | 0.034 | 0.056 | 0.072 | 0.085 | 0.109 | 0.133 | 0.164 |
|  | 80 | 0.034 | 0.056 | 0.072 | 0.085 | 0.109 | 0.134 | 0.166 |
|  | 100 | 0.034 | 0.056 | 0.072 | 0.086 | 0.109 | 0.134 | 0.167 |
|  | $\infty$ | 0.034 | 0.058 | 0.072 | 0.086 | 0.110 | 0.136 | 0.171 |
|  | 10 | 0.055 | 0.086 | 0.107 | 0.126 | 0.156 | 0.187 | 0.229 |
|  | 20 | 0.055 | 0.086 | 0.110 | 0.130 | 0.167 | 0.203 | 0.253 |
|  | 40 | 0.055 | 0.087 | 0.111 | 0.131 | 0.167 | 0.203 | 0.253 |
| $p = 0.8$ | 60 | 0.055 | 0.087 | 0.112 | 0.132 | 0.168 | 0.205 | 0.256 |
|  | 80 | 0.055 | 0.087 | 0.112 | 0.132 | 0.169 | 0.206 | 0.257 |
|  | 100 | 0.055 | 0.087 | 0.112 | 0.132 | 0.169 | 0.206 | 0.258 |
|  | $\infty$ | 0.055 | 0.087 | 0.113 | 0.133 | 0.170 | 0.209 | 0.261 |
|  | 10 | 0.065 | 0.100 | 0.126 | 0.147 | 0.182 | 0.219 | 0.265 |
|  | 20 | 0.065 | 0.102 | 0.132 | 0.155 | 0.194 | 0.238 | 0.289 |
|  | 40 | 0.064 | 0.102 | 0.129 | 0.152 | 0.193 | 0.229 | 0.290 |
| $p = 0.9$ | 60 | 0.064 | 0.101 | 0.130 | 0.153 | 0.195 | 0.234 | 0.294 |
|  | 80 | 0.065 | 0.101 | 0.131 | 0.154 | 0.196 | 0.236 | 0.297 |
|  | 100 | 0.065 | 0.101 | 0.131 | 0.155 | 0.196 | 0.238 | 0.298 |
|  | $\infty$ | 0.065 | 0.101 | 0.132 | 0.156 | 0.199 | 0.243 | 0.303 |

(continued)

TABLE 4.16 (continued)

| Statistic | n | Significance level $\alpha$ | | | | | | |
|---|---|---|---|---|---|---|---|---|
| | | 0.50 | 0.25 | 0.15 | 0.10 | 0.05 | 0.025 | 0.01 |
| $W^2$ | 10 | 0.070 | 0.109 | 0.136 | 0.160 | 0.200 | 0.239 | 0.292 |
| | 20 | 0.070 | 0.110 | 0.142 | 0.166 | 0.209 | 0.251 | 0.313 |
| | 40 | 0.069 | 0.108 | 0.138 | 0.161 | 0.205 | 0.246 | 0.304 |
| $p = 0.95$ | 60 | 0.069 | 0.108 | 0.139 | 0.163 | 0.207 | 0.250 | 0.313 |
| | 80 | 0.069 | 0.108 | 0.139 | 0.164 | 0.208 | 0.252 | 0.318 |
| | 100 | 0.070 | 0.108 | 0.140 | 0.164 | 0.209 | 0.254 | 0.321 |
| | $\infty$ | 0.070 | 0.109 | 0.141 | 0.166 | 0.212 | 0.259 | 0.333 |
| | 10 | 0.075 | 0.116 | 0.147 | 0.171 | 0.216 | 0.259 | 0.319 |
| | 20 | 0.073 | 0.115 | 0.148 | 0.175 | 0.221 | 0.265 | 0.328 |
| | 40 | 0.074 | 0.115 | 0.147 | 0.172 | 0.218 | 0.267 | 0.331 |
| $p = 1.0$ | 60 | 0.074 | 0.115 | 0.147 | 0.173 | 0.219 | 0.267 | 0.334 |
| | 80 | 0.074 | 0.115 | 0.147 | 0.173 | 0.220 | 0.268 | 0.336 |
| | 100 | 0.074 | 0.115 | 0.147 | 0.173 | 0.220 | 0.268 | 0.337 |
| | $\infty$ | 0.074 | 0.116 | 0.148 | 0.175 | 0.222 | 0.271 | 0.338 |
| $A^2$ | 20 | 0.080 | 0.127 | 0.161 | 0.188 | 0.232 | 0.271 | 0.325 |
| | 40 | 0.078 | 0.126 | 0.161 | 0.189 | 0.241 | 0.292 | 0.355 |
| $p = 0.2$ | 60 | 0.077 | 0.126 | 0.164 | 0.192 | 0.244 | 0.300 | 0.373 |
| | 80 | 0.077 | 0.126 | 0.164 | 0.194 | 0.249 | 0.306 | 0.385 |
| | 100 | 0.078 | 0.126 | 0.163 | 0.195 | 0.252 | 0.311 | 0.394 |
| | $\infty$ | 0.078 | 0.128 | 0.161 | 0.200 | 0.274 | 0.336 | 0.438 |
| | 10 | 0.158 | 0.248 | 0.312 | 0.363 | 0.445 | 0.528 | 0.671 |
| | 20 | 0.157 | 0.248 | 0.319 | 0.379 | 0.477 | 0.582 | 0.719 |
| | 40 | 0.157 | 0.250 | 0.322 | 0.382 | 0.485 | 0.584 | 0.736 |
| $p = 0.4$ | 60 | 0.156 | 0.251 | 0.324 | 0.382 | 0.493 | 0.605 | 0.753 |
| | 80 | 0.156 | 0.252 | 0.326 | 0.385 | 0.496 | 0.611 | 0.762 |
| | 100 | 0.157 | 0.252 | 0.326 | 0.388 | 0.497 | 0.614 | 0.767 |
| | $\infty$ | 0.158 | 0.255 | 0.330 | 0.407 | 0.501 | 0.614 | 0.788 |
| | 10 | 0.243 | 0.373 | 0.474 | 0.549 | 0.684 | 0.835 | 1.058 |
| | 20 | 0.241 | 0.375 | 0.482 | 0.568 | 0.721 | 0.875 | 1.104 |
| | 40 | 0.243 | 0.385 | 0.492 | 0.580 | 0.733 | 0.892 | 1.126 |
| $p = 0.6$ | 60 | 0.244 | 0.382 | 0.491 | 0.580 | 0.730 | 0.892 | 1.126 |
| | 80 | 0.244 | 0.382 | 0.491 | 0.580 | 0.731 | 0.894 | 1.128 |
| | 100 | 0.244 | 0.383 | 0.492 | 0.581 | 0.733 | 0.897 | 1.130 |
| | $\infty$ | 0.244 | 0.390 | 0.494 | 0.584 | 0.746 | 0.914 | 1.145 |

(continued)

TABLE 4.16 (continued)

| Statistic | n | Significance level $\alpha$ | | | | | | |
|---|---|---|---|---|---|---|---|---|
| | | 0.50 | 0.25 | 0.15 | 0.10 | 0.05 | 0.025 | 0.01 |
| $A^2$ | 10 | 0.337 | 0.510 | 0.636 | 0.740 | 0.929 | 1.130 | 1.434 |
| | 20 | 0.337 | 0.518 | 0.662 | 0.773 | 0.979 | 1.195 | 1.512 |
| | 40 | 0.344 | 0.529 | 0.669 | 0.782 | 0.985 | 1.186 | 1.465 |
| p = 0.8 | 60 | 0.341 | 0.527 | 0.670 | 0.782 | 0.989 | 1.207 | 1.516 |
| | 80 | 0.341 | 0.530 | 0.670 | 0.783 | 0.991 | 1.214 | 1.529 |
| | 100 | 0.341 | 0.532 | 0.671 | 0.785 | 0.993 | 1.218 | 1.533 |
| | $\infty$ | 0.345 | 0.549 | 0.675 | 0.793 | 1.003 | 1.222 | 1.521 |
| | 10 | 0.391 | 0.580 | 0.732 | 0.852 | 1.059 | 1.289 | 1.584 |
| | 20 | 0.396 | 0.611 | 0.768 | 0.894 | 1.117 | 1.360 | 1.706 |
| | 40 | 0.399 | 0.608 | 0.771 | 0.893 | 1.117 | 1.330 | 1.633 |
| p = 0.9 | 60 | 0.398 | 0.609 | 0.766 | 0.897 | 1.127 | 1.367 | 1.699 |
| | 80 | 0.398 | 0.612 | 0.766 | 0.900 | 1.132 | 1.380 | 1.720 |
| | 100 | 0.399 | 0.615 | 0.768 | 0.902 | 1.135 | 1.386 | 1.728 |
| | $\infty$ | 0.407 | 0.630 | 0.781 | 0.914 | 1.149 | 1.394 | 1.729 |
| | 10 | 0.433 | 0.653 | 0.800 | 0.928 | 1.176 | 1.422 | 1.738 |
| | 20 | 0.431 | 0.657 | 0.822 | 0.959 | 1.205 | 1.466 | 1.811 |
| | 40 | 0.437 | 0.657 | 0.824 | 0.958 | 1.195 | 1.432 | 1.779 |
| p = 0.95 | 60 | 0.434 | 0.654 | 0.824 | 0.959 | 1.202 | 1.447 | 1.803 |
| | 80 | 0.434 | 0.657 | 0.827 | 0.962 | 1.208 | 1.456 | 1.812 |
| | 100 | 0.435 | 0.660 | 0.830 | 0.965 | 1.212 | 1.462 | 1.817 |
| | $\infty$ | 0.444 | 0.680 | 0.850 | 0.983 | 1.232 | 1.490 | 1.830 |
| | 10 | 0.485 | 0.746 | 0.886 | 1.017 | 1.278 | 1.524 | 1.894 |
| | 20 | 0.488 | 0.723 | 0.904 | 1.052 | 1.315 | 1.570 | 1.924 |
| | 40 | 0.494 | 0.732 | 0.907 | 1.049 | 1.299 | 1.565 | 1.933 |
| p = 1.0 | 60 | 0.491 | 0.728 | 0.905 | 1.051 | 1.303 | 1.570 | 1.964 |
| | 80 | 0.491 | 0.728 | 0.906 | 1.053 | 1.306 | 1.574 | 1.971 |
| | 100 | 0.492 | 0.729 | 0.907 | 1.054 | 1.308 | 1.576 | 1.973 |
| | $\infty$ | 0.496 | 0.736 | 0.916 | 1.062 | 1.321 | 1.591 | 1.959 |

Some asymptotic points taken, from Pettitt (1977b), with permission of the author and of the Biometrika Trustees.

(c) Refer $_2W^2_{r,n}$ or $_2A^2_{r,n}$ to the percentage points given in Table 4.16.

Pettitt (1977b) gave asymptotic theory and points for this test: some of the points have been used in Table 4.16. Tables for $_2U^2_{r,n}$ are given by Stephens (1986).

For a test with Type 1 censored data, the test statistic, say, $_1W^2_{t,n}$, can be found by setting $\hat{p} = Z_{(r+1)} = 1 - \exp(-t/\hat{\beta})$, where t is the censoring value and $\hat{\beta}$ is found as above, and then using the formulas of Section 4.7.3, with sample size r + 1. For large samples, an approximate test may be made by referring the statistic to Table 4.16, with entries p = $\hat{p}$ and n, but for smaller samples, entering the table at an estimate of p instead of the true value can produce a considerable error in significance level; see the comments in Section 4.8.4 on tests for normality.

Another method of treating right-censored data is to use the N-transformation of Chapter 10 (see Section 10.5.6). This converts a right-censored exponential sample to a complete exponential sample, and the above tests of exponentiality for complete samples, or others given in Chapter 10, may then be used to test $H_0$.

## 4.10  EDF TESTS FOR THE EXTREME-VALUE DISTRIBUTION

One form of the extreme-value distribution is

$$F(x) = \exp\left[-\exp\left\{-\frac{(x-\alpha)}{\beta}\right\}\right], \quad -\infty < x < \infty \tag{4.6}$$

where $-\infty < \alpha < \infty$, and $\beta > 0$.

The distribution of $X' = -X$ gives a second form of the extreme-value distribution:

$$F(x') = 1 - \exp\left\{-\exp\left(\frac{x'-\alpha'}{\beta}\right)\right\}, \quad -\infty < x' < \infty \tag{4.7}$$

(here $\alpha' = -\alpha$ above).

The first version (4.6) has a long tail to the right, and (4.7) has a long tail to the left.

In this section we discuss EDF tests of the null hypothesis

$H_0$: a random sample $X_1, \ldots, X_n$ comes from distribution (4.6)
   with one or both of parameters $\alpha$ and $\beta$ unknown

Three test situations can again be distinguished (Stephens, 1977):

Case 1: $\beta$ known, $\alpha$ to be estimated.
Case 2: $\alpha$ known, $\beta$ to be estimated.
Case 3: $\alpha, \beta$ both unknown, and to be estimated.

We suppose the parameters will be estimated by maximum likelihood; the estimates, for Case 3, are given by the equations (Johnson and Kotz (1970), p. 283):

$$\hat{\beta} = \Sigma_j X_j / n - [\Sigma_j X_j \exp(-X_j/\hat{\beta})]/[\Sigma_j \exp(-X_j/\hat{\beta})] \qquad (4.8)$$

and

$$\hat{\alpha} = -\hat{\beta} \log[\Sigma_j \exp(-X_j/\hat{\beta})/n] \qquad (4.9)$$

Equation (4.8) is solved iteratively for $\hat{\beta}$, and then (4.9) can be solved for $\hat{\alpha}$. In Case 1, $\beta$ is known; then $\hat{\alpha}$ is given by (4.9) with $\beta$ replacing $\hat{\beta}$. In Case 2,

TABLE 4.17 Modifications and Upper Tail Percentage Points for Statistics $W^2$, $U^2$, and $A^2$ for the Extreme-Value or Weibull Distributions (Sections 4.10, 4.11)

| Statistic | Modification | Significance level $\alpha$ | | | | |
|---|---|---|---|---|---|---|
| | | .25 | .10 | .05 | .025 | .01 |
| $W^2$ | | | | | | |
| Case 1 | $W^2(1 + 0.16/n)$ | .116 | .175 | .222 | .271 | .338 |
| Case 2 | None | .186 | .320 | .431 | .547 | .705 |
| Case 3 | $W^2(1 + 0.2/\sqrt{n})$ | .073 | .102 | .124 | .146 | .175 |
| $U^2$ | | | | | | |
| Case 1 | $U^2(1 + 0.16/n)$ | .090 | .129 | .159 | .189 | .230 |
| Case 2 | $U^2(1 + 0.15/\sqrt{n})$ | .086 | .123 | .152 | .181 | .220 |
| Case 3 | $U^2(1 + 0.2/\sqrt{n})$ | .070 | .097 | .117 | .138 | .165 |
| $A^2$ | | | | | | |
| Case 1 | $A^2(1 + 0.3/n)$ | .736 | 1.062 | 1.321 | 1.591 | 1.959 |
| Case 2 | None | 1.060 | 1.725 | 2.277 | 2.854 | 3.640 |
| Case 3 | $A^2(1 + 0.2/\sqrt{n})$ | .474 | .637 | .757 | .877 | 1.038 |

TABLE 4.18 Upper Tail Percentage Points for Statistics $\sqrt{n}D^+$, $\sqrt{n}D^-$, $\sqrt{n}D$, and $\sqrt{n}V$, for Tests for the Extreme-Value or Weibull Distributions (Sections 4.10, 4.11)

| Statistic | n | Significance level $\alpha$ | | | |
|---|---|---|---|---|---|
| | | .10 | .05 | .025 | .01 |
| $\sqrt{n}D^+$ | 10 | .872 | .969 | 1.061 | 1.152 |
| Case 1 | 20 | .878 | .979 | 1.068 | 1.176 |
| | 50 | .882 | .987 | 1.070 | 1.193 |
| | $\infty$ | .886 | .996 | 1.094 | 1.211 |
| $\sqrt{n}D^-$ | 10 | .773 | .883 | .987 | 1.103 |
| Case 1 | 20 | .810 | .921 | 1.013 | 1.142 |
| | 50 | .840 | .950 | 1.031 | 1.171 |
| | $\infty$ | .886 | .996 | 1.094 | 1.211 |
| $\sqrt{n}D$ | 10 | .934 | 1.026 | 1.113 | 1.206 |
| Case 1 | 20 | .954 | 1.049 | 1.134 | 1.239 |
| | 50 | .970 | 1.067 | 1.148 | 1.263 |
| | $\infty$ | .995 | 1.094 | 1.184 | 1.298 |
| $\sqrt{n}V$ | 10 | 1.43 | 1.55 | 1.65 | 1.77 |
| Case 1 | 20 | 1.46 | 1.58 | 1.69 | 1.81 |
| | 50 | 1.48 | 1.59 | 1.72 | 1.84 |
| | $\infty$ | 1.53 | 1.65 | 1.77 | 1.91 |
| $\sqrt{n}D^+$ | 10 | .99 | 1.14 | 1.27 | 1.42 |
| Case 2 | 20 | 1.00 | 1.15 | 1.28 | 1.43 |
| | 50 | 1.01 | 1.17 | 1.29 | 1.44 |
| | $\infty$ | 1.02 | 1.17 | 1.30 | 1.46 |
| $\sqrt{n}D^-$ | 10 | 1.01 | 1.16 | 1.28 | 1.41 |
| Case 2 | 20 | 1.01 | 1.15 | 1.28 | 1.43 |
| | 50 | 1.00 | 1.14 | 1.29 | 1.45 |
| | $\infty$ | 1.02 | 1.17 | 1.30 | 1.46 |

(continued)

TABLE 4.18 (continued)

| Statistic | n | Significance level $\alpha$ | | | |
|---|---|---|---|---|---|
| | | .10 | .05 | .025 | .01 |
| $\sqrt{n}D$ | 10 | 1.14 | 1.27 | 1.39 | 1.52 |
| Case 2 | 20 | 1.15 | 1.28 | 1.40 | 1.53 |
| | 50 | 1.16 | 1.29 | 1.41 | 1.53 |
| | $\infty$ | 1.16 | 1.29 | 1.42 | 1.53 |
| $\sqrt{n}V$ | 10 | 1.39 | 1.49 | 1.60 | 1.72 |
| Case 2 | 20 | 1.42 | 1.54 | 1.64 | 1.76 |
| | 50 | 1.45 | 1.56 | 1.67 | 1.79 |
| | $\infty$ | 1.46 | 1.58 | 1.69 | 1.81 |
| $\sqrt{n}D^+$ | 10 | .685 | .755 | .842 | .897 |
| Case 3 | 20 | .710 | .780 | .859 | .926 |
| | 50 | .727 | .796 | .870 | .940 |
| | $\infty$ | .734 | .808 | .877 | .957 |
| $\sqrt{n}D^-$ | 10 | .700 | .766 | .814 | .892 |
| Case 3 | 20 | .715 | .785 | .843 | .926 |
| | 50 | .724 | .796 | .860 | .944 |
| | $\infty$ | .733 | .808 | .877 | .957 |
| $\sqrt{n}D$ | 10 | .760 | .819 | .880 | .944 |
| Case 3 | 20 | .779 | .843 | .907 | .973 |
| | 50 | .790 | .856 | .922 | .988 |
| | $\infty$ | .803 | .874 | .939 | 1.007 |
| $\sqrt{n}V$ | 10 | 1.287 | 1.381 | 1.459 | 1.535 |
| Case 3 | 20 | 1.323 | 1.428 | 1.509 | 1.600 |
| | 50 | 1.344 | 1.453 | 1.538 | 1.639 |
| | $\infty$ | 1.372 | 1.477 | 1.557 | 1.671 |

Taken from Chandra, Singpurwalla, and Stephens (1981), with permission of the authors and of the American Statistical Association.
The table for $\sqrt{n}D$, Case 2, has been corrected.

$\alpha$ is known; suppose then that $Y_i = X_i - \alpha$; $\hat{\beta}$ is given by solving

$$\hat{\beta} = \{\Sigma_j Y_j - \Sigma_j Y_j \exp(-Y_j/\hat{\beta})\}/n$$

The steps in making the test are then:

(a) Estimate unknown parameters as above.
(b) Calculate $Z_{(i)} = F(X_{(i)})$, $i = 1, \ldots, n$, where $F(x)$ is given by equation (4.6), using estimated parameters when necessary.
(c) Use formulas (4.2) to calculate the EDF statistics.
(d) Modify the test statistics as shown in Table 4.17, or use Table 4.18 and compare with the upper tail percentage points given.

Table 4.17 is taken from Stephens (1977), and Table 4.18 from Chandra, Singpurwalla, and Stephens (1981).

Case 1 above is equivalent to a test for the exponential distribution on the transformed variable $Y = \exp(-X/\beta)$. This transformation in (4.6) gives, for Y, the distribution $F(y) = 1 - \exp(-\delta y)$, $y > 0$, with $\delta = \exp(\alpha/\beta)$. When $\beta$ is known, the transformation can be made, and the Y values are then tested to come from the exponential distribution with origin zero and unknown scale parameter (Section 4.9.3). The test statistics for the exponential test will take the same values as those for the Case 1 test in the present section, except that $D^+$ becomes $D^-$ and vice versa.

## 4.11 EDF TESTS FOR THE WEIBULL DISTRIBUTION

### 4.11.1 Test Situations

The general form of the Weibull distribution $W(x;\alpha,\beta,m)$ is

$$F(x) = 1 - \exp\left\{-\left(\frac{x-\alpha}{\beta}\right)^m\right\}, \quad x > \alpha; \beta > 0, m > 1$$

Here $\alpha$ and $\beta$ are location and scale parameters, respectively, and m is a shape parameter; $\alpha$ is called the **origin** of the distribution. The Weibull density function is

$$f(x) = \frac{m}{\beta}\left(\frac{x-\alpha}{\beta}\right)^{m-1} \exp\left\{-\left(\frac{x-\alpha}{\beta}\right)^m\right\}$$

The null hypothesis in this section is

$H_0$: a random sample $X_1, \ldots, X_n$ comes from the Weibull distribution $W(x;\alpha,\beta,m)$

4.11.2  Tests When the Location Parameter is Known;
         Reduction to a Test for the Extreme-Value Distribution

We consider the case where $\alpha$ is known. Suppose its value is zero, so that $H_0$ becomes

$$H_{0\alpha}: \text{ the set X comes from } W(x;0,\beta,m)$$

This distribution is often called the two-parameter Weibull distribution. If $\alpha$ is not zero, but has value $\alpha_0$, say, the transformation $X' = X - \alpha_0$, $i = 1, \ldots, n$, gives a set $X'$, for which $H_{0\alpha}$ will be true when $H_0$ is true for X; thus $H_{0\alpha}$ is tested for $X'$.

In considering $H_{0\alpha}$ we distinguish three cases:

Case 1: m is known and $\beta$ is unknown;
Case 2: $\beta$ is known and m is unknown;
Case 3: both m and $\beta$ are unknown.

For the test of $H_{0\alpha}$, the tables for the extreme-value distribution tests may be used. Let $Y = -\log X$ in the distribution $W(x;0,\beta,m)$; the distribution for Y becomes

$$F(y) = \exp\left[-\exp\left\{-\frac{(y-\phi)}{\theta}\right\}\right], \quad -\infty < y < \infty \tag{4.10}$$

with $\theta = 1/m$ and $\phi = -\log \beta$. This distribution is the extreme-value distribution of Section 4.10, and a test of $H_{0\alpha}$ for X may be made by testing that Y has the extreme-value distribution, with one or both of $\theta$ and $\phi$ unknown.

The test procedure therefore becomes:

(a)  Make the transformation $Y_i = -\log X_i$, $i = 1, \ldots, n$.
(b)  Arrange the $Y_i$ in ascending order (note that if the $X_i$ were given in ascending order the $Y_i$ will be in descending order).
(c)  Test that the Y-sample is from the extreme-value distribution (4.6) as described in Section 4.10.

In Case 1, m will be known, and so $\theta$ will be known in distribution (4.10) for Y. The test is therefore a Case 1 test as described in Section 4.10. In Case 2, $\beta$ is known, and so $\phi$ is known in distribution (4.10) for Y. The test will be a Case 2 test of Section 4.10. In Case 3, both parameters $\theta$ and $\phi$ in (4.10) will be unknown, so the test will be a Case 3 test of Section 4.10. Tables for the rather more unusual cases where $\alpha$ is unknown have been given by Lockhart and Stephens (1985a).

## 4.12  EDF TESTS FOR THE GAMMA DISTRIBUTION

In this section we discuss the tests of the null hypothesis

$H_0$: a random sample $X_1, \ldots, X_n$ comes from the Gamma distribution, $G(x;\alpha,\beta,m)$ with density

$$f(x) = \frac{1}{\beta\Gamma(m)}\left(\frac{x-\alpha}{\beta}\right)^{m-1} \exp\{-(x-\alpha)/\beta\}, \; x > \alpha; \; \beta > 0, \; m > 0$$

The location parameter $\alpha$ will be called the <u>origin</u> of the distribution; $\beta$ and m are, respectively, scale and shape parameters.

### 4.12.1  Tests with Known Origin, Cases 1, 2, and 3

We consider the case where $\alpha$ is known. If $\alpha = 0$, $H_0$ becomes

$H_{0\alpha}$: set X comes from $G(x;0,\beta,m)$

If $\alpha$ is not zero, but has value $\alpha_0$, say, the transformation $X_i' = X_i - \alpha_0$, $i = 1, \ldots, n$, is made to give a set X': then the null hypothesis $H_0$ for set X reduces to $H_{0\alpha}$ for set X', and $H_{0\alpha}$ is tested using the set X'.
  In considering $H_{0\alpha}$ we can distinguish three cases:

Case 1: m is known, and $\beta$ is unknown;
Case 2: $\beta$ is known, and m is unknown;
Case 3: both m and $\beta$ are unknown.

For Cases 2 and 3, distribution theory, even asymptotic theory, when m is estimated by maximum likelihood or another efficient method, will depend on the true m; this is because m is not a location or scale parameter (Section 4.3.3). However, useful approximate tests can still be made as follows.

### 4.12.2  Tests for Case 1

The steps for making this test are:

(a)  Put the sample in ascending order $X_{(1)} < \cdots < X_{(n)}$.
(b)  Let $\bar{X}$ be the sample mean, and estimate $\beta$ by $\hat{\beta} = \bar{X}/m$; $\hat{\beta}$ is the maximum likelihood estimator of $\beta$.
(c)  Define

$$I(X;m,\beta) = \frac{1}{\beta^m\Gamma(m)}\int_0^X x^{m-1}\exp(-x/\beta)\,dx$$

TABLE 4.19 Upper Tail Asymptotic Percentage Points for $W^2$, $U^2$, and $A^2$ in Tests for the Gamma Distribution (Section 4.12)[a]

| Statistic | m | Significance level $\alpha$ | | | |
|---|---|---|---|---|---|
| | | .10 | .05 | .025 | .01 |
| $W^2$ | 1 | .175 | .222 | .271 | .338 |
| | 2 | .156 | .195 | .234 | .288 |
| | 3 | .149 | .185 | .222 | .271 |
| | 4 | .146 | .180 | .215 | .262 |
| | 5 | .144 | .177 | .211 | .257 |
| | 6 | .142 | .175 | .209 | .254 |
| | 8 | .140 | .173 | .205 | .250 |
| | 10 | .139 | .171 | .204 | .247 |
| | 12 | .138 | .170 | .202 | .245 |
| | 15 | .138 | .169 | .201 | .244 |
| | 20 | .137 | .169 | .200 | .243 |
| | ∞ | .135 | .165 | .196 | .237 |
| $U^2$ | 1 | .129 | .159 | .189 | .230 |
| | 2 | .129 | .158 | .188 | .228 |
| | 3 | .128 | .158 | .187 | .227 |
| | 4 | .128 | .158 | .187 | .227 |
| | 5 | .128 | .158 | .187 | .227 |
| | 6 | .128 | .157 | .187 | .227 |
| | 8 | .128 | .157 | .187 | .227 |
| | 10 | .128 | .157 | .187 | .227 |
| | 12 | .128 | .157 | .187 | .227 |
| | 15 | .128 | .157 | .187 | .227 |
| | 20 | .128 | .157 | .187 | .227 |
| | ∞ | .128 | .157 | .187 | .227 |
| $A^2$ | 1 | 1.062 | 1.321 | 1.591 | 1.959 |
| | 2 | .989 | 1.213 | 1.441 | 1.751 |
| | 3 | .959 | 1.172 | 1.389 | 1.683 |
| | 4 | .944 | 1.151 | 1.362 | 1.648 |
| | 5 | .935 | 1.139 | 1.346 | 1.627 |
| | 6 | .928 | 1.130 | 1.335 | 1.612 |
| | 8 | .919 | 1.120 | 1.322 | 1.595 |
| | 10 | .915 | 1.113 | 1.314 | 1.583 |
| | 12 | .911 | 1.110 | 1.310 | 1.578 |
| | 15 | .908 | 1.106 | 1.304 | 1.570 |
| | 20 | .905 | 1.101 | 1.298 | 1.562 |
| | ∞ | .893 | 1.087 | 1.281 | 1.551 |

[a]Parameters: location $\alpha$ known; scale $\beta$ unknown; shape m known.

Accurate computer routines now exist for this expression (the incomplete gamma function). Calculate $Z_{(i)} = I(X_{(i)}; m, \hat{\beta})$, for $i = 1, \ldots, n$.

(d)  Calculate the EDF statistics from the $Z_{(i)}$, using formulas (4.2).

(e)  Modify the statistics as follows:

For $m = 1$, calculate

$$W^* = W^2(1 + 0.16/n); \quad U^* = U^2(1 + 0.16/n); \quad A^* = A^2(1 + 0.6/n)$$

For $m \geq 2$, calculate

$$W^* = \frac{(1.8nW^2 - 0.14)}{1.8n - 1}; \quad U^* = \frac{(1.8nU^2 - 0.14)}{1.8n - 1}; \quad A^* = A^2 + \frac{1}{n}\left(0.2 + \frac{0.3}{m}\right)$$

The modified statistics are then referred to the upper tail percentage points given in Table 4.19 for the appropriate known value of m. These points are the asymptotic points for the various distributions; they were given by Pettitt and Stephens (1983).

The modifications given above are based on Monte Carlo studies for finite n, and have been designed to be as comprehensive as possible, covering all values of m and n; when the given percentage points are used at level $\alpha$ it is believed that the true level of significance will not differ by more than 0.5% for $n \geq 5$.

## 4.12.3  Application to a Test for the Chi-Square Distribution

The Gamma distribution, with $m = r/2$ and $\beta = 2$, becomes the chi-square distribution with r degrees of freedom. Thus this Case 1 test can be used to test that observations $X_i$, multiplied by an unknown constant, come from a chi-square distribution with known degrees of freedom. For example, it may be used to test $H_0$: n independent sample variances $s_1^2, \ldots, s_n^2$, each calculated from a sample of size k, come from parent populations which are normal with the same (unknown) variance $\sigma^2$. An application might be to test for constant variance in an Analysis of Variance with cells each containing k observations. Other applications are given by Pettitt and Stephens (1983).

## 4.12.4  Test for Case 2

For this case, the steps are as follows:

(a)  Put the sample in ascending order $X_{(1)} < \cdots < X_{(n)}$.

(b)  Estimate m by solving for $\hat{m}$ the equation $\{\Sigma_i \log X_i\}/n = \psi(m) - \log \beta$, where $\psi(m)$ is the digamma function $\frac{d}{dm} \log \Gamma(m)$; $\hat{m}$ is the maximum likelihood estimator of m.

(c)  Calculate $Z_{(i)} = I(X_{(i)}; \hat{m}, \beta)$, for $i = 1, \ldots, n$.

TABLE 4.20 Upper Tail Asymptotic Percentage Points for $W^2$, $U^2$, and $A^2$ in Tests for the Gamma Distribution (Section 4.12)[a]

| Statistic | m | Significance level $\alpha$ | | | | | |
|---|---|---|---|---|---|---|---|
| | | .25 | .10 | .05 | .025 | .01 | .005 |
| $W^2$ | 1 | .103 | .150 | .186 | .223 | .273 | .311 |
| | 2 | .099 | .143 | .176 | .210 | .256 | .291 |
| | 3 | .097 | .140 | .172 | .205 | .250 | .283 |
| | 4 | .096 | .138 | .171 | .203 | .247 | .280 |
| | 5 | .096 | .138 | .169 | .202 | .245 | .278 |
| | 6 | .095 | .137 | .169 | .201 | .244 | .276 |
| | 8 | .095 | .136 | .168 | .200 | .242 | .275 |
| | 10 | .095 | .136 | .167 | .199 | .241 | .274 |
| | 12 | .095 | .136 | .167 | .199 | .241 | .273 |
| | 15 | .094 | .135 | .167 | .198 | .240 | .272 |
| | 20 | .094 | .135 | .166 | .198 | .240 | .272 |
| | ∞ | .094 | .134 | .165 | .197 | .238 | .270 |
| $U^2$ | 1 | .090 | .129 | .159 | .189 | .230 | .262 |
| | 2 | .089 | .128 | .158 | .189 | .229 | .261 |
| | 3 | .089 | .128 | .158 | .188 | .229 | .260 |
| | 4 | .089 | .128 | .158 | .188 | .229 | .260 |
| | 5 | .089 | .128 | .158 | .188 | .229 | .260 |
| | 6 | .089 | .128 | .158 | .188 | .228 | .260 |
| | 8 | .089 | .128 | .157 | .188 | .228 | .260 |
| | 10 | .089 | .128 | .157 | .188 | .228 | .260 |
| | 12 | .089 | .128 | .157 | .188 | .228 | .260 |
| | 15 | .089 | .128 | .157 | .188 | .228 | .260 |
| | 20 | .089 | .127 | .157 | .187 | .228 | .260 |
| | ∞ | .090 | .127 | .157 | .187 | .228 | .259 |
| $A^2$ | 1 | .680 | .956 | 1.170 | 1.390 | 1.687 | 1.916 |
| | 2 | .661 | .926 | 1.130 | 1.338 | 1.619 | 1.836 |
| | 3 | .655 | .915 | 1.115 | 1.320 | 1.596 | 1.809 |
| | 4 | .651 | .909 | 1.108 | 1.310 | 1.584 | 1.795 |
| | 5 | .649 | .906 | 1.103 | 1.305 | 1.577 | 1.787 |
| | 6 | .648 | .904 | 1.101 | 1.301 | 1.572 | 1.781 |
| | 8 | .646 | .901 | 1.097 | 1.297 | 1.567 | 1.775 |
| | 10 | .645 | .899 | 1.095 | 1.294 | 1.563 | 1.771 |
| | 12 | .644 | .898 | 1.094 | 1.293 | 1.561 | 1.768 |
| | 15 | .644 | .897 | 1.092 | 1.291 | 1.559 | 1.766 |
| | 20 | .643 | .896 | 1.091 | 1.289 | 1.557 | 1.763 |
| | ∞ | .644 | .894 | 1.087 | 1.285 | 1.551 | 1.756 |

[a]Parameters: location $\alpha$ known; scale $\beta$ known; shape m unknown.

TABLE 4.21 Upper Tail Asymptotic Percentage Points for $W^2$, $U^2$, and $A^2$ in Tests for the Gamma Distribution (Section 4.12)[a]

| Statistic | m | Significance level $\alpha$ | | | | | |
|---|---|---|---|---|---|---|---|
| | | .25 | .10 | .05 | .025 | .01 | .005 |
| $W^2$ | 1 | .079 | .111 | .136 | .162 | .196 | .222 |
| | 2 | .076 | .107 | .131 | .155 | .187 | .211 |
| | 3 | .075 | .106 | .129 | .153 | .184 | .208 |
| | 4 | .075 | .105 | .128 | .152 | .183 | .207 |
| | 5 | .075 | .105 | .128 | .151 | .182 | .206 |
| | 6 | .075 | .105 | .128 | .151 | .181 | .205 |
| | 8 | .074 | .104 | .127 | .150 | .181 | .204 |
| | 10 | .074 | .104 | .127 | .150 | .180 | .204 |
| | 12 | .074 | .104 | .127 | .150 | .180 | .203 |
| | 15 | .074 | .104 | .127 | .149 | .180 | .203 |
| | 20 | .074 | .104 | .126 | .149 | .180 | .203 |
| | $\infty$ | .074 | .104 | .126 | .148 | .178 | .201 |
| $U^2$ | 1 | .071 | .098 | .119 | .141 | .169 | .190 |
| | 2 | .070 | .097 | .118 | .139 | .166 | .187 |
| | 3 | .070 | .097 | .118 | .138 | .165 | .186 |
| | 4 | .070 | .097 | .117 | .138 | .165 | .186 |
| | 5 | .069 | .097 | .117 | .138 | .165 | .185 |
| | 6 | .069 | .097 | .117 | .138 | .165 | .185 |
| | 8 | .069 | .096 | .117 | .137 | .164 | .185 |
| | 10 | .069 | .096 | .117 | .137 | .164 | .185 |
| | 12 | .069 | .096 | .117 | .137 | .164 | .185 |
| | 15 | .069 | .096 | .117 | .137 | .164 | .185 |
| | 20 | .069 | .096 | .117 | .137 | .164 | .185 |
| | $\infty$ | .069 | .096 | .117 | .136 | .164 | .183 |
| $A^2$ | 1 | .486 | .657 | .786 | .917 | 1.092 | 1.227 |
| | 2 | .477 | .643 | .768 | .894 | 1.062 | 1.190 |
| | 3 | .475 | .639 | .762 | .886 | 1.052 | 1.178 |
| | 4 | .473 | .637 | .759 | .883 | 1.048 | 1.173 |
| | 5 | .472 | .635 | .758 | .881 | 1.045 | 1.170 |
| | 6 | .472 | .635 | .757 | .880 | 1.043 | 1.168 |
| | 8 | .471 | .634 | .755 | .878 | 1.041 | 1.165 |
| | 10 | .471 | .633 | .754 | .877 | 1.040 | 1.164 |
| | 12 | .471 | .633 | .754 | .876 | 1.039 | 1.163 |
| | 15 | .470 | .632 | .754 | .876 | 1.038 | 1.162 |
| | 20 | .470 | .632 | .753 | .875 | 1.037 | 1.161 |
| | $\infty$ | .470 | .631 | .752 | .873 | 1.035 | 1.159 |

[a]Parameters: location $\alpha$ known; scale $\beta$ unknown; shape m unknown.

(d) Calculate the EDF statistics from the $Z_{(i)}$, using formulas (4.2).
(e) Reject $H_{0\alpha}$ if the value of the statistic used is greater than the value in Table 4.20 for desired significance level $\alpha$ and for appropriate $\hat{m}$.

### 4.12.5  Tests for Case 3

The steps in the test are as follows:

(a) Estimate m by solving for $\hat{m}$ the equation

$$\{\Sigma_i \log X_i\}/n - \log \bar{X} = \psi(m) - \log m$$

where $\psi(m)$ is the digamma function as above, and estimate $\beta$ by
$\hat{\beta} = \bar{X}/\hat{m}$.

(b) Calculate EDF statistics from $Z_{(i)} = I(X_{(i)}; \hat{m}, \hat{\beta})$, $i = 1, \ldots, n$.
(c) Reject $H_{0\alpha}$ if the value of the statistic used is greater than the value in Table 4.21, for desired significance level $\alpha$, and appropriate $\hat{m}$.

#### 4.12.5.1  Comment

The points in Table 4.21 remain remarkably stable as m changes, especially for $U^2$, and accurate results can be expected when $\hat{m}$ is used for m, except possibly for small values of m. Note that only asymptotic points are given; experience with $W^2$, $U^2$, and $A^2$ suggests these will be very good approximations to the points for finite n, even for quite small n. The points in Tables 4.20 and 4.21 are taken from Lockhart and Stephens (1985b), where the asymptotic theory is also developed. A somewhat different treatment was given much earlier in an unpublished report by Mickey, Mundle, Walker, and Glinski (1963). The various cases when the origin is not known are much more unlikely; furthermore, it is often difficult to estimate parameters efficiently. Tests for these cases have been given by Lockhart and Stephens (1985b). Tables for the Kolmogorov statistic D, for n = 4(1)10(5)30, have been given for Cases 1, 2, and 3 above (a different estimate of m is used in Case 3) by Schneider and Clickner (1976).

### 4.13  EDF TESTS FOR THE LOGISTIC DISTRIBUTION

In this section is discussed the test of

$H_0$: a random sample $X_1, \ldots, X_n$ comes from the logistic distribution

$$F(x; \alpha, \beta) = 1/[1 + \exp\{-(x - \alpha)/\beta\}], \quad -\infty < x < \infty; \ \beta < 0$$

with parameters $\alpha$ or $\beta$, or both, unknown

TABLE 4.22 Modifications and Upper Tail Percentage Points for $W^2$, $U^2$, $A^2$ in Tests for the Logistic Distribution (Section 4.13)[a]

| Statistic | Modification | Significance level $\alpha$ | | | | | |
|---|---|---|---|---|---|---|---|
| | | .25 | .10 | .05 | .025 | .01 | .005 |
| $W^2$ | | | | | | | |
| Case 1 | $(1.9nW^2 - 0.15)/(1.9n - 1.0)$ | .083 | .119 | .148 | .177 | .218 | .249 |
| Case 2 | $(0.95nW^2 - 0.45)/(0.95n - 1.0)$ | .184 | .323 | .438 | .558 | .721 | .847 |
| Case 3 | $(nW^2 - 0.08)/(n - 1.0)$ | .060 | .081 | .098 | .114 | .136 | .152 |
| $U^2$ | | | | | | | |
| Case 2 | $(1.6nU^2 - 0.16)/(1.6n - 1.0)$ | .080 | .116 | .145 | .174 | .214 | .246 |
| $A^2$ | | | | | | | |
| Case 1 | $A^2 + 0.15/n$ | .615 | .857 | 1.046 | 1.241 | 1.505 | 1.710 |
| Case 2 | $(0.6nA^2 - 1.8)/(0.6n - 1.0)$ | 1.043 | 1.725 | 2.290 | 2.880 | 3.685 | 4.308 |
| Case 3 | $A^2(1.0 + 0.25/n)$ | .426 | .563 | .660 | .769 | .906 | 1.010 |

[a]For $U^2$ Cases 1 and 3 use modifications and percentage points for $W^2$ Cases 1 and 3, respectively (see Section 4.13). Taken from Stephens (1979), with permission of the Biometrika Trustees.

TABLE 4.23  Upper Tail Percentage Points for Statistics $D^+\sqrt{n}$, $D\sqrt{n}$, and $V\sqrt{n}$, for Tests for the Logistic Distribution (Section 4.14)

| Case | n | Significance level $\alpha$ | | | |
|------|------|--------|--------|--------|--------|
| | | 0.10 | 0.05 | 0.025 | 0.01 |
| | | Statistic $D^+\sqrt{n}$ | | | |
| 1 | 5 | 0.702 | 0.758 | 0.805 | 0.854 |
| | 10 | 0.730 | 0.792 | 0.846 | 0.913 |
| | 20 | 0.744 | 0.809 | 0.867 | 0.944 |
| | 50 | 0.752 | 0.819 | 0.880 | 0.962 |
| | ∞ | 0.757 | 0.826 | 0.888 | 0.974 |
| 2 | 5 | 0.971 | 1.120 | 1.239 | 1.380 |
| | 10 | 0.990 | 1.143 | 1.268 | 1.423 |
| | 20 | 0.999 | 1.150 | 1.282 | 1.444 |
| | 50 | 1.005 | 1.161 | 1.290 | 1.456 |
| | ∞ | 1.009 | 1.166 | 1.297 | 1.464 |
| 3 | 5 | 0.603 | 0.650 | 0.690 | 0.735 |
| | 10 | 0.636 | 0.687 | 0.736 | 0.789 |
| | 20 | 0.653 | 0.705 | 0.758 | 0.816 |
| | 50 | 0.663 | 0.716 | 0.773 | 0.832 |
| | ∞ | 0.669 | 0.723 | 0.781 | 0.842 |
| | | Statistic $D\sqrt{n}$ | | | |
| 1 | 5 | 0.736 | 0.791 | 0.845 | 0.883 |
| | 10 | 0.777 | 0.837 | 0.895 | 0.653 |
| | 20 | 0.800 | 0.865 | 0.926 | 0.997 |
| | 50 | 0.808 | 0.874 | 0.937 | 1.011 |
| | ∞ | 0.816 | 0.883 | 0.947 | 1.025 |
| 2 | 5 | 1.108 | 1.236 | 1.349 | 1.474 |
| | 10 | 1.148 | 1.274 | 1.388 | 1.521 |
| | 20 | 1.167 | 1.294 | 1.406 | 1.545 |
| | 50 | 1.179 | 1.305 | 1.419 | 1.559 |
| | ∞ | 1.187 | 1.313 | 1.427 | 1.568 |
| 3 | 5 | 0.643 | 0.679 | 0.723 | 0.751 |
| | 10 | 0.679 | 0.730 | 0.774 | 0.823 |
| | 20 | 0.698 | 0.755 | 0.800 | 0.854 |
| | 50 | 0.708 | 0.770 | 0.817 | 0.873 |
| | ∞ | 0.715 | 0.780 | 0.827 | 0.886 |

(continued)

TABLE 4.23 (continued)

| Case | n | Significance level $\alpha$ | | | |
|------|-----|--------|--------|--------|--------|
| | | 0.10 | 0.05 | 0.025 | 0.01 |
| | | Statistic $V\sqrt{n}$ | | | |
| 1 | 5 | 1.369 | 1.471 | 1.580 | 1.658 |
| | 10 | 1.410 | 1.520 | 1.630 | 1.741 |
| | 20 | 1.433 | 1.550 | 1.659 | 1.790 |
| | 50 | 1.447 | 1.564 | 1.675 | 1.815 |
| | $\infty$ | 1.454 | 1.574 | 1.685 | 1.832 |
| 2 | 5 | 1.314 | 1.432 | 1.547 | 1.674 |
| | 10 | 1.372 | 1.483 | 1.587 | 1.711 |
| | 20 | 1.400 | 1.510 | 1.607 | 1.730 |
| | 50 | 1.417 | 1.525 | 1.619 | 1.741 |
| | $\infty$ | 1.429 | 1.535 | 1.627 | 1.748 |
| 3 | 5 | 1.170 | 1.246 | 1.299 | 1.373 |
| | 10 | 1.230 | 1.311 | 1.381 | 1.466 |
| | 20 | 1.260 | 1.344 | 1.422 | 1.514 |
| | 50 | 1.277 | 1.364 | 1.448 | 1.542 |
| | $\infty$ | 1.289 | 1.376 | 1.463 | 1.560 |

As in earlier sections, we distinguish three cases:

Case 1: $\beta$ known, $\alpha$ unknown;
Case 2: $\alpha$ known, $\beta$ unknown;
Case 3: $\alpha$ and $\beta$ both unknown.

The parameters are estimated from the data by maximum likelihood. For Case 3, when both $\alpha$ and $\beta$ are unknown, the equations for the estimates $\hat{\alpha}$, $\hat{\beta}$ are

$$n^{-1} \Sigma_i [1 + \exp\{(X_i - \hat{\alpha})/\hat{\beta}\}]^{-1} = 0.5 \tag{4.11}$$

$$n^{-1} \Sigma_i \left(\frac{X_i - \hat{\alpha}}{\hat{\beta}}\right) \frac{1 - \exp\{(X_i - \hat{\alpha})/\hat{\beta}\}}{1 + \exp\{(X_i - \hat{\alpha})/\hat{\beta}\}} = -1 \tag{4.12}$$

These equations may be solved iteratively; good starting values for $\hat{\alpha}$ and $\hat{\beta}$ are the sample mean $\bar{X}$ and the sample standard deviation s. In Case 1, $\hat{\alpha}$ is the solution of equation (4.11), with $\beta$ replacing $\hat{\beta}$. In Case 2, $\beta$ is the solution of (4.12), with $\alpha$ replacing $\hat{\alpha}$. The steps in making the test are then as follows:

(a)  Find estimates of any unknown parameters.
(b)  Calculate $Z_{(i)} = 1/[1 + \exp\{-(X_{(i)} - \alpha)/\beta\}]$, $i = 1, \ldots, n$, with $\alpha$ and $\beta$ replaced by estimates where necessary.
(c)  Calculate the EDF statistics from (4.2).
(d)  For $W^2$, $U^2$, and $A^2$ modify the statistic as in Table 4.22; reject $H_0$ if the statistic exceeds the percentage point given for desired significance level $\alpha$. For $D^+$, $D^-$, D, and V, multiply by $\sqrt{n}$ and use Table 4.23. The table for $D^+\sqrt{n}$ can also be used for $D^-\sqrt{n}$.

The asymptotic percentage points for $W^2$, $U^2$, and $A^2$ given in Table 4.22 are based on theoretical work of Stephens (1979), and the modifications have been derived, as in previous sections, from extensive Monte Carlo studies for finite n. For each case, 10,000 samples were used to give the percentage points, for n = 5, 8, 10, 20, and 50. The percentage points for $D^+\sqrt{n}$, $D\sqrt{n}$, and $V\sqrt{n}$ given in Table 4.23 were derived from the same Monte Carlo studies.

## 4.14  EDF TESTS FOR THE CAUCHY DISTRIBUTION

The Cauchy distribution has density

$$f(x;\alpha,\beta) = \frac{1}{\pi} \frac{\beta}{\beta^2 + (x - \alpha)^2}, \quad -\infty < x < \infty; \beta > 0 \tag{4.13}$$

and distribution function

$$F(x;\alpha,\beta) = \frac{1}{2} + \frac{1}{\pi} \tan^{-1}\left(\frac{x - \alpha}{\beta}\right), \quad -\infty < x < \infty; \beta > 0 \tag{4.14}$$

In this section we discuss tests of the null hypothesis

$H_0$: a random sample $X_1, \ldots, X_n$ comes from the Cauchy distribution, with one or both of parameters $\alpha$ and $\beta$ unknown

As with previous tests, we consider three cases:

Case 1: $\beta$ known, $\alpha$ unknown;
Case 2: $\alpha$ known, $\beta$ unknown;
Case 3: both $\alpha$ and $\beta$ unknown.

For other distributions, the parameters have been estimated by maximum likelihood; however, for the Cauchy distribution, the likelihood may have

TABLE 4.24  Upper Tail Percentage Points for $W^2$ and $A^2$ for Tests for the Cauchy Distribution (Section 4.14)

| | Significance level $\alpha$ | | | | | |
|---|---|---|---|---|---|---|
| n | .25 | .15 | .10 | .05 | .025 | .01 |
| | Case 1. Statistic $W^2$ | | | | | |
| 5 | .208 | .382 | .667 | 1.26 | 1.51 | 1.61 |
| 8 | .227 | .480 | .870 | 1.68 | 2.30 | 2.55 |
| 10 | .227 | .460 | .840 | 1.80 | 2.60 | 3.10 |
| 12 | .220 | .430 | .770 | 1.76 | 2.85 | 3.65 |
| 15 | .205 | .372 | .670 | 1.59 | 2.88 | 4.23 |
| 20 | .189 | .315 | .520 | 1.25 | 2.65 | 4.80 |
| 25 | .175 | .275 | .420 | .870 | 2.10 | 4.70 |
| 30 | .166 | .250 | .360 | .710 | 1.60 | 4.10 |
| 40 | .153 | .220 | .290 | .510 | 1.50 | 3.05 |
| 50 | .145 | .200 | .260 | .400 | .70 | 2.05 |
| 100 | .130 | .170 | .210 | .270 | .35 | .60 |
| $\infty$ | .115 | .146 | .173 | .216 | .260 | .319 |
| | Case 1. Statistic $A^2$ | | | | | |
| 5 | 1.19 | 2.22 | 3.83 | 8.00 | 12.75 | 17.980 |
| 8 | 1.33 | 2.62 | 4.7 | 10.0 | 17.4 | 25.0 |
| 10 | 1.34 | 2.52 | 4.5 | 10.6 | 18.2 | 29.0 |
| 12 | 1.31 | 2.42 | 4.1 | 9.9 | 18.8 | 32.0 |
| 15 | 1.30 | 2.15 | 3.5 | 8.2 | 17.2 | 31.2 |
| 20 | 1.17 | 1.86 | 2.8 | 6.5 | 14.4 | 27.5 |
| 25 | 1.12 | 1.68 | 2.3 | 4.7 | 10.8 | 23.0 |
| 30 | 1.08 | 1.55 | 2.1 | 3.8 | 8.2 | 20.0 |
| 40 | 1.02 | 1.38 | 1.8 | 2.9 | 5.2 | 15.5 |
| 50 | .970 | 1.29 | 1.6 | 2.4 | 3.8 | 10 |
| 100 | .890 | 1.16 | 1.4 | 1.8 | 2.2 | 3.5 |
| $\infty$ | .834 | 1.02 | 1.219 | 1.519 | 1.812 | 2.212 |

TABLE 4.25. Upper Tail Percentage Points for $W^2$ and $A^2$ for Tests for the Cauchy Distribution (Section 4.14)

| | Significance level $\alpha$ | | | | | |
|---|---|---|---|---|---|---|
| n | .25 | .15 | .10 | .05 | .025 | .01 |

Case 2. Statistic $W^2$

| | | | | | | |
|---|---|---|---|---|---|---|
| 5 | .199 | .236 | .261 | .338 | .437 | .590 |
| 8 | .211 | .273 | .321 | .389 | .463 | .564 |
| 10 | .212 | .279 | .332 | .414 | .501 | .626 |
| 12 | .212 | .281 | .337 | .433 | .525 | .661 |
| 15 | .206 | .279 | .339 | .444 | .537 | .684 |
| 20 | .199 | .273 | .333 | .442 | .547 | .698 |
| 25 | .194 | .268 | .328 | .437 | .551 | .704 |
| 30 | .189 | .265 | .326 | .435 | .553 | .708 |
| 40 | .185 | .260 | .323 | .434 | .555 | .712 |
| 50 | .183 | .258 | .321 | .433 | .557 | .714 |
| 100 | .179 | .254 | .319 | .432 | .559 | .715 |
| $\infty$ | .176 | .250 | .316 | .131 | .560 | .714 |

Case 2. Statistic $A^2$

| | | | | | | |
|---|---|---|---|---|---|---|
| 5 | .974 | 1.131 | 1.239 | 1.59 | 2.08 | 2.84 |
| 8 | 1.085 | 1.360 | 1.560 | 1.88 | 2.18 | 2.55 |
| 10 | 1.110 | 1.414 | 1.653 | 2.04 | 2.38 | 2.89 |
| 12 | 1.117 | 1.443 | 1.710 | 2.14 | 2.55 | 3.15 |
| 15 | 1.117 | 1.449 | 1.728 | 2.22 | 2.65 | 3.31 |
| 20 | 1.101 | 1.444 | 1.728 | 2.24 | 2.73 | 3.44 |
| 25 | 1.083 | 1.432 | 1.727 | 2.25 | 2.77 | 3.50 |
| 30 | 1.064 | 1.422 | 1.724 | 2.25 | 2.80 | 3.53 |
| 40 | 1.051 | 1.41 | 1.723 | 2.26 | 2.82 | 3.56 |
| 50 | 1.045 | 1.405 | 1.722 | 2.27 | 2.83 | 3.59 |
| 100 | 1.038 | 1.40 | 1.718 | 2.28 | 2.86 | 3.64 |
| $\infty$ | 1.034 | 1.409 | 1.716 | 2.283 | 2.872 | 3.677 |

TABLE 4.26  Upper Tail Percentage Points for $W^2$ and $A^2$
for Tests for the Cauchy Distribution (Section 4.14)

| n | Significance level $\alpha$ | | | | | |
|---|------|------|------|------|------|------|
|   | .25 | .15 | .10 | .05 | .025 | .01 |
| | Case 3.  Statistic $W^2$ | | | | | |
| 5 | .167 | .242 | .305 | .393 | .445 | .481 |
| 8 | .192 | .315 | .441 | .703 | .940 | 1.13 |
| 10 | .197 | .331 | .481 | .833 | 1.201 | 1.571 |
| 12 | .194 | .329 | .487 | .896 | 1.391 | 1.901 |
| 15 | .185 | .317 | .472 | .904 | 1.54 | 2.33 |
| 20 | .169 | .281 | .419 | .835 | 1.63 | 2.96 |
| 25 | .154 | .253 | .366 | .726 | 1.47 | 3.08 |
| 30 | .143 | .225 | .319 | .615 | 1.25 | 2.90 |
| 40 | .126 | .195 | .263 | .460 | .850 | 2.17 |
| 50 | .117 | .175 | .235 | .381 | .642 | 1.56 |
| 60 | .1097 | .160 | .211 | .330 | .508 | 1.07 |
| 100 | .098 | .135 | .174 | .2378 | .331 | .544 |
| ∞ | .080 | .108 | .130 | .170 | .212 | .270 |
| | Case 3.  Statistic $A^2$ | | | | | |
| 5 | .835 | 1.14 | 1.40 | 1.77 | 2.00 | 2.16 |
| 8 | .992 | 1.52 | 2.06 | 3.20 | 4.27 | 5.24 |
| 10 | 1.04 | 1.63 | 2.27 | 3.77 | 5.58 | 7.50 |
| 12 | 1.04 | 1.65 | 2.33 | 4.14 | 6.43 | 9.51 |
| 15 | 1.02 | 1.61 | 2.28 | 4.25 | 7.20 | 11.50 |
| 20 | .975 | 1.51 | 2.13 | 4.05 | 7.58 | 14.57 |
| 25 | .914 | 1.40 | 1.94 | 3.57 | 6.91 | 14.96 |
| 30 | .875 | 1.30 | 1.76 | 3.09 | 5.86 | 13.80 |
| 40 | .812 | 1.16 | 1.53 | 2.48 | 4.23 | 10.20 |
| 50 | .774 | 1.08 | 1.41 | 2.14 | 3.37 | 7.49 |
| 60 | .743 | 1.02 | 1.30 | 1.92 | 2.76 | 5.32 |
| 100 | .689 | .927 | 1.14 | 1.52 | 2.05 | 3.30 |
| ∞ | .615 | .780 | .949 | 1.225 | 1.52 | 1.90 |

local maxima, and it may be difficult to find the true maximum. We therefore find estimates using sums of weighted order statistics. Chernoff, Gastwirth, and Johns (1967) have given the estimate $\hat{\alpha} = \Sigma_i c_i X_{(i)}$ with

$$c_i = \frac{\sin 4\pi\{i/(n+1) - 0.5\}}{n \tan \pi\{i/(n+1) - 0.5\}}$$

The estimate of $\beta$ is $\hat{\beta} = \Sigma_i d_i X_{(i)}$ with

$$d_i = \frac{8 \tan \pi\{i/(n+1) - 0.5\}}{n \sec^4 \pi\{i/(n+1) - 0.5\}}$$

These estimates are asymptotically efficient, and asymptotic distributions of $W^2$, $U^2$, and $A^2$ can be found. The test of $H_0$ is then as follows:

(a)  Estimate parameter $\alpha$ or $\beta$ or both, as described above.
(b)  Calculate $Z_{(i)} = F(X_{(i)}; \alpha, \beta)$, given in (4.14), with estimates replacing unknown parameters.
(c)  Use the formulas (4.2) to calculate EDF statistics.
(d)  Refer to Tables 4.24, 4.25, or 4.26 to make the test; reject $H_0$ if the test statistic is greater than the value given for n and for the desired significance level $\alpha$.

The points are taken from Stephens (1985), where the asymptotic theory, and tables for $U^2$, are given.

## 4.15  EDF TESTS FOR THE VON MISES DISTRIBUTION

The von Mises distribution is used to describe unimodal data on the circumference of a circle. Suppose the circle has center O and radius 1, and let a radius OP be measured by the polar coordinate $\theta$, from ON as origin. Let $\theta_0$ be the coordinate of a radius OA, and let $\kappa$ be a positive constant. The von Mises density is

$$f(\theta; \theta_0, \kappa) = \frac{1}{2\pi I_0(\kappa)} \exp\{\kappa \cos(\theta - \theta_0)\}, \quad 0 < \theta < 2\pi$$

Here $I_0(\kappa)$ is the imaginary Bessel function of order zero. The distribution has a mode along OA (that is, at $\theta = \theta_0$) and is symmetric around OA; for $\kappa = 0$ the distribution reduces to the uniform distribution around the circle. Suppose a random sample of values $\theta_1$, $\theta_2$, ..., $\theta_n$ is given, denoting locations on the circumference of points $P_1$, $P_2$, ..., $P_n$. We discuss the test of

$H_0$: the random sample of $\theta$-values comes from the von Mises distribution

$$F(\theta; \theta_0, \kappa) = \int_0^\theta f(\theta; \theta_0, \kappa) d\theta$$

TABLE 4.27  Upper Tail Percentage Points for $U^2$ for Tests
of the von Mises Distribution (Section 4.15)

| True shape $\kappa$ | Significance level $\alpha$ | | | | | | | |
|---|---|---|---|---|---|---|---|---|
| | 0.500 | 0.250 | 0.150 | 0.100 | 0.050 | 0.025 | 0.010 | 0.005 |
| | | | | Case 1 | | | | |
| 0.0 | 0.047 | 0.071 | 0.089 | 0.105 | 0.133 | 0.163 | 0.204 | 0.235 |
| 0.50 | 0.048 | 0.072 | 0.091 | 0.107 | 0.135 | 0.165 | 0.205 | 0.237 |
| 1.00 | 0.051 | 0.076 | 0.095 | 0.111 | 0.139 | 0.169 | 0.209 | 0.241 |
| 1.50 | 0.053 | 0.080 | 0.099 | 0.115 | 0.144 | 0.173 | 0.214 | 0.245 |
| 2.00 | 0.055 | 0.082 | 0.102 | 0.119 | 0.147 | 0.177 | 0.217 | 0.248 |
| 2.50 | 0.056 | 0.084 | 0.104 | 0.121 | 0.150 | 0.180 | 0.220 | 0.251 |
| 3.00 | 0.057 | 0.085 | 0.106 | 0.122 | 0.152 | 0.181 | 0.222 | 0.253 |
| 3.50 | 0.058 | 0.086 | 0.107 | 0.123 | 0.153 | 0.182 | 0.223 | 0.254 |
| 4.00 | 0.058 | 0.086 | 0.107 | 0.124 | 0.153 | 0.183 | 0.224 | 0.255 |
| 10.00 | 0.059 | 0.088 | 0.109 | 0.126 | 0.155 | 0.186 | 0.227 | 0.258 |
| $\infty$ | 0.059 | 0.089 | 0.110 | 0.127 | 0.157 | 0.187 | 0.228 | 0.259 |
| | | | | Case 2 | | | | |
| 0.0 | 0.047 | 0.071 | 0.089 | 0.105 | 0.133 | 0.163 | 0.204 | 0.235 |
| 0.50 | 0.048 | 0.072 | 0.091 | 0.107 | 0.135 | 0.165 | 0.205 | 0.237 |
| 1.00 | 0.051 | 0.076 | 0.095 | 0.111 | 0.139 | 0.169 | 0.209 | 0.241 |
| 1.50 | 0.053 | 0.080 | 0.100 | 0.116 | 0.144 | 0.174 | 0.214 | 0.245 |
| 2.00 | 0.055 | 0.082 | 0.103 | 0.119 | 0.148 | 0.177 | 0.218 | 0.249 |
| 2.50 | 0.056 | 0.084 | 0.105 | 0.121 | 0.150 | 0.180 | 0.220 | 0.251 |
| 3.00 | 0.057 | 0.085 | 0.105 | 0.122 | 0.151 | 0.181 | 0.221 | 0.252 |
| 3.50 | 0.057 | 0.085 | 0.106 | 0.122 | 0.151 | 0.181 | 0.221 | 0.253 |
| 4.00 | 0.057 | 0.085 | 0.106 | 0.122 | 0.151 | 0.181 | 0.221 | 0.253 |
| 10.00 | 0.057 | 0.085 | 0.105 | 0.122 | 0.151 | 0.180 | 0.221 | 0.252 |
| $\infty$ | 0.057 | 0.085 | 0.105 | 0.122 | 0.151 | 0.180 | 0.221 | 0.252 |
| | | | | Case 3 | | | | |
| 0.0 | 0.030 | 0.040 | 0.046 | 0.052 | 0.061 | 0.069 | 0.081 | 0.090 |
| 0.50 | 0.031 | 0.042 | 0.050 | 0.056 | 0.065 | 0.077 | 0.090 | 0.100 |
| 1.00 | 0.035 | 0.049 | 0.059 | 0.066 | 0.079 | 0.092 | 0.110 | 0.122 |
| 1.50 | 0.039 | 0.056 | 0.067 | 0.077 | 0.092 | 0.108 | 0.128 | 0.144 |
| 2.00 | 0.043 | 0.061 | 0.074 | 0.084 | 0.101 | 0.119 | 0.142 | 0.159 |
| 2.50 | 0.045 | 0.064 | 0.078 | 0.089 | 0.107 | 0.125 | 0.150 | 0.168 |
| 3.00 | 0.046 | 0.066 | 0.080 | 0.091 | 0.110 | 0.129 | 0.154 | 0.173 |
| 3.50 | 0.047 | 0.067 | 0.081 | 0.093 | 0.112 | 0.131 | 0.157 | 0.176 |
| 4.00 | 0.047 | 0.067 | 0.082 | 0.093 | 0.113 | 0.132 | 0.158 | 0.178 |
| 10.00 | 0.048 | 0.068 | 0.083 | 0.095 | 0.115 | 0.135 | 0.162 | 0.182 |
| $\infty$ | 0.048 | 0.069 | 0.084 | 0.096 | 0.117 | 0.137 | 0.164 | 0.184 |

As for other distributions, there are three cases:

Case 1: $\theta_0$ unknown, $\kappa$ known;
Case 2: $\theta_0$ known, $\kappa$ unknown;
Case 3: both $\theta_0$ and $\kappa$ unknown.

Maximum likelihood estimates of $\theta_0$ and of $\kappa$ are found as follows. Let $\underline{R}$ be the vector sum or resultant of vectors $OP_i$, $i = 1, \ldots, n$, and let R be its length. The estimate $\hat{\theta}_0$ of $\theta_0$ is the direction of $\underline{R}$, and the estimate $\hat{\kappa}$ of $\kappa$ is given by solving

$$\frac{I_1(\kappa)}{I_0(\kappa)} = \frac{R}{N} \tag{4.15}$$

where $I_1(\kappa)$ is the imaginary Bessel function of order 1. Tables for solving (4.15) are given in, for example, Biometrika Tables for Statisticians, Vol. 2 (Pearson and Hartley, 1972), and by Mardia (1972).

When OA is known, let X be the component of $\underline{R}$ on OA; then the estimate of $\kappa$ is now given by $\hat{\kappa}_1$, obtained by replacing R by X in (4.15).

Since the distribution is on a circle, only $U^2$ or V are valid EDF statistics, of those we have been considering (see Section 4.5.3). Asymptotic null distributions can be found for $U^2$; because $\kappa$ is not a scale parameter, the distribution depends on $\kappa$. However, as for the gamma and Weibull distributions, useful tests are still available.

The steps in making a test of $H_0$ are then as follows:

(a) For the appropriate case, estimate unknown parameters as described above.
(b) Calculate $Z_{(i)} = F(\theta_{(i)}; \theta_0, \kappa)$, where $\theta_0$ and $\kappa$ are replaced by estimates if necessary.
(c) Calculate $U^2$ from formula (4.2).

Refer $U^2$ to the part of Table 4.27 appropriate for the given case, using $\kappa$ or $\hat{\kappa}$; reject $H_0$ if $U^2$ exceeds the point given for $\alpha$. The test is approximate, since asymptotic points are used; however, these are likely to be accurate, for practical purposes, for $n \geq 20$. The points are taken from Lockhart and Stephens (1985), where asymptotic theory is also given.

## 4.16 EDF TESTS FOR CONTINUOUS DISTRIBUTIONS: MISCELLANEOUS TOPICS

### 4.16.1 Power of EDF Statistics when Parameters Are Estimated

In Section 4.6 some comments were made on the power of different EDF statistics for Case 0, using complete samples, where essentially the final test is a test for uniformity of the Z-values given by the Probability Integral Transformation. Different statistics were found to detect different types of departure from uniformity. When unknown parameters are estimated from the same sample as is used for the goodness-of-fit test, the differences in the powers of the statistics tend to become smaller. It appears that fitting the parameter or parameters makes it possible to adjust the tested distribution to the sample in such a way that the statistics can detect a departure from the null distribution with roughly the same efficiency; nevertheless, $A^2$ tends to lead the others, probably because it is effective at detecting departures in the tails.

Some asymptotic theory is available to examine power, at least for quadratic statistics. Durbin and Knott (1972) demonstrated a method by which asymptotic power results could be obtained, and applied it to tests for the normal distribution with mean 0 and variance 1, that is, Case 0 tests, against normal alternatives with a shift in mean or a shift in variance. Stephens (1974a) extended the results to shifts in both mean and variance. The technique rests on a partition of the appropriate statistics into components (see also Section 8.12). Durbin, Knott, and Taylor (1975) showed how the decomposition into components could be done also for the test for normality with mean and variance unknown (Case 3), or for the exponential test with scale parameter unknown, and used their method to discuss the asymptotic power of the components. Stephens (1976b) followed the method and applied it to tests for the statistics $W^2$, $U^2$, and $A^2$ for these situations. The overall result when tests for normality or exponentiality are made with unknown parameters, is that $A^2$ is slightly better than $W^2$ for the alternatives discussed, with $U^2$ not far behind $W^2$.

The superiority of $A^2$ has also been documented by various power studies based on Monte Carlo sampling. Some of these, in comparisons of tests of uniformity and normality, are by Stephens (1974b). These power studies also included the statistics $D^+$, $D^-$, $D$, and $V$.

The most famous statistic, the Kolmogorov-Smirnov $D$, tends to be weak in power. Statistics $D^+$ and $D^-$, on the other hand, often have good power but each one against only certain classes of alternatives. For example, in tests of exponentiality (see Table 10.6, results for Group 1 statistics) $D^+$ appears to be powerful against alternatives with decreasing failure rate and $D^-$ is powerful against alternatives with increasing failure rate. In some applications the alternative of interest may be clearly identified, and then it will be possible to identify which statistic to use. However, $D^+$ and $D^-$ will be biased when used against the wrong alternatives, so these statistics must be used with caution. From the power studies for tests for normality and exponentiality it appears that $A^2$ (or $W^2$ as second choice) should be the recommended omnibus test statistic for EDF tests with unknown parameters, with good power against a wide range of alternatives.

## 4.16.2  The Effect on Power of Knowing
         Certain Parameters

In Section 4.9 above, as an illustration of the test for exponentiality, the
example was worked in the case when the parameter $\beta$ was known, and also
when it was necessary to estimate it. It is clear from the example (or from
a comparison of Tables 4.11 and 4.14) that when the estimate is very close
to the true value, one has a much more sensitive test using the tables for
the parameters unknown than using the tables for Case 0; in general, the
critical values for rejection are much smaller when parameters must be
estimated. It would quite frequently happen that the estimated value of $\beta$
would be close to the true value, and then the practitioner who does not know
$\beta$ will obtain greater power than if $\beta$ were known. This appears somewhat
paradoxical, in that usually in statistical testing one assumes that the more
knowledge the better. However, the tests are (ideally) intended as tests for
distributional form, not as tests for parameter values, and some knowledge
of parameters may not be very important in assessing distributional form.
For example, it may be unhelpful to know, and to use, the mean of the true
distribution, when this is not the one tested. Stephens (1974b) and Dyer (1974)
have noted these effects in tests for normality; being given means and variances changes the test from Case 3 to Case 0, with a consequent loss of
power. On the other hand, Spinelli and Stephens (1987) have shown that in
tests for exponentiality it is better to use the value of the origin, when this
is known, than to estimate it. Note also that in Example E4.9.3 the exponential form, when $\beta$ was given, was acceptable, but when the test focused more
on the exponential shape (the main point of the test) and less on the parameter,
the exponential form was rejected. Further work is still needed on what
parametric information is useful and what is not.

## 4.16.3  Other Techniques for Unknown Parameters

### 4.16.3.1  Use of Sufficient Statistics

Some other interesting methods have been proposed to deal with unknown
parameters. When sufficient statistics are available for $\theta$, Srinivasan (1970,
1971) has suggested using the Kolmogorov statistic D calculated from a comparison of $F_n(x)$ with the estimate $\tilde{F}(x;\hat{\theta})$ obtained by applying the Rao-
Blackwell theorem to $F(x, \hat{\theta})$, where $\hat{\theta}$ is, say, the maximum likelihood
estimator of $\theta$. The resulting tests are asymptotically equivalent to the tests
given in previous sections using $F(x;\hat{\theta})$ itself (Moore, 1973) and can be expected to have similar properties for finite n. The method will usually lead
to complicated calculations, and has been developed only for tests for normality (Srinivasan, 1970; see also Kotz, 1973) and for tests of exponentiality (see
Section 10.8.1).

### 4.16.3.2 The Half-Sample and Related Methods

Another method of eliminating unknown parameters is called the half-sample method; this can be useful when unknown parameters are not location or scale. Unknown components in $\theta$ are estimated by asymptotically efficient methods (for example, maximum likelihood) using only half the given sample, randomly chosen. The estimates, together with any known components, give an estimate $\hat{\theta}$ of the vector $\theta$. The transformation $Z_{(i)} = F(X_{(i)}; \hat{\theta})$, $i = 1$, ..., n, is made, and EDF statistics are calculated from formulas (4.2), now using the whole sample. A remarkable result is that, asymptotically, the EDF statistics will have their Case 0 distributions (Section 4.4), although this will not be true for finite n. Stephens (1978) has examined the half-sample method applied to tests for normality and exponentiality, to compare with the techniques given in Sections 4.8 and 4.9. Several points can be made:

(a) The quadratic statistics $W^2$, $U^2$, and $A^2$, as in other situations, appear to converge fairly rapidly to their asymptotic distributions: this is probably the case for tests for other distributions also, so that for reasonably large (say $n \geq 20$) samples, the half-sample method could be used with the Case 0 asymptotic points.

(b) The half-sample technique is not invariant; different statisticians will obtain different values of the estimates, according to the different possible random half-samples chosen for estimation, and so will get different values of the test statistics.

(c) There is considerable loss in power when the half-sample method is used for tests of normality and exponentiality, compared with using EDF statistics with parameter estimates obtained from the whole sample, as described in Sections 4.8 and 4.9. The powers also tend to vary among the different statistics.

Braun (1980) has also suggested a technique for dealing with unknown parameters. These are first estimated using the whole sample; then the sample itself is randomly divided into several groups and a Case 0 test made on each group separately, using the estimates as though they were true values. This technique can be expected to be valuable only for large samples; see Braun (1980).

It seems clear that the above methods should not be preferred to the techniques previously presented for tests involving unknown location and scale parameters, where the complete sample is used to estimate parameters, but they might be useful for tests for distributions involving shape parameters. More information would be helpful on how the methods compare with other tests with unknown shape parameters, for example, use of $\chi^2$ or its improvements discussed in Chapter 3, or with EDF tests such as the tests for the Weibull or gamma distributions given in Sections 4.11 or 4.12 above.

### 4.16.4 Tests for Symmetry

Tests have been derived for $H_0$: X has a symmetric distribution about a specified median. If the median is a, the transformation $X' = X - a$ gives a sample set which, on $H_0$, will be symmetric with median zero. Hence only this situation need be considered. The test is not strictly a goodness-of-fit test, but a test of the very general hypothesis $F(x) = 1 - F(-x)$. A basic technique is to compare the EDF's of X and -X; the statistics are then based on ranks. Smirnov (1947) and Butler (1969) suggested a modification of the Kolmogorov statistic for this problem. Distribution theory of the Butler-Smirnov test was given by Chatterjee and Sen (1973), and power results were discussed by Koul and Stoudte (1976). Other variations, and methods of obtaining confidence bands, were illustrated by Doksum, Fenstad, and Aaberge (1977); these authors find versions of Kolmogorov statistics which are competitive with EDF statistics and with the Shapiro-Wilk statistic (Section 5.10.3) when used as tests for normality against gamma and log-normal alternatives. Review articles on Kolmogorov-type statistics for symmetry are given by Niederhausen (1982) and Gibbons (1983).

Rothman and Woodroofe (1972) and Srinivasan and Godio (1974) have given test statistics $S_n$ of Cramér-von Mises type for $H_0$; Hill and Rao (1977) showed connections between the two statistics (called there $R_n$ and $S_n$), have generalized them, and finally have proposed a statistic $T_n^{(0)}$ which is based on the generalizations. The statistic has the property that it takes the same value if $X_i$ is replaced by $-X_i$ or if it is replaced by $1/X_i$ for all i. Hill and Rao (1977) gave tables of probabilities in the upper tail for $n^2 T_n^{(0)}/4$, for n from 10 to 24. Lockhart and McLaren (1985) have given asymptotic points for this test.

Use of the EDF to estimate the center of symmetry was discussed by Butler (1969) and by Rao, Schuster, and Littell (1975).

### 4.16.5 Tests Based on the Empirical Characteristic Function

Some authors have proposed goodness-of-fit tests based on the empirical characteristic function (ECF). This is defined, for a random sample $X_1$, $X_2$, ..., $X_n$, by $\phi_n(t) = \{\Sigma_{j=1}^n \exp(it\,X_j)\}/n$, where here $i^2 = -1$; $\phi_n(t)$ converges, as $n \to \infty$, to $\phi(t)$, the characteristic function of the distribution $F(x)$ of X, and the real and imaginary parts of $\phi_n(t)$, say $C_n(t)$ and $S_n(t)$, suitably normalized, are asymptotically jointly normal. Tests of fit can be based on how well $\phi_n(t)$, $C_n(t)$, or $S_n(t)$ correspond to hypothesized values (corresponding to a given distribution) at particular values of t; or they can be of Kolmogorov-Smirnov type or Cramér-von Mises type, based on

$$\sup_t \mid \phi_n(t) - \phi(t) \mid \quad \text{or on} \quad \int_{-\infty}^{\infty} \mid \phi_n(t) - \phi(t) \mid^2 dG(t) ,$$

where G is a suitable measure. Many practical questions remain for such tests, such as the choice of t-values or of G(t); also, tables rarely exist for finite n and power studies are often limited, so that more work is needed in this area. Epps and Pulley (1983) have given a test for normality, with tables, and references to earlier work.

## 4.17 EDF TESTS FOR DISCRETE DISTRIBUTIONS

### 4.17.1 Introduction

The tests given in previous sections have all been developed for various cases in which the tested distribution F(x) was continuous. Historically, the test statistics were introduced with this intention, and the field was left clear to the Pearson chi-square statistic for testing for discrete distributions. However, an EDF can also be drawn for discrete data and it can be compared with the cumulative distribution from which the data are supposed to be drawn; it is then natural to define measures of discrepancy analogous to the statistics given for continuous distributions. Here we examine tests based on such measures. A general review of goodness-of-fit tests for discrete distributions was given by Horn (1977).

Data may appear to be discrete either because the sample genuinely arises from a discrete distribution like the Binomial or Poisson, for example, in measurements of counts, or alternatively because originally continuous data may have been grouped. The grouping may occur because the unit of measurement is very coarse, for example, when angles are measured to the nearest 5 degrees, or weights to the nearest pound or gram. This occurs in the data on leghorn chicks in Table 4.1; the two chicks which are recorded as having weight 190 gm obviously do not possess exactly equal weight, but each weighs somewhere between 189.5 and 190.5 gm. With large amounts of data, grouping may also be done to facilitate display or handling of the data, and the original values, and therefore some information, may be lost before a goodness-of-fit test is to be made. This happens with Monte Carlo sampling, when very many observations will be recorded, and for ease of tabulation will often be graded into a histogram as they are collected.

Of course, in practice all continuous data are subject to the limits of accurate measurement, but the inherent grouping may be so fine as to have negligible effect. This was assumed to be so for the data on chicks when they were tested for normality in Sections 4.4 and 4.8.

### 4.17.2  The EDF for Discrete Data: Case 0

Suppose that for discrete data the possible outcomes are divided into k cells and the null hypothesis is

$H_0$:  P(an observation falls in cell i) = $p_i$,  i = 1, $\ldots$, k

The $p_i$ are assumed given, so that $H_0$ is completely specified, and the situation is Case 0 for discrete distributions. The cell boundaries may be determined by the actual values taken by a random variable X, especially if there are exactly k of these, or some values may be grouped together, as in the tail of a Poisson distribution, to give k cells overall. Suppose n independent observations are taken, and let $O_i$ be the observed number and $E_i$ be the expected number ($E_i = np_i$), in the i-th cell. Define the statistic S by

$$S = \max_{1 \leq j \leq k} \left| \sum_{i=1}^{j} (O_i - E_i) \right|$$

For grouped continuous data, let the cell boundaries, in ascending order, be $c_0$, $c_1$, $\ldots$, $c_k$; cell i contains values X such that $c_{i-1} \leq X < c_i$. If $O_i$ and $E_i$ are the observed and expected values in cell i, the statistic S can be defined as above. Also, an EDF may be defined as

$$F_n(c_j) = \left( \sum_{i=1}^{j} O_i \right) \Big/ n, \quad j = 1, \ldots, k$$

$$F_n(x) = F_n(c_j), \quad c_j \leq x < c_{j+1};$$

$F_n(x)$ is the cumulative histogram of the data. The grouped distribution function $F_g(x)$ may be defined in the same way, by replacing $O_i$ by $E_i$. Then the statistic S is equal to

$$S = n \sup_x |F_n(x) - F_g(x)|$$

and there is an obvious parallel with the Kolmogorov statistic nD. Similarly, a statistic parallel to $W^2$ would be

$$W_d^2 = n^{-1} \sum_{j=1}^{k} \left\{ \sum_{i=1}^{j} (O_i - E_i) \right\}^2$$

and it is possible to construct parallels to the other statistics for continuous distributions.

The value of the statistic S depends on the ordering of the cells so that a different ordering will produce a different value for the same data. It is therefore recommended that S be used when there is a natural ordering of the categories.

Several authors have discussed the statistic S or the statistics $S^+$ and $S^-$ defined by

$$S^+ = \max_{1 \le j \le k} \left\{ \sum_{i=1}^{j} (O_i - E_i) \right\} \quad \text{and} \quad S^- = \max_{1 \le j \le k} \left\{ \sum_{i=1}^{j} (E_i - O_i) \right\}$$

which are analogous to $nD^+$ and $nD^-$, and we confine ourselves to tests for discrete data based on these three statistics.

Pettitt and Stephens (1977) have given exact probabilities for the distribution of S for equal cell probabilities. They also showed how the tables can be used as good approximations for probability distributions of S for unequal probabilities per cell, and also to deduce approximate probabilities for $S^+$ or $S^-$ (see also Conover, 1972). Table 4.28 is taken from Table 1 of Pettitt and Stephens (1977). The table gives values of $P(S \ge m)$, for values of m which give probabilities near the usual test levels. Thus a test of $H_0$ is made as follows:

(a) Record the observed number of observations $O_i$ and the expected number $E_i$, for all i, i = 1, ..., k.

(b) Calculate $T_j = \sum_{i=1}^{j} (O_i - E_i)$, j = 1, ..., k.

(c) Calculate $S^+ = \max_j T_j$, or $S^- = \max_j (-T_j)$, or $S = \max_j |T_j|$; let m be the value of the test statistic used.

(d) Use Table 4.28 to find p-levels, that is $P(S \ge m)$. The p-levels for $S^+$ or $S^-$, that is, $P(S^+ \ge m)$ or $P(S^- \ge m)$ are each approximately $\frac{1}{2} P(S \ge m)$.

(e) If the p-level for the statistic used is less than the test level $\alpha$, reject $H_0$ at significance level.

Statistic S gives a two-sided test and statistics $S^+$ and $S^-$ give one-sided tests.

E4.17.2 Example

The data given in Table 4.29, used by Pettitt and Stephens (1977), are taken from Siegel (1956). Each of ten subjects was presented with five photographs of himself, varying in tone (grades 1-5), and was asked to choose the photograph he liked best. The hypothesis tested was that there was no overall preference for any tone, that is, each tone was equally likely to be chosen. The values of $T_j = \sum_{i=1}^{j} (O_i - E_i)$ are given in the table. The values of $S^+$

TABLE 4.28 Table of Probabilities for EDF Statistic S for a Fully
Specified Discrete Distribution with k Classes (Section 4.17)

k = 3

| n = 6 | m : | 4 | 3 | |
|---|---|---|---|---|
| | | .00274 | .03567 | |

| n = 9 | m : | 5 | 4 | 3 |
|---|---|---|---|---|
| | | .00193 | .01656 | .12361 |

| n = 12 | m : | 6 | 5 | 4 |
|---|---|---|---|---|
| | | .00109 | .00771 | .04994 |

| n = 15 | m : | 6 | 5 | 4 |
|---|---|---|---|---|
| | | .00361 | .02089 | .09181 |

| n = 18 | m : | 6 | 5 | 4 |
|---|---|---|---|---|
| | | .00902 | .04005 | .13579 |

| n = 21 | m : | 7 | 6 | 5 |
|---|---|---|---|---|
| | | .00402 | .01760 | .06308 |

| n = 24 | m : | 7 | 6 | 5 |
|---|---|---|---|---|
| | | .00792 | .02897 | .08824 |

| n = 27 | m : | 7 | 6 | 5 |
|---|---|---|---|---|
| | | .01325 | .04245 | .011433 |

| n = 30 | m : | 8 | 7 | 6 |
|---|---|---|---|---|
| | | .00609 | .02015 | .05757 |

k = 4

| n = 8 | m : | 4 | 3 | |
|---|---|---|---|---|
| | | .01514 | .10791 | |

| n = 12 | m : | 5 | 4 | |
|---|---|---|---|---|
| | | .01115 | .05974 | |

| n = 16 | m : | 6 | 5 | 4 |
|---|---|---|---|---|
| | | .00706 | .03299 | .12611 |

| n = 20 | m : | 7 | 6 | 5 |
|---|---|---|---|---|
| | | .00424 | .01826 | .06598 |

| n = 24 | m : | 7 | 6 | 5 |
|---|---|---|---|---|
| | | .01014 | .03526 | .10519 |

| n = 28 | m : | 8 | 7 | 6 |
|---|---|---|---|---|
| | | .00566 | .01914 | .05689 |

k = 5

| n = 10 | m : | 5 | 4 | |
|---|---|---|---|---|
| | | .00477 | .04162 | |

| n = 15 | m : | 6 | 5 | 4 |
|---|---|---|---|---|
| | | .00584 | .03202 | .12322 |

| n = 20 | m : | 7 | 6 | 5 |
|---|---|---|---|---|
| | | .00496 | .02203 | .07617 |

| n = 25 | m : | 8 | 6 | 5 |
|---|---|---|---|---|
| | | .00368 | .04717 | .13083 |

| n = 30 | m : | 8 | 7 | 6 |
|---|---|---|---|---|
| | | .00946 | .02930 | .07924 |

k = 6

| n = 12 | m : | 6 | 5 | 4 |
|---|---|---|---|---|
| | | .00173 | .01422 | .08064 |

| n = 18 | m : | 7 | 6 | 5 |
|---|---|---|---|---|
| | | .00308 | .01599 | .06435 |

| n = 24 | m : | 7 | 6 | 5 |
|---|---|---|---|---|
| | | .01375 | .04695 | .13203 |

| n = 30 | m : | 8 | 7 | 6 |
|---|---|---|---|---|
| | | .01071 | .13317 | .08836 |

k = 7

| n = 14 | m : | 6 | 5 | 4 |
|---|---|---|---|---|
| | | .00511 | .02996 | .12856 |

| n = 21 | m : | 7 | 6 | 5 |
|---|---|---|---|---|
| | | .00807 | .03242 | .10550 |

| n = 28 | m : | 8 | 7 | 6 |
|---|---|---|---|---|
| | | .00853 | .02828 | .08047 |

k = 8

| n = 16 | m : | 6 | 5 | |
|---|---|---|---|---|
| | | .01122 | .05166 | |

| n = 24 | m : | 8 | 7 | 6 |
|---|---|---|---|---|
| | | .00410 | .01641 | .05477 |

(continued)

TABLE 4.28 (continued)

---

$k = 9$

| $n = 18$ m : | 7 | 6 | 5 |
|---|---|---|---|
| | .00406 | .02043 | .07840 |

| $n = 27$ m : | 8 | 7 | 6 |
|---|---|---|---|
| | .00833 | .02831 | .08210 |

$k = 10$

| $n = 20$ m : | 7 | 6 | 5 |
|---|---|---|---|
| | .00781 | .03276 | .10909 |

| $n = 30$ m : | 9 | 7 | 6 |
|---|---|---|---|
| | .00421 | .04365 | .11333 |

---

[a]For given n and k, the table gives values of $P(S \geq m)$ beneath values of m. The probabilities given are exact for cells of equal probability. Half the tabulated probability is a good approximation to $P(S^+ \geq m) = P(S^- \geq m)$. Taken from Pettitt and Stephens (1977), with permission of the American Statistical Association.

TABLE 4.29 Data for EDF Test for a Discrete Distribution[a]

---

| Tone grade j of chosen photograph | Number choosing grade j: $O_j$: | Expected number: $E_j$: | $T_j = \sum_{i=1}^{j} (O_i - E_i)$ |
|---|---|---|---|
| 1 | 0 | 2 | −2 |
| 2 | 1 | 2 | −3 |
| 3 | 0 | 2 | −5 |
| 4 | 5 | 2 | −2 |
| 5 | 4 | 2 | 0 |

---

[a]The first column gives the five grades of tone of a photograph, and the data in Column 2 are the numbers out of 10 persons in an experiment who chose the different tone grades. (See Section 4.17.)

and $S^-$ are respectively 0 and 5, and the value of S is 5. From Table 4.29 for $n = 10$, $k = 5$, we have $P(S \geq 5) = 0.00477$, so S is highly significant, with p-level less than .005, and $H_0$ will be rejected. The Pearson statistic

$$X^2 = \sum_{i=1}^{k} (O_i - E_i)^2 / E_i$$

has the value 11. Using the usual $\chi_4^2$ approximation, $P(X^2 \geq 11) = 0.024$, while by exact enumeration the probability is 0.04. The S statistic thus gives a much more extreme value than does $X^2$, and appears to be more sensitive in this instance. Pettitt and Stephens have investigated the power of S, espe-cially against alternatives representing a trend in cell probability as the cell index i increases, and it appears that for such alternatives, S will often be more powerful than $X^2$.

Note that Case 0 tables for nD should not be used for S, despite the parallel between the two statistics. Noether (1963) suggested that use of the nD tables would give a conservative test; Pettitt and Stephens have given several examples to show this to be true, with the true $\alpha$-value very different from the supposed value.

The test for S has been given above for Case 0 where the null hypothesis is completely specified. The analogue of S is not available for the various cases where probabilities for each cell must be estimated, for example, in a test for a Poisson or binomial distribution, where an unknown parameter must be estimated from the data.

Wood and Altavela (1978) have discussed asymptotic properties of Kolmogorov-Smirnov statistics $D^+$, $D^-$, and D when used with discrete dis-tributions, and have shown how asymptotic percentage points may be simulated.

## 4.18  COMBINATIONS OF TESTS

### 4.18.1  Introduction

Suppose k independent statistical tests are made. It may be that the p-levels are quite small, but not small enough to be significant. If the k tests are all tests of similar type—for example, all tests for normality of similar data, with small samples of each—the results may suggest, overall, that the data are non-normal, but the samples are too small to detect this. It then becomes desirable to combine the tests. The general problem of combining tests, even of different types, has been discussed by many authors; see, for example, Fisher (1967), Birnbaum (1954), and Volodin (1965). Fisher (1967) suggested an easy method of combination, based on the p-levels of the k separate test statistics. In effect, the p-levels are tested for uniformity. This method is discussed in Section 8.15 and has been used to combine various tests for normality by Wilk and Shapiro (1968) and by Pettitt (1977a). Volodin (1965) has discussed tests for one distribution (the normal, exponential, Poisson, or Weibull) against specific alternatives which are close to the one tested.

### 4.18.2 Combining EDF Test Statistics: Case 0

In this section we give a method of combining test statistics obtained from EDF tests. Suppose $A^2$ is the test statistic in a Case 0 test, and let test j give value $A_j^*$: the proposed test statistic is $Z_k = \sum_{j=1}^{k} A_j^*/k$. For Case 0, $A^2$ is chosen as test statistic since, as was stated in Section 4.4, its distribution function almost does not depend on n. A table of percentage points for $Z_k$, for Case 0 tests, is given in Table 4.30. When k is too large for the table, a good approximation to the percentage point of $Z_k$ is given by the corresponding point for a normal distribution with mean $\mu$ and variance $\sigma^2/k$, where $\mu$ and $\sigma^2$ are given in the last line of the table.

### E 4.18.2 Example

Suppose six Case 0 tests for normality are made, and the values of $A^2$ are 2.353, 1.526, 0.550, 0.252, 2.981, 2.309. The p-levels of the tests are, from Table 4.3: 0.06, 0.17, 0.70, 0.97, 0.03, 0.06. The average is $Z_6 = 1.662$, and reference to Table 4.30 shows $Z_6$ to be significant at the 5% level. Although only one component test is significant at the 5% level, the overall combination suggests a total picture of non-normality.

TABLE 4.30  Table for Combining Tests for k Samples, Case 0 (Section 4.18.2)[a]

| No. of samples k | Significance level $\alpha$ | | | | |
|---|---|---|---|---|---|
|  | .25 | .10 | .05 | .025 | .01 |
| 2 | 1.242 | 1.705 | 2.047 | 2.387 | 2.838 |
| 3 | 1.219 | 1.582 | 1.842 | 2.096 | 2.427 |
| 4 | 1.200 | 1.506 | 1.721 | 1.928 | 2.195 |
| 6 | 1.173 | 1.414 | 1.578 | 1.735 | 1.934 |
| 8 | 1.155 | 1.358 | 1.495 | 1.624 | 1.786 |
| 10 | 1.142 | 1.320 | 1.439 | 1.550 | 1.689 |
| $\infty$ | | $\mu = 1.000$ | $\sigma^2 = 0.57974/k$ | | |

[a]Upper tail percentage points of $Z = \sum_j A_j^*/k$, where $A_j^*$ is the value of $A^2$ for sample j, case 0. For k > 10, Z is approximated by a normal distribution with $\mu$ and $\sigma^2$ as shown.

TABLE 4.31   Table for Combining Tests for k Samples,
Parameters Unknown (Section 4.18.3)[a]

| Test | No. of samples k | Significance level $\alpha$ | | | | |
|------|------|------|------|------|------|------|
| | | .25 | .10 | .05 | .025 | .01 |
| Normal | 2 | 1.081 | 1.537 | 1.878 | 2.220 | 2.674 |
| Case 2 | 3 | 1.065 | 1.422 | 1.680 | 1.934 | 2.265 |
| | 4 | 1.050 | 1.350 | 1.562 | 1.768 | 2.035 |
| | 6 | 1.027 | 1.262 | 1.424 | 1.579 | 1.777 |
| | 8 | 1.010 | 1.208 | 1.343 | 1.470 | 1.631 |
| | 10 | 0.998 | 1.172 | 1.289 | 1.398 | 1.535 |
| | $\infty$ | (See Table 4.30: $\mu = 0.8649$, $\sigma^2 = 0.5303/k$) | | | | |
| Normal | 2 | 0.455 | 0.562 | 0.638 | 0.710 | 0.805 |
| Case 3 | 3 | 0.446 | 0.530 | 0.588 | 0.643 | 0.713 |
| | 4 | 0.439 | 0.510 | 0.559 | 0.604 | 0.661 |
| | 6 | 0.431 | 0.487 | 0.524 | 0.559 | 0.602 |
| | 8 | 0.425 | 0.473 | 0.504 | 0.533 | 0.569 |
| | 10 | 0.421 | 0.463 | 0.490 | 0.515 | 0.546 |
| | $\infty$ | (See Table 4.30: $\mu = 0.3843$, $\sigma^2 = 0.3615/k$) | | | | |
| Exponential | 2 | 0.723 | 0.942 | 1.102 | 1.260 | 1.469 |
| Case 1 | 3 | 0.708 | 0.881 | 1.003 | 1.122 | 1.276 |
| | 4 | 0.698 | 0.843 | 0.945 | 1.042 | 1.166 |
| | 6 | 0.683 | 0.798 | 0.876 | 0.950 | 1.043 |
| | 8 | 0.673 | 0.771 | 0.836 | 0.897 | 0.973 |
| | 10 | 0.666 | 0.752 | 0.809 | 0.861 | 0.927 |
| | $\infty$ | (See Table 4.30: $\mu = 0.5959$, $\sigma^2 = 0.1392/k$) | | | | |

[a]Upper tail percentage points of $Z_k = \Sigma_j A_j^*/k$ where $A_j^*$ is the modified $A^2$ for sample j. Values are given for the test of normality with mean zero and variance unknown (Case 2 of Section 4.8), the test for normality with mean and variance unknown (Case 3 of Section 4.8), and for the test for exponentiality (Section 4.9, Case 2).

### 4.18.3 Combining EDF Test Statistics: Other Cases

The same technique can be applied to combine tests of fit when parameters are estimated. Here each value of $A^2$ will be modified, as described in previous sections, for the appropriate test, and then the mean $Z_k$ of the modified values $A_j^*$, $j = 1, \ldots, k$, will be taken as the overall test statistic. Upper tail percentage points of $Z_k$, for tests of normality and for tests of exponentiality, are in Table 4.31. For k too large for the table, follow the same procedure as described in Section 4.18.2, for Table 4.30.

### E 4.18.3 Example

Proschan (1963) has given a number of sets of failure times of airconditioning equipment for several aircraft. The data for aircraft 7910 have already been listed in Table 4.13. When EDF tests of exponentiality are made on the sets for the first 6 aircraft, with $\alpha = 0$ and $\beta$ estimated separately for each aircraft, the values of A* (that is, $A^2$ modified as in Table 4.11) are: 0.543, 0.722, 0.763, 1.187, 0.499, 1.175, and the mean A* is 0.814; reference to Table 4.31 shows this value to be near the 10% point, for k = k, although only two of the individual values are significant at the 10% level.

### 4.18.4 Combining the Standardized Values
###          from Several Tests

Pierce (1978) has suggested the following method of combining tests based on k samples, for testing $H_0$: the sample comes from a distribution $F(x; \theta)$, with $\theta$ containing unknown location and/or scale parameters $\alpha$ and $\beta$. The true values of these parameters may be different for each test. For sample i, let $\hat{\alpha}_i$ and $\hat{\beta}_i$ be the maximum likelihood estimates. Define <u>standardized values</u> $W_{ri} = (X_{ri} - \hat{\alpha}_i)/\hat{\beta}_i$, $r = 1, \ldots, n_i$, where $X_{ri}$, $r = 1, \ldots, n_i$, are the observations in sample i. The proposal of Pierce is that the $W_{ri}$ for all the k samples should be pooled to form one large sample of size $n = \sum_{i=1}^{k} n_i$.

Pierce showed that the limiting distribution of any EDF statistic calculated from the combined sample will be the same as its limiting distribution for one sample. Similar results apply if only one parameter is not known. Pierce gave results of a Monte Carlo study on tests for normality (Case 3), comparing the method, using $A^2$ calculated from the k pooled samples of standardized values, with Fisher's method using the p-levels of k values of $A^2$. The alternative, that is, the true distribution, was a Weibull distribution, and the pooled method is more sensitive in this study. Quesenberry, Whitaker, and Dickens (1976) have given another method, somewhat similar but less direct, of combining tests for normality.

## 4.19  EDF STATISTICS AS INDICATORS OF PARENT POPULATIONS

Statisticians sometimes use goodness-of-fit statistics to decide which population appears best suited to describe a data set. EDF statistics may be used for this purpose; when the different parent populations $F_i(x)$ are fully specified (Case 0), a typical statistic, say $A^2$, is calculated assuming $F_i(x)$ to be correct, giving value $A_i^2$, and values of $A_i^2$ may be directly compared. A smaller value of $A_i^2$ will indicate a better fit than a larger value. However, when parameters are estimated from the data, the value of $A_i^2$ must not be the indicator, since the distribution of $A_i^2$ will now vary with $F_i(x)$; instead, the $p_i$-value attached to $A_i^2$ will be a suitable indicator, with a larger $p_i$-value (measured from the upper-tail) indicating a better fit.

## 4.20  TESTS BASED ON NORMALIZED SPACINGS

### 4.20.1  Normalized Spacings and the EDF Statistic $A_s^2$

In this section tests are discussed based on the spacings of a sample. Each spacing is normalized by division of a constant and a transformation made to produce z-values between 0 and 1. We give the test based on the EDF statistic $A_s^2$, the Anderson-Darling $A^2$ calculated from these z-values, and compare it with tests based on the median or the mean of the values. The technique affords an interesting method of testing by eliminating location and scale parameters, rather than directly estimating them, and is based on tests for the exponential distribution. The tests can be used for censored data. The general case will be treated of a sample which has been censored at both ends. Suppose $t + 2$ successive observations $X_{(k)}, X_{(k+1)}, \ldots,$ $X_{(k+t+1)}$ are given, and the test is of the null hypothesis:

> $H_0$: the original X-sample comes from a continuous distribution $F(x; \alpha, \beta)$, where $\alpha$ and $\beta$ are unknown location and scale parameters

Then values X can be viewed as being constructed from a random sample W from $F(x; 0, 1)$, by the relation $X_{(i)} = \alpha + \beta W_{(i)}$, $i = k, k + 1, \ldots, k + t + 1$. Suppose $m_i = E(W_{(i)})$, and define spacings $E_i = X_{(k+i)} - X_{(k+i-1)}$, $i = 1, \ldots,$ $t + 1$; these are analogous to the spacings $D_i$ of Section 8.2 or the spacings $E_i$ of Section 10.5.2. Then <u>normalized spacings</u> are given by

$$y_i = E_i / (m_{k+i} - m_{k+i-1}), \quad i = 1, \ldots, t + 1$$

When the $X_{(i)}$ are from the exponential distribution $Exp(\alpha, \beta)$, given in Section 4.9.1, the $y_i$ will be i.i.d. $Exp(0, \beta)$. The J transformation of

Section 10.5.4 can then be applied to give values $z_{(i)}$:

$$z_{(i)} = \sum_{j=1}^{i} y_j \bigg/ \sum_{j=1}^{t+1} y_j, \quad i = 1, \ldots, t$$

and the $z_{(i)}$ will be ordered uniforms, that is, they will be distributed as an ordered sample of size t from the uniform distribution with limits 0 and 1.

A special case of normalized spacings is the set X' derived, as in Section 10.5.2, by the N-transformation applied to a complete sample of X-values from $Exp(0, \beta)$; here an extra spacing is available between $X_{(1)}$ and the known lower endpoint $\alpha = 0$ of the distribution of X. The spacings are $E_1 = X_{(1)}$, $E_2 = X_{(2)} - X_{(1)}$, $E_3 = X_{(3)} - X_{(2)}$, etc., and, for the $Exp(0, 1)$ distribution, $m_i - m_{i-1} = 1/(n + 1 - i)$, with $m_0 \equiv 0$; thus the normalized spacings are the values $X_i'$ in Section 10.5.2. Tests that the original sample is exponential can then be based on statistics used for testing that the $z_{(i)}$ are ordered uniforms, for example, EDF statistics, Case 0 (Section 4.4), or any of the test statistics described in Chapter 8. Tests of this type, for exponentiality of the original X, are discussed in Chapter 10.

These tests can be adapted to test that X came from a more general $F(x; \alpha, \beta)$ as follows. The spacings are calculated as described above, and, provided values of $k_i = m_i - m_{i-1}$ are known, normalized spacings $y_i$ can be found; then the transformation to $z_{(i)}$ can be made. Then, subject to important conditions particularly affecting the extreme spacings, suitably separated normalized spacings from any continuous distribution are asymptotically independent and exponentially distributed with mean $\beta$ (see Pyke, 1965, for rigorous and detailed results). The conditions on this result are sufficiently strong, however, that the transformed values $z_{(i)}$ must not be assumed to be distributed as uniform order statistics, even asymptotically, for the purpose of finding distributions of test statistics. However, asymptotic distribution theory of three statistics calculated from the z-values has been given by Lockhart, O'Reilly, and Stephens (1985, 1986).

### 4.20.2 Tests for the Normal, Logistic, and Extreme-Value Distributions Based on $A_s^2$

We first discuss statistic $A_s^2$; this is the Anderson-Darling statistic $A^2$ defined in equation (4.2), but using the $z_{(i)}$ above derived from the spacings. Later we discuss statistics based on the median and the mean of the z-values.

Asymptotic percentage points for $A_s^2$, for use with tests that the $X_i$ are from one of the normal, logistic, or extreme-value distributions, are given in Table 4.32; these are taken from Lockhart, O'Reilly, and Stephens (1985, 1986). Thus to make a test for normality, $p = k/n$ and $q = (k + t + 1)/n$ are found, and Table 4.32 is entered on the line corresponding to p and q; the null hypothesis that the original $X_i$ are normal is rejected if $A_s^2$ exceeds the percentage point given for the desired test level $\alpha$.

TABLE 4.32  Asymptotic Percentage Points for $A_s^2$, for Tests for Normal, Logistic, or Extreme-Value Populations (Section 4.20). The table is entered at $p = k/n$ and $q = (k + t + 1)/n$.

| Left censoring point, p | Right censoring point, q | Significance level $\alpha$ | | | | | | |
|---|---|---|---|---|---|---|---|---|
| | | 0.25 | 0.20 | 0.15 | 0.10 | 0.05 | 0.025 | 0.01 |
| | | | Normal Distribution | | | | | |
| 0 | 1 | 0.955 | 1.066 | 1.211 | 1.422 | 1.798 | 2.191 | 2.728 |
| 0 | 0.75 | 1.056 | 1.183 | 1.350 | 1.592 | 2.026 | 2.479 | 3.100 |
| 0 | 0.50 | 1.098 | 1.232 | 1.409 | 1.667 | 2.129 | 2.612 | 3.273 |
| 0 | 0.25 | 1.133 | 1.273 | 1.459 | 1.730 | 2.215 | 2.722 | 3.416 |
| 0.25 | 0.75 | 1.178 | 1.324 | 1.518 | 1.800 | 2.306 | 2.835 | 3.559 |
| 0.25 | 0.50 | 1.225 | 1.381 | 1.587 | 1.889 | 2.430 | 2.996 | 3.770 |
| | | | Logistic Distribution | | | | | |
| 0 | 1 | 1.123 | 1.263 | 1.448 | 1.720 | 2.206 | 2.716 | 3.413 |
| 0 | 0.75 | 1.141 | 1.281 | 1.468 | 1.741 | 2.230 | 2.741 | 3.441 |
| 0 | 0.50 | 1.178 | 1.325 | 1.521 | 1.806 | 2.318 | 2.852 | 3.584 |
| 0 | 0.25 | 1.215 | 1.369 | 1.574 | 1.873 | 2.409 | 2.969 | 3.736 |
| 0.25 | 0.75 | 1.177 | 1.323 | 1.517 | 1.801 | 2.308 | 2.838 | 3.564 |
| 0.25 | 0.50 | 1.223 | 1.378 | 1.584 | 1.885 | 2.424 | 2.989 | 3.761 |
| | | | Extreme-Value Distribution | | | | | |
| 0 | 1 | 1.016 | 1.138 | 1.300 | 1.535 | 1.957 | 2.398 | 3.000 |
| 0.0 | 0.75 | 1.159 | 1.302 | 1.492 | 1.770 | 2.267 | 2.787 | 3.498 |
| 0.0 | 0.50 | 1.202 | 1.354 | 1.555 | 1.849 | 2.376 | 2.927 | 3.682 |
| 0.0 | 0.25 | 1.229 | 1.386 | 1.594 | 1.898 | 2.444 | 3.015 | 3.797 |
| 0.25 | 1.0 | 1.027 | 1.150 | 1.312 | 1.549 | 1.972 | 2.413 | 3.018 |
| 0.25 | 0.75 | 1.187 | 1.336 | 1.532 | 1.819 | 2.333 | 2.870 | 3.605 |
| 0.25 | 0.50 | 1.231 | 1.388 | 1.596 | 1.901 | 2.447 | 3.018 | 3.800 |
| 0.50 | 1.0 | 1.051 | 1.177 | 1.345 | 1.589 | 2.025 | 2.481 | 3.105 |
| 0.50 | 0.75 | 1.224 | 1.380 | 1.586 | 1.887 | 2.428 | 2.993 | 3.767 |
| 0.75 | 1.0 | 1.081 | 1.213 | 1.387 | 1.641 | 2.096 | 2.571 | 3.222 |

[a]The table is entered at $p = k/n$ and $q = (k + t + 1)/n$.

Monte Carlo studies suggest that, as in previous sections, the points may be used to give a good approximate test for, say, $n \geq 20$. The tables can be interpolated for values of p and q not given.

The test for the extreme-value distribution above refers to distribution (4.7); for a test that X comes from (4.6) the substitution $X' = -X$ must be made, and the $X'$ tested to come from (4.7). The test may be adapted to a test that X comes from the two-parameter Weibull distribution $W(x;0,\beta,m)$ of Section 4.11 by taking $X' = \log X$ and testing that $X'$ comes from (4.7).

In the above tests, values of $m_i$ (or more precisely, of $k_i = m_{i+1} - m_i$) are needed. For the normal distribution values of $m_i$ have been extensively tabulated or can be accurately calculated; see Section 5.7.2. For the logistic distribution, $k_i = n/\{(i-1)(n-i+1)\}$, $i = 2, \ldots, n$. For the extreme-value distribution, values of $m_i$ have been given by Mann (1968) and by White (1968)} also values of $k_i$ are tabulated, for $3 \leq n \leq 25$, by Mann, Scheuer, and Fertig (1973). Alternatively the approximation $m_i = \log[-\log\{1 - (i - 0.5)/(n + 0.25)\}]$ is quite accurate for $n \geq 10$ (Lawless, 1982).

### 4.20.3 Tests Based on the Median and the Mean of the Transformed Spacings

Although they are not properly EDF statistics, it is convenient to discuss here tests based on the median and the mean of the $z_{(i)}$. The test based on the median was introduced by Mann, Scheuer, and Fertig (1973) for the special problem of testing that an original right-censored sample of t-values comes from the two parameter Weibull distribution $W(t;0,\beta,m)$ (Section 4.11) against the three-parameter alternative $W(t;\alpha,\beta,m)$, with $\alpha > 0$. The test situation arises in reliability theory. To make the test, the transformation $X = \log t$ is made as described above, and the right-censored sample of X-values is tested to come from the extreme-value distribution (4.7). Suppose therefore that the r smallest order statistics of an X-sample of size n are available, giving $r - 1$ normalized spacings $y_i$; the statistic proposed is $S = \sum_{i=s}^{r-1} y_i/T$, where $T = \sum_{i=1}^{r-1} y_i$, and where $s = (r+1)/2$ if r is odd and $s = (r+2)/2$ if r is even. It is easily shown that $S = 1 - z_{(w)}$, where $w = s - 1$, so that S is equivalent to $z_{(w)}$, which is essentially the median of the z-values. Mann, Scheuer, and Fertig (1973) give tables of percentage points for S, based on Monte Carlo studies, for $n = 3(1)25$, and for $r = 3(1)n$. For the problem they were studying, the authors proposed a one-tail test: $H_0$ is rejected if S is too large (so the median is too small). An example of the S-test, with censored data, is given in Section 12.4.1. Mann and Fertig (1975) later modified the S statistic, by choosing a smaller value of s. They also showed how to obtain confidence intervals for the origin of a three-parameter Weibull distribution by choosing those values which gave non-significant test results.

Another possible test statistic for $H_0$ is $\bar{Z}$, the mean of the $z_{(i)}$. Tiku and Singh (1981) gave a test, also for the extreme-value distribution, with double censoring; the statistic is their $Z^*$. On $H_0$, the authors gave an accurate normal approximation for $Z^*$ with mean 1, and with variance dependent on the variances and covariances of standard extreme-value order statistics. It is easily shown that $Z^* = 2\bar{Z}$, so that the Tiku-Singh approximation can easily be adapted to give percentage points for $\bar{Z}$. The distributions of S and $\bar{Z}$, when the X come from a suitably regular parent population, have been investigated by Lockhart, O'Reilly, and Stephens (1985, 1986). Both

statistics are asymptotically normal; $S_1 = C_1 t^{\frac{1}{2}}(S - 0.5)$ and $Z_1 = C_2 t^{\frac{1}{2}}(\bar{Z} - 0.5)$ are asymptotically $N(0, 1)$, where $C_1$ and $C_2$ are constants dependent on the parent population and on the censoring levels.

For example, for the extreme-value test, for complete samples, $C_1 = 2.233$, and $C_2 = 4.0098$. If the $z_{(i)}$ could be assumed to be ordered uniforms, S would have a beta distribution and $\bar{Z}$ would have a distribution tabulated in Chapter 8; asymptotically these also tend to normal distributions of the type given above, with $C_1 = 2$ for S and $C_2 = \sqrt{12} = 3.464$ for $\bar{Z}$. Mann, Scheuer, and Fertig found the beta distribution to give a good approximation to their Monte Carlo points for small n, and suggested its use for n beyond their tables; however, comparison of the normal approximations above shows that use of the beta distribution for large n will give a test which is too conservative.

## 4.20.4 Power Properties

Monte Carlo power studies by Mann, Scheuer, and Fertig showed that S has good power properties for the problem they were considering where S was used with one tail. Other power studies for the extreme-value test have been given by Littell, McClave, and Offen (1979), Tiku and Singh (1981), and Lockhart, O'Reilly, and Stephens (1986). These show that $A_8^2$ and $\bar{Z}$ have high power, often better than S. For some alternatives, S will be biased; we return to this point below.

Tiku (1981) has investigated $S^*$, equivalent to $\bar{Z}$, in testing for normality. Lockhart, O'Reilly, and Stephens (1985) have made further comparisons, involving $A_8^2$, S, $\bar{Z}$, the $A^2$ (Case 3) test of Section 4.8, and the Shapiro-Wilk W test discussed in Chapter 5; here S must be used with two tails to cover reasonable alternatives. The most powerful tests for normality are given by W, $A^2$ (Case 3), and $A_8^2$; these three give quite similar results.

Lockhart, O'Reilly, and Stephens have also shown that statistics S and $\bar{Z}$ can give non-consistent tests for some alternatives to the null; for example, this is the case in testing for normality. Statistic S may also be biased, as was observed by Tiku and Singh (1981), although for the problem discussed by Mann, Scheuer, and Fertig (1973) this does not appear to be the case. It would seem that statistic $A_S^2$, which is consistent, should be preferred to $\bar{Z}$ and S except perhaps for some situations in which the alternatives to the null are very carefully specified to avoid problems of bias and non-consistency. $A_S^2$ has good power in studies so far reported. It can be easily calculated without direct estimation of parameters, can be used for censored data, and is consistent. These properties suggest that $A_S^2$, and possibly other EDF statistics found from normalized spacings, might prove useful in other test situations, rivaling the regular use of EDF statistics as described in the rest of this chapter.

REFERENCES

Ali, M. M. and Chan, L. K. (1964). On Gupta's estimates of the parameters of the normal distribution. Biometrika 51, 498-501.

Anderson, T. W. and Darling, D. A. (1952). Asymptotic theory of certain goodness-of-fit criteria based on stochastic processes. Ann. Math. Statist. 23, 193-212.

Anderson, T. W. and Darling, D. A. (1954). A test of goodness-of-fit. J. Amer. Statist. Assoc. 49, 765-769.

Anscombe, F. J. and Tukey, J. W. (1963). The examination and analysis of residuals. Technometrics 5, 141-160.

Barr, D. R. and Davidson, T. (1973). A Kolmogorov-Smirnov test for censored samples. Technometrics 15, 739-757.

Berk, R. H. and Jones, D. H. (1979). Goodness-of-fit test statistics that dominate the Kolmogorov statistics. Z. Wahrsch. Verw. Gebiete 47, 47-59.

Birnbaum, A. (1954). Combining independent tests of significance. J. Amer. Statist. Assoc. 49, 559-574.

Birnbaum, Z. W. and Lientz, B. P. (1969a). Exact distributions for some Renyi type statistics. Applicationes Mathematicae 10, 179-192.

Birnbaum, Z. W. and Lientz, B. P. (1969b). Tables of critical values of some Renyi type statistics for finite sample sizes. J. Amer. Statist. Assoc. 64, 870-877.

Bliss, C. (1967). Statistics in Biology: Statistical Methods for Research in the Natural Sciences. New York: McGraw-Hill.

Blom, G. (1958). Statistical Estimates and Transformed Beta Variates. New York: Wiley.

Braun, H. (1980). A simple method for testing goodness of fit in the presence of nuisance parameters. J. Roy. Statist. Soc., B 42, 53-63.

Butler, C. C. (1969). A test for symmetry using the sample distribution function. Ann. Math. Statist. 40, 2209-2210.

Chandra, M., Singpurwalla, N. D., and Stephens, M. A. (1981). Kolmogorov statistics for tests of fit for the extreme-value and Weibull distributions. J. Amer. Statist. Assoc. 76, 729-731.

Chatterjee, S. K. and Sen, P. K. (1973). On Kolmogorov-Smirnov type tests for symmetry. Ann. Inst. Statist. Math. 25, 287-300.

Chernoff, H., Gastwirth, J. L., and Johns, M. V. (1967). Asymptotic distribution of linear combinations of functions of order statistics with applications to estimation. Ann. Math. Statist. 38, 52-73.

Conover, W. J. (1972). A Kolmogorov goodness of fit test of discontinuous distributions. J. Amer. Statist. Assoc. 67, 591-596.

Csörgő, S. and Horvath, L. (1981). On the Koziol-Green model for random censorship. Biometrika 68, 391-401.

Doksum, K. A. (1975). Measures of location and asymmetry. Scand. J. Statist. 2, 11-22.

Doksum, K. A., Fenstad, G., and Aaberge, R. (1977). Plots and tests for symmetry. Biometrika 64, 473-487.

Draper, N. R. and Smith, H. (1966). Applied Regression Analysis. New York: Wiley.

Dufour, R. and Maag, U. R. (1978). Distribution results for modified Kolmogorov-Smirnov statistics for truncated or censored data. Technometrics 20, 29-32.

Durbin, J. (1973). Distribution Theory for Tests Based on the Sample Distribution Function. Regional Conference Series in Appl. Math., No. 9, SIAM, Philadelphia, Pa.

Durbin, J. (1975). Kolmogorov-Smirnov tests when parameters are estimated with applications to tests of exponentiality and tests on spacings. Biometrika 62, 5-22.

Durbin, J. and Knott, M. (1972). Components of Cramér-Von Mises statistics, I. J. Roy. Statist. Soc., B 34, 290-307.

Durbin, J., Knott, M., and Taylor, C. C. (1975). Components of Cramér-Von Mises statistics, II. J. Roy. Statist. Soc., B 37, 216-237.

Dyer, A. R. (1974). Comparisons of tests for normality with a cautionary note. Biometrika 61, 185-189.

Easterling, R. G. (1976). Goodness-of-fit and parameter estimation. Technometrics 18, 1-9.

Epps, T. W. and Pulley, L. B. (1983). A test for normality based on the empirical characteristic function. Biometrika 70, 723-728.

Finkelstein, J. M. and Schafer, R. E. (1971). Improved goodness-of-fit tests. Biometrika 58, 641-645.

Fisher, R. A. (1967). Statistical Methods for Research Workers, 4th Edition. New York: Stechert.

Gibbons, J. D. (1983). Kolmogorov-Smirnov symmetry test. In Encyclopaedia of Statistical Sciences, S. Kotz, N. L. Johnson, and C. B. Read, Eds., 4, 396-398.

Gillespie, M. J. and Fisher, L. (1979). Confidence band for the Kaplan-Meier survival curve estimate. Ann. Statist. 4, 920-924.

Green, J. R. and Hegazy, Y. A. S. (1976). Powerful modified EDF goodness-of-fit tests. J. Amer. Statist. Assoc. 71, 204-209.

Gupta, A. K. (1952). Estimation of the mean and standard deviation of a normal population from a censored sample. Biometrika 39, 266-273.

Hall, W. J. and Wellner, J. A. (1980). Confidence band for a survival curve from censored data. Biometrika 67, 133-143.

Hegazy, Y. A. S. and Green, J. R. (1975). Some new goodness-of-fit tests using order statistics. Appl. Statist. 24, 299-308.

Hill, D. L. and Rao, P. V. (1977). Tests of symmetry based on Cramér-Von Mises statistics. Biometrika 64, 489-494.

Horn, S. D. (1977). Goodness-of-fit tests for discrete data; a review and an application to a health impairment scale. Biometrics 33, 237-248.

Johnson, N. L. and Kotz, S. (1970). Distributions in Statistics, Volume 1: Continuous Univariate Distributions. Boston: Houghton Mifflin.

Kaplan, E. L. and Meier, P. (1958). Non-parametric estimation from incomplete observations. J. Amer. Statist. Assoc. 53, 457-481.

Kleinbaum, D. G. and Kupper, L. L. (1978). Applied Regression Analysis and Other Multivariable Methods. Belmont, Calif.: Duxbury Press.

Kolmogorov, A. N. (1933). Sulla determinazione empirica di una legge di distibuziane. Giorna. Ist. Attuari. 4, 83-91.

Kotz, S. (1973). Normality versus lognormality with applications. Comm. Statist. 1, 113-132.

Koul, H. L. and Staudte, R. G., Jr. (1976). Power bounds for a Smirnov statistic in testing the hypothesis of symmetry. Ann. Statist. 4, 924-935.

Koziol, J. A. (1980). Goodness-of-fit tests for randomly censored data. Biometrika 67, 693-696.

Koziol, J. A. and Byar, D. P. (1975). Percentage points of the asymptotic distributions of one and two sample k-s statistics for truncated or censored data. Technometrics 17, 507-510.

Koziol, J. A. and Green, S. B. (1976). A Cramér-von Mises statistic for randomly censored data. Biometrika 63, 465-474.

Kuiper, N. H. (1960). Tests concerning random points on a circle. Proc. Koninkl. Neder. Akad. van. Wetenschappen, A 63, 38-47.

Lawless, J. F. (1982). Statistical Models and Methods of Lifetime Data. New York: Wiley.

Lewis, P. A. W. (1961). Distribution of the Anderson–Darling statistic. Ann. Math. Statist. 32, 1118-1124.

Lilliefors, H. W. (1967). On the Kolmogorov-Smirnov tests for normality with mean and variance unknown. J. Amer. Statist. Assoc. 62, 399-402.

Lilliefors, H. W. (1969). On the Kolmogorov-Smirnov test for the exponential distribution with mean unknown. J. Amer. Statist. Assoc. 64, 387-389.

Littell, R. C. and Rao, P. V. (1978). Confidence regions for location and scale parameters based on the Kolmogorov-Smirnov goodness-of-fit statistic. Technometrics 20, 23-27.

Littell, R. C., McClave, J. T., and Offen, W. W. (1979). Goodness of fit test for the two parameter Weibull distribution. Commun. Statist. Simulation Comput. B,3, 257-269.

Lockhart, R. A. and McLaren, C. G. (1985). Asymptotic points for a test of symmetry about a specified median. Biometrika 72, 208-210.

Lockhart, R. A. and Stephens, M. A. (1985a). Tests for the Weibull distribution based on the empirical distribution function. Technical Report, Department of Mathematics and Statistics, Simon Fraser University.

Lockhart, R. A. and Stephens, M. A. (1985b). Goodness-of-fit tests for the gamma distribution. Technical Report, Department of Mathematics and Statistics, Simon Fraser University.

Lockhart, R. A. and Stephens, M. A. (1985c). Tests of fit for the Von Mises distribution. Biometrika 72 (To appear).

Lockhart, R. A., O'Reilly, F. J., and Stephens, M. A. (1985). Tests of fit based on normalized spacings. Technical report, Department of Statistics, Stanford University.

Lockhart, R. A., O'Reilly, F. J., and Stephens, M. A. (1986). Tests of fit for the extreme value and Weibull distributions based on normalized spacings. Nav. Res. Logist. Quart. (To appear).

Louter, A. S. and Koerts, J. (1970). On the Kuiper test for normality with mean and variance unknown. Statistica Neerlandica 24, 83-87.

Loynes, R. A. (1980). The empirical distribution function of residuals from generalised regression. Ann. Statist. 8, 285-298.

Mann, N. R. (1968). Results on statistical estimation and hypothesis testing with application to the Weibull and extreme value distributions. Technical report no. ARL 68-0068, Office of Aerospace Research, United States Air Force, Wright-Patterson Air Force Base, Ohio.

Mann, N. R. and Fertig, K. W. (1975). A goodness-of-fit test for the two parameters versus three parameters Weibull; confidence bounds for threshold. Technometrics 17, 237-245.

Mann, N. R., Scheuer, E. M., and Fertig, K. (1973). A new goodness-of-fit test for the two parameter Weibull or extreme-value distribution. Commun. Statist. 2, 383-400.

Mardia, K. V. (1972). Statistics of Directional Data. New York: Academic Press.

Margolin, B. H. and Maurer, W. (1976). Tests of the Kolmogorov-Smirnov type for exponential data with unknown scale, and related problems. Biometrika 63, 149-160.

Martynov, G. V. (1976). Computation of limit distributions of statistics for normality tests of type w squared. Theor. Probability Appl. 21, 1-13.

Michael, J. R. and Schucany, W. R. (1979). A new approach to testing goodness-of-fit for censored samples. Technometrics 21, 435-441.

Mickey, M. R., Mundle, P. B., Walker, D. N., and Glinski, A. M. (1963). Test criteria for Pearson type III distributions. Technical report ARL 63-100, Aeronautical Research Laboratories, United States Air Force.

Moore, D. S. (1973). A note on Srinivasan's goodness-of-fit test. Biometrika 60, 209-211.

Mukantseva, L. A. (1977). Testing normality in one dimensional and multidimensional linear regression. Theor. Probability Appl. 22, 591-601.

Nesenko, G. A. and Tjurin, Ju. N. (1978). Asymptotics of Kolmogorov's statistic for a parametric family. Soviet Math. Dokl. 19, 516-519.

Neuhaus, G. (1979). Asymptotic theory of goodness-of-fit tests when parameters are present: A survey. Math. Op. Stat. S. 10, 479-494.

Niederhausen, H. (1981). Tables of significance points for the variance-weighted Kolmogorov-Smirnov statistics. Technical report, Department of Statistics, Stanford University.

Niederhausen, H. (1982). Butler-Smirnov test. In Encyclopaedia of Statistical Sciences, S. Kotz, N. L. Johnson, and C. B. Read, Eds., 1, 340-344.

Noether, G. E. (1963). A note on the Kolmogorov-Smirnov statistic in the discrete case. Metrika 7, 115-116.

Pearson, E. S. (1963). Comparison of tests for randomness of points on a line. Biometrika 50, 315-325.

Pearson, E. S. and Hartley, H. O. (1972). Biometrika Tables for Statisticians, Vol. 2. New York: Cambridge University Press.

Pettitt, A. N. (1976). Cramér-von Mises statistics for testing normality with censored samples. Biometrika 63, 475-481.

Pettitt, A. N. (1977a). Testing the normality of several independent samples using the Anderson-Darling statistic. J. Roy. Statist. Soc. C 26, 156-161.

Pettitt, A. N. (1977b). Tests for the exponential distribution with censored data using Cramér-von Mises statistics. Biometrika 64, 629-632.

Pettitt, A. N. and Stephens, M. A. (1976). Modified Cramér-von Mises statistics for censored data. Biometrika 63, 291-298.

Pettitt, A. N. and Stephens, M. A. (1977). The Kolmogorov-Smirnov goodness-of-fit statistic with discrete and grouped data. Technometrics 19, 205-210.

Pettitt, A. N. and Stephens, M. A. (1983). EDF statistics for testing for the Gamma distribution. Technical Report, Department of Statistics, Stanford University.

Pierce, D. A. (1978). Combining evidence from several samples for testing goodness-of-fit to a location-scale family. Technical Report, Department of Statistics, Stanford University.

Pierce, D. A. and Gray, R. J. (1982). Testing normality of errors in regression models. Biometrika 69, 233-236.

Pierce, D. A. and Kopecky, K. J. (1978). Testing goodness-of-fit for the distribution of errors in regression models. Technical Report Symp. 16, Department of Statistics, Stanford University.

Proschan, F. (1963). Theoretical explanation of observed decreasing failure rate. Technometrics 5, 375-383.

Pyke, R. (1965). Spacings. J. Roy. Statist. Soc., B 27, 395-449.

Quesenberry, C. P., Whitaker, T. B., and Dickens, J. W. (1976). On testing normality using several samples: Analysis of peanut aflatoxin data. Biometrics 32, 753-754.

Rao, P. V., Schuster, E. F., and Littell, R. C. (1975). Estimation of shift and center of symmetry based on Kolmogorov-Smirnov statistics. Ann. Statist. 3, 862-873.

Renyi, A. (1953). On the theory of order statistics. Acta Math. Acad. Sci. Hung. 4, 191-227.

Riedwyl, H. (1967). Goodness-of-fit. J. Amer. Statist. Assoc. 62, 390-398.

Rothman, E. D. and Woodroofe, M. (1972). A Cramér-von Mises type statistic for testing symmetry. Ann. Math. Statist. 43, 2035-2038.

Sahler, W. (1968). A survey of distribution-free statistics based on distances between distribution functions. Metrika 13, 149-169.

Schey, H. M. (1977). The asymptotic distribution of the one-sided Kolmogorov-Smirnov statistic for truncated data. Comm. Statist. Theory Methods, A 6, 1361-1366.

Schneider, B. E. and Clickner, R. P. (1976). On the distribution of the Kolmogorov-Smirnov statistic for the gamma distribution with unknown parameters. Mimeo Series No. 36, Department of Statistics, School of Business Administration, Temple University, Philadelphia, Pa.

Serfling, R. J. and Wood, C. L. (1975). On null-hypothesis limiting distributions of Kolmogorov-Smirnov type statistics with estimated location and scale parameters. Technical Report: Florida State University.

Siegel, S. (1956). Non-parametric Statistics for Behavioral Scientists. New York: McGraw-Hill.

Smirnov, N. V. (1947). Akad. Nauk SSR., C. R. (Dokl.) Acad. Sci. URSS 56, 11-14.

Smith, R. M. and Bain, L. J. (1976). Correlation-type goodness-of-fit statistics with censored sampling. Comm. Statist. Theory Methods, A 5, 119-132.

Spinelli, J. J. and Stephens, M. A. (1983). Tests for exponentiality when origin and scale parameters are unknown. Technometrics 29, 471-476.

Srinivasan, R. (1970). An approach to testing the goodness-of-fit of incompletely specified distributions. Biometrika 57, 605-611.

Srinivasan, R. (1971). Tests for exponentiality. Statist. Hefte 12, 157-160.

Srinivasan, R. and Godio, L. B. (1974). A Cramér-von Mises type statistic for testing symmetry. Biometrika 61, 196-198.

Stephens, M. A. (1970). Use of the Kolmogorov-Smirnov, Cramér-von Mises and related statistics without extensive tables. J. Roy. Statist. Soc., B 32, 115-122.

Stephens, M. A. (1971). Asymptotic results for goodness-of-fit statistics when parameters must be estimated. Technical Reports 159, 180, Department of Statistics, Stanford University.

Stephens, M. A. (1974a). Components of goodness-of-fit statistics. Ann. Inst. H. Poincare, B 10, 37-54.

Stephens, M. A. (1974b). EDF statistics for goodness-of-fit and some comparisons. J. Amer. Statist. Assoc. 69, 730-737.

Stephens, M. A. (1976a). Asymptotic results for goodness-of-fit statistics with unknown parameters. Ann. Statist. 4, 357-369.

Stephens, M. A. (1976b). Asymptotic power of EDF statistics for exponentiality against Gamma and Weibull alternatives. Technical Report No. 297: Department of Statistics, Stanford University.

Stephens, M. A. (1977). Goodness-of-fit for the extreme value distribution. Biometrika 64, 583-588.

Stephens, M. A. (1978). On the half-sample method for goodness-of-fit. J. Roy. Statist. Soc., B 40, 64-70.

Stephens, M. A. (1979). Tests of fit for the logistic distribution based on the empirical distribution function. Biometrika 66, 591-595.

Stephens, M. A. (1985). Tests for the Cauchy distribution based on the EDF. Technical Report, Department of Statistics, Stanford University.

Stephens, M. A. (1986). Goodness-of-fit for censored data. Technical Report, Department of Statistics, Stanford University.

Stephens, M. A. and Wagner, D. (1986). On a replacement method for goodness-of-fit with randomly censored data. Technical report, Department of Mathematics and Statistics, Simon Fraser University.

Tiku, M. L. (1981). A goodness of fit statistic based on the sample spacings for testing a symmetric distribution against symmetric alternatives. Australian J. Statist. 23, 149-158.

Tiku, M. L. and Singh, M. (1981). Testing the two parameter Weibull distribution. Comm. Statist. A, 10, 907-918.

Van Soest, J. (1967). Some experimental results concerning tests of normality. Statistica Neerlandica 21, 91-97.

Van Soest, J. (1969). Some goodness-of-fit tests for exponential distributions. Statistica Neerlandica 23, 41-51.

Van Tilmann-Deutler, D., Griesenbrock, H. P., and Schwensfeier, H. E. (1975). Der Kolmogorov-Smirnov Eisntichprobentest auf Normalitat bei unbekanntem Mittelwert und undbekannter Varianz. Allegemeins Statistishe Archiv. A 59, 228-250.

Volodin, I. N. (1965). Testing of statistical hypothesis on the type of distribution by small sample. Kazan. Gos. Univ. Ucen. Zap. 125, 3-23 (Russian). Also in Selected translations in mathematical statistics and probability, AMS (1973), 11, 1-26.

Watson, G. S. (1961). Goodness-of-fit tests on a circle. Biometrika 48, 109-114.

White, H. and MacDonald, G. M. (1980). Some large-sample tests for non-normality in the linear regression model (with discussion). J. Amer. Statist. Assoc. 75, 16-28.

White, J. S. (1969). The moments of log-Weibull order statistics. Technometrics 11, 373-386.

Wilk, M. B. and Shapiro, S. S. (1968). The joint assessment of normality of several independent samples. Technometrics 10, 825-839.

Wood, C. L. (1978a). On null-hypothesis limiting distributions of Kolmogorov-Smirnov type statistics with estimated location and scale parameters. Comm. Statist., A 7, 1181-1198.

Wood, C. L. (1978b). A large sample Kolmogorov-Smirnov test for normality of experimental error in a randomised block design. Biometrika 65, 673-676.

Wood, C. L. and Altavela, M. M. (1978). Large-sample results for Kolmogorov-Smirnov statistics for discrete distributions. Biometrika 65, 235-239.

# 5

# Tests Based on Regression and Correlation

Michael A. Stephens  Simon Fraser University, Burnaby, B.C., Canada

## 5.1 INTRODUCTION

In previous chapters it has been shown how a random sample can be used in
a graphical display (for example, on probability paper, or by drawing the
EDF) which is then used to indicate whether the sample comes from a given
distribution. The techniques make use of the sample values arranged in
ascending order, that is, they use the order statistics. In this chapter we
examine another graphical method, related to probability plots, in which the
order statistics $X_{(i)}$ are plotted on the vertical axis of the graph, against $T_i$,
a suitable function of i, on the horizontal axis. A straight line is then fitted
to the points, and tests are based on statistics associated with this line. This
type of test will be called a regression test; when the test statistic used is
the correlation coefficient between X and T, the test will be called a correlation test.

### 5.1.1 Notation

The following notation is used: $X_1$, ..., $X_n$ is a random sample from a
continuous distribution $F_0(x)$, known or hypothesized; $X_{(1)} < X_{(2)} < \cdots < X_{(n)}$
are the order statistics. X may refer to the n-vector with components $X_i$ or
$X_{(i)}$ depending on context. $F_0(x)$ is often of the form $F(w)$ with $w = (x - \alpha)/\beta$;
$\alpha$ is the location parameter and $\beta$ is the scale parameter. If $\alpha = 0$, $F_0(x)$ is
said to contain a scale parameter only.

If $u = F(w)$, $w = F^{-1}(u)$, so that $F^{-1}(\cdot)$ is the <u>inverse function</u> of $F(\cdot)$;
$f(w)$ is the density function corresponding to $F(w)$.

$W_1, \ldots, W_n$ is a random sample from $F(w)$; $W_{(1)} < W_{(2)} < \cdots < W_{(n)}$ are the order statistics. $T_i$, $i = 1, \ldots, n$, is used to describe a set of constants; two important sets are $T_i \equiv m_i = E(W_{(i)})$, where E denotes expectation, and $T_i \equiv H_i = F^{-1}\{i/(n+1)\}$, $i = 1, \ldots, n$; m, H, T are column vectors of length n with components $m_i$, $H_i$, $T_i$, respectively.
$\Sigma$ or $\Sigma_i$ denotes the sum over i from 1 to n unless other limits are given.
$\bar{X} = \Sigma_i X_i/n$, $\bar{T} = \Sigma_i T_i/n$, etc.; log X means $\log_e X$.

### 5.1.2  Definitions: Correlation Coefficient

Let X refer to the vector $X_{(1)}, \ldots, X_{(n)}$, and T to vector $T_1, \ldots, T_n$; define the sums

$$S(X, T) = \Sigma(X_{(i)} - \bar{X})(T_i - \bar{T}) = \Sigma X_{(i)} T_i - n\bar{X}\bar{T}$$

$$S(X, X) = \Sigma(X_{(i)} - \bar{X})^2 = \Sigma(X_i - \bar{X})^2$$

$$S(T, T) = \Sigma(T_i - \bar{T})^2$$

and let $v(X, T) = S(X, T)/(n - 1)$; similarly define $v(X, X)$ and $v(T, T)$. The correlation coefficient between X and T is $R(X, T) = v(X, T)/\{v(X, X) v(T, T)\}^{\frac{1}{2}}$ $= S(X, T)/\{S(X, X) S(T, T)\}^{\frac{1}{2}}$. The statistic $Z(X, T) = n\{1 - R^2(X, T)\}$ is often used in subsequent sections. $S(X, X)$ may be referred to as $S^2$; similarly $v(X, T)$ or $R(X, T)$ may be referred to as v or R when there is no ambiguity in context. The usual meaning of (sample) correlation, defined for two sets of random variables, is here being extended to define "correlation" between a set of random variables X and a set of constants T, following the same definition.

### 5.1.3  The Correlation Coefficient for Censored Data

Suppose only a subset of the $X_{(i)}$ is available, because the data have been censored. Provided the ranks i of the known $X_{(i)}$ are known also, the corresponding $T_i$ can be paired with the $X_{(i)}$, and the correlation coefficient calculated as above, with the sums running only over the available values i. The calculation of $R(X, T)$ is thus very easily adapted to all types of censored data.

### 5.2  REGRESSION TESTS: MODELS

Regression tests arise most naturally when unknown parameters in the tested distribution $F_0(x)$ are location and scale parameters. Suppose $F_0(x)$ is $F(w)$ with $w = (x - \alpha)/\beta$, so that $\alpha$ is a location parameter and $\beta$ a scale parameter, and suppose any other parameters in $F(w)$ are known. If a sample of values $W_i$ were taken from $F(w)$ with $\alpha = 0$ and $\beta = 1$, we could construct a sample $X_i$ from $F_0(x)$ by calculating

$$X_i = \alpha + \beta W_i, \quad i = 1, \ldots, n \tag{5.1}$$

Let $m_i = E(W_{(i)})$; then

$$E(X_{(i)}) = \alpha + \beta m_i \tag{5.2}$$

and a plot of $X_{(i)}$ against $m_i$ should be approximately a straight line with intercept $\alpha$ on the vertical axis and slope $\beta$. The values $m_i$ are the most natural values to plot along the horizontal axis, but for most distributions they are difficult to calculate. Various authors have therefore proposed alternatives $T_i$ which are convenient functions of $i$; then (5.2) can be replaced by the model

$$X_{(i)} = \alpha + \beta T_i + \epsilon_i \tag{5.3}$$

where $\epsilon_i$ is an "error" which for $T_i = m_i$ will have mean zero. A frequent choice for $T_i$ is $H_i$ defined in Section 5.1.1.

## 5.3 MEASURES OF FIT

Three main approaches to testing how well the data fit (5.3) can be identified:

(a)  A test is based on the correlation coefficient $R(X, T)$ between the paired sets $X_i$ and $T_i$ as defined in Section 5.1.2.
(b)  Estimates $\hat{\alpha}$ and $\hat{\beta}$ of the parameters in the line $\alpha + \beta T_i$ are found by a suitable method, and a test is based on the sum of squares of the residuals $X_{(i)} - \hat{X}_{(i)}$, where $\hat{X}_{(i)} = \hat{\alpha} + \hat{\beta} T_i$. In order to give a scale-free statistic, this must be divided by another quadratic form in the $X_{(i)}$.
(c)  The scale parameter $\beta$ is estimated as in (b), and the squared value compared with another estimate of $\beta^2$, for example that obtained from the sample variance.

These three methods are closely connected. In particular, when the method of estimation in (b) and (c) is ordinary least squares, the techniques often lead to test statistics equivalent to $R^2$. We discuss $R^2(X, m)$ in this chapter, for various distributions, beginning with the uniform. $R^2(X, m)$ is consistent against all alternatives; methods (b) and (c) can yield statistics which are not consistent against certain classes of alternative. We shall discuss statistics based on methods (b) and (c) when they arise in connection with tests for the normal and exponential distributions.

## 5.4 TESTS BASED ON THE CORRELATION COEFFICIENT

### 5.4.1 The Correlation Coefficient and the ANOVA Table

When ordinary least squares is used to estimate $\alpha$ and $\beta$ in the line $X_{(i)} = \alpha + \beta T_i$, the estimate of $\beta$ is $\hat{\beta} = S(X, T)/S(T, T)$, and the standard ANOVA table for the model is, with $\hat{X}_{(i)} = \hat{\alpha} + \hat{\beta} T_i$,

---

Regression sum of squares: $S^2(X, T)/S(T, T)$

Error sum of squares (ESS): $\Sigma_i \{ X_{(i)} - \hat{X}_{(i)} \}^2$

---

Total sum of squares (TSS): $S^2(X, X)$

---

Define

$$Z(X, T) = n \{ 1 - R^2(X, T) \} \tag{5.4}$$

It is easily shown that $Z(X, T)$, or simply $Z$, is $n(ESS/TSS)$. Thus $Z$ is equivalent to $R^2$, but can also be regarded as based on ESS (Section 5.3, method (b)), or on the ratio of two quadratic forms (method (c)). The interconnections exist because ordinary least squares has been used to estimate $\alpha$ and $\beta$.

When $T = m$, $R^2$ is an appealing statistic for measuring the fit, since if a "perfect" sample is given, that is, a sample whose ordered values fall exactly at their expected values, $R^2(X, m)$ will be 1. A test based on $R^2(X, m)$ will be one-tail, with small values of $R^2$ indicating a poor fit. There are then some advantages in using $Z$ instead of $R^2$; first, high values of $Z$ lead to rejection of a good fit, corresponding to many other goodness-of-fit statistics (such as EDF statistics or Pearson's $X^2$), and second, $Z$ often has an asymptotic distribution, which can be found, and interpolation in tables is easier. We therefore tabulate $Z$, and insert asymptotic points in the tables where these have been calculated.

$R^2(X, T)$ is naturally suited to the model $X_{(i)} = \alpha + \beta T_i$ where both $\alpha$ and $\beta$ are unknown. When one or more of these parameters are known (for example, in many tests for exponentiality $\alpha = 0$, and in some tests for uniformity $\alpha = 0$ and $\beta = 1$), $R^2$ will not necessarily be a good statistic for measuring the fit, and a test based on the residuals (method (b)) will be more natural. These points will be discussed in Sections 5.6.1 and 5.11.4 below.

### 5.4.2 Consistency of the R(X, m) Test

Sarkadi (1975) showed the consistency of the text based on $R(X, m)$ for testing normality, and more recently Gerlach (1979) has shown consistency for correlation tests based on $R(X, m)$, or equivalently on $Z = n\{1 - R^2(X, m)\}$, for a wide class of distributions including all the usual continuous distributions. This is to be expected, since for large n we expect a sample to become perfect in the sense of Section 5.4.1. We can expect the consistency property to extend to $R(X, T)$ provided T approaches m sufficiently rapidly for large samples. We now give tests based on $R^2(X, m)$, for the uniform distribution, with unknown limits (next section) and with limits 0 and 1 (Section 5.6).

## 5.5 THE CORRELATION TEST FOR THE UNIFORM DISTRIBUTION WITH UNKNOWN LIMITS

### 5.5.1 Complete Samples

Suppose $U(a, b)$ refers to the uniform distribution between limits a and b. The test for uniformity is a test of

$H_0$: a random sample $X_1$, $X_2$, ..., $X_n$ comes from $U(a, b)$

with a, b unknown.

The ordered values $X_{(i)}$ will be plotted against $m_i = i/(n + 1)$; $m_i = E(W_{(i)})$, where $W_1$, $W_2$, ..., $W_n$ is a random sample from $U(0, 1)$, and $\bar{m} = 0.5$; for this distribution $m_i = H_i$. The correlation test then takes the following steps:

(a)  Calculate $Z = n\{1 - R^2(X, m)\}$.
(b)  Compare Z to the percentage points in Table 5.1; reject $H_0$ if Z exceeds the appropriate value for given n and $\alpha$. Z has an asymptotic null distribution, found and tabulated by Lockhart and Stephens (1986). The points given for finite n, obtained by Monte Carlo studies, approach the asymptotic points smoothly and rapidly, and interpolation in Table 5.1 is straightforward. Note that since $R^2(X, T)$ is scale-free, $R^2(X, m)$ is here the same as $R^2(X, T)$ with $T_i = i$.

### 5.5.2 Tests for Type 2 Censored Samples

Suppose the sample is Type 2 right-censored (see Chapter 11), and only the subset $X_{(i)}$, $i = 1, ..., r$, is available. This subset, on $H_0$, will be uniform with unknown limits a* and b* if a and b are unknown. Thus the subset may be treated as a complete sample of size r and the test above may be used; $m_i$ is then $i/(r + 1)$. Alternatively, $R^2(X, m)$ can be calculated as described in Section 5.1.2; since the $X_{(i)}$ are the first r values out of n, the values $m_i$

TABLE 5.1  Upper Tail Percentage Points for $Z = n\{1 - R^2(X, m)\}$ for a
Test for Uniformity, Full Sample

| n | Significance level $\alpha$ | | | | | | |
|---|-------|-------|-------|-------|-------|-------|-------|
|   | 0.5   | 0.25  | 0.15  | 0.10  | 0.05  | 0.025 | 0.01  |
| 4  | 0.344 | 0.559 | 0.734 | 0.888 | 1.089 | 1.238 | 1.388 |
| 6  | 0.441 | 0.703 | 0.901 | 1.053 | 1.325 | 1.590 | 1.918 |
| 8  | 0.495 | 0.792 | 1.000 | 1.163 | 1.474 | 1.739 | 2.100 |
| 10 | 0.535 | 0.833 | 1.068 | 1.245 | 1.532 | 1.846 | 2.294 |
| 12 | 0.560 | 0.864 | 1.093 | 1.280 | 1.628 | 1.938 | 2.360 |
| 18 | 0.605 | 0.940 | 1.147 | 1.385 | 1.716 | 2.083 | 2.503 |
| 20 | 0.610 | 0.960 | 1.200 | 1.420 | 1.760 | 2.140 | 2.550 |
| 40 | 0.640 | 0.980 | 1.215 | 1.420 | 1.762 | 2.140 | 2.550 |
| 60 | 0.648 | 0.988 | 1.227 | 1.420 | 1.765 | 2.140 | 2.550 |
| 80 | 0.658 | 0.997 | 1.228 | 1.420 | 1.770 | 2.140 | 2.550 |
| ∞  | 0.666 | 0.993 | 1.234 | 1.430 | 1.774 | 2.129 | 2.612 |

are $i/(n + 1)$, $i = 1, \ldots, r$, with mean $\bar{m} = (r + 1)/\{2(n + 1)\}$. Because
$R^2(X, m)$ is scale-free, both these procedures are equivalent to finding the
correlation between $X_{(i)}$ and $i$, and the two values of $R^2$ are identical. Similarly, a left-censored or double-censored sample can be treated; if the ranks
of the known $X_{(i)}$ are $s \leq i \leq r$, the correlation R between $X_i$ and $i$ may be
found, and Table 5.1 used as though the data were a complete sample of size
$n = r + 1 - s$. Note that this technique cannot be used for a randomly censored
sample. Chen (1984) has examined correlation tests for randomly censored
data, especially in connection with testing exponentiality.

### 5.5.3  Test for Type 1 Censored Samples

For a Type 1 censored sample, the procedure above does not make use of
the censoring values, say A for the lower value and B for the upper. Suppose,
for simplicity, A is zero and B is 11. Then it is possible to have, say, five
values out of ten, $X_{(1)}$ to $X_{(5)}$, which are 0.9, 2.1, 3.1, 3.9, 5.2; these
could give a large correlation coefficient, but would be suspect as being a
uniform sample because of the large gap between the maximum $X_{(5)} = 5.2$
and $B = 11$. One way to make use of A and B is to include these in the sample,
making it now a sample of size 7. On $H_0$, all 7 values will be uniform between

unknown limits a*, b* and again the complete sample test in Section 5.5.1 may be used. The test as now made combines a test for uniformity of the $X_{(1)}$ with a test that they are spread over the range (A, B).

## 5.6 THE CORRELATION TEST FOR U(0,1)

### 5.6.1 Complete Samples

In many goodness-of-fit situations, a transformation reduces the original test to a test of $H_0$: that a set of X-values is U(0,1). Examples are the Probability Integral Transformation discussed in Chapter 4, and the J and K transformations in tests for exponentiality discussed in Chapter 10. For this null hypothesis, equation (5.2) reduces to

$$E(X_{(i)}) = m_i$$

where $m_i = i/(n+1)$; that is, $\alpha$ and $\beta$ are known to be 0 and 1, respectively. As in Section 5.5.3, statistic $R^2(X, m)$, taken alone, is not a suitable test statistic for $H_0$; for example, $R^2$ could be very close to 1 even if the X-values were uniform on only a small subset of [0,1]. $R^2(X, m)$ can be modified for this situation by including 0, 1 as values in the sample, but a more direct use of the known range is to base a test on the residuals $v_i = X_{(i)} - m_i$, for example, using $M_u^2 = \Sigma_i v_i^2$; then $M_u^2/n$ is statistic $T_1$ of Section 8.8.1. Other tests based on $v_i$ are given in Section 8.8.1.

### 5.6.2 Censored Samples from U(0,1)

Suppose the sample is Type 1 left-censored at A and right-censored at B; a test for uniformity should not only test for linearity of the values given in this interval, but also should test that the number of values is reasonable. $R^2(X, m)$ alone will not do this, for reasons similar to those given in Section 5.5.3; similar objections would apply if the given values were Type 2 censored. Thus the correlation coefficient alone will not usually be appropriate for testing for U(0,1) with censored samples.

Further illustrations of these difficulties are given in Chapter 11, where other methods are proposed for censored U(0,1) samples.

## 5.7 REGRESSION TESTS FOR THE NORMAL DISTRIBUTION 1

### 5.7.1 Tests Based on the Correlation Coefficient

In this section correlation tests are discussed for

$H_0$: a random sample $X_1, \ldots, X_n$ comes from $N(\mu, \sigma^2)$
with both $\mu$ and $\sigma$ unknown

$$(5.5)$$

In sections 5.8 and 5.9 other regression tests for (5.5) are developed, based on residuals, or on the ratio of two estimates of scale.

For the normal distribution $N(\mu, \sigma^2)$, $f(w) = (2\pi)^{-\frac{1}{2}} \exp(-w^2/2)$, with $w = (x - \mu)/\sigma$; thus $\alpha = \mu$ and $\beta = \sigma$, and $m_i$ are the expected values of standard normal order statistics. Equation (5.2) becomes

$$E(X_{(i)}) = \mu + \sigma m_i \qquad (5.6)$$

For the normal distribution $\bar{m} = 0$, and $R^2(X, m)$ can conveniently be written in vector notation. Let X be the vector $\{X_{(1)}, X_{(2)}, \ldots, X_{(n)}\}$ as before, and let m be the vector $(m_1, m_2, \ldots, m_n)$; let a prime, for example, in $X'$ and $m'$, denote the transpose of a vector or of a matrix. Then, for a complete sample,

$$R^2(X, m) = \frac{(X'm)^2}{(m'm)S^2} \qquad (5.7)$$

This statistic will be seen later to be identical to the Shapiro-Francia statistic $W'$, so that, for testing normality, we can refer to $R^2(X, m)$ also as $W'$.

## 5.7.2 Expected Values of Standard Normal Order Statistics

Values of $m_i$, for the calculation of $R^2(X, m)$, must be calculated numerically: extensive tables have been given by Harter (1961). The approximation

$$m_i = \Phi^{-1}\left(\frac{i - 0.375}{n + 0.125}\right) \qquad (5.8)$$

first suggested by Blom (1958) can also be used; $\Phi^{-1}(\cdot)$ is the inverse of the standard normal cdf and computer routines exist for this function. Other formulas given by Hastings (1955) are quoted in Abramowitz and Stegun (1965, p. 933); an algorithm has been given by Milton and Hotchkiss (1969), and simpler formulas, suitable for use on pocket calculators, have since been given by Page (1977) and by Hamaker (1978). Weisberg and Bingham (1975) have shown that use of Blom's formula and the Milton-Hotchkiss algorithm for $\Phi^{-1}(\cdot)$ to approximate $m_i$ makes negligible difference to the distribution of $R^2(X, m) = W'$, which in any case must be found by Monte Carlo methods. Shapiro and Francia (1972) originally introduced $W'$ as a replacement for the Shapiro-Wilk statistic below, for use when $n \geq 50$, but the above articles suggest that $W' = R^2(X, m)$ will be useful, because of the

TABLE 5.2  Upper Tail Percentage Points for $Z = n\{1 - R^2(X,m)\}$ for a Test for Normality with Complete or Type 2 Censored Data; p = censoring ratio r/n

| | n | Significance level $\alpha$ | | | | | | |
|---|---|---|---|---|---|---|---|---|
| | | 0.50 | 0.25 | 0.15 | 0.10 | 0.05 | 0.025 | 0.01 |
| p = 0.2 | 20 | 1.52 | 2.86 | 3.65 | 4.16 | 5.26 | 6.65 | 8.13 |
| | 40 | 2.68 | 4.25 | 5.40 | 6.27 | 7.69 | 9.07 | 10.76 |
| | 60 | 3.25 | 5.05 | 6.34 | 7.46 | 9.25 | 11.00 | 13.01 |
| | 80 | 3.65 | 5.64 | 7.08 | 8.32 | 10.38 | 12.50 | 15.17 |
| | 100 | 3.94 | 6.07 | 7.64 | 8.95 | 11.22 | 13.64 | 16.96 |
| p = 0.4 | 10 | 0.77 | 1.42 | 1.80 | 2.03 | 2.51 | 3.17 | 3.90 |
| | 20 | 1.33 | 2.09 | 2.64 | 3.07 | 3.76 | 4.46 | 5.38 |
| | 40 | 1.77 | 2.72 | 3.42 | 3.98 | 4.92 | 5.97 | 7.30 |
| | 60 | 2.02 | 3.04 | 3.84 | 4.49 | 5.65 | 6.79 | 8.51 |
| | 80 | 2.18 | 3.30 | 4.16 | 4.84 | 6.13 | 7.44 | 9.38 |
| | 100 | 2.30 | 3.51 | 4.39 | 5.09 | 6.46 | 7.92 | 10.01 |
| p = 0.6 | 10 | 0.76 | 1.19 | 1.50 | 1.76 | 2.18 | 2.61 | 3.15 |
| | 20 | 1.03 | 1.57 | 1.97 | 2.28 | 2.83 | 3.38 | 4.09 |
| | 40 | 1.30 | 1.93 | 2.42 | 2.81 | 3.47 | 4.22 | 5.31 |
| | 60 | 1.42 | 2.13 | 2.66 | 3.08 | 3.86 | 4.70 | 5.97 |
| | 80 | 1.52 | 2.27 | 2.82 | 3.29 | 4.12 | 5.04 | 6.36 |
| | 100 | 1.60 | 2.38 | 2.94 | 3.43 | 4.30 | 5.27 | 6.63 |
| p = 0.8 | 10 | 0.65 | 1.01 | 1.26 | 1.45 | 1.79 | 2.14 | 2.59 |
| | 20 | 0.84 | 1.25 | 1.55 | 1.79 | 2.20 | 2.63 | 3.27 |
| | 40 | 1.01 | 1.48 | 1.81 | 2.09 | 2.61 | 3.14 | 3.85 |
| | 60 | 1.08 | 1.58 | 1.95 | 2.25 | 2.80 | 3.38 | 4.29 |
| | 80 | 1.15 | 1.67 | 2.06 | 2.38 | 2.95 | 3.56 | 4.50 |
| | 100 | 1.19 | 1.78 | 2.14 | 2.47 | 3.07 | 3.68 | 4.61 |
| p = 0.9 | 10 | 0.62 | 0.94 | 1.17 | 1.35 | 1.67 | 2.01 | 2.42 |
| | 20 | 0.76 | 1.14 | 1.40 | 1.60 | 1.97 | 2.33 | 2.86 |
| | 40 | 0.90 | 1.31 | 1.61 | 1.86 | 2.27 | 2.70 | 3.34 |
| | 60 | 0.96 | 1.40 | 1.70 | 1.96 | 2.38 | 2.87 | 3.54 |
| | 80 | 1.01 | 1.46 | 1.78 | 2.05 | 2.50 | 2.99 | 3.68 |
| | 100 | 1.05 | 1.50 | 1.84 | 2.12 | 2.60 | 3.07 | 3.78 |
| p = 0.95 | 10 | 0.60 | 0.92 | 1.13 | 1.30 | 1.62 | 1.95 | 2.34 |
| | 20 | 0.74 | 1.11 | 1.36 | 1.55 | 1.89 | 2.26 | 2.73 |
| | 40 | 0.87 | 1.26 | 1.54 | 1.76 | 2.14 | 2.52 | 3.14 |
| | 60 | 0.92 | 1.33 | 1.61 | 1.85 | 2.26 | 2.69 | 3.27 |
| | 80 | 0.97 | 1.39 | 1.68 | 1.92 | 2.35 | 2.80 | 3.39 |
| | 100 | 1.01 | 1.43 | 1.74 | 1.98 | 2.41 | 2.88 | 3.48 |
| p = 1.0 | 10 | 0.59 | 0.89 | 1.10 | 1.26 | 1.58 | 1.90 | 2.27 |
| | 20 | 0.76 | 1.11 | 1.36 | 1.56 | 1.91 | 2.31 | 2.81 |
| | 40 | 0.91 | 1.32 | 1.60 | 1.83 | 2.23 | 2.66 | 3.30 |
| | 60 | 1.01 | 1.42 | 1.72 | 1.96 | 2.39 | 2.83 | 3.43 |
| | 80 | 1.06 | 1.50 | 1.80 | 2.05 | 2.49 | 2.94 | 3.54 |
| | 100 | 1.08 | 1.55 | 1.86 | 2.11 | 2.56 | 3.01 | 3.61 |
| | 400 | 1.34 | 1.83 | 2.17 | 2.45 | 2.89 | 3.36 | 3.95 |
| | 600 | 1.43 | 1.94 | 2.26 | 2.52 | 2.98 | 3.42 | 4.04 |
| | 1000 | 1.50 | 2.00 | 2.33 | 2.60 | 3.11 | 3.61 | 4.25 |

ready availability of good approximations to $m_i$, for smaller n. We have therefore constructed tables for this test based on $R^2(X, m)$, using (5.8) for m.

### 5.7.3 Correlation Test Based on $R^2(X, m)$

The correlation test using $R^2(X, m)$ then takes the following form:

(a) Calculate $Z = n\{1 - R^2(X, m)\}$, with $m_i$ given by (5.8).
(b) Refer Z to Table 5.2. $H_0$ is rejected if Z is greater than the percentage point for given values of n and of $p = r/n$, and significance level $\alpha$.
Table 5.2 contains percentage points for use with Type 2 censored data. They have been found by extensive Monte Carlo samuling, using 10,000 samples for each n. Shapiro and Francia (1972) have given points for $W' = R^2(X, m)$, for complete samples only, for $n = 35$ and for all n between 50 and 99. Fotopoulos, Leslie, and Stephens (1984) have produced asymptotic theory for $Z(X, m)$, for complete samples.

Tables are not given for Type 1 right-censored data since objections apply similar to those against EDF statistics. If the upper censoring value were t, and if $p = \Phi\{t - \mu)/\sigma\}$, tables of Z could be given for selected p and n; however, they would have to be entered, in practice, at $\hat{p} = \Phi\{(t - \hat{\mu})/\hat{\sigma}\}$, and this could cause an error in the apparent significance level of Z. For large samples, Z can be calculated from the available observations, and Table 5.2 can be used to give an approximate test.

### E 5.7.3 Example

Table 5.3 contains 20 values of weights of chicks, taken from Bliss (1976), already used in Chapter 4 in tests of normality. When these are correlated

### TABLE 5.3 Data for Tests of Normality

$X^a$: 156, 162, 168, 182, 186, 190, 190, 196, 202, 210, 214, 220, 226, 230, 230, 236, 236, 242, 246, 270

$m_{20}, \ldots, m_{11}{}^b$: 1.86748, 1.40760, 1.13095, 0.92098, 0.74538, 0.59030, 0.44833, 0.31493, 0.18696, 0.06200

$m'_{20}, \ldots, m'_{11}{}^c$: 1.8682, 1.4034, 1.1281, 0.9191, 0.7441, 0.5895, 0.4478, 0.3146, 0.1868, 0.0619

[a]The values X are weights of 20 chicks, in grams.
[b]Values m are the exact expected values of standard normal order statistics.
[c]Values m' are those given by (5.8).

with $m_i$ calculated from (5.8), also given in the table, the value of R is 0.9907, leading to $Z = 0.3719$. When the weights are correlated against the exact $m_i$ (also given), $Z = 0.3720$, a negligible difference. Reference to Table 5.2 shows that Z is not significant at $p = 0.50$, so that normality is not rejected (the same conclusion as in Section 4.8.1). An example with censored data is in Section 11.4.1.2.

### 5.7.4 Correlation Test Based on $R^2(X, H)$

Let $H_i = \Phi^{-1}\{i/(n + 1)\}$; a test could be based on statistic $Z(X, H) = n\{1 - R^2(X, H)\}$, following the procedure given above. This test statistic was suggested by de Wet and Venter (1972). Use of the $H_i$ instead of $m_i$ makes distribution theory of $Z(X, H)$ easier than that of $Z(X, m)$, and de Wet and Venter have given the asymptotic theory for $Z(X, H)$, for full samples. Stephens (1986) has given tables for $Z(X, H)$, for finite n and for complete and Type 2 censored samples. These are not included here since computation of $Z(X, H)$ is no easier than that of $Z(X, m)$ and the corresponding tests can be expected to have similar properties.

### 5.7.5 Other Correlation Statistics for Normality

Smith and Bain (1976) have proposed the correlation statistic $R(X, K)$, where $K_i$ is a close approximation to $H_i$, given by Abramowitz and Stegun (1965, p. 933). Smith and Bain have given tables for use when $R^2(X, K)$ has been calculated from Type 2 censored data. Filliben (1975) investigated tests using $T_i = \tilde{m}_i$ the median of the distribution of $W_{(i)}$; $\tilde{m}_i$ is given by $\Phi^{-1}(\tilde{u}_i)$, where $\tilde{u}_i$ is the median of the i-th order statistic of a uniform sample. Filliben has given an empirical approximation for $\tilde{u}_i$ which gives a formula for $\tilde{m}_i$ similar to that for $m_i$ given in (5.8) above; thus $R^2(X, \tilde{m})$ is close to $R^2(X, m) = W'$ and has similar power properties (Filliben, 1975). Filliben also gave tables of critical values of $R^2(X, \tilde{m})$.

## 5.8 REGRESSION TESTS BASED ON RESIDUALS

We next turn to the second method of testing described in Section 5.3, in which parameters $\alpha$ and $\beta$ in the model $X_{(i)} = \alpha + \beta T_i + \epsilon_i$ are estimated and a test is then based on the residuals. We consider estimates given by generalized least squares, and suppose $T = m$. In the notation of Section 5.2, let $v_{ij} = E\{W_{(i)} - m_i\}\{W_{(j)} - m_j\}$, the covariance of $W_{(i)}$ and $W_{(j)}$. Then, as before, let X be the column vector with components $X_{(1)}, \ldots, X_{(n)}$, let m be the column vector with components $m_1, \ldots, m_n$, and let 1 be a column vector with each component 1. Let V be the matrix $V \equiv (v_{ij})$ and let X' be the transpose of X; similarly for other vectors and matrices. The generalized least squares estimates of $\alpha$ and $\beta$ are

$$\hat{\alpha} = -m'GX \quad \text{and} \quad \hat{\beta} = l'GX \tag{5.9}$$

where

$$G = \frac{V^{-1}(lm' - ml')V^{-1}}{(l'V^{-1}l)(m'V^{-1}m) - (l'V^{-1}m)^2}$$

For particular distributions, for example, for the normal and exponential distributions, these equations simplify considerably.

The estimate of $X_{(i)}$ given by the line is again $\hat{X}_{(i)} = \hat{\alpha} + \hat{\beta}m_i$, and the sum of squares of residuals, corresponding to the Error Sum of Squares in the ANOVA table of Section 5.4, is $ESS_1 = \Sigma_i (X_{(i)} - \hat{X}_{(i)})^2$; a test for fit may then be based on $Z_1(X, m) = ESS_1/S^2$, where $S^2 = \Sigma_i (X_{(i)} - \bar{X})^2$ as before. Alternatively, since in generalized least squares analysis the quantity $ESS_2 = (X - \hat{X})'V^{-1}(X - \hat{X})$, where $\hat{X}$ is the column vector with components $\hat{X}_{(1)}, \ldots, \hat{X}_{(n)}$, is minimized by the parameter estimates $\hat{\alpha}$ and $\hat{\beta}$, the test might be based on $Z_2(X, m) = ESS_2/S^2$. The test based on $Z_1$ can be shown to be consistent. Some examination of such tests has been made by Spinelli (1980), for the exponential and the extreme-value distributions, but otherwise they have not been much developed.

## 5.9 TESTS BASED ON THE RATIO OF
##       TWO ESTIMATES OF SCALE

Finally we turn to the third method of testing the fit of the model (5.2), one which has been developed by Shapiro and Wilk for testing normality and exponentiality. The procedure used is to compare $\hat{\beta}^2$, where $\hat{\beta}$ is the generalized least squares estimate given in equation (5.9), with the estimate of $\beta^2$ given by the sample variance, and the test statistic is essentially $\hat{\beta}^2/S^2$. Tests of this type are closely related to those in the previous section.

For the normal test, this statistic works very well, but in other test situations, for example, for the exponential test, the statistic is inconsistent; in practical terms there will be certain alternative distributions which will not be detected even with very large samples (see Section 5.12).

In the case of tests for normality, modifications of the first estimate of $\beta^2$ above have also been suggested, since the estimate is complicated to calculate. It is not, of course, necessary to use the particular estimates of scale given above, and a test can be developed using the ratio of any two convenient estimates for $\beta^2$. Some comments on the choice of these estimates are in Section 5.11 below.

In the next section we give tests for normality based on residuals and on ratios of scale estimates. Regression tests for exponentiality of all types are included with other tests in Section 5.11, and some general comments on the techniques are given in Section 5.12.

## 5.10  REGRESSION TESTS FOR THE
## NORMAL DISTRIBUTION 2

### 5.10.1  Tests for Normality Based on Residuals

The model (5.2) can be extended to provide further tests of fit. Suppose the $m_i$ are the expected values of order statistics from a $N(0,1)$ distribution; the model expressed by (5.2) is then correct only when the sample $X_i$ comes from a normal distribution. A wide class of alternative distributions can be specified by supposing that the order statistics in a sample of size n from a distribution $F(x)$ have expectations which satisfy the model

$$E(X_{(i)}) = \mu + \sigma m^i + \beta_2 w_2(m_i) + \beta_3 w_3(m_i) \cdots \qquad (5.10)$$

where $\beta_2$, $\beta_3$, $\cdots$ are constants and $w_2(m_i)$, $w_3(m_i)$, $\cdots$ are functions of $m_i$. By different choices of these functions, the normal model for X is embedded in various classes of densities. For the appropriate class, for given $w_j(\cdot)$, the estimates $\hat{\beta}_j$ of the constants $\beta_j$ can be derived by generalized least squares. A test for normality can be based on these estimates by using them to test $H_0 : \beta_2 = \beta_3 = \cdots = 0$. Equivalently, tests may be based on the reduction in error sum of squares between a fit of model (5.2) and the more general model (5.10) above. LaBrecque (1973, 1977) has developed such tests, for $w_j(m)$ a j-th order polynomial in m, chosen so that the covariance matrix of the estimates $\hat{\mu}$, $\hat{\sigma}$, $\hat{\beta}_2$, etc., should be diagonal, and has given the necessary tables of coefficients and significance points. Stephens (1976) has suggested use of Hermite polynomials for $w_j(\cdot)$ and has given asymptotic theory of the tests based on $\hat{\beta}_j$, for $j \geq 2$.

### 5.10.2  Test for N(0,1)

When the parameters in $N(\mu, \sigma^2)$ are specified, the test based on residuals takes a simple form. Let $X'_{(i)} = (X_{(i)} - \mu)/\sigma$; the null hypothesis then reduces to a test that the $X'_{(i)}$ are $N(0,1)$, and a natural statistic based on residuals is

$$M_N^2 = \Sigma (X'_{(i)} - m_i)^2$$

De Wet and Venter (1972) have investigated the asymptotic theory of this statistic (which they call $L_n^0$). However, no tables are available for small sample sizes, and the power properties of the test are not known. The test based on $M_N^2$ would be a rival to the many tests available for this case, called Case 0 in Chapter 4, when parameters are fully specified; for example, the $X_i$ could be transformed to uniforms (on the null hypothesis) by the Probability Integral Transformation (Section 4.2.3) and any of the many tests for $U(0,1)$ could be used.

### 5.10.3 Tests for Normality Based on the Ratio of Estimates of Scale; the Shapiro-Wilk Test

For the normal distribution, $\alpha$ and $\beta$ in (5.2) are $\mu$ and $\sigma$, respectively; the estimates of these parameters given by (5.9) then become

$$\hat{\mu} = \bar{X} \quad \text{and} \quad \hat{\sigma} = \frac{m' V^{-1} X}{m' V^{-1} m} \tag{5.11}$$

The test statistic proposed by Shapiro and Wilk (1965) is

$$W = \frac{\hat{\sigma}^2 R^4}{S^2 C^2} \tag{5.12}$$

where $S^2 = \Sigma (X_{(i)} - \bar{X})^2 = \Sigma (X_i - \bar{X})^2$ as before, and where here $R^2 = m'V^{-1}m$ and $C^2 = m'V^{-1}V^{-1}m$. The factors $R^4$ and $C^2$ ensure that W always takes values between 0 and 1.

#### 5.10.3.1 Computing Formulas

When $\hat{\sigma}$ is substituted in the formula for W, it may be reduced to the following computing formula. First, let vectors a* and a be defined by

$$a^* = V^{-1}m \quad \text{and} \quad a = a^*/C \tag{5.13}$$

then

$$W = \frac{(a'X)^2}{S^2} = \frac{\left\{ \Sigma_i a_i X_{(i)} \right\}^2}{S^2} \tag{5.14}$$

In order to calculate W, the vector a is needed, and this in turn requires values of m and $V^{-1}$, derived from V. At the time of the introduction of W, exact values of V were published only for sample sizes up to n = 20 (Teichroew, 1956; Sarhan and Greenberg, 1962). They have since been calculated for larger sample sizes (Tietjen, Kahaner, and Beckman, 1978), and an algorithm for approximating V has been given by Davis and Stephens (1977). For values of n between 21 and 50, Shapiro and Wilk used approximations for the components $a_i$ of a, and gave a table for sample sizes from n = 3 to 50. These are given in Table 5.4. The $a_i$ are symmetric about 0, that is, $a_i = -a_{n+1-i}$, i = 1, ..., r, where r = (n - 1)/2 if n is odd and r = n/2 if n is even, so that only the positive values are given.

#### 5.10.3.2 Test Procedures

The steps in making the W test for normality, that is, for testing $H_0$: the $X_i$ are a random sample from $N(\mu, \sigma^2)$, with $\mu$, $\sigma$ unknown, are:

TABLE 5.4  Coefficients $\{a_{n+1-i}\}$ for the W Test for Normality

| i | | | | | n | | | | | |
|---|---|---|---|---|---|---|---|---|---|---|
| | 2 | 3 | 4 | 5 | 6 | 7 | 8 | 9 | 10 | |
| 1 | .7071 | .7071 | .6872 | .6646 | .6431 | .6233 | .6052 | .5888 | .5739 | |
| 2 | — | .0000 | .1677 | .2413 | .2806 | .3031 | .3164 | .3244 | .3291 | |
| 3 | — | — | — | .0000 | .0875 | .1401 | .1743 | .1976 | .2141 | |
| 4 | — | — | — | — | — | .0000 | .0561 | .0947 | .1224 | |
| 5 | — | — | — | — | — | — | — | .0000 | .0399 | |
| | 11 | 12 | 13 | 14 | 15 | 16 | 17 | 18 | 19 | 20 |
| 1 | .5601 | .5475 | .5359 | .5251 | .5150 | .5056 | .4968 | .4886 | .4808 | .4734 |
| 2 | .3315 | .3325 | .3325 | .3318 | .3306 | .3290 | .3273 | .3253 | .3232 | .3211 |
| 3 | .2260 | .2347 | .2412 | .2460 | .2495 | .2521 | .2540 | .2553 | .2561 | .2565 |
| 4 | .1429 | .1586 | .1707 | .1802 | .1878 | .1939 | .1988 | .2027 | .2059 | .2085 |
| 5 | .0695 | .0922 | .1099 | .1240 | .1353 | .1447 | .1524 | .1587 | .1641 | .1686 |
| 6 | .0000 | .0303 | .0539 | .0727 | .0880 | .1005 | .1109 | .1197 | .1271 | .1334 |
| 7 | — | — | .0000 | .0240 | .0433 | .0593 | .0725 | .0837 | .0932 | .1013 |
| 8 | — | — | — | — | .0000 | .0196 | .0359 | .0496 | .0612 | .0711 |
| 9 | — | — | — | — | — | — | .0000 | .0163 | .0303 | .0422 |
| 10 | — | — | — | — | — | — | — | — | .0000 | .0140 |
| | 21 | 22 | 23 | 24 | 25 | 26 | 27 | 28 | 29 | 30 |
| 1 | .4643 | .4590 | .4542 | .4493 | .4450 | .4407 | .4366 | .4328 | .4291 | .4254 |
| 2 | .3185 | .3156 | .3126 | .3098 | .3069 | .3043 | .3018 | .2992 | .2968 | .2944 |
| 3 | .2578 | .2571 | .2563 | .2554 | .2543 | .2533 | .2522 | .2510 | .2499 | .2487 |
| 4 | .2119 | .2131 | .2139 | .2145 | .2148 | .2151 | .2152 | .2151 | .2150 | .2148 |
| 5 | .1736 | .1764 | .1787 | .1807 | .1822 | .1836 | .1848 | .1857 | .1864 | .1870 |
| 6 | .1399 | .1443 | .1480 | .1512 | .1539 | .1563 | .1584 | .1601 | .1616 | .1630 |
| 7 | .1092 | .1150 | .1201 | .1245 | .1283 | .1316 | .1346 | .1372 | .1395 | .1415 |
| 8 | .0804 | .0878 | .0941 | .0997 | .1046 | .1089 | .1128 | .1162 | .1192 | .1219 |
| 9 | .0530 | .0618 | .0696 | .0764 | .0823 | .0876 | .0923 | .0965 | .1002 | .1036 |
| 10 | .0263 | .0368 | .0459 | .0539 | .0610 | .0672 | .0728 | .0778 | .0822 | .0862 |
| 11 | .0000 | .0122 | .0228 | .0321 | .0403 | .0476 | .0540 | .0598 | .0650 | .0697 |
| 12 | — | — | .0000 | .0107 | .0200 | .0284 | .0358 | .0424 | .0483 | .0537 |
| 13 | — | — | — | — | .0000 | .0094 | .0178 | .0253 | .0320 | .0381 |
| 14 | — | — | — | — | — | — | .0000 | .0084 | .0159 | .0227 |
| 15 | — | — | — | — | — | — | — | — | .0000 | .0076 |

(continued)

TABLE 5.4 (continued)

| i | n | | | | | | | | | |
|---|---|---|---|---|---|---|---|---|---|---|
|    | 31 | 32 | 33 | 34 | 35 | 36 | 37 | 38 | 39 | 40 |
| 1  | .4220 | .4188 | .4156 | .4127 | .4096 | .4068 | .4040 | .4015 | .3989 | .3964 |
| 2  | .2921 | .2898 | .2876 | .2854 | .2834 | .2813 | .2794 | .2774 | .2755 | .2737 |
| 3  | .2475 | .2463 | .2451 | .2439 | .2427 | .2415 | .2403 | .2391 | .2380 | .2368 |
| 4  | .2145 | .2141 | .2137 | .2132 | .2127 | .2121 | .2116 | .2110 | .2104 | .2098 |
| 5  | .1874 | .1878 | .1880 | .1882 | .1883 | .1883 | .1883 | .1881 | .1880 | .1878 |
| 6  | .1641 | .1651 | .1660 | .1667 | .1673 | .1678 | .1683 | .1686 | .1689 | .1691 |
| 7  | .1433 | .1449 | .1463 | .1475 | .1487 | .1496 | .1505 | .1513 | .1520 | .1526 |
| 8  | .1243 | .1265 | .1284 | .1301 | .1317 | .1331 | .1344 | .1356 | .1366 | .1376 |
| 9  | .1066 | .1093 | .1118 | .1140 | .1160 | .1179 | .1196 | .1211 | .1225 | .1237 |
| 10 | .0899 | .0931 | .0961 | .0988 | .1013 | .1036 | .1056 | .1075 | .1092 | .1108 |
| 11 | .0739 | .0777 | .0812 | .0844 | .0873 | .0900 | .0924 | .0947 | .0967 | .0986 |
| 12 | .0585 | .0629 | .0669 | .0706 | .0739 | .0770 | .0798 | .0824 | .0848 | .0870 |
| 13 | .0435 | .0485 | .0530 | .0572 | .0610 | .0645 | .0677 | .0706 | .0733 | .0759 |
| 14 | .0289 | .0344 | .0395 | .0441 | .0484 | .0523 | .0559 | .0592 | .0622 | .0651 |
| 15 | .0144 | .0206 | .0262 | .0314 | .0361 | .0404 | .0444 | .0481 | .0515 | .0546 |
| 16 | .0000 | .0068 | .0131 | .0187 | .0239 | .0287 | .0331 | .0372 | .0409 | .0444 |
| 17 | — | — | .0000 | .0062 | .0119 | .0172 | .0220 | .0264 | .0305 | .0343 |
| 18 | — | — | — | — | .0000 | .0057 | .0110 | .0158 | .0203 | .0244 |
| 19 | — | — | — | — | — | — | .0000 | .0053 | .0101 | .0146 |
| 20 | — | — | — | — | — | — | — | — | .0000 | .0049 |

| i | 41 | 42 | 43 | 44 | 45 | 46 | 47 | 48 | 49 | 50 |
|---|---|---|---|---|---|---|---|---|---|---|
| 1  | .3940 | .3917 | .3894 | .3872 | .3850 | .3830 | .3808 | .3789 | .3770 | .3751 |
| 2  | .2719 | .2701 | .2684 | .2667 | .2651 | .2635 | .2620 | .2604 | .2589 | .2574 |
| 3  | .2357 | .2345 | .2334 | .2323 | .2313 | .2302 | .2291 | .2281 | .2271 | .2260 |
| 4  | .2091 | .2085 | .2078 | .2072 | .2065 | .2058 | .2052 | .2045 | .2038 | .2032 |
| 5  | .1876 | .1874 | .1871 | .1868 | .1865 | .1862 | .1859 | .1855 | .1851 | .1847 |
| 6  | .1693 | .1694 | .1695 | .1695 | .1695 | .1695 | .1695 | .1693 | .1692 | .1691 |
| 7  | .1531 | .1535 | .1539 | .1542 | .1545 | .1548 | .1550 | .1551 | .1553 | .1554 |
| 8  | .1384 | .1392 | .1398 | .1405 | .1410 | .1415 | .1420 | .1423 | .1427 | .1430 |
| 9  | .1249 | .1259 | .1269 | .1278 | .1286 | .1293 | .1300 | .1306 | .1312 | .1317 |
| 10 | .1123 | .1136 | .1149 | .1160 | .1170 | .1180 | .1189 | .1197 | .1205 | .1212 |

(continued)

TABLE 5.4 (continued)

| i | 41 | 42 | 43 | 44 | 45 | 46 | 47 | 48 | 49 | 50 |
|---|---|---|---|---|---|---|---|---|---|---|
|    |    |    |    |    | n  |    |    |    |    |    |
| 11 | .1004 | .1020 | .1035 | .1049 | .1062 | .1073 | .1085 | .1095 | .1105 | .1113 |
| 12 | .0891 | .0909 | .0927 | .0943 | .0959 | .0972 | .0986 | .0998 | .1010 | .1020 |
| 13 | .0782 | .0804 | .0824 | .0842 | .0860 | .0876 | .0892 | .0906 | .0919 | .0932 |
| 14 | .0677 | .0701 | .0724 | .0745 | .0765 | .0783 | .0801 | .0817 | .0832 | .0846 |
| 15 | .0575 | .0602 | .0628 | .0651 | .0673 | .0694 | .0713 | .0731 | .0748 | .0764 |
| 16 | .0476 | .0506 | .0534 | .0560 | .0584 | .0607 | .0628 | .0648 | .0667 | .0685 |
| 17 | .0379 | .0411 | .0442 | .0471 | .0497 | .0522 | .0546 | .0568 | .0588 | .0608 |
| 18 | .0283 | .0318 | .0352 | .0383 | .0412 | .0439 | .0465 | .0489 | .0511 | .0532 |
| 19 | .0188 | .0227 | .0263 | .0296 | .0328 | .0357 | .0385 | .0411 | .0436 | .0459 |
| 20 | .0094 | .0136 | .0175 | .0211 | .0245 | .0277 | .0307 | .0335 | .0361 | .0386 |
| 21 | .0000 | .0045 | .0087 | .0126 | .0163 | .0197 | .0229 | .0259 | .0288 | .0314 |
| 22 | — | — | .0000 | .0042 | .0081 | .0118 | .0153 | .0185 | .0215 | .0244 |
| 23 | — | — | — | — | .0000 | .0039 | .0076 | .0111 | .0143 | .0174 |
| 24 | — | — | — | — | — | — | .0000 | .0037 | .0071 | .0104 |
| 25 | — | — | — | — | — | — | — | — | .000 | .0035 |

Taken from Shapiro and Wilk (1965), with permission of the authors and of the Biometrika Trustees.

(a)  Calculate Y from $Y = \Sigma_{i=1}^{r} a_{n+1-i}(X_{(n+1-i)} - X_{(i)})$ where $r = (n - 1)/2$, if n is odd, and $r = n/2$ if n is even.

(b)  Calculate $W = Y^2/S^2$.

(c)  If W is less than the value given in the lower tail in Table 5.5 for appropriate values of n and $\alpha$, reject $H_0$ at level $\alpha$.

The exact distribution of W under the null hypothesis will depend on n, but not on the true values of $\mu$ and $\sigma$. This distribution is not known, and Shapiro and Wilk provided Monte Carlo percentage points for use with the test. Table 5.5 contains these percentage points of W, for sample sizes $n \leq 50$. Shapiro and Wilk (1968) gave an approximation to the null distribution of W.

The test is made in the lower tail of W, because extensive Monte Carlo studies by Shapiro and Wilk suggested that when the sample is not from a normal distribution, low values of W will usually result; this is so because W is approximately $R^2(X, m)$ for the normal case (see Section 5.10.4).

TABLE 5.5 Percentage Points for the W Test for Normality

| | Significance level $\alpha$ | | | | | | | | |
|---|---|---|---|---|---|---|---|---|---|
| | Lower tail | | | | | | | Upper tail | |
| n | 0.01 | 0.02 | 0.05 | 0.10 | 0.50 | 0.10 | 0.05 | 0.02 | 0.01 |
| 3 | 0.753 | 0.756 | 0.767 | 0.789 | 0.959 | 0.998 | 0.999 | 1.000 | 1.000 |
| 4 | .687 | .707 | .748 | .792 | .935 | .987 | .992 | .996 | .997 |
| 5 | .686 | .715 | .762 | .806 | .927 | .979 | .986 | .991 | .993 |
| 6 | 0.713 | 0.743 | 0.788 | 0.826 | 0.927 | 0.974 | 0.981 | 0.986 | 0.989 |
| 7 | .730 | .760 | .803 | .838 | .928 | .972 | .979 | .985 | .988 |
| 8 | .749 | .778 | .818 | .851 | .932 | .972 | .978 | .984 | .987 |
| 9 | .764 | .791 | .829 | .859 | .935 | .972 | .978 | .984 | .986 |
| 10 | .781 | .806 | .842 | .869 | .938 | .972 | .978 | .983 | .986 |
| 11 | 0.792 | 0.817 | 0.850 | 0.876 | 0.940 | 0.973 | 0.979 | 0.984 | 0.986 |
| 12 | .805 | .828 | .859 | .883 | .943 | .973 | .979 | .984 | .986 |
| 13 | .814 | .837 | .866 | .889 | .945 | .974 | .979 | .984 | .986 |
| 14 | .825 | .846 | .874 | .895 | .947 | .975 | .980 | .984 | .986 |
| 15 | .835 | .855 | .881 | .901 | .950 | .975 | .980 | .984 | .987 |
| 16 | 0.844 | 0.863 | 0.887 | 0.906 | 0.952 | 0.976 | 0.981 | 0.985 | 0.987 |
| 17 | .851 | .869 | .892 | .910 | .954 | .977 | .981 | .985 | .987 |
| 18 | .858 | .874 | .897 | .914 | .956 | .978 | .982 | .986 | .988 |
| 19 | .863 | .879 | .901 | .917 | .957 | .978 | .982 | .986 | .988 |
| 20 | .868 | .884 | .905 | .920 | .959 | .979 | .983 | .986 | .988 |
| 21 | 0.873 | 0.888 | 0.908 | 0.923 | 0.960 | 0.980 | 0.983 | 0.987 | 0.989 |
| 22 | .878 | .892 | .911 | .926 | .961 | .980 | .984 | .987 | .989 |
| 23 | .881 | .895 | .914 | .928 | .962 | .981 | .984 | .987 | .989 |
| 24 | .884 | .898 | .916 | .930 | .963 | .981 | .984 | .987 | .989 |
| 25 | .888 | .901 | .918 | .931 | .964 | .981 | .985 | .988 | .989 |
| 26 | 0.891 | 0.904 | 0.920 | 0.933 | 0.965 | 0.982 | 0.985 | 0.988 | 0.989 |
| 27 | .894 | .906 | .923 | .935 | .965 | .982 | .985 | .988 | .990 |
| 28 | .896 | .908 | .924 | .936 | .966 | .982 | .985 | .988 | .990 |
| 29 | .898 | .910 | .926 | .937 | .966 | .982 | .985 | .988 | .990 |
| 30 | .900 | .912 | .927 | .939 | .967 | .983 | .985 | .988 | .900 |
| 31 | 0.902 | 0.914 | 0.929 | 0.940 | 0.967 | 0.983 | 0.986 | 0.988 | 0.990 |
| 32 | .904 | .915 | .930 | .941 | .968 | .983 | .986 | .988 | .990 |
| 33 | .906 | .917 | .931 | .942 | .968 | .983 | .986 | .989 | .990 |
| 34 | .908 | .919 | .933 | .943 | .969 | .983 | .986 | .989 | .990 |
| 35 | .910 | .920 | .934 | .944 | .969 | .984 | .986 | .989 | .990 |

(continued)

TABLE 5.5 (continued)

| | Significance level $\alpha$ | | | | | | | | |
|---|---|---|---|---|---|---|---|---|---|
| | Lower tail | | | | | | | | Upper tail |
| n | 0.01 | 0.02 | 0.05 | 0.10 | 0.50 | 0.10 | 0.05 | 0.02 | 0.01 |
| 36 | 0.912 | 0.922 | 0.935 | 0.945 | 0.970 | 0.984 | 0.986 | 0.989 | 0.990 |
| 37 | .914 | .924 | .936 | .946 | .970 | .984 | .987 | .989 | .990 |
| 38 | .916 | .925 | .938 | .947 | .971 | .984 | .987 | .989 | .990 |
| 39 | .917 | .927 | .939 | .948 | .971 | .984 | .987 | .989 | .991 |
| 40 | .919 | .928 | .940 | .949 | .972 | .985 | .987 | .989 | .991 |
| 41 | 0.920 | 0.929 | 0.941 | 0.950 | 0.972 | 0.985 | 0.987 | 0.989 | 0.991 |
| 42 | .922 | .930 | .942 | .951 | .972 | .985 | .987 | .989 | .991 |
| 43 | .923 | .932 | .943 | .951 | .973 | .985 | .987 | .990 | .991 |
| 44 | .924 | .933 | .944 | .952 | .973 | .985 | .987 | .990 | .991 |
| 45 | .926 | .934 | .945 | .953 | .973 | .985 | .988 | .990 | .991 |
| 46 | 0.927 | 0.935 | 0.945 | 0.953 | 0.974 | 0.985 | 0.988 | 0.990 | 0.991 |
| 47 | .928 | .936 | .946 | .954 | .974 | .985 | .988 | .990 | .991 |
| 48 | .929 | .937 | .947 | .954 | .974 | .985 | .988 | .990 | .991 |
| 49 | .929 | .937 | .947 | .955 | .974 | .985 | .988 | .990 | .991 |
| 50 | .930 | .938 | .947 | .955 | .974 | .985 | .988 | .990 | .991 |

Taken from Shapiro and Wilk (1965), with permission of the authors and of the Biometrika Trustees.

E 5.10.3.3 Example

For the data of Table 5.3, Y in Step (a) above is (using coefficients from Table 5.4) $Y = 0.4734(270 - 156) + 0.3211(246 - 162) + \cdots = 131.95$. $S^2$ is 17845 and $W = 0.976$. Reference to Table 5.5 shows W to be significant at approximately the 10% level, upper tail, indicating a very good normal fit.

### 5.10.4 The Shapiro-Francia Test

A test similar to W, but for use with $n \geq 50$, was later suggested by Shapiro and Francia (1972). This is based on the observation of Gupta (1952), that the estimate $\hat{\sigma}$ is almost the same if $V^{-1}$ is ignored in (5.11); the test statistic then given by Shapiro and Francia is

$$W' = \frac{(m'X)^2}{(m'm)S^2}$$

As has already been observed, W' is equivalent to the sample correlation statistic $R^2(X, m)$ given by equation (5.7) of Section 5.7.1. A justification for the equivalence, for large n, of W and W' has been given by Stephens (1975), who showed that, for large n, $V^{-1}m \sim 2m$; then $R^2/n$ and $C^2/n$ in Section 5.10.3 will have limits 2 and 4 and W reduces to $W' = R^2(X, m)$.

### 5.10.5 The d'Agostino Test

In the above tests, $\hat{\sigma}$ is given by a linear combination of the order statistics. The coefficients are difficult to calculate for W, and in W' the formula is replaced by an easier one. Other linear combinations of order statistics can also be used to estimate $\sigma$, and one in particular, proposed by Downton (1966), has been used by d'Agostino to provide a further adaption of the Shapiro-Wilk statistic. d'Agostino's statistic is given by

$$D_A = [\Sigma X_{(i)}\{i - \tfrac{1}{2}(n + 1)\}]/(S n^{3/2}) \qquad (5.15)$$

The statistic $D_A$ is easier to calculate than W or W' but must be used with both tails; d'Agostino has shown by Monte Carlo studies that alternative distributions may produce large or small values of the statistic. D'Agostino (1971, 1972, 1973) tabled percentage points for a standardized value of $D_A$, given by $Y = \{D_A - (2\sqrt{\pi})^{-1}\} \sqrt{n}/0.02998598$, and gave power studies (see Chapter 9). Note that $D_A$ can be regarded as a correlation statistic, based on the correlation $R(X, T)$ where now $T_i = \{i - \tfrac{1}{2}(n + 1)\}$; however, the fact that $T_i$ is not near $m_i$ means that $D_A$ can take significantly large or small values when the sample is not normal.

### 5.10.6 Power Studies for Regression Tests of Normality

Shapiro and Wilk (1965) and Shapiro, Wilk, and Chen (1968) have given extensive Monte Carlo studies to compare W with other test statistics for normality. Their studies indicate that W is a powerful omnibus test, particularly when compared with the Pearson chi-square tests and against other older tests, for example, those based on skewness $b_1$ and kurtosis $b_2$, or u = range/(standard deviation). Stephens (1974) also gave comparisons, particularly of W with EDF statistics, and pointed out that low power results for the latter, given in the papers quoted above, are based on non-comparable test situations. Nevertheless, over a wide range of alternative distributions, W gives slightly better power than EDF statistics $A^2$ and $W^2$, and considerably better than the Kolmogorov D, or the Pearson $X^2$ or Pearson-Fisher $X^2$ discussed in Chapter 3. For large samples, Stephens (1974) also compared W, W', and $D_A$ for power over a wide range of alternatives. The power of W' is marginally less than that of W when W is available, and that of $D_A$ is smaller again. Thus the power drops as the statistic becomes more easy to calculate. Dyer (1974) has shown that $W'$ has good power properties. For large samples these studies are effectively showing the value of the correlation coefficients $R^2(X, m)$ or $R^2(X, H)$ as test statistics for normality. Of these statistics, it would appear to be best to use the Shapiro-Wilk $W$ for small samples, and $Z(X, m)$ for larger samples ($n \geq 50$), but further comparisons would be useful.

## 5.11 REGRESSION TESTS FOR THE EXPONENTIAL DISTRIBUTION

### 5.11.1 Introduction

In this and the following sections we discuss tests of

$H_0$: a random sample of X-values comes from $\text{Exp}(\alpha, \beta)$

that is, the distribution

$$F_0(x) = 1 - \exp\{-(x - \alpha)/\beta\}, \quad x > \alpha, \; \beta > 0 \qquad (5.16)$$

Parameter $\alpha$ is the origin of the distribution, and $\beta$ is the scale parameter. First suppose that both $\alpha$ and $\beta$ are unknown.

### 5.11.2 Correlation Coefficient Tests

For the exponential distribution, $m_i = \Sigma^i_{j=1} (n - j + 1)^{-1}$ and so can be calculated without numerical integration. The test statistic $Z(X, m) = n\{1 - R^2(X, m)\}$, for either complete or right-censored samples of Type 2, is referred to Table 5.6. Points in this table were found from Monte Carlo samples, using 10,000 samples for each n.

Also, for the exponential distribution, $H_i = -\log\{1 - i/(n + 1)\}$; the correlation test statistic using H is then $Z(X, H) = n\{1 - R^2(X, H)\}$ and is referred to Table 5.7. The points differ only slightly from those for $Z(X, m)$ except when p, the censoring ratio, is near 1. Although Spinelli and Stephens (1983; see Section 5.11.6) found $Z(X, H)$ less sensitive than $Z(X, m)$ for complete samples we include the table because $H_i$ can be calculated more easily than $m_i$. It might also be true that the power of $Z(X, H)$ improves for right-censored samples where the influential tail observations are not available. Smith and Bain (1976) have given a table of points for $Z(X, H)/n$, and Lockhart and Stephens (1986) have discussed asymptotic theory for both $Z(X, m)$ and $Z(X, H)$. Examples of the use of these statistics are given in Chapter 10.

TABLE 5.6 Upper Tail Percentage Points for $Z = n\{1 - R^2(X, m)\}$ for a Test for Exponentiality, Parameters Unknown, for Complete or Type-2 Right-Censored Data; $p = r/n$ is the censoring ratio

| | n | \multicolumn{7}{c}{Significance level $\alpha$} | | | | | | |
|---|---|---|---|---|---|---|---|---|
| | | 0.50 | 0.25 | 0.15 | 0.10 | 0.05 | 0.025 | 0.01 |
| p = 0.2 | 20 | 1.74 | 2.83 | 3.68 | 4.40 | 5.44 | 6.24 | 6.97 |
| | 40 | 2.53 | 3.95 | 5.00 | 5.87 | 7.32 | 8.81 | 10.71 |
| | 60 | 2.80 | 4.32 | 5.44 | 6.34 | 7.90 | 9.47 | 11.64 |
| | 80 | 2.95 | 4.51 | 5.68 | 6.58 | 8.20 | 9.81 | 12.13 |
| | 100 | 3.05 | 4.63 | 5.82 | 6.74 | 8.39 | 10.03 | 12.44 |
| p = 0.4 | 10 | 0.86 | 1.41 | 1.82 | 2.16 | 2.68 | 3.09 | 3.50 |
| | 20 | 1.25 | 1.98 | 2.51 | 2.96 | 3.73 | 4.49 | 5.46 |
| | 40 | 1.48 | 2.26 | 2.84 | 3.34 | 4.13 | 4.96 | 6.15 |
| | 60 | 1.54 | 2.33 | 2.93 | 3.42 | 4.23 | 5.08 | 6.27 |
| | 80 | 1.57 | 2.38 | 2.98 | 3.47 | 4.28 | 5.14 | 6.34 |
| | 100 | 1.59 | 2.40 | 3.01 | 3.49 | 4.31 | 5.18 | 6.37 |
| p = 0.6 | 10 | 0.75 | 1.22 | 1.55 | 1.81 | 2.25 | 2.65 | 3.19 |
| | 20 | 0.95 | 1.47 | 1.86 | 2.17 | 2.69 | 3.25 | 4.03 |
| | 40 | 1.06 | 1.60 | 2.02 | 2.36 | 2.93 | 3.54 | 4.36 |
| | 60 | 1.09 | 1.65 | 2.05 | 2.39 | 2.96 | 3.56 | 4.37 |
| | 80 | 1.11 | 1.66 | 2.07 | 2.40 | 2.98 | 3.58 | 4.39 |
| | 100 | 1.12 | 1.66 | 2.08 | 2.41 | 2.99 | 3.59 | 4.40 |
| p = 0.8 | 10 | 0.65 | 1.04 | 1.32 | 1.55 | 1.93 | 2.31 | 2.83 |
| | 20 | 0.79 | 1.22 | 1.55 | 1.81 | 2.27 | 2.76 | 3.46 |
| | 40 | 0.87 | 1.32 | 1.67 | 1.94 | 2.43 | 2.85 | 3.56 |
| | 60 | 0.90 | 1.35 | 1.69 | 1.97 | 2.45 | 2.92 | 3.63 |
| | 80 | 0.91 | 1.37 | 1.71 | 1.99 | 2.46 | 2.95 | 3.66 |
| | 100 | 0.92 | 1.37 | 1.72 | 2.00 | 2.47 | 2.97 | 3.68 |
| p = 0.9 | 10 | 0.62 | 1.00 | 1.27 | 1.49 | 1.85 | 2.23 | 2.68 |
| | 20 | 0.79 | 1.21 | 1.53 | 1.79 | 2.25 | 2.74 | 3.45 |
| | 40 | 0.86 | 1.32 | 1.66 | 1.94 | 2.42 | 2.96 | 3.67 |
| | 60 | 0.89 | 1.35 | 1.69 | 1.97 | 2.46 | 2.99 | 3.70 |
| | 80 | 0.91 | 1.36 | 1.70 | 1.99 | 2.49 | 3.01 | 3.71 |
| | 100 | 0.92 | 1.37 | 1.71 | 2.00 | 2.50 | 3.02 | 3.72 |
| p = 0.95 | 10 | 0.63 | 1.03 | 1.31 | 1.53 | 1.89 | 2.24 | 2.67 |
| | 20 | 0.81 | 1.26 | 1.59 | 1.89 | 2.40 | 2.89 | 3.73 |
| | 40 | 0.93 | 1.41 | 1.80 | 2.11 | 2.71 | 3.26 | 4.06 |
| | 60 | 0.97 | 1.47 | 1.85 | 2.16 | 2.73 | 3.30 | 4.14 |
| | 80 | 0.99 | 1.50 | 1.88 | 2.18 | 2.74 | 3.32 | 4.18 |
| | 100 | 1.00 | 1.52 | 1.89 | 2.20 | 2.74 | 3.34 | 4.20 |
| p = 1.0 | 10 | 0.64 | 1.05 | 1.34 | 1.56 | 1.92 | 2.25 | 2.67 |
| | 20 | 0.92 | 1.46 | 1.87 | 2.20 | 2.77 | 3.35 | 4.10 |
| | 40 | 1.26 | 2.00 | 2.58 | 3.05 | 3.94 | 4.92 | 6.42 |
| | 60 | 1.47 | 2.32 | 2.95 | 3.52 | 4.67 | 5.94 | 8.01 |
| | 80 | 1.64 | 2.58 | 3.30 | 3.96 | 5.25 | 6.81 | 9.33 |
| | 100 | 1.78 | 2.78 | 3.57 | 4.30 | 5.70 | 7.49 | 10.35 |

TABLE 5.7 Upper Tail Percentage Points for $Z = n\{1 - R^2(X, H)\}$ for a Test for Exponentiality, Parameters Unknown, for Complete or Type 2 Right-Censored Data; $p = r/n$ is the censoring ratio

| | n | 0.50 | 0.25 | 0.15 | 0.10 | 0.05 | 0.025 | 0.01 |
|---|---|---|---|---|---|---|---|---|
| | | | | | Significance level $\alpha$ | | | |
| p = 0.2 | 20 | 2.13 | 3.38 | 4.33 | 5.12 | 6.40 | 7.59 | 8.94 |
| | 40 | 2.68 | 4.14 | 5.25 | 6.15 | 7.61 | 9.06 | 11.10 |
| | 60 | 2.89 | 4.42 | 5.57 | 6.47 | 8.04 | 9.63 | 11.87 |
| | 80 | 3.00 | 4.57 | 5.73 | 6.64 | 8.27 | 9.92 | 12.27 |
| | 100 | 3.06 | 4.66 | 5.83 | 6.74 | 8.40 | 10.11 | 12.52 |
| p = 0.4 | 10 | 0.86 | 1.41 | 1.82 | 2.16 | 2.69 | 3.08 | 3.48 |
| | 20 | 1.25 | 1.98 | 2.51 | 2.96 | 3.73 | 4.49 | 5.39 |
| | 40 | 1.48 | 2.25 | 2.84 | 3.33 | 4.13 | 4.96 | 6.14 |
| | 60 | 1.54 | 2.33 | 2.93 | 3.42 | 4.23 | 5.08 | 6.27 |
| | 80 | 1.57 | 2.38 | 2.98 | 3.46 | 4.28 | 5.14 | 6.33 |
| | 100 | 1.59 | 2.40 | 3.00 | 3.49 | 4.31 | 5.18 | 6.37 |
| p = 0.6 | 10 | 0.75 | 1.22 | 1.54 | 1.80 | 2.25 | 2.65 | 3.18 |
| | 20 | 0.95 | 1.47 | 1.85 | 2.17 | 2.70 | 3.27 | 4.01 |
| | 40 | 1.06 | 1.61 | 2.02 | 2.36 | 2.93 | 3.53 | 4.36 |
| | 60 | 1.09 | 1.64 | 2.05 | 2.39 | 2.96 | 3.56 | 4.37 |
| | 80 | 1.11 | 1.66 | 2.07 | 2.40 | 2.98 | 3.58 | 4.39 |
| | 100 | 1.11 | 1.67 | 2.08 | 2.41 | 2.99 | 3.58 | 4.40 |
| p = 0.8 | 10 | 0.65 | 1.03 | 1.31 | 1.53 | 1.93 | 2.31 | 2.81 |
| | 20 | 0.79 | 1.22 | 1.55 | 1.81 | 2.28 | 2.78 | 3.48 |
| | 40 | 0.86 | 1.32 | 1.67 | 1.94 | 2.44 | 2.96 | 3.56 |
| | 60 | 0.90 | 1.35 | 1.69 | 1.97 | 2.46 | 2.96 | 3.64 |
| | 80 | 0.92 | 1.37 | 1.71 | 1.99 | 2.47 | 2.96 | 3.69 |
| | 100 | 0.93 | 1.37 | 1.71 | 2.00 | 2.48 | 2.97 | 3.71 |
| p = 0.9 | 10 | 0.62 | 0.99 | 1.25 | 1.47 | 1.86 | 2.25 | 2.75 |
| | 20 | 0.78 | 1.20 | 1.53 | 1.81 | 2.31 | 2.80 | 3.57 |
| | 40 | 0.85 | 1.32 | 1.66 | 1.94 | 2.43 | 2.99 | 3.75 |
| | 60 | 0.89 | 1.35 | 1.69 | 1.98 | 2.48 | 3.03 | 3.77 |
| | 80 | 0.91 | 1.36 | 1.71 | 1.99 | 2.50 | 3.05 | 3.77 |
| | 100 | 0.92 | 1.37 | 1.72 | 2.00 | 2.51 | 3.06 | 3.78 |
| p = 0.95 | 10 | 0.62 | 1.02 | 1.29 | 1.51 | 1.90 | 2.28 | 2.74 |
| | 20 | 0.80 | 1.25 | 1.61 | 1.88 | 2.43 | 3.05 | 3.96 |
| | 40 | 0.92 | 1.41 | 1.82 | 2.15 | 2.78 | 3.41 | 4.29 |
| | 60 | 0.96 | 1.48 | 1.87 | 2.19 | 2.79 | 3.40 | 4.30 |
| | 80 | 0.98 | 1.51 | 1.89 | 2.20 | 2.79 | 3.40 | 4.31 |
| | 100 | 1.00 | 1.53 | 1.90 | 2.22 | 2.79 | 3.40 | 4.31 |
| p = 1.0 | 10 | 0.63 | 1.01 | 1.30 | 1.54 | 1.94 | 2.31 | 2.74 |
| | 20 | 0.88 | 1.44 | 1.89 | 2.27 | 2.99 | 3.71 | 4.76 |
| | 40 | 1.19 | 1.99 | 2.69 | 3.33 | 4.57 | 5.90 | 7.86 |
| | 60 | 1.39 | 2.33 | 3.20 | 3.96 | 5.52 | 7.33 | 9.85 |
| | 80 | 1.55 | 2.63 | 3.62 | 4.49 | 6.35 | 8.52 | 11.52 |
| | 100 | 1.67 | 2.86 | 3.94 | 4.90 | 7.00 | 9.44 | 12.83 |

### 5.11.3 Tests Based on the Residuals

As was described before, tests may also be based on the ESS of Section 5.4.1, divided by a quadratic form in the $X_{(i)}$. When simple least squares is used, and when the divisor is $S^2$, regression on m yields $n(ESS/S^2) = Z(X, m)$ as test statistic, and regression on H yields $Z(X, H)$. The divisor $S^2$ is an estimate of $n\beta^2$, but for the exponential distribution it might be better to use the estimate $n\bar{X}^2$, since $\bar{X}$ is a sufficient estimator of $\beta$. The corresponding test statistic $ESS/(n\bar{X}^2)$ does not appear to have been investigated.

### 5.11.4 Tests Based on Residuals, When the Origin Is Known

Situations often arise in testing for exponentiality where $\alpha$ is known; usually $\alpha = 0$. If $\alpha$ is $\alpha_0$, the substitution $X' = X - \alpha_0$ will reduce the test to a test of $H_0$, with $\alpha = 0$, on the $X'$ values. Thus we can suppose the null hypothesis is

$H_0$: a set of values of X is from $Exp(0, \beta)$

In the test statistics $R^2(X, m)$ or $R^2(X, H)$ the fact that $\alpha$ is 0 is not used; however, the line $E(X_{(i)}) = \beta m_i$ can be fitted to the pairs $\{m_i, X_{(i)}\}$ and a test statistic can be constructed using the ESS divided by a suitable estimate of $\beta^2$, similar to those in Section 5.11.3 above. If $\alpha$ and $\beta$ were both known, a natural statistic on these lines would be $M_E^2 = \Sigma(X'_{(i)} - m_i)^2$ where $X'_{(i)} = (X_{(i)} - \alpha)/\beta$; this is analogous to $M_N^2$ of Section 5.10.2. Lockhart and Stephens (1986) have studied statistics based on residuals.

### 5.11.5 Tests Based on the Ratio of Two Estimates of Scale

#### 5.11.5.1 The Shapiro-Wilk Test, for Origin and Scale Unknown

For the exponential distribution, the estimates in (5.9) become

$$\hat{\alpha} = X_{(1)} \quad \text{and} \quad \hat{\beta} = \frac{n(\bar{X} - X_{(1)})}{n-1} \tag{5.17}$$

and the comparison of $\hat{\beta}^2$ with the sample variance leads to the Shapiro-Wilk (1972) statistic

$$W_E = \frac{n(\bar{X} - X_{(1)})^2}{(n-1)S^2} \tag{5.18}$$

Thus the test for exponentiality with origin and scale unknown is as follows:

(a)  Calculate $W_E$ from (5.18).
(b)  Refer $W_E$ to Table 5.8, using, in general, a two-tail test.

Shapiro and Wilk pointed out that, in general, $W_E$ will give a two-tail test, since for alternative distributions $W_E$ may take either low or high values. Shapiro and Wilk (1972) gave a table of percentage points for $W_E$ based on Monte Carlo studies; Table 5.8 is adapted from their table. Currie (1980) has since calculated the points by numerical methods. Points for $W_E$ can also be found from those of Greenwood's statistic based on uniform spacings (see Section 10.9.3.2). The test is discussed in Section 5.12.

### 5.11.5.2  Adaptation of the Shapiro-Wilk Statistic for Known Origin

It is often required to test for $F(x)$ in (5.16) with $\alpha$ known; for the present, suppose $\alpha = 0$. The estimate of $\beta$ in the new model $X_{(i)} = \beta m_i$ now becomes $\hat{\beta} = \bar{X}$, the same as the maximum likelihood estimate, and the corresponding Shapiro-Wilk statistic would be based on $\bar{X}^2/S^2$; Hahn and Shapiro (1967) have proposed $WE_0 = S^2/(n\bar{X})^2$, and have given percentage points derived from Monte Carlo methods. Stephens (1978) later gave a test statistic, here called $W_S$, which, for sample size n, has the same distribution as $W_E$ for sample size $n + 1$. Statistics $WE_0$ and $W_S$ are in fact equivalent, and both are equivalent to Greenwood's statistic based on spacings (see Section 10.9.3.2). The test based on $W_S$ will be given here since no new tables are necessary for its use. Suppose, returning to the general situation, that the known origin has value $\alpha = \alpha_0$. The steps in the test are as follows:

(a)  Let $z_i = X_i - \alpha_0$, for $i = 1, 2, \ldots, n$.
(b)  Calculate

$$A = \sum_{i=1}^{n} z_i \quad \text{and} \quad B = \sum_{i=1}^{n} z_i^2$$

(c)  Calculate $W_S = A^2/[n\{(n + 1)B - A^2\}]$.
(d)  Refer $W_S$ to Table 5.8, using the percentage points given for sample size $n + 1$, and using a two-tail test.

$W_S$ can also be calculated by adding one extra value $X_{n+1}$ equal to $\alpha_0$, to the given sample of X-values of size n, and then using all $n + 1$ values to calculate $W_E$ from (5.18). This is a useful device if a computer program is already available for $W_E$. Stephens (1978) showed that, for most alternatives, use of $W_S$ gives greater power than using $W_E$ as though $\alpha$ were not known.

Shapiro and Wilk (1972) also discussed the situation when it is desired to test the exponential distributional form, and at the same time to test that

TABLE 5.8   Percentage Points for $W_E$ and $W_S$ for Testing Exponentiality

| n | Lower tail | | | | | | Upper tail | | | | |
|---|---|---|---|---|---|---|---|---|---|---|---|
| | **Significance level $\alpha$** | | | | | | | | | | |
| | 0.005 | 0.01 | 0.025 | 0.05 | 0.10 | 0.50 | 0.10 | 0.05 | 0.025 | 0.01 | 0.005 |
| 3 | .252 | .254 | .260 | .270 | .292 | .571 | .971 | .993 | .998 | .9997 | .9999 |
| 4 | .124 | .130 | .143 | .160 | .189 | .377 | .751 | .858 | .924 | .968 | .984 |
| 5 | .085 | .091 | .105 | .119 | .144 | .288 | .555 | .668 | .759 | .860 | .919 |
| 6 | .061 | .067 | .080 | .096 | .117 | .228 | .429 | .509 | .584 | .678 | .750 |
| 7 | .051 | .059 | .070 | .081 | .097 | .187 | .347 | .416 | .485 | .571 | .643 |
| 8 | .045 | .051 | .061 | .071 | .085 | .163 | .293 | .350 | .403 | .485 | .543 |
| 9 | .040 | .044 | .054 | .063 | .075 | .142 | .255 | .301 | .345 | .402 | .443 |
| 10 | .037 | .040 | .049 | .057 | .068 | .123 | .218 | .253 | .288 | .339 | .370 |
| 12 | .031 | .036 | .041 | .049 | .057 | .101 | .172 | .202 | .236 | .272 | .298 |
| 14 | .027 | .033 | .036 | .043 | .050 | .085 | .142 | .165 | .186 | .213 | .232 |
| 16 | .023 | .028 | .033 | .037 | .044 | .073 | .119 | .136 | .154 | .177 | .193 |
| 18 | .021 | .025 | .029 | .033 | .039 | .064 | .102 | .116 | .131 | .148 | .167 |

| n | | | | | | | | | | | | |
|-----|------|------|------|------|------|------|------|------|------|------|------|------|
| 20 | .020 | .023 | .026 | .030 | .035 | .057 | .088 | .100 | .112 | .129 | .137 |
| 25 | .017 | .019 | .022 | .025 | .029 | .045 | .067 | .075 | .084 | .093 | .100 |
| 30 | .015 | .016 | .019 | .021 | .024 | .036 | .054 | .059 | .064 | .072 | .079 |
| 35 | .013 | .014 | .017 | .019 | .021 | .031 | .044 | .049 | .054 | .059 | .064 |
| 40 | .012 | .013 | .015 | .016 | .019 | .027 | .038 | .041 | .045 | .050 | .051 |
| 45 | .011 | .012 | .013 | .015 | .017 | .024 | .033 | .036 | .039 | .042 | .045 |
| 50 | .010 | .011 | .012 | .014 | .015 | .021 | .029 | .032 | .034 | .036 | .039 |
| 55 | .009 | .010 | .012 | .013 | .014 | .019 | .026 | .028 | .030 | .032 | .034 |
| 60 | .009 | .010 | .011 | .012 | .013 | .018 | .023 | .025 | .027 | .029 | .031 |
| 65 | .008 | .009 | .010 | .011 | .012 | .016 | .022 | .023 | .025 | .027 | .028 |
| 70 | .008 | .008 | .009 | .010 | .011 | .015 | .019 | .021 | .022 | .024 | .026 |
| 75 | .007 | .008 | .009 | .010 | .011 | .014 | .018 | .019 | .021 | .022 | .023 |
| 80 | .007 | .008 | .008 | .009 | .010 | .013 | .017 | .018 | .019 | .021 | .021 |
| 85 | .007 | .007 | .008 | .009 | .009 | .012 | .016 | .017 | .017 | .019 | .019 |
| 90 | .006 | .007 | .008 | .008 | .009 | .012 | .015 | .016 | .016 | .018 | .018 |
| 95 | .006 | .007 | .007 | .008 | .008 | .011 | .014 | .015 | .015 | .016 | .017 |
| 100 | .006 | .006 | .007 | .007 | .008 | .010 | .013 | .014 | .015 | .015 | .016 |

Adapted from Shapiro and Wilk (1972), with permission of the authors and of the American Statistical Association.

$\alpha = \alpha_0$, where $\alpha_0$ is a given constant. The test statistic for $\alpha$ is based on

$$U = (X_{(1)} - \alpha_0)/(\bar{X} - X_{(1)}) \qquad (5.20)$$

When the observations are exponential, U is distributed independently of $W_E$, and Shapiro and Wilk combine the two statistics by Fisher's method (Section 8.15).

### 5.11.5.3 The Jackson Statistic

A statistic suggested by Jackson (1967) is effectively a comparison between the slope of the regression line through the origin, when $X_{(i)}$ is plotted against $m_i$ and the covariance of the X-values is ignored, and $\bar{X}$, the maximum likelihood estimator of $\beta$. The statistic is

$$J = \Sigma m_i X_{(i)}/(n\bar{X})$$

In general, J is used as a two-tail test. Jackson has given moments of J, and percentage points for n = 5, 10, 15, 20. These are based on curve-fitting, using the moments.

### 5.11.5.4 The de Wet-Venter Statistic

De Wet and Venter (1973) have devised a statistic dependent on the ratio of two asymptotically efficient estimators of $\beta$. It is straightforward to apply their method to the exponential test with $\alpha = 0$, for a complete sample; the test statistic is $V_E = n\bar{X}/\{\Sigma X_{(i)}^2/H_i\}$ where $H_i = -\log\{1 - i/(n + 1)\}$. De Wet and Venter have given asymptotic null distribution theory for $V_E$.

### 5.11.6 Comparison of Regression Tests for Exponentiality

Spinelli and Stephens (1987) have reported results on power studies to compare $R^2(X, m)$ and $R^2(X, H)$ in tests for exponentiality, with EDF statistics $W^2$ and $A^2$, and with $W_E$; for these studies both $\alpha$ and $\beta$ were assumed unknown. On the whole, the correlation statistics $R^2(X, m)$ and $R^2(X, H)$ were less effective than EDF statistics or than $W_E$, particularly for large sample sizes. For tests for this distribution (in contrast to tests for normality) we might also expect some difference between $R^2(X, m)$ and $R^2(X, H)$; this emerges clearly from the studies, with $R^2(X, H)$ less powerful overall than $R^2(X, m)$. Statistic $W_E$ has good power over a wide range of alternatives, although it will have lower power against alternatives with coefficient of variation equal to 1 (see Section 5.12).

## 5.12  TESTS BASED ON THE RATIO OF TWO ESTIMATES
## OF SCALE: FURTHER COMMENTS

In Section 5.3 it was noted that Sarkadi (1975), for normality, and Gerlach (1979) more generally, have proved the consistency of tests based on the correlation coefficient $R(X, m)$. The value of $R(X, m)$ will, loosely speaking, approach 1 as the fit gets better. The test based on $R(X, m)$ is then a one-tail test; the null hypothesis is rejected only for small values of R, or for large values of $Z = n(1 - R^2(X, m))$.

This consistency does not necessarily extend to correlation statistics $R(X, T)$ when T is not m. For example, tests based on the ratio of the generalized least squares estimate of $\beta$ with an estimate obtained from the sample variance can be put in terms of correlation statistics, but will not generally be consistent. For the Shapiro-Wilk test for normality in Section 5.10.3, $\hat{\sigma}/S$ is equivalent to the correlation coefficient $R(X, T)$ where $T_i$ is the i-th component of $T = V^{-1}m$. Then if T is proportional to m, or very nearly so, the graph of $X_{(i)}$ against $T_i$ will be approximately a straight line; $R(X, T)$ will be a good measure of fit, and low values of $R(X, T)$ will lead to rejection of the normal model. For the normal distribution, T is nearly proportional to m, since $Vm \approx m/2$ implying $V^{-1}m \approx 2m$ (Stephens 1975), with the approximation becoming better for large n; then the Shapiro-Wilk W approaches the Shapiro-Francia W', and this is the same as the correlation statistic $R^2(X, m)$, which is consistent.

However, the normal case is exceptional. Even for other symmetric distributions $V^{-1}m$ will not as a rule be proportional to m, even asymptotically, and the situation is more complicated for non-symmetric distributions, such as the exponential. For such distributions the test based on the ratio of $\hat{\beta}^2$ to $S^2$ is equivalent to the correlation $R^2(X, T)$ with the vector $T = 1'V^{-1}(1m' - m1')V^{-1}$, and for most distributions this vector will not even be close to m. For example, for the exponential distribution, with a sample of size n, T is proportional to a vector with one component equal to $-(n - 1)$, and the other n - 1 components all equal to 1. A plot of $X_{(i)}$ against $T_i$, even for a "perfect" exponential sample with $X_{(i)} = m_i$, would not be close to a straight line, and the value of $R(X, T)$ would not be close to 1. The statistic $R(X, T)$ then gives no indication of the fit in the sense that a large value indicates a good fit and a small value a bad fit. In practical terms, this means that for $W_E$, equivalent to $R^2(X, T)$, both tails are needed to make the test. Also, the test statistic will not be consistent.

For the exponential distribution, the coefficient of variation $C_V = \sigma/\mu$ is 1, and for large samples $nW_E$ converges in probability to $1/C_V^2 = 1$; however, there are other distributions for which $nW_E$ would also converge to 1 (for example, the Beta distribution Beta $(x; p, q)$ defined in Section 8.8.2, with $p < 1$, and $q = p\{p + 1\}/\{1 - p\}$), and $W_E$ will not detect these alternatives. Nevertheless, when the alternatives to exponentiality are identified and exclude such distributions, $W_E$ can be powerful, as was reported in Section 5.11.6; see also Section 10.14.

For the uniform distribution, the Shapiro-Wilk method gives $W_u = (n - 1) \{ X_{(n)} - X_{(1)} \}^2/S^2$; this also can be shown to be inconsistent. Similar objections apply to the corresponding test for the extreme-value distribution, which was examined by Spinelli (1980).

In general, consistency of a test based on the ratio of two estimates of scale must depend critically on how these are chosen, and this question deserves closer examination. It is intuitively reasonable that efficient estimates, or estimates which are at least asymptotically efficient, will give better tests. The ESS, after fitting the line (5.3) may not be asymptotically efficient for the chosen T vector, nor will be the sample variance in the denominator except in the test for normality; this will affect the consistency of tests which use these estimates. De Wet and Venter (1973) have devised a general procedure for tests of distributions when only the scale is unknown, in which the ratio of two asymptotically efficient estimates is used. The authors gave asymptotic theory, and illustrated with tests for the Gamma distribution. The specific case for the exponential distribution is given in Section 5.11.5 above. However, computing formulas for the test statistic for other Gamma distributions, and tables for finite n are not available. Practical aspects of the method also remain to be explored, such as how the power is influenced by which estimates are used in the test ratio.

## 5.13   REGRESSION TESTS FOR OTHER DISTRIBUTIONS: GENERAL COMMENTS

The various techniques above for tests of normality and exponentiality have not been so extensively developed for other distributions, and the tests to follow, for the extreme-value, logistic, and Cauchy distributions, are all based on the simple correlation coefficients $R^2(X, H)$. H is used for computational simplicity rather than m, although for some distributions (for example, the extreme-value, see Lawless, 1982) good approximations to m have been found and so tests could be developed; there may be some difference in sensitivity between tests based on $R^2(X, H)$ and those based on $R^2(X, m)$ as was found for the exponential distribution (Section 5.11.6). On this question more work needs to be done; also more comparisons are needed between correlation tests and others. The tables given in connection with the tests below were found from Monte Carlo studies with 10,000 samples for each n. The tests are given for Type 2 censored data: objections to tests for Type 1 censored data are similar to those given in Section 5.7.3 in connection with tests for normality.

## 5.14  CORRELATION TESTS FOR THE EXTREME-VALUE DISTRIBUTION

### 5.14.1  Version 1

The null hypothesis is

$H_0$: a random sample of X-values comes from

$$F_0(x;\alpha,\beta) = \exp[-\exp\{-(x - \alpha)/\beta\}], \quad -\infty < x < \infty, \ \beta > 0 \qquad (5.21)$$

The test given is for complete or Type 2 right-censored samples; for left-censored data see Section 5.14.2. For this distribution $H_i = -\log[-\log\{i/(n + 1)\}]$. This version of the extreme-value distribution has a long tail to the right and is used to model data in, for example, reliability studies. The test statistic $Z = n\{1 - R^2(X,H)\}$ can be calculated from the r smallest observations; $H_0$ is rejected if Z exceeds the appropriate value in Table 5.9. The table is entered, for Type 2 censored data, at $p = r/n$ and at n.

### 5.14.2  Version 2

Another version of the extreme-value distribution is

$$F_0(x';\alpha,\beta) = 1 - \exp[-\exp\{(x' - \alpha)/\beta\}], \quad -\infty < x < \infty, \ \beta > 0 \qquad (5.22)$$

This is the distribution of $X' = -X$, where X has distribution (5.21). For this distribution $H_i = \log[-\log\{1 - i/(n + 1)\}]$. The test statistic $Z = n\{1 - R^2(X,H)\}$, calculated from the r smallest observations (Type 2 censoring) is referred to Table 5.10. For left-censored data from (5.21), the signs can be changed and tested as right-censored data from (5.22), and vice-versa. These tables can also be used to give a test for the two-parameter Weibull distribution (density $W(x;0,\beta,m)$ of Section 4.11, also given in Equation (10.4)). This is a distribution with a long right tail, also used in reliability and survival studies. Here, left-censored data values are transformed by $Y_{(n+1-i)} = -\log(X_{(i)})$ and the Y-values tested to come from (5.21), since they will be right-censored; similarly, right-censored Weibull test data are transformed by $Y_{(i)} = \log(X_{(i)})$ and tested to come from (5.22). Gerlach (1979) has considered the test for (5.22) based on $R^2(X,m)$; the test is shown to be consistent and tables are given to make the test; tables for $Z(X,m)$ are given by Stephens (1986). An example is given in Section 11.4.1.3.

## 5.15  CORRELATION TESTS FOR OTHER DISTRIBUTIONS

### 5.15.1  The Logistic Distribution

For the logistic distribution, $F(x) = 1/[1 + \exp\{-(x - \alpha)/\beta\}]$, $(-\infty < x < \infty, \beta > 0)$ and $H_i = \log\{i/(n + 1 - i)\}$, $i = 1, \ldots, n$. The statistic

TABLE 5.9 Upper Tail Percentage Points for $Z = n\{1 - R^2(X, H)\}$ for a
Test for the Extreme-Value Distribution, Equation (5.21), Parameters
Unknown, for Complete or Type 2 Right-Censored Data; $p = r/n$ is the
censoring ratio

|  | n | .50 | .25 | .15 | .10 | .05 | .025 | .01 |
|---|---|---|---|---|---|---|---|---|
|  |  |  |  | Significance level $\alpha$ |  |  |  |  |
|  | 20 | 1.62 | 2.81 | 3.52 | 4.07 | 4.90 | 5.69 | 7.19 |
|  | 40 | 2.58 | 4.12 | 5.18 | 6.09 | 7.48 | 8.92 | 10.47 |
| $p = 0.2$ | 60 | 3.01 | 4.68 | 5.89 | 6.90 | 8.75 | 10.45 | 13.02 |
|  | 80 | 3.33 | 5.11 | 6.45 | 7.59 | 9.71 | 11.79 | 14.50 |
|  | 100 | 3.56 | 5.42 | 6.88 | 8.12 | 10.42 | 12.84 | 15.47 |
|  | 10 | 0.81 | 1.38 | 1.75 | 2.04 | 2.52 | 2.93 | 3.39 |
|  | 20 | 1.27 | 1.98 | 2.51 | 2.94 | 3.76 | 4.56 | 5.46 |
| $p = 0.4$ | 40 | 1.62 | 2.47 | 3.09 | 3.66 | 4.66 | 5.59 | 6.85 |
|  | 60 | 1.77 | 2.71 | 3.39 | 3.93 | 4.92 | 5.88 | 7.35 |
|  | 80 | 1.88 | 2.86 | 3.58 | 4.16 | 5.19 | 6.25 | 7.72 |
|  | 100 | 1.95 | 2.95 | 3.72 | 4.33 | 5.41 | 6.55 | 7.99 |
|  | 10 | 0.74 | 1.17 | 1.49 | 1.75 | 2.16 | 2.61 | 3.18 |
|  | 20 | 0.98 | 1.49 | 1.88 | 2.20 | 2.72 | 3.28 | 4.03 |
| $p = 0.6$ | 40 | 1.15 | 1.73 | 2.15 | 2.49 | 3.08 | 3.77 | 4.66 |
|  | 60 | 1.23 | 1.82 | 2.25 | 2.61 | 3.26 | 3.92 | 4.77 |
|  | 80 | 1.28 | 1.89 | 2.34 | 2.71 | 3.35 | 4.04 | 4.91 |
|  | 100 | 1.32 | 1.94 | 2.41 | 2.78 | 3.41 | 4.12 | 5.03 |
|  | 10 | 0.64 | 0.99 | 1.25 | 1.46 | 1.79 | 2.14 | 2.58 |
|  | 20 | 0.79 | 1.19 | 1.48 | 1.71 | 2.14 | 2.58 | 3.29 |
| $p = 0.8$ | 40 | 0.90 | 1.32 | 1.63 | 1.85 | 2.27 | 2.70 | 3.32 |
|  | 60 | 0.94 | 1.37 | 1.68 | 1.94 | 2.38 | 2.79 | 3.37 |
|  | 80 | 0.97 | 1.40 | 1.72 | 1.98 | 2.41 | 2.82 | 3.37 |
|  | 100 | 0.99 | 1.42 | 1.74 | 1.99 | 2.41 | 2.84 | 3.35 |
|  | 10 | 0.61 | 0.93 | 1.24 | 1.37 | 1.71 | 2.08 | 2.51 |
|  | 20 | 0.74 | 1.13 | 1.42 | 1.64 | 2.03 | 2.44 | 3.05 |
| $p = 0.9$ | 40 | 0.84 | 1.23 | 1.53 | 1.77 | 2.17 | 2.59 | 3.14 |
|  | 60 | 0.88 | 1.28 | 1.57 | 1.80 | 2.19 | 2.59 | 3.17 |
|  | 80 | 0.91 | 1.31 | 1.59 | 1.81 | 2.20 | 2.60 | 3.18 |
|  | 100 | 0.92 | 1.32 | 1.60 | 1.82 | 2.20 | 2.60 | 3.18 |
|  | 10 | 0.61 | 0.94 | 1.23 | 1.41 | 1.76 | 2.13 | 2.60 |
|  | 20 | 0.75 | 1.14 | 1.44 | 1.68 | 2.11 | 2.57 | 3.20 |
|  | 40 | 0.85 | 1.28 | 1.60 | 1.84 | 2.28 | 2.73 | 3.33 |
| $p = 0.95$ | 60 | 0.90 | 1.33 | 1.63 | 1.88 | 2.30 | 2.74 | 3.39 |
|  | 80 | 0.93 | 1.35 | 1.65 | 1.89 | 2.31 | 2.75 | 3.43 |
|  | 100 | 0.95 | 1.36 | 1.66 | 1.90 | 2.32 | 2.75 | 3.45 |
|  | 10 | 0.61 | 0.95 | 1.23 | 1.45 | 1.81 | 2.18 | 2.69 |
|  | 20 | 0.82 | 1.30 | 1.69 | 2.03 | 2.65 | 3.36 | 4.15 |
| $p = 1.0$ | 40 | 1.04 | 1.67 | 2.23 | 2.66 | 3.63 | 4.78 | 6.42 |
|  | 60 | 1.20 | 1.93 | 2.57 | 3.18 | 4.33 | 5.69 | 7.79 |
|  | 80 | 1.32 | 2.14 | 2.87 | 3.55 | 4.92 | 6.54 | 8.86 |
|  | 100 | 1.41 | 2.30 | 3.09 | 3.82 | 5.38 | 7.22 | 9.67 |

TABLE 5.10  Upper Tail Percentage Points for $Z = n\{1 - R^2(X, H)\}$ for a Test for the Extreme-Value Distribution, Equation (5.22), Parameters Unknown, for Complete or Type 2 Right-Censored Data; $p = r/n$ is the censoring ratio

| | n | .50 | .25 | .15 | .10 | .05 | .025 | .01 |
|---|---|---|---|---|---|---|---|---|
| | | | | | Significance level $\alpha$ | | | |
| p = 0.2 | 20 | 1.52 | 2.79 | 3.54 | 4.02 | 4.69 | 6.24 | 7.72 |
| | 40 | 2.76 | 4.43 | 5.71 | 6.66 | 8.22 | 9.82 | 11.33 |
| | 60 | 3.45 | 5.58 | 7.21 | 8.41 | 10.64 | 12.83 | 15.79 |
| | 80 | 3.98 | 6.48 | 8.46 | 10.08 | 13.03 | 16.06 | 20.00 |
| | 100 | 4.37 | 7.16 | 9.45 | 11.46 | 15.06 | 18.91 | 23.56 |
| p = 0.4 | 10 | 0.76 | 1.42 | 1.77 | 2.01 | 2.40 | 3.15 | 3.96 |
| | 20 | 1.38 | 2.19 | 2.80 | 3.28 | 4.12 | 4.89 | 5.69 |
| | 40 | 1.92 | 3.11 | 4.05 | 4.86 | 6.44 | 7.94 | 9.72 |
| | 60 | 2.32 | 3.79 | 5.04 | 6.06 | 7.99 | 10.09 | 13.05 |
| | 80 | 2.63 | 4.32 | 5.80 | 7.10 | 9.65 | 12.39 | 15.93 |
| | 100 | 2.86 | 4.72 | 6.37 | 7.92 | 11.09 | 14.39 | 18.25 |
| p = 0.6 | 10 | 0.77 | 1.23 | 1.56 | 1.83 | 2.25 | 2.63 | 3.10 |
| | 20 | 1.11 | 1.80 | 2.31 | 2.73 | 3.48 | 4.27 | 5.28 |
| | 40 | 1.49 | 2.44 | 3.25 | 3.93 | 5.33 | 6.67 | 8.57 |
| | 60 | 1.77 | 2.94 | 3.94 | 4.79 | 6.46 | 8.51 | 11.21 |
| | 80 | 1.99 | 3.30 | 4.48 | 5.53 | 7.68 | 10.13 | 13.23 |
| | 100 | 2.15 | 3.57 | 4.90 | 5.12 | 8.74 | 11.44 | 14.74 |
| p = 0.8 | 10 | 0.68 | 1.08 | 1.38 | 1.61 | 2.02 | 2.43 | 2.90 |
| | 20 | 0.93 | 1.50 | 1.95 | 2.35 | 3.04 | 3.81 | 4.88 |
| | 40 | 1.22 | 2.00 | 2.69 | 3.27 | 4.46 | 5.69 | 7.51 |
| | 60 | 1.42 | 2.35 | 3.19 | 3.96 | 5.35 | 7.15 | 9.43 |
| | 80 | 1.59 | 2.54 | 3.59 | 4.49 | 6.21 | 8.31 | 11.04 |
| | 100 | 1.71 | 2.85 | 3.89 | 4.88 | 6.92 | 9.20 | 12.30 |
| p = 0.9 | 10 | 0.64 | 1.02 | 1.30 | 1.54 | 1.93 | 2.27 | 2.80 |
| | 20 | 0.86 | 1.38 | 1.81 | 2.17 | 2.87 | 3.59 | 4.51 |
| | 40 | 1.11 | 1.82 | 2.44 | 2.99 | 4.10 | 5.27 | 7.04 |
| | 60 | 1.29 | 2.13 | 2.86 | 3.57 | 4.88 | 6.45 | 8.63 |
| | 80 | 1.43 | 2.37 | 3.21 | 4.01 | 5.59 | 7.46 | 9.97 |
| | 100 | 1.53 | 2.56 | 3.48 | 4.35 | 6.17 | 8.24 | 11.01 |
| p = 0.95 | 10 | 0.62 | 0.98 | 1.26 | 1.49 | 1.86 | 2.22 | 2.74 |
| | 20 | 0.84 | 1.33 | 1.75 | 2.10 | 2.76 | 3.51 | 4.36 |
| | 40 | 1.07 | 1.73 | 2.32 | 2.82 | 3.93 | 5.05 | 6.78 |
| | 60 | 1.23 | 2.01 | 2.73 | 3.38 | 4.61 | 6.11 | 8.22 |
| | 80 | 1.36 | 2.25 | 3.04 | 3.79 | 5.27 | 7.04 | 9.43 |
| | 100 | 1.46 | 2.43 | 3.28 | 4.08 | 5.82 | 7.77 | 10.38 |
| p = 1.0 | 10 | 0.61 | 0.95 | 1.23 | 1.45 | 1.81 | 2.18 | 2.69 |
| | 20 | 0.82 | 1.30 | 1.69 | 2.03 | 2.65 | 3.36 | 4.15 |
| | 40 | 1.04 | 1.67 | 2.23 | 2.66 | 3.63 | 4.78 | 6.42 |
| | 60 | 1.26 | 1.93 | 2.57 | 3.18 | 4.33 | 5.69 | 7.79 |
| | 80 | 1.35 | 2.14 | 2.87 | 3.55 | 4.92 | 6.54 | 8.86 |
| | 100 | 1.40 | 2.30 | 3.09 | 3.82 | 5.38 | 7.22 | 9.67 |

TABLE 5.11 Upper Tail Percentage Points for $Z = n\{1 - R^2(X, H)\}$ for a Test for the Logistic Distribution, Parameters Unknown, for Complete or Type 2 Right-Censored Data; $p = r/n$ is the censoring ratio

| | n | Significance level $\alpha$ | | | | | | |
| | | .50 | .25 | .15 | .10 | .05 | .025 | .01 |
|---|---|---|---|---|---|---|---|---|
| p = 0.2 | 20 | 1.56 | 2.85 | 3.57 | 4.03 | 4.78 | 6.09 | 7.35 |
| | 40 | 2.75 | 4.47 | 5.70 | 6.65 | 8.29 | 9.73 | 11.13 |
| | 60 | 3.37 | 5.40 | 6.92 | 8.21 | 10.37 | 14.48 | 15.32 |
| | 80 | 3.88 | 6.25 | 8.08 | 9.71 | 12.67 | 15.40 | 19.35 |
| | 100 | 4.29 | 6.94 | 9.03 | 10.95 | 14.70 | 17.96 | 22.78 |
| p = 0.4 | 10 | 0.79 | 1.43 | 1.78 | 2.00 | 2.37 | 3.01 | 3.76 |
| | 20 | 1.37 | 2.18 | 2.78 | 3.26 | 4.10 | 4.90 | 5.71 |
| | 40 | 1.94 | 3.08 | 4.01 | 4.79 | 6.25 | 7.93 | 9.87 |
| | 60 | 2.23 | 3.62 | 4.75 | 5.73 | 7.66 | 9.68 | 12.99 |
| | 80 | 2.50 | 4.08 | 5.43 | 6.66 | 9.01 | 11.70 | 15.61 |
| | 100 | 2.72 | 4.44 | 5.97 | 7.43 | 10.12 | 13.48 | 17.68 |
| p = 0.6 | 10 | 0.77 | 1.23 | 1.57 | 1.84 | 2.26 | 2.68 | 3.16 |
| | 20 | 1.10 | 1.76 | 2.24 | 2.63 | 3.41 | 4.19 | 5.02 |
| | 40 | 1.46 | 2.34 | 3.07 | 3.72 | 5.07 | 6.37 | 8.38 |
| | 60 | 1.66 | 2.69 | 3.56 | 4.40 | 5.99 | 7.72 | 10.43 |
| | 80 | 1.84 | 2.99 | 4.00 | 4.95 | 6.86 | 9.07 | 12.14 |
| | 100 | 1.98 | 3.24 | 4.35 | 5.38 | 7.57 | 10.20 | 13.48 |
| p = 0.8 | 10 | 0.68 | 1.07 | 1.36 | 1.58 | 1.99 | 2.34 | 2.81 |
| | 20 | 0.91 | 1.43 | 1.85 | 2.20 | 2.86 | 3.54 | 4.43 |
| | 40 | 1.16 | 1.84 | 2.42 | 2.94 | 3.99 | 5.15 | 6.99 |
| | 60 | 1.30 | 2.09 | 2.76 | 3.39 | 4.62 | 6.01 | 8.15 |
| | 80 | 1.42 | 2.29 | 3.05 | 3.76 | 5.21 | 6.90 | 0.27 |
| | 100 | 1.51 | 2.44 | 3.27 | 4.05 | 5.68 | 7.64 | 10.18 |
| p = 0.9 | 10 | 0.66 | 1.04 | 1.33 | 1.54 | 1.96 | 2.32 | 2.82 |
| | 20 | 0.85 | 1.34 | 1.71 | 2.05 | 2.63 | 3.29 | 4.20 |
| | 40 | 1.07 | 1.68 | 2.18 | 2.64 | 3.50 | 4.59 | 6.14 |
| | 60 | 1.18 | 1.89 | 2.48 | 3.02 | 4.02 | 5.31 | 7.17 |
| | 80 | 1.28 | 2.04 | 2.70 | 3.30 | 4.47 | 5.97 | 8.07 |
| | 100 | 1.35 | 2.15 | 2.86 | 3.51 | 4.82 | 6.50 | 8.79 |
| p = 0.95 | 10 | 0.65 | 1.03 | 1.31 | 1.52 | 1.94 | 2.31 | 2.82 |
| | 20 | 0.85 | 1.33 | 1.71 | 2.03 | 2.57 | 3.17 | 4.00 |
| | 40 | 1.05 | 1.64 | 2.12 | 2.51 | 3.30 | 4.27 | 5.71 |
| | 60 | 1.17 | 1.83 | 2.38 | 2.84 | 3.76 | 4.89 | 6.49 |
| | 80 | 1.25 | 1.96 | 2.56 | 3.10 | 4.12 | 5.42 | 7.28 |
| | 100 | 1.31 | 2.06 | 2.89 | 3.28 | 4.39 | 5.83 | 7.92 |
| p = 1.0 | 10 | 0.65 | 1.02 | 1.29 | 1.51 | 1.93 | 2.31 | 2.84 |
| | 20 | 0.90 | 1.42 | 1.84 | 2.19 | 2.78 | 3.42 | 4.20 |
| | 40 | 1.20 | 1.90 | 2.46 | 2.94 | 3.76 | 4.64 | 5.94 |
| | 60 | 1.38 | 2.20 | 2.88 | 3.40 | 4.38 | 5.37 | 6.99 |
| | 80 | 1.52 | 2.42 | 3.15 | 3.74 | 4.83 | 5.93 | 7.73 |
| | 100 | 1.62 | 2.59 | 3.35 | 3.99 | 5.16 | 6.34 | 8.26 |

TABLE 5.12 Upper Tail Percentage Points for $Z = n\{1 - R^2(X, H)\}$ for a Test for the Cauchy Distribution, Parameters Unknown, for Complete or Type 2 Right-Censored Data; $p = r/n$ is the censoring ratio

| | n | Significance level $\alpha$ | | | | | | |
|---|---|---|---|---|---|---|---|---|
| | | .50 | .25 | .15 | .10 | .05 | .025 | .01 |
| | 20 | 1.19 | 2.02 | 3.94 | 5.20 | 6.64 | 7.39 | 9.39 |
| | 40 | 3.23 | 5.85 | 6.99 | 8.10 | 10.45 | 12.61 | 15.35 |
| p = 0.2 | 60 | 4.90 | 8.85 | 10.78 | 12.03 | 14.22 | 17.40 | 20.91 |
| | 80 | 6.76 | 12.00 | 14.91 | 16.70 | 18.99 | 22.12 | 26.71 |
| | 100 | 8.50 | 14.85 | 18.76 | 21.18 | 23.60 | 26.19 | 31.84 |
| | 10 | 0.67 | 1.17 | 2.01 | 2.57 | 3.21 | 3.84 | 4.83 |
| | 20 | 1.65 | 2.99 | 3.64 | 3.97 | 5.11 | 6.15 | 7.44 |
| p = 0.4 | 40 | 3.48 | 6.12 | 7.64 | 8.64 | 9.71 | 10.66 | 13.09 |
| | 60 | 5.17 | 9.09 | 11.73 | 13.25 | 15.10 | 16.20 | 18.39 |
| | 80 | 7.05 | 12.39 | 15.92 | 18.11 | 20.95 | 22.35 | 24.16 |
| | 100 | 8.79 | 15.48 | 19.68 | 22.55 | 26.27 | 28.11 | 29.42 |
| | 10 | 0.91 | 1.61 | 1.94 | 2.12 | 2.80 | 3.45 | 4.12 |
| | 20 | 1.83 | 3.16 | 3.94 | 4.45 | 5.06 | 5.63 | 6.74 |
| p = 0.6 | 40 | 3.66 | 6.35 | 8.13 | 9.29 | 10.74 | 11.52 | 12.61 |
| | 60 | 5.40 | 9.51 | 12.30 | 14.10 | 16.51 | 17.78 | 18.51 |
| | 80 | 7.27 | 12.90 | 16.62 | 19.15 | 22.49 | 24.20 | 25.24 |
| | 100 | 8.99 | 16.00 | 20.51 | 23.73 | 27.88 | 29.97 | 31.56 |
| | 10 | 1.10 | 1.86 | 2.37 | 2.68 | 3.05 | 3.72 | 4.67 |
| | 20 | 2.07 | 3.57 | 4.58 | 5.27 | 6.12 | 6.59 | 7.27 |
| p = 0.8 | 40 | 3.94 | 6.96 | 8.97 | 10.47 | 12.28 | 13.21 | 13.78 |
| | 60 | 5.69 | 10.19 | 13.40 | 15.48 | 18.19 | 19.78 | 20.68 |
| | 80 | 7.63 | 13.65 | 17.74 | 20.51 | 24.43 | 26.47 | 27.54 |
| | 100 | 9.42 | 16.81 | 21.53 | 24.96 | 30.08 | 32.45 | 33.59 |
| | 10 | 1.34 | 2.48 | 2.99 | 3.35 | 3.97 | 4.36 | 5.08 |
| | 20 | 2.39 | 4.26 | 5.52 | 6.43 | 7.45 | 8.06 | 9.56 |
| p = 0.9 | 40 | 4.45 | 7.94 | 10.38 | 12.11 | 14.11 | 15.24 | 15.87 |
| | 60 | 6.39 | 11.39 | 14.92 | 17.48 | 20.42 | 21.93 | 22.85 |
| | 80 | 8.36 | 14.92 | 19.57 | 23.03 | 26.87 | 28.94 | 30.17 |
| | 100 | 10.10 | 18.07 | 23.71 | 27.99 | 32.59 | 35.24 | 36.76 |
| | 10 | 1.50 | 2.78 | 3.45 | 3.86 | 4.49 | 4.86 | 5.32 |
| | 20 | 2.84 | 5.24 | 6.77 | 7.81 | 8.98 | 9.78 | 12.35 |
| p = 0.95 | 40 | 5.27 | 9.66 | 12.43 | 14.32 | 16.61 | 17.78 | 20.82 |
| | 60 | 7.48 | 13.31 | 17.89 | 20.15 | 23.30 | 24.97 | 26.12 |
| | 80 | 9.55 | 17.02 | 22.34 | 26.09 | 30.13 | 32.25 | 33.42 |
| | 100 | 11.31 | 20.28 | 26.68 | 31.33 | 36.16 | 38.65 | 40.43 |
| | 10 | 1.74 | 3.19 | 4.02 | 4.51 | 5.08 | 5.42 | 5.58 |
| | 20 | 4.08 | 7.35 | 9.08 | 10.19 | 11.42 | 12.05 | 12.42 |
| p = 1.0 | 40 | 8.85 | 15.78 | 19.56 | 21.67 | 24.00 | 25.18 | 25.82 |
| | 60 | 13.79 | 24.29 | 30.00 | 33.13 | 36.33 | 38.06 | 38.92 |
| | 80 | 18.58 | 32.50 | 39.91 | 44.19 | 46.46 | 50.66 | 51.97 |
| | 100 | 22.72 | 39.66 | 48.46 | 53.80 | 59.12 | 61.70 | 63.50 |

$Z = n\{1 - R^2(X, H)\}$ is found from a complete or Type 2 right-censored sample, and referred to Table 5.11. The hypothesis that the sample comes from $F(x)$ is rejected at the level $\alpha$ if $Z$ exceeds the given percentage point.

### 5.15.2 The Cauchy Distribution

For the Cauchy distribution, $F(x) = 0.5 + [\tan^{-1}\{(x - \alpha)/\beta\}]/\pi$, $(-\infty < x < \infty$, $\beta > 0)$, and $H_i = \tan(\pi[\{i/(n + 1)\} - 0.5])$, $i = 1, \ldots, n$. The statistic $Z = n\{1 - R^2(X, H)\}$ is found from a complete or Type 2 right-censored sample and referred to Table 5.12.

### 5.15.3 The Exponential Power Distribution

Smith and Bain (1976) have given tables of null critical values of $1 - R^2(X, H)$ for the exponential power distribution, $F(x) = 1 - \exp[1 - \exp\{(x - \alpha)/\beta\}]$, $(\alpha < x < \infty$, $\beta > 0)$. These are for Type 2 right-censored data, with $r/n = 0.5$, 0.75, and 1, and for $n = 8, 20, 40, 60$, and 80.

### REFERENCES

Abramowitz, M. and Stegun, I. A. (eds.) (1965). Handbook of Mathematical Functions (National Bureau of Standards). New York: Dover Publications.

Bliss, C. (1967). Statistics in Biology: Statistical Methods for Research in the Natural Sciences. New York: McGraw-Hill.

Blom, G. (1958). Statistical Estimates and Transformed Beta Variates. New York: Wiley.

Chen, C-H. (1984). A correlation goodness-of-fit test for randomly censored data. Biometrika 71, 315-322.

Currie, I. D. (1980). The upper tail of the distribution of W-exponential. Scand. J. Statist. 7, 147-149.

D'Agostino, R. B. (1971). An omnibus test of normality for moderate and large sample sizes. Biometrika 58, 341-348.

D'Agostino, R. B. (1972). Small sample probability points for D test of normality. Biometrika 59, 219-221.

D'Agostino, R. B. (1973). Monte Carlo comparison of the W' and D test of normality for N = 100. Commun. Statist. 1, 545-551.

Davis, C. S. and Stephens, M. A. (1977). The covariance matrix of normal order statistics. Commun. Statist.-Simula. Computa. B6, 135-149.

De Wet, T. (1974). Rates of convergence of linear combinations of order statistics. S. Afr. Statist. J. 8, 35-43.

De Wet, T. and Venter, J. H. (1972). Asymptotic distributions of certain test criteria of normality. S. Afr. Statist. J. 6, 135-149.

De Wet, T. and Venter, J. H. (1973). A goodness-of-fit test for a scale parameter family of distributions. S. Afr. Statist. J. 7, 35-46.

Downton, F. (1966). Linear estimates with polynomial coefficients. Biometrika 53, 129-141.

Dyer, A. R. (1974). Comparisons of tests for normality with a cautionary note. Biometrika 61, 185-189.

Filliben, J. J. (1975). The probability plot correlation coefficient test for normality. Technometrics 17, 111-117.

Fotopoulos, S., Leslie, J. R., and Stephens, M. A. (1984). Approximations for expected values of normal order statistics with an application to goodness-of-fit. Technical report, Department of Statistics, Stanford University.

Gerlach, B. (1979). A consistent correlation-type goodness-of-fit test; with application to the two-parameter Weibull distribution. Math. Operationsforsch. Statist. Ser. Statist. 10, 427-452.

Gupta, A. K. (1952). Estimation of the mean and standard deviation of a normal population from a censored sample. Biometrika 39, 266-273.

Hahn, G. J. and Shapiro, S. S. (1967). Statistical Models in Engineering. New York: Wiley.

Hamaker, H. C. (1978). Approximating the cumulative normal distribution and its inverse. Appl. Statist. 27, 76-77.

Harter, H. L. (1961). Expected values of normal order statistics. Biometrika 48, 151-165.

Hastings, C. Jr. (1955). Approximations for Digital Computers. Princeton: Princeton University Press.

Jackson, O. A. Y. (1967). An analysis of departure from the exponential distribution. J. Roy. Statist. Soc. B 29, 540-549.

Labrecque, J. F. (1973). New goodness-of-fit procedures for the case of normality. Ph.D. Thesis, SUNY at Buffalo, Department of Statistics.

Labrecque, J. F. (1977). Goodness-of-fit tests based on nonlinearity in probability plots. Technometrics 19, 292-306.

Lawless, J. F. (1982). Statistical Models and Methods of Lifetime Data. New York: Wiley.

Lockhart, R. A. and Stephens, M. A. (1986). Correlation tests of fit. Technical Report, Department of Mathematics and Statistics, Simon Fraser University.

Milton, R. C. and Hotchkiss, R. (1969). Computer evaluation of the normal and inverse normal distribution function. Technometrics 11, 817-822.

Page, E. (1977). Approximations to the cumulative normal function and its inverse for use on a pocket calculator. App. Statist. 26, 75-76.

Sarhan, A. E. and Greenberg, B. G. (1962). Contributions to Order Statistics. New York: Wiley.

Sarkadi, K. (1975). The consistency of the Shapiro-Francia test. Biometrika 62, 445-450.

Shapiro, S. S. and Francia, R. S. (1972). Approximate analysis of variance test for normality. J. Amer. Statist. Assoc. 67, 215-216.

Shapiro, S. S. and Wilk, M. B. (1965). An analysis of variance test for normality (complete samples). Biometrika 52, 591-611.

Shapiro, S. S. and Wilk, M. B. (1968). Approximations for the null distribution of the W statistic. Technometrics 10, 861-866.

Shapiro, S. S. and Wilk, M. B. (1972). An analysis of variance test for the exponential distribution (complete samples). Technometrics 14, 355-370.

Shapiro, S. S., Wilk, M. B. and Chen, H. J. (1968). A comparative study of various tests for normality. J. Amer. Statist. Assoc. 63, 1343-1372.

Smith, R. M. and Bain, L. J. (1976). Correlation-type goodness-of-fit statistics with censored sampling. Comm. Statist. Theory Methods A 5, 119-132.

Spinelli, J. (1980). Contributions to goodness-of-fit. M.Sc. Thesis, Department of Mathematics, Simon Fraser University.

Spinelli, J. J. and Stephens, M. A. (1987). Tests for exponentiality when origin and scale parameters are unknown. Technometrics 29, 471-476.

Stephens, M. A. (1974). EDF statistics for goodness-of-fit and some comparisons. J. Amer. Statist. Assoc. 69, 730-737.

Stephens, M. A. (1975). Asymptotic properties for covariance matrices of order statistics. Biometrika 62, 23-28.

Stephens, M. A. (1976). Extensions to the W-test for normality. Technical Report, Department of Statistics, Stanford University.

Stephens, M. A. (1978). On the W test for exponentiality with origin known. Technometrics 20, 33-35.

Stephens, M. A. (1986). Goodness-of-fit for censored data. Technical report, Department of Statistics, Stanford University.

Teichroew, D. (1956). Tables of expected values of order statistics and products of order statistics for samples of size 20 and less from the normal distribution. Ann. Math. Statist. 27, 410–426.

Tietjen, G. L., Kahaner, D. K. and Beckman, R. J. (1978). Variances and covariances of the normal order statistics for sample sizes 2–50. Selected Tables in Mathematical Statistics 5 (D. B. Owen and R. E. Odeh, eds.), Providence: American Mathematical Society.

Weisberg, S. and Bingham, C. (1975). An approximate analysis of variance test for non-normality suitable for machine calculation. Technometrics 17, 133–134.

# 6
# Some Transformation Methods in Goodness-of-Fit*

C. P. Quesenberry   North Carolina State University, Raleigh, North Carolina

## 6.1 INTRODUCTION

### 6.1.1 Hypothesis Testing Problems

In this chapter let $X_1$, $X_2$, $\cdots$, $X_n$ be identically and independently distributed (i.i.d.) real-valued random variables with a common continuous distribution function (df) F. The classic <u>simple goodness-of-fit problem</u> is to test

$$H_0: F = F_0$$
$$H_1: F \neq F_0 \tag{6.1}$$

where $F_0$ is a specific continuous distribution function. The hypothesis testing problem of (6.1) is often not a very useful model in practice. It is more meaningful in many instances to test that the distribution function F has some specified functional form without assuming that the values of all parameters are known.

Let $\theta = (\theta_1, \cdots, \theta_p)$ be a vector of real-valued parameters and $\mathscr{F}_0 = \{F_\theta: \theta \in \Omega\}$ be a parametric class of probability distribution functions. Moreover, we assume that $\Omega$ is a natural parameter space, i.e., that it contains all points $\theta$ for which $F_\theta$ is a continuous probability distribution function. Then the <u>composite goodness-of-fit problem</u> is to test

$$H_0: F \in \mathscr{F}_0$$
$$H_1: F \notin \mathscr{F}_0 \tag{6.2}$$

---

*Work supported in part by National Science Foundation Grant MCS76-82652.

235

Let $\tau = (\tau_1, \ldots, \tau_{p_1})$ be another vector of real-valued parameters and $\mathcal{F}_1 = \{F_\tau : \tau \in \Omega_1\}$ be another parametric class of continuous distribution functions. The classes $\mathcal{F}_0$ and $\mathcal{F}_1$ are called <u>separate</u> families (Cox 1961) if the density of an arbitrary member of either class cannot be obtained as the limit of a sequence of densities from the other class. If $\mathcal{F}_0$ and $\mathcal{F}_1$ are separate families, then the <u>separate families testing problem</u> is to test

$$H_0: F \in \mathcal{F}_0$$
$$H_1: F \in \mathcal{F}_1 \qquad\qquad\qquad (6.3)$$

In words, this testing problem is to test that the sample is from a member of one class of distributions against the composite alternative hypothesis that it is from a certain alternative class. As a particular example let $\mathcal{F}_0$ be the df's of the scale parameter exponential class of densities given by (6.24) for $\theta$ unknown and $\mu = 0$, and let $\mathcal{F}_1$ be the df's of the lognormal class of densities given by (6.35) for $\mu = 0$ and $\sigma^2$ unknown. For this choice of $\mathcal{F}_0$ and $\mathcal{F}_1$, (6.3) is to test that the sample is from a scale-parameter exponential distribution against the alternative that it is from a shape-parameter lognormal distribution.

Suppose there are available k independent samples:

$$
\begin{aligned}
&X_{11},\ X_{12},\ \ldots,\ X_{1n_1}\\
&X_{21},\ X_{22},\ \ldots,\ X_{2n_2}\\
&\ \vdots\qquad\ \ \vdots\qquad\qquad\ \vdots \qquad\qquad\qquad (6.4)\\
&X_{k1},\ X_{k2},\ \ldots,\ X_{kn_k}
\end{aligned}
$$

The <u>several samples goodness-of-fit problem</u> is to test

$$H_0: X_{ij} \sim F_{\theta_i} \in \mathcal{F}_0; \quad i = 1, \ldots, k; \ j = 1, \ldots, n_i$$
$$H_1: \text{Negation of } H_0 \qquad\qquad\qquad (6.5)$$

In words, this null hypothesis is that all random variables (rv's) have df's of the same functional form; however, the parameters may change from sample to sample. The testing problem of (6.5) is an important generalization of the classical single sample goodness-of-fit composite null hypothesis testing problem of (6.2). The several samples null hypothesis can also be considered against a corresponding <u>several samples separate families</u> hypothesis testing problem as follows:

$$H_0: X_{ij} \sim F_{\theta_i} \in \mathcal{F}_0 \, ; \; i = 1, \, \ldots, \, k; \; j = 1, \, \ldots, \, n_i$$

$$H_1: X_{ij} \sim F_{\tau_i} \in \mathcal{F}_1 \, ; \; i = 1, \, \ldots, \, k; \; j = 1, \, \ldots, \, n_i$$

(6.6)

In the following sections we shall consider some techniques for testing the five types of problems described above. The approach used here may be described as follows for the classical goodness-of-fit problem of display (6.2). The sample $X_1, \, \ldots, \, X_n$ is transformed to a set of values $U_1, \, \ldots, \, U_N$ ($N = n - p$, the number of observations minus the number of parameters) in such a way that when $H_0$ is true, then $U_1, \, \ldots, \, U_N$ are independently and identically distributed uniform random variables on the $(0, 1)$ interval—i.i.d. $U(0, 1)$ rv's. The composite null hypothesis $H_0$ of (6.2) is then replaced by the surrogate simple null hypothesis that the U's are i.i.d. $U(0, 1)$ rv's.

The reader who is interested only in the methodology of this approach may wish to go directly to Section 6.4, which considers studying the uniformity of the transformed values; and then to Section 6.5, where the formulas are given for the transformations for a number of families of distributions. Numerical examples are given in Section 6.6 to illustrate the application of this approach to some data sets.

### 6.1.2 The Transformations Approach

In this section we consider again a single sample $X_1, \, \ldots, \, X_n$ with parent df F that is assumed under a null hypothesis to be a member of the parametric class $\mathcal{F}_0$ of the last subsection. Consider a set of transformations of the structure

$$U_1 = h_1(X_1, \ldots, X_n)$$

$$U_2 = h_2(X_1, \ldots, X_n)$$

$$\vdots$$

$$U_N = h_N(X_1, \ldots, X_n)$$

(6.7)

where $N \leq n$. Each $h_i$ for $i \in \{1, \ldots, N\}$ is a real-valued measurable function. Let $\bar{P}_F$ denote the probability distribution of $(U_1, \ldots, U_N)$ induced from the parent distribution with df F, and recall that a test is <u>similar</u> on $H_0$ of (6.2) if it has constant probability of rejecting for every $F \in \mathcal{F}_0$. The following are three properties of the U-transformations of (6.7) which we list for consideration.

(1)   There exists a probability distribution Q such that $P_F = Q$ for every $F \in \mathcal{F}_0$, i.e., $(U_1, \ldots, U_N)$ has the same distribution for every $F \in \mathcal{F}_0$.

(2)  If $P_F = Q$, then $F \in \mathscr{F}_0$, i.e., $(U_1, \ldots, U_N)$ has the characterizing
     distribution $Q$ only if $F$ is a member of the class $\mathscr{F}_0$.
(3)  There exists a test based on $(U_1, \ldots, U_N)$ for testing the null hypothesis
     $H_0$ of (6.2) against a particular simple alternative hypothesis that has
     the same power as the most powerful similar test based on $(X_1, \ldots, X_n)$
     for this same testing problem.

Condition (1) is very important because it assures that every size $\alpha$ test
based on $(U_1, \ldots, U_N)$ is also a similar $\alpha$ test for the same null hypothesis.
Condition (2) is a theoretically interesting property of transformations
of the structure of (6.7); however, it should be pointed out that this type of
characterization property of transformations has not played an important
role in the goodness-of-fit field. The apparent reason why this is so involves
the following considerations. The actual test statistic will itself be a real-
valued function of the values $(U_1, \ldots, U_N)$ of (6.7), i.e., the test statistic is
obtained by composition of a real-valued function with the transformations
of (6.7). Unless the distribution of the test statistic is also a characterization
of the distribution $Q$ in property (2), then property (2) for the transformations
of (6.7) is of little relevance. In other words, it matters little whether characterization is lost in the first or second step of the transformations. In this
context, it should be observed that most of the goodness-of-fit statistics that
are important in applied statistics do not characterize a null class of distributions. As a particular example, consider the chi-squared test statistic.
Although chi-squared test statistics do not characterize null hypothesis
classes of distributions, they have, of course, been and are of great importance in applied statistics (see Chapter 3).

A number of transformations of the form of (6.7) have been given in the
literature for particular parametric families. David and Johnson (1948) considered the probability integral transformation when parameters are replaced
by estimates. They showed that the transformed values are dependent, and
for many location-scale parameter families that the transformed random
variables have distributions that do not depend upon the values of the parameters.

A number of writers have given transformations for particular families
of the structure of (6.7) that satisfy condition (1) for transformations. Sarkadi
(1960, 1965) gives transformations for the three univariate normal families.
Durbin (1961) has proposed a transformation approach that eliminates the
nuisance parameters by introducing a further randomization step. Störmer
(1964) gives a method for transforming a sample from a $N(\mu, \sigma^2)$ distribution
to a sample of n - 2 values from a $N(0, 1)$ distribution. A number of transformations of the structure of (6.7) and satisfying property (1) have been
considered in the literature for one and two parameter exponential classes.
Two of these transformations are considered by Seshadri, Csörgö and
Stephens (1969), and one is shown to have property (2), also. Csörgö and
Seshadri (1970), Csörgö and Seshadri (1971), and Csörgö, Seshadri, and

Yalovsky (1973) have considered this transformations approach, and have given transformations for some particular normal, exponential, and gamma families of distributions with properties (1) and (2). A number of writers have considered the "recursive residuals" for normal regression models, which are of the structure of (6.7) when the variance is known. We shall discuss these in subsection 6.5.7 below.

A general theory for obtaining transformations such that the transformed U's are i.i.d. U(0, 1) random variables is given by O'Reilly and Quesenberry (1973) and extended to additional classes by Quesenberry (1975). These authors call the transformations involved <u>conditional probability integral transformations</u>—CPIT's, and we shall in this chapter consider only transformations obtained by this approach. We do not claim that the transformations and resulting tests and other analyses obtained by this approach have advantages over all other transformations for particular classes of distributions. Indeed, in some cases they will be found to give statistics equivalent to those of some other approaches.

## 6.2 PROBABILITY INTEGRAL TRANSFORMATIONS

### 6.2.1 Classical Probability Integral Transformations

The well-known probability integral transformation theorem due to R. A. Fisher (1930) can be stated as follows.

### Theorem 6.1

If X is a real-valued random variable with continuous df F, then $U = F(X)$ has a uniform distribution on the interval $(0, 1)$, i.e., U is a U(0, 1) rv.

(Historical aside: Actually, Fisher did not explicitly discuss the transformation in this paper; rather, he used the fact that a continuous distribution function $F(T; \theta)$ of a statistic T is a U(0, 1) rv in deriving fiducial limits for a parameter. He also used the result in Theorem 6.1 in <u>Statistical Methods for Research Workers</u> (Fisher 1932), in his method for combining tests of significance—again, without explicit discussion of the transformation itself.)

Thus if $X_1, \ldots, X_n$ are i.i.d. with continuous common df F, then $U_1 = F(X_1), \ldots, U_n = F(X_n)$ are i.i.d. U(0, 1) random variables. An important generalization of this basic theorem due to Rosenblatt (1952) is given in the next theorem.

If F is a multivariate distribution function we denote by $F(\cdot)$, $F(\cdot \mid \cdot)$, etc., the usual marginal and conditional df's.

### Theorem 6.2

If $(Y_1, \ldots, Y_m)$ is a vector of m random variables with absolutely continuous multivariate distribution function F, then the m random variables

$$U_1 = F(Y_1), \quad U_2 = F(Y_2|Y_1), \quad \ldots, \quad U_m = F(Y_m|Y_1, \ldots, Y_{m-1})$$

are i.i.d. $U(0,1)$ rv's.

As a simple example to illustrate Theorem 6.2, let $m = 2$ and $(Y_1, Y_2)$ have joint density function

$$f(y_1, y_2) = \exp(-y_2) \quad \text{for } 0 < y_1 < y_2 < +\infty$$

Then

$$F(y_1) = 1 - \exp(-y_1) \quad \text{for } y_1 > 0$$

and

$$F(y_2|y_1) = 1 - \exp(y_1 - y_2), \quad 0 < y_1 < y_2 < \infty$$

Theorem 6.2 says that

$$U_1 = F(Y_1) = 1 - \exp(-Y_1) \quad \text{and} \quad U_2 = F(Y_2|Y_1) = 1 - \exp(Y_1 - Y_2)$$

for $0 < Y_1 < Y_2 < \infty$, are i.i.d. $U(0,1)$ rv's. This result is easily verified directly. Applying the standard transformation of densities gives

$$h(u_1, u_2) = \exp(-y_2) \cdot \exp(y_2) = 1 \quad \text{for } 0 < u_1 < 1, \ 0 < u_2 < 1$$

The generality with which both of the preceding theorems hold should be carefully noted. We shall apply these in particular for cases when F is already explicitly a conditional distribution function.

## 6.2.2  Conditional Probability Integral Transformations: CPIT's

The model assumptions we make in this chapter are more restrictive than those in O'Reilly and Quesenberry (1973) (O–Q), but they are sufficiently general to cover many important cases. We assume that the parametric class $\mathcal{F}_0$ of distribution functions corresponds to a continuous exponential class (cf. Zacks, 1971, Section 2.5), and that $T_n$ is a p-component vector that is a sufficient and complete statistic for $\theta = (\theta_1, \ldots, \theta_p)$. Denote by $\tilde{F}_n(x_1, \ldots, x_n)$ the distribution function of $(X_1, \ldots, X_n)$ given the statistic $T_n$. Then $\tilde{F}_n(x_1)$, $\tilde{F}_n(x_2|x_1), \ldots, \tilde{F}_n(x_n|x_1, \ldots, x_{n-1})$ are the marginal and conditional distribution functions obtained from $\tilde{F}_n(x_1, \ldots, x_n)$. The following theorem is a direct consequence of Theorem 2.3 of O–Q and can be obtained from Theorem 6.2 above.

Theorem 6.3

The $(n - p)$ random variables

$$U_1 = \tilde{F}_n(X_1), \; U_2 = \tilde{F}_n(X_2 | X_1), \; \cdots$$

$$U_{n-p} = \tilde{F}_n(X_{n-p} | X_1, \ldots, X_{n-p-1})$$

(6.8)

are i.i.d. $U(0,1)$ rv's.

We note that the assumption that $(X_1, \ldots, X_n)$ are i.i.d. is not necessary to obtain Theorem 6.3. It is sufficient to require that $(X_1, \ldots, X_n)$ have a full-rank absolutely continuous distribution. We give results below in subsection 6.5.3 obtained by applying Theorem 6.3 to the order statistics of a sample from an exponential distribution.

A sequence $(T_n)_{n \geq 1}$ of statistics is said to be <u>doubly transitive</u> if each of the pairs of values $(T_n, X_{n+1})$ and $(T_{n+1}, X_{n+1})$ can be computed from the other. For example, if $T_n = \bar{X}_n = (X_1 + \cdots + X_n)/n$ then $(\bar{X}_{n+1}, X_{n+1}) = ((n\bar{X}_n + X_{n+1})/(n + 1), \; X_{n+1})$, and $(\bar{X}_n, X_{n+1}) = (((n + 1)\bar{X}_{n+1})/n, \; X_{n+1})$; and the sample mean $\bar{X}_n$ is doubly transitive.

The i.i.d. assumptions on $(X_1, \ldots, X_n)$ are necessary, in general, for the next theorem, which is obtained from Corollary 2.1 of O-Q. Now, it is often rather difficult to apply Theorem 6.3 directly to obtain explicit formulas for particular parametric families. The next theorem greatly simplifies the task of deriving the actual transformations for many important parametric families.

Theorem 6.4

If $(T_n)_{n \geq 1}$ is doubly transitive, then the $(n - p)$ random variables

$$U_1 = \tilde{F}_{p+1}(X_{p+1}), \; U_2 = \tilde{F}_{p+2}(X_{p+2}), \; \ldots, \; U_{n-p} = \tilde{F}_n(X_n)$$

(6.9)

are i.i.d. $U(0,1)$ rv's.

## 6.2.3 Conditional Probability Integral Transformations, Truncation Parameter Families

Some important classes of distributions which are not covered by the transformation theory of the preceding section are the truncation parameter families considered by Quesenberry (1975). For these families we assume that the parent density defined on an interval $(a, b)$, finite or infinite, is of one of the three forms:

$$f(x; \mu_1, \mu_2, \theta) = c(\mu_1, \mu_2, \theta)h(x, \theta); \quad a < \mu_1 < x < \mu_2 < b \qquad (6.10)$$

$$f_1(x; \mu, \theta) = c_1(\mu, \theta)h_1(x; \theta), \quad a < \mu < x < b \qquad (6.11)$$

$$f_2(x; \mu, \theta) = c_2(\mu, \theta)h_2(x; \theta), \quad a < x < \mu < b \qquad (6.12)$$

Here a and b are known constants; $\mu$, $\mu_1$ and $\mu_2$ are truncation parameters; $\theta$ is a p-component parameter vector; and $h(x, \theta)$, $h_1(x, \theta)$, and $h_2(x, \theta)$ are positive, continuous, and integrable functions over the intervals $(\mu_1, \mu_2)$, $(\mu, b)$, and $(a, \mu)$, respectively.

For $X_1, \ldots, X_n$ a sample we now set out a particular transformation to another set of values. Let r denote the antirank of $X_{(n)}$, i.e., r must satisfy $X_r = X_{(n)}$; and put

$$W_1 = X_1, \ldots, W_{r-1} = X_{r-1}, W_r = X_{r+1}, \ldots, W_{n-1} = X_n$$

Then we shall call $W_1, \ldots, W_{n-1}$ the <u>sample with $X_{(n)}$ deleted</u>. In words: $W_1, \ldots, W_{n-1}$ are the sample members that are less than the largest order statistic and subscripted in the same order.

Next, let r denote the antirank of $X_{(1)}$, i.e., r must satisfy $X_r = X_{(1)}$. Define $W_1, \ldots, W_{n-1}$ in terms of this $X_r$ in the same manner as above, and these values will be called the <u>sample with $X_{(1)}$ deleted</u>.

Finally, let $r_1$ and $r_2$ denote the antiranks of $X_{(1)}$ and $X_{(n)}$, respectively; and put $m_1 = \min\{r_1, r_2\}$, and $m_2 = \max\{r_1, r_2\}$. Put

$$W_1 = X_1, \ldots, W_{m_1-1} = X_{m_1-1}, W_{m_1} = X_{m_1+1}, \ldots, W_{m_2-2} = X_{m_2-1}$$

$$W_{m_2-1} = X_{m_2+1}, \ldots, W_{n-2} = X_n$$

These values $W_1, \ldots, W_{n-2}$ will be called the <u>sample with $X_{(1)}$ and $X_{(n)}$ deleted</u>.

For $X_1, \ldots, X_n$ a sample from the density f of (6.10) and $W_1, \ldots, W_{n-2}$ the sample with $X_{(1)}$ and $X_{(n)}$ deleted, the next theorem is from Quesenberry (1975).

<u>Theorem 6.5</u>

For fixed $(X_{(1)}, X_{(n)}) = (x_{(1)}, x_{(n)})$, the members $W_1, \ldots, W_{n-2}$ of the deleted sample are conditionally independent, identically distributed, continuous rv's with common density function

$$g(w, \theta) = \frac{h(w, \theta)I_{(x_{(1)}, x_{(n)})}(w)}{\int_{x_{(1)}}^{x_{(n)}} h(w, \theta) \, dw} \qquad (6.13)$$

where $I_{(x_{(1)}, x_{(n)})}(w) = 1$ if $x_{(1)} < w < x_{(n)}$, and is otherwise zero.

Let $T_{n-2}$ be a complete sufficient statistic [a function of $(W_1, \ldots, W_{n-2})$] for $\theta$ in the family of $g(w, \theta)$, and let $\tilde{G}_{n-2}(w) = P(W_1 \leq w | T_{n-2})$. That is, $\tilde{G}_{n-2}$ is the Rao-Blackwell (MVU) estimating df of the df $G(w, \theta)$ corresponding to the density function $g(w, \theta)$.

From Theorem 6.4 above or from Theorem 2 of Quesenberry (1975), the next result follows.

### Theorem 6.6

If $(T_j)_{j \geq 1}$ is doubly transitive then the $(n - p - 2)$ rv's

$$U_{j-r} = \tilde{G}_j(W_j) \quad j = p + 1, \ldots, n - 2 \tag{6.14}$$

are i.i.d. $U(0, 1)$ rv's.

Transformations of samples from distributions with densities of the form of $f_1$ of (6.11) or $f_2$ of (6.12) can be obtained in a similar fashion. If the sample is from $f_1$ we fix $X_{(1)} = x_{(1)}$ and let $W_1, \ldots, W_{n-1}$ be the sample with $X_{(1)}$ deleted. Then for fixed $X_{(1)} = x_{(1)}$ these values are conditionally independent, identically distributed, continuous rv's with common density function

$$g_1(w) = \frac{h_1(w, \theta) I_{(x_{(1)}, b)}(w)}{\int_{x_{(1)}}^{b} h_1(w, \theta) \, dw} \tag{6.15}$$

Finally, if the sample is from $f_2$ we fix $X_{(n)} = x_{(n)}$ and let $W_1, \ldots, W_{n-1}$ be the sample with $X_{(n)}$ deleted. For fixed $X_{(n)} = x_{(n)}$ these values are conditionally independent, identically distributed, continuous rv's with common density function

$$g_2(w) = \frac{h_2(w, \theta) I_{(a, x_{(n)})}(w)}{\int_{a}^{x_{(n)}} h_2(w, \theta) \, dw} \tag{6.16}$$

In each of these two cases we apply Theorem 6.4 to the $W_j$'s to produce $(n - p - 1)$ i.i.d. $U(0, 1)$ rv's. The transformations for some particular families of truncation parameter distributions will be given in Section 6.5.

## 6.3 SOME PROPERTIES OF CPIT'S

### 6.3.1 Notation and Terminology

In this section we give a largely descriptive account of some of the more important properties of CPIT's, and for this we shall use some notation and terminology that is not used elsewhere in this chapter.

Denote by $\mathscr{P}_j$ a parametric class of probability measures corresponding to the class $\mathscr{F}_j$ of df's of Section 6.1 for $j = 0, 1$. The members of these classes are probability measures on a Borel set $\mathscr{X}$ of real numbers and let $\mathscr{C}$ denote the Borel subsets of $X$. Let $(\mathscr{X}^n, \mathscr{C}^n)$ be the usual product space, and

$$\mathscr{P}_j^n = \{ P^n ; P^n = P \times \cdots \times P, \; P \in \mathscr{P}_j \}, \quad j = 0, 1$$

Let $g : \mathscr{X} \to \mathscr{X}$ be a strictly increasing 1-to-1 function and $g^n : \mathscr{X}^n \to \mathscr{X}^n$ be defined by $g^n(x_1, \ldots, x_n) = (gx_1, \ldots, gx_n)$. For each function $g^n$ assume there exists a function $\overline{g} : \mho \to \mho$ such that $P_{\overline{g}\theta}(X \in g^n A) = P_\theta(X \in A)$ for every $A \in \mathscr{A}^n$ and $X' = (X_1, \ldots, X_n)$. Let $G$ denote a transformation group on $\mathscr{X}$, $G^n$ the corresponding transformation group on $\mathscr{X}^n$, and $\overline{G}$ the corresponding transformation group on $\Omega$. A transformation group on a space is said to be <u>transitive</u> if its maximal invariant is constant on the space. The U-values of Theorem 6.4 can be expressed as functions on $\mathscr{X}^n$. The next three theorems are from Quesenberry and Starbuck (1976) (Q-S). In the following $\underline{U}' = (U_1, \ldots, U_{n-p})$.

### Theorem 6.7

If $G$ is a transformation group of strictly increasing functions on $\mathscr{X}$ that induces a transitive group $\overline{G}$ on $\Omega$, and conditions for Theorem 6.4 are satisfied, then $\underline{U}$ is equivalent to an invariant statistic, i.e.,

$$\underline{U}(x_1, \ldots, x_n) = \underline{U}(gx_1, \ldots, gx_n) \quad \text{a.s.} \quad \mathscr{P}^n \; \forall \; g \in G$$

From the distributional result of Theorem 6.3 and Basu (1955, 1960), the next result follows.

### Theorem 6.8

For T a complete sufficient statistic for $\Omega$ and $\underline{U}$ as given in Theorem 6.3, T and $\underline{U}$ are independent vectors.

The following result from Q-S shows that the U-transformations of Section 6.2 are efficient from the power viewpoint.

Theorem 6.9

A most powerful similar test exists for testing $H_0$ of (6.2) against a simple
alternative $F = F_1$, and this test can be expressed as a function of $(U_1, \ldots, U_{n-p})$ only.

The import of this result is that all information in the sample about the
class $\mathcal{F}_0$ of df's is also in the U-values $U_1, \ldots, U_{n-p}$. This result and
Theorem 6.8 shows that the CPIT transformations may be regarded as a
technique whereby the information in the sample $(X_1, \ldots, X_n)$ can be partitioned into two vectors $(\underline{U}';T)$, and the vector $T$ contains all information
about the parameters $(\theta_1, \ldots, \theta_p)$, the vector $\underline{U}$ contains all information
about the class $\mathcal{F}_0$, and $T$ and $\underline{U}$ are independent. Thus the values $(U_1, \ldots, U_{n-p})$ may be used to make inferences about the class of distributions (is
it normal? exponential?), the statistic(s) $T$ may be used to make inferences
within the class about parameter values (estimate the mean?), and the independence of $\underline{U}$ and $T$ can be exploited to assess overall error rates.

## 6.3.2 Sequential Nature of CPIT's

From the nature of the transformations in (6.8) and (6.9) it is apparent that,
in general, the vector $\underline{U}$ of transformed values is not a symmetric function
of the X sample. That is, the vector $\underline{U}$ is not invariant under permutations
of the observations. For those cases when the transformations are not permutation invariant, this property requires consideration on a number of
points.

One point of concern when the u's are not permutationally invariant is
that a goodness-of-fit test or other analysis may lead to a conclusion that
depends upon the presumably irrelevant indexing of the X's. If two randomly
selected orderings are used to compute $\underline{U}$ and then a particular goodness-of-
fit test (such as the Neyman smooth test discussed below) is computed on the
two $\underline{U}$ vectors, the statistics are identically distributed and dependent, but
nonidentical. If we consider these two test statistics, then the situation is
similar to that when more than one competing test statistic is computed for
the same testing problem. In particular, it is common practice today to compute a number of the goodness-of-fit test statistics, discussed elsewhere in
this book, for each sample. A quantity of relevance here is the probability
that two tests will agree in their conclusions. Quesenberry and Dietz (1983)
considered this probability for Neyman smooth tests made on the U's from
random permutations of a sample. They gave empirical evidence that these
agreement probabilities are very high in many cases of interest and are
bounded below by the value 2/3 for all cases considered.

It is possible to obtain CPIT transforms that are invariant under permutations of the original sample. To obtain these transforms recall that in
order to apply the approach of Rosenblatt it was necessary to assume only
that the rv's $(X_1, \ldots, X_n)$ had a full-rank continuous distribution. Thus to

obtain permutation invariant transformations we can first transform an i.i.d.
sample $(X_1, \ldots, X_n)$ to its order statistics, say $(X_{(1)}, \ldots, X_{(n)})$, and then
make CPIT transformations using Theorem 6.3 on these order statistics.
However, for this case Theorem 6.4 is not applicable and it is a difficult
task to find the transforms in practice. They have been found only for two
rather simple cases of exponential and uniform distributions, that will be
considered further below.

The discussion thus far has assumed that the sample is entirely symmetric of given size n. In practice, we often have data that can be naturally
ordered by some variable. Perhaps the most common ordering variable is
time, and when this is so the observations themselves may arrive sequen-
tially in time. When the data are ordered by time, or some other variable,
then the CPIT transformations approach has an especially strong appeal for
it will allow the analyst to design tests and other analysis techniques specifically to detect misspecifications in the i.i.d. model that are related to the
ordering of the data. For example, one or more of the parameters might
change with time. Some problems of this type are classical ones in statistics—there is a large literature concerned with slippage of normal means
and with detecting heteroscedasticity of normal variances. However, we
shall not consider these problems in this chapter.

## 6.4  TESTING SIMPLE UNIFORMITY

### 6.4.1  Introductory Remarks

The transformations of Section 6.2 can be used to construct similar $\alpha$ tests
for the testing problems of (6.2) and (6.3) by making size $\alpha$ tests of the
surrogate null hypothesis that the U-values are themselves independently
and identically distributed as uniform random variables on the interval $(0,1)$,
i.e., are i.i.d. $U(0,1)$ rv's. These transformed values of the U's should be
studied with care because they contain all test information in the sense of
Theorem 6.9. It should also be observed that when the null hypothesis
$H_0 : P \in \mathscr{P}_0$ fails, the distribution of the U's may fail to be i.i.d. $U(0,1)$ in
many ways. They may no longer be independent, nor identically distributed,
nor uniformly distributed. Moreover, if the model properties of independence
or identical distributions of the observations as assumed by the formal
goodness-of-fit and separate hypothesis testing problems of (6.2)-(6.6) are
violated, then this will also result in transformed values that are not, in
general, i.i.d. $U(0,1)$ rv's. Thus we can use these transformed values to
study the validity of other model specification, in addition to violations of an
assumed parametric density functional form. Now, the choice of tests to be
made will, of course, depend upon the type of violation of model assumptions
of concern. In some problems, we may not have reason to be concerned about
particular types of model violations, and we would like to perform an analysis
with good sensitivity against a wide range of alternatives. Such an analysis

can be made by studying the transformed values to determine if these are feasible values for i.i.d. $U(0,1)$ rv's. The analyses which we propose here for this purpose include graphical methods, and two omnibus goodness-of-fit tests on the transformed U values.

Subsequently, we shall use lower case u's and write $\underline{u} = (u_1, \ldots, u_N)'$ for both the random variables and their observed values. Here $N = n - p$, the number of observations minus the number of parameters in the model. Also, when we consider models with normal distributions of the errors in subsections 6.5.5 and 6.5.7 (linear regression models) below, it will be seen that the values $u_j$ are closely related to quantities usually called <u>residuals</u> from the least squares fitted lines from these models. For this reason we shall call $\underline{u}$ the vector of <u>uniform residuals</u> from the parametric model, in general.

Next put

$$z_j = \Phi^{-1}(u_j), \quad j = 1, \ldots, N \qquad (6.17)$$

for $\Phi$ a $N(0,1)$ df, and $\Phi^{-1}$ its inverse. By the inverse of Theorem 6.1, when the $u_j$'s are i.i.d. $U(0,1)$ rv's, then the $z_j$'s are i.i.d. $N(0,1)$ rv's. Thus we can also test (6.3) by testing the surrogate null hypothesis that the $z_j$'s are i.i.d. $N(0,1)$ rv's. We shall consider further tests based on the $z_j$'s below when we consider particular parametric classes. We shall call the values $z_j$ normal uniform (NU) residuals. A principal reason for considering these NU residuals is that the problem of testing normality is the most extensively studied problem in the goodness-of-fit area. See Chapter 9 of this book, and we will recommend tests of normality below. Hester and Quesenberry (1984) have found that a test based on these NU residuals has attractive power properties for testing for heteroscedasticity, i.e., for either increasing or decreasing variance for ordered regression data.

### 6.4.2 Graphical Methods for Symmetric Samples

Graphical methods are very useful for studying the uniformity of the values $u_1, \ldots, u_N$ on the unit interval. As a first step the data can be plotted on the $(0,1)$ interval; or, if N is large it will be more convenient to partition the $(0,1)$ interval into a number of subintervals of equal length and to construct a histogram on these subintervals. The data pattern on the unit interval, or the shape of the histogram constructed, conveys important information about the shape of the parent density from which the data were drawn relative to the shapes of the densities of the null hypothesis class. We next consider this in more detail.

In order to study the significance of particular patterns of the u-values on the unit interval, we first suppose that the u's were obtained from the classical probability integral transformation of Theorem 6.1 by transforming a sample $X_1, \ldots, X_N$ using a continuous df $F_0(x)$ with corresponding density function $f_0(x)$. However, if $X_1, \ldots, X_N$ is actually a sample from a parent distribution with df $F_1(x)$ and density function $f_1(x)$, then the u's will not constitute a set of i.i.d. $U(0,1)$ rv's, unless $F_0 = F_1$. Let $(a,b)$ be an interval on the real line where $f_0(x) < f_1(x)$ for every $x$ in $(a,b)$. Then the expected number of u's in the interval $(F_0(a), F_0(b))$ will be greater than under uniformity. Conversely, if $f_0(x) > f_1(x)$ for $x$ in the interval $(c,d)$, then the expected pattern of points in $(F_0(c), F_0(d))$ will be more sparse than under uniformity. Thus the splatter pattern of $u_1, \ldots, u_N$ on the unit interval should be interpreted as follows. If the data are too dense on an interval $(F_0(a), F_0(b))$, this implies that the true density function $f_1$ exceeds the density function $f_0$ on the interval $(a,b)$ and, conversely, when the data are too sparse on $(F_0(a), F_0(b))$ the true density $f_1$ is less than $f_0$ on $(a,b)$.

Next, suppose that the u's were obtained from a sample $X_1, \ldots, X_n$ by the CPIT transformations of Theorem 6.4 for a parametric class $\mathcal{F}_0$. Then the data patterns have the same interpretations as for the classical probability integral transformations just discussed. This is true because the transforming function used for the ith observation is an estimator of the parent df. Indeed, this transforming function is a minimum variance unbiased (MVU) estimator of the parent df based on $(X_1, \ldots, X_i)$ whenever the parent df is a member of $\mathcal{F}_0$. (See Seheult and Quesenberry (1971).)

If $u_{(1)}, \ldots, u_{(N)}$ are the order statistics of a sample from a $U(0,1)$ parent, then $u_{(i)}$ is a beta random variable with parameters $(i, n - i + 1)$. Thus $u_{(i)}$ has mean and variance given by

$$i/(N+1) \quad \text{and} \quad i(N+i-1)/(N+1)^2(N+2)$$

respectively, and its distribution function is the incomplete beta-function. If the points $(u_{(i)}, i/(N+1))$ are plotted on Cartesian axes, these points should approximate the line $g(u) = u$ for $0 < u < 1$. Quesenberry and Hales (1980) have given graphs called concentration bands for N = 2, 5, 10, 15, 20, 30, 40, 50, 60, 80, 100, 150, 200, 300, 500 that are helpful guides for judging the significance of these uniform probability plots.

The discussion made above in this subsection allows us to anticipate the pattern of points that the uniform residuals will make on the unit interval when a particular alternative density is considered. For a given data set this allows a direct interpretation of the data patterns observed in uniform and NU residuals both in histograms and in uniform probability plots. Some examples of data patterns in uniform probability plots will be given in the numerical examples of Section 6.6.

### 6.4.3 Test Statistics for Uniformity and Normality

The statistical literature contains a large number of goodness-of-fit test statistics which can be used to test that a set of values $u_1, \ldots, u_N$ consti-

TABLE 6.1  Empirical Critical Values for $U^2_{MOD}$

| $N/\alpha$ | 0.1 | 0.05 | 0.01 |
|---|---|---|---|
| 2 | 0.164 | 0.181 | 0.195 |
| 3 | 0.158 | 0.191 | 0.242 |
| 4 | 0.152 | 0.187 | 0.254 |
| 5 | 0.152 | 0.185 | 0.256 |
| 6 | 0.151 | 0.187 | 0.260 |
| 7 | 0.151 | 0.187 | 0.262 |
| 8 | 0.152 | 0.188 | 0.262 |
| 9 | 0.152 | 0.189 | 0.270 |
| 10 | 0.151 | 0.185 | 0.265 |
| $\infty$ | 0.152 | 0.187 | 0.267 |

tutes a sample from a $U(0, 1)$ distribution. We are here interested in tests which have good power against a wide range of alternative shapes. Monte Carlo power studies of tests of uniformity on which we shall base our choice of test statistics have been given by Stephens (1974), Quesenberry and Miller (1977) (Q-M), and Miller and Quesenberry (1979). We recommend two test statistics for use in testing uniformity of the u-values, and our choices are the $U^2$ statistic of Watson (1961) and the smooth test of Neyman (1937).

The Watson $U^2$ statistic was found in Q-M to have good power against a number of classes of alternatives, even for small sample sizes. This statistic is given by

$$U^2 = (1/12N) + \sum_{i=1}^{N} \{(2i - 1)/2N - u_{(i)}\}^2 - N(\bar{u} - 0.5)^2 \qquad (6.18)$$

where $\bar{u} = (u_1 + \cdots + u_N)/N$. Stephens (1970) found empirically that a linear function of $U^2$ given by

$$U^2_{MOD} = \{U^2 - 0.1/N + 0.1/N^2\}\{1 + 0.8/N\} \qquad (6.19)$$

has critical values that are approximately constant in N for $N > 10$. Table 6.1 gives approximate significance points for $U^2_{MOD}$ that were obtained by Q-M by Monte Carlo methods, and the Stephens approximation for $N > 10$.

In a classic paper Neyman (1937) proposed a test statistic that is defined as follows. Let $\pi_r$ denote the rth degree Legendre polynomial, the first five of which are, for $0 < u < 1$,

$$\pi_0(u) = 1$$

$$\pi_1(u) = \sqrt{12}(u - 1/2)$$

$$\pi_2(u) = \sqrt{5}(6(u - 1/2)^2 - 1/2)$$

$$\pi_3(u) = \sqrt{7}(20(u - 1/2)^3 - 3(u - 1/2))$$

$$\pi_4(u) = 210(u - 1/2)^4 - 45(u - 1/2)^2 + 9/8$$

Then put

$$t_r = \sum_{j=1}^{N} \pi_r(u_j), \quad r = 1, \ldots, k$$

and

$$p_k^2 = (1/N) \sum_{r=1}^{k} t_r^2$$

Miller and Quesenberry (1979) (M-Q) showed that each of the tests $p_k^2$ for $k = 2, 3, 4$ have good power against a number of classes of alternatives. Neyman showed that $p_k^2$ has an asymptotic $\chi^2(k)$ distribution under the uniformity null hypothesis and a noncentral $\chi^2$ distribution under the alternative.

Following recommendations of M-Q we shall use $p_4^2$ as a general omnibus test statistic for the uniformity null hypothesis. Table 6.2 gives some significance points for $p_4^2$ obtained by M-Q by Monte Carlo methods.

We shall consider only the statistics $U^2$ and $p_4^2$ for testing uniformity because they appear on the basis of the above noted power studies to be two of the best general omnibus tests of uniformity. In those studies it was noted that one weakness shared by many goodness-of-fit tests (including $p_4^2$) is that for small sample sizes (sometimes for N as large as 15), and some important alternatives, the tests are <u>biased</u>. We have not observed a case where $U^2$ gives a biased test for any sample size, nor one for which $p_4^2$ gives a biased test for N larger than 10. Thus we recommend only $U^2$ if N < 10, and we will compute both $U^2$ and $p_4^2$ for larger sample sizes. A practical advantage of $p_4^2$ is that it has an approximate $\chi^2(4)$ distribution as N increases, and so its observed significance level, p-value, is easily evaluated (see Table 6.2). These points will be illustrated in the numerical examples of Section 6.6.

Instead of (or in addition to) the graphs and tests using the uniform residuals, we can make graphs and tests using the NU residuals defined in (6.17) above. For most purposes, we feel that the graphs of uniform residuals are more easily interpreted than those of NU residuals. It seems to this writer that it is a bit easier to judge if a histogram agrees with an assumed uniform parent than with a normal-shaped parent. A probability plot of the

TABLE 6.2  Empirical Critical Values for $p_4^2$

| $N/\alpha$ | 0.1 | 0.05 | 0.01 |
|---|---|---|---|
| 2 | 7.19 | 9.52 | 16.14 |
| 3 | 7.34 | 9.51 | 15.80 |
| 4 | 7.46 | 9.50 | 15.43 |
| 5 | 7.53 | 9.49 | 15.12 |
| 6 | 7.57 | 9.48 | 14.86 |
| 7 | 7.60 | 9.47 | 14.65 |
| 8 | 7.62 | 9.47 | 14.47 |
| 9 | 7.63 | 9.46 | 14.32 |
| 10 | 7.64 | 9.46 | 14.19 |
| 11 | 7.65 | 9.45 | 14.09 |
| 12 | 7.65 | 9.45 | 14.00 |
| 13 | 7.66 | 9.44 | 13.93 |
| 14 | 7.66 | 9.44 | 13.87 |
| 15 | 7.66 | 9.43 | 13.82 |
| 16 | 7.66 | 9.43 | 13.78 |
| 17 | 7.66 | 9.43 | 13.74 |
| 18 | 7.67 | 9.42 | 13.71 |
| 19 | 7.67 | 9.42 | 13.69 |
| 20 | 7.67 | 9.42 | 13.67 |
| 30 | 7.68 | 9.40 | 13.58 |
| 40 | 7.68 | 9.40 | 13.52 |
| 50 | 7.69 | 9.40 | 13.48 |
| $\infty$ | 7.78 | 9.49 | 13.28 |

NU residuals requires normal probability paper, whereas the uniform residuals requires no special paper. Thus the uniform residuals are more convenient to plot with widely available software.

Aside from graphs of NU residuals, we can also make omnibus tests of a wide variety of parametric models by testing the normality of these residuals. There are a number of reasons to consider tests on NU residuals in place of, or in addition to, tests on the uniform residuals. The problem of testing normality for a simple random sample is the most extensively studied problem in the goodness-of-fit field, and excellent tests for this problem are readily available (see Chapter 9 of this handbook). Also, for at least some of the many important problems of testing null hypothesis parametric models with normally distributed errors—such as the multiple samples problems, regression models of Section 6.5.7, or the ANOVA model of Section 6.5.8— the NU residuals have a useful tendency to retain data patterns from the original data. For example, Hester and Quesenberry (1984) have exploited this property to construct efficient tests for heteroscedasticity for normal regression models.

Any of the multitude of goodness-of-fit tests for the completely specified null hypothesis testing problem of (6.1) may be used to test normality of the NU residuals. However, it has been shown by both Stephens (1974) and Dyer (1974) that the test for composite normality usually has better power against most alternatives, even when the parameters are known. We have done some simulation work in studying the efficiency of the Anderson-Darling (AD) test on NU residuals, and it appears that this characteristic of goodness-of-fit tests to show better power when parameters are not assumed known can be expected to obtain here, also. Finally, we must choose the particular tests of normality to make on the NU residuals. At the present state of knowledge, we feel that reasonable choices of test statistics would be the Shapiro-Wilk test or the AD test.

Another point that should be recalled in interpreting the analysis of residuals discussed here is the following. We generally consider testing the goodness-of-fit hypothesis testing problem as stated in display (6.2), and this means, of course, that we assume that the observations are i.i.d. However, with real data it will often be the case that we cannot really be sure that these assumptions are valid, and thus in practice in these cases the classical goodness-of-fit null hypothesis of (6.2) should be expanded to include the i.i.d. assumptions. That is, it is desirable to validate the entire model including these sampling assumptions.

## 6.5  TRANSFORMATIONS FOR PARTICULAR FAMILIES

### 6.5.1  Introduction

In this section the CPIT transformations are given for a number of parametric families of probability distributions. Lower case letters are used to denote both random variables and their observed values here and in the next section, which considers numerical examples. In all cases we assume that a sample $x_1, \ldots, x_n$ is available from an unspecified member of the class of distributions under consideration. We shall denote by $(w_1, \ldots, w_{n-1})$ the sample with $x_{(n)}$ deleted, and use similar notation for the other cases when $x_{(1)}$ is deleted or both $x_{(1)}$ and $x_{(n)}$ are deleted. (See Section 6.2.3 for details of this transformation.)

### 6.5.2  Uniform Distributions

Densities:

$$f(x; \mu_1, \mu_2) = (1/(\mu_2 - \mu_1)) I_{(\mu_1, \mu_2)}(x) \qquad (6.20)$$

for $-\infty < \mu_1 < \mu_2 < +\infty$.

Case I, $\mu_1 = \mu_{10}$ known, $\mu_2$ unknown.

   For $w_1, \ldots, w_{n-1}$ the sample with $x_{(n)}$ deleted put

$$u_j = (w_j - \mu_{10})/(x_{(n)} - \mu_{10}), \quad j = 1, \ldots, n-1 \tag{6.21}$$

Case II, $\mu_1$ unknown, $\mu_2 = \mu_{20}$ known.

   For $w_1, \ldots, w_{n-1}$ the sample with $x_{(1)}$ deleted put

$$u_j = (\mu_{20} - w_j)/(\mu_{20} - x_{(1)}), \quad j = 1, \ldots, n-1 \tag{6.22}$$

Case III, both $\mu_1$ and $\mu_2$ unknown.

   For $w_1, \ldots, w_{n-2}$ the sample with both $x_{(1)}$ and $x_{(n)}$ deleted put

$$u_j = (w_j - x_{(1)})/(x_{(n)} - x_{(1)}), \quad j = 1, \ldots, n-2 \tag{6.23}$$

   For these transformations for uniform distributions, we note that the lack of invariance with respect to permutations of the X's cited in subsection 6.3.2 does not obtain. That is, two permutations of the X's will give the same values for the elements of the $\underline{\mu}$ vector, though not in the same order. That is, the ordered values of the components of $\underline{u}$ will be the same, and any analysis or test that is symmetric in the components of $\underline{u}$ will be unaffected by a permutation of the X's.


## 6.5.3  Exponential Distributions

Densities:

$$f(x; \mu, \theta) = (1/\theta) \exp\{-(x - \mu)/\theta\} I_{(\mu, \infty)}(x) \tag{6.24}$$

for $-\infty < \mu < \infty$, and $\theta > 0$.

Case I, $\mu$ unknown, $\theta = \theta_0$ known.

   Let $w_1, \ldots, w_{n-1}$ be the sample with $x_{(1)}$ deleted.

$$u_j = 1 - \exp\{-(w_j - x_{(1)})/\theta_0\}, \quad j = 1, \ldots, n-1 \tag{6.25}$$

   Note that the remarks about the effect of permuting the X's on the u's above also obtain to these u's, viz., if the X's are permuted we still get the same $\underline{set}$ of values $\{u_1, \ldots, u_{n-1}\}$ from (6.25). For cases II and III that follow, we give the transforms obtained from the sample $X_1, \ldots, X_n$ from Theorem 6.4, as (6.26), and, also, transforms obtained by applying Theorem 6.3 directly to the order statistics $(x_{(1)}, \ldots, x_{(n)})$, as (6.27); these latter formulas were obtained by O'Reilly and Stephens (1984).

Case II, $\theta$ unknown, $\mu = \mu_0$ known.

Put $x_i' = x_i - \mu_0$, $i = 1, \ldots, n$.

$$u_{j-1} = 1 - \left\{ \sum_{i=1}^{j-1} x_i' \Big/ \sum_{i=1}^{j} x_i' \right\}^{j-1}, \quad j = 2, \ldots, n \qquad (6.26)$$

Next, put $y_i = x_{(i)} - \mu_0$; $i = 1, \ldots, n$; $y_0 = 0$.

$$u_j = 1 - \left\{ \frac{1 - (n - j + 1)y_j / (y_j + \cdots + y_n)}{1 - (n - j + 1)y_{j-1} / (y_j + \cdots + y_n)} \right\}^{n-j}, \quad j = 1, \ldots, n-1 \quad (6.27)$$

Case III, $\mu$ and $\theta$ unknown.

First, let $w_1, \ldots, w_{n-1}$ be the sample with $x_{(1)}$ deleted and put $w_i' = w_i - x_{(1)}$, $i = 1, \ldots, n - 1$. Then the transformations for this case are given by (6.26) with $x_i'$ replaced by $w_i'$ and n by n - 1.

Next, put $y_{i-1} = x_{(i)} - x_{(1)}$, $i = 1, \ldots, n$. Then the transformations for this case are given by (6.27) with these y's, and n replaced by n - 1.

### 6.5.4 Pareto Distributions

Densities:

$$f(x; \alpha, \gamma) = (\alpha \gamma^\alpha / x^{1+\alpha}) I_{(\gamma, \infty)}(x), \quad \alpha > 0, \ \gamma > 0 \qquad (6.28)$$

If the transformed values $\ln x_i$, $i = 1, \ldots, n$, are considered, then these transformed values are a sample from the exponential family of (6.23) where $\theta = 1/\alpha$ and $\mu = \ln \gamma$.

Case I, $\gamma$ unknown, $\alpha = \alpha_0$ known.

Let $w_1, \ldots, w_{n-1}$ be the sample with $x_{(1)}$ deleted.

$$u_j = 1 - (x_{(1)} / w_j)^{\alpha_0} \qquad (6.29)$$

for $j = 1, \ldots, n - 1$.

Case II, $\alpha$ unknown, $\gamma = \gamma_0$ known.

Put $x_i' = \ln x_i - \ln \gamma_0$; $i = 1, \ldots, n$. Then the CPIT transforms are given by (6.26) using these $x_i'$'s.

Put $y_i = \ln x_{(i)} = \ln \gamma_0$, $i = 1, \ldots, n$. Then the order statistics CPIT is given by (6.27) using these y's.

Case III, $\alpha$ and $\gamma$ unknown.

Let $w_1, \ldots, w_{n-1}$ be the sample with $x_{(1)}$ deleted, as above in Case I, and put $x_i' = \ln x_{(1)}$ for $i = 1, \ldots, n - 1$. Then the CPIT transforms are given by (6.26) using these $x_i$'s and n replaced by n - 1.

To obtain the order statistics transforms for this case put $y_{i-1} = \ln x_{(i)} - \ln x_{(1)}$, $n = 1, \ldots, n$, and use these y's in (6.27) with n replaced by n - 1.

## 6.5.5 Normal Distributions

Densities:

$$f(x; \mu, \sigma^2) = (1/\sigma\sqrt{2\pi}) \exp\left\{-(x - \mu)^2/2\sigma^2\right\} \tag{6.30}$$

for $-\infty < \mu < \infty$, and $\sigma^2 > 0$.

Case I, $\mu$ unknown, $\sigma^2 = \sigma_0^2$ known.

Let $\bar{x}_j = (x_1 + \cdots + x_j)/j$ for $j = 1, \ldots, n$ and $\Phi$ denote the df of a $N(0, 1)$ distribution.

$$u_{j-1} = \Phi\left\{(j - 1)^{\frac{1}{2}}(X_j - \bar{X}_{j-1})/j^{\frac{1}{2}}\sigma_0\right\} \tag{6.31}$$

for $j = 2, \ldots, n$.

Case II, $\sigma$ unknown, $\mu = \mu_0$ known.

For this case only we put

$$s_j^2 = \sum_{i=1}^{j} (X_i - \mu_0)^2/j, \quad j = 1, \ldots, n$$

and let $G_\nu$ denote a Student-t distribution function with $\nu$ degrees of freedom. Then

$$u_{j-1} = G_{j-1}\left\{(x_j - \mu_0)/S_{j-1}\right\}, \quad j = 2, \ldots, n \tag{6.32}$$

This can be generalized somewhat, as follows. Suppose that $S_\nu^2$ is a mean square estimator of $\sigma^2$ that is independent of the X's, and that $\nu S_\nu^2/\sigma^2$ has a $\chi^2(\nu)$ distribution. Then we put

$$S_j^* = \left\{[(j - 1)S_{j-1}^2 + \nu S_\nu^2]/(\nu + j - 1)\right\}^{\frac{1}{2}}$$

and

$$u_{j-1} = G_{\nu+j-1}\left\{(x_j - \mu_0)/S_j^*\right\}, \quad j = 2, \ldots, n \tag{6.32'}$$

Case III, $\mu$ and $\sigma^2$ unknown.

For this case put

$$\bar{x}_j = (1/j) \sum_{i=1}^{j} x_i, \quad S_j^2 = (1/(j-1)) \sum_{i=1}^{j} (x_i - \bar{x}_j)^2$$

Then

$$u_{j-2} = G_{j-2} \left\{ [(j-1)/j]^{\frac{1}{2}}(x_j - \bar{x}_{j-1})/S_{j-1} \right\}, \quad j = 3, \ldots, n \qquad (6.33)$$

Again, let $S_\nu^2$ be an independent mean square estimator of $\sigma^2$ such that $\nu S_\nu^2/\sigma^2$ is a $\chi^2(\nu)$ rv. Then put

$$S_j^* = \left\{ [(j-2)S_{j-1}^2 + \nu S_\nu^2]/(\nu + j - 2) \right\}^{\frac{1}{2}}$$

and

$$u_{j-2} = G_{\nu+j-2} \left\{ [(j-1)/j]^{\frac{1}{2}}(x_j - \bar{x}_{j-1})/S_j^* \right\} \qquad (6.33')$$

In computing the above quantities it is helpful to use the following updating formulas. For $\bar{x}_j$ as defined above, and for

$$(SS_j) = \sum_{i=1}^{j} (x_i - \bar{x}_j)^2$$

then

$$\bar{x}_j = [(j-1)\bar{x}_{j-1} + x_j]/j, \quad (SS_j) = (SS_{j-1}) + [j/(j-1)](x_j - \bar{x}_j)^2 \qquad (6.34)$$

Youngs and Cramer (1972) and Chan, Golub, and Le Veque (1983) discussed these formulas; the latter authors were primarily concerned with numerical accuracy.

## 6.5.6 Lognormal Distributions

Densities:

$$f(x; \mu, \sigma^2) = (x\sigma\sqrt{2\pi})^{-1} \exp \left\{ -(\ln x - \mu)^2/2\sigma^2 \right\} \qquad (6.35)$$

for $-\infty < \mu < \infty$, and $\sigma^2 > 0$.

The values $\ln x_1, \ldots, \ln x_n$ constitute a sample for the normal family of (6.30), and transformations for the three cases can be obtained by replacing $(x_1, \ldots, x_n)$ by $(\ln x_1, \ldots, \ln x_n)$ in the transformations for the corresponding normal cases above.

### 6.5.7 Normal Linear Regression Models

For $\underset{\sim}{y}_n = (y_1, \ldots, y_n)'$ a vector of independent rv's, $X_n$ an $n \times p$ matrix of full rank $(n > p)$, and $\underset{\sim}{\beta} = (\beta_1, \ldots, \beta_p)'$ a vector of p parameters, we consider the family of normal distributions

$$y_j \sim N(\underset{\sim}{x}_j' \underset{\sim}{\beta}, \sigma^2), \quad j = 1, \ldots, n \tag{6.36}$$

for $\underset{\sim}{x}_j'$ the jth row of $X_n$. The matrix $X_j$ denotes the matrix consisting of the first j rows of $X_n$. We shall give transformations here for two cases, viz., for $\sigma^2$ known and unknown.

Case I, $\sigma^2 = \sigma_0^2$ known.

For this case it is readily verified that the statistic $\underset{\sim}{t}_n' = X_n' \underset{\sim}{y}_n$ is a complete sufficient statistic for $\underset{\sim}{\beta}$. Moreover, the UMVU estimating distribution function is itself the df of a normal distribution with

$$\text{mean} = \underset{\sim}{x}_j'(X_j'X_j)^{-1}X_j'\underset{\sim}{y}_j, \quad \text{variance} = \sigma_0^2[1 - \underset{\sim}{x}_j'(X_j'X_j)^{-1}\underset{\sim}{x}_j]$$

Using this result with Theorem 6.4 gives the following results. Put

$$A_j = \frac{y_j - \underset{\sim}{x}_j'(X_j'X_j)^{-1}X_j'\underset{\sim}{y}_j}{\sigma_0[1 - \underset{\sim}{x}_j'(X_j'X_j)^{-1}\underset{\sim}{x}_j]^{\frac{1}{2}}} \tag{6.37}$$

and

$$u_{j-p} = \Phi(A_j), \quad j = p+1, \ldots, n \tag{6.38}$$

Certain well-known updating formulas [Placket (1950), Bartlett (1951)] are very convenient for the computation of $A_j$ as well as for the development below. For

$$\underset{\sim}{b}_j = (X_j'X_j)^{-1}X_j'\underset{\sim}{y}_j , \text{ the least squares estimate of } \underset{\sim}{\beta} \text{ from the first j observations}$$

$$s_j^2 = \underset{\sim}{y}_j'[I - X_j(X_j'X_j)^{-1}X_j']\underset{\sim}{y}_j , \text{ the usual least squares sum of squares for residuals from the first j observations,}$$

then the updating formulas are

$$(X_j' X_j)^{-1} = (X_{j-1}' X_{j-1})^{-1} - \frac{(X_{j-1}' X_{j-1})^{-1} x_j x_j' (X_{j-1}' X_{j-1})^{-1}}{1 + x_j'(X_{j-1}' X_{j-1})^{-1} x_j} \qquad (6.39)$$

$$b_j = b_{j-1} + (X_j' X_j)^{-1} x_j (y_j - x_j' b_{j-1}) \qquad (6.40)$$

$$S_j^2 = S_{j-1}^2 + \sigma_0^2 A_j^2 \qquad (6.41)$$

Using these relations, $A_j$ can be written in the alternative form

$$A_j = \frac{y_j - x_j' b_{j-1}}{\sigma_0 [1 + x_j'(X_{j-1}' X_{j-1})^{-1} x_j]^{\frac{1}{2}}} \qquad (6.37')$$

Put $w_j = \sigma_0 A_j$, and these $w_j$'s are the quantities sometimes called **recursive residuals**, and have been considered by a number of writers including Hedayat and Robson (1970), Brown, Durbin, and Evans (1975), and, recently, by Galpin and Hawkins (1984). These writers have generally not assumed that $\sigma^2$ is known. Then in the form (6.37') the $w_j$'s can be shown to be i.i.d. $N(0, \sigma^2)$, which also follows easily from the CPIT results given above.

Case II, $\sigma^2$ unknown.
From O-Q, Example 4.3, we put

$$B_j = \frac{(j - p - 1)^{\frac{1}{2}}(y_j - x_j' b_j)}{\{[1 - x_j'(X_j' X_j)^{-1} x_j]S_j^2 - (y_j - x_j' b_j)^2\}^{\frac{1}{2}}} \qquad (6.42)$$

and

$$u_{j-p-1} = G_{j-p-1}(B_j), \quad j = p + 2, \ldots, n \qquad (6.43)$$

Using the updating formulas again, we express $B_j$ in the alternative form

$$B_j = \frac{(j - p - 1)^{\frac{1}{2}}(y_j - x_j' b_{j-1})}{S_{r-1}[1 + x_j'(X_{j-1}' X_{j-1})^{-1} x_j]^{\frac{1}{2}}} \qquad (6.42')$$

Note that the quantity $(y_j - x_j' b_{j-1})$ in the numerator of (6.37') and (6.42') is the residual of $y_j$ from the least squares line fitted using the first $j - 1$ points. This is a normal rv with mean zero, and by examining (6.37') and (6.42'), we see that $A_j$ and $B_j$ are the standardized and Studentized forms of this rv, respectively. Moreover, by again using the updating formulas, we can show that

$$B_j = (j - p - 1)^{\frac{1}{2}} A_j \Big/ \Big( \sum_{i=1}^{j-1} A_i^2 \Big)^{\frac{1}{2}} = (j - p - 1)^{\frac{1}{2}} w_j \Big/ \Big( \sum_{i=1}^{j-1} w_i^2 \Big)^{\frac{1}{2}} \qquad (6.44)$$

for $w_j$ the recursive residual defined above.

It should be noted that the formulas given above in Section 6.5.5 for the univariate normal distribution for the cases when $\sigma^2$ is known are, of course, special cases of the formulas given in this section for the univariate regression model.

## 6.5.8 Normal Analysis of Variance Model

The material in this section is largely from Quesenberry, Giesbrecht, and Burns (1983) (QGB). We consider k mutually independent samples, $x_{ij}$, $i = 1, \ldots, k$, $j = 1, \ldots, n_i$, as in (6.4), and the family of distributions

$$x_{ij} \sim N(\mu_i, \sigma^2), \quad i = 1, \ldots, k \qquad (6.45)$$

In words, this is a fixed-effects, one-way, normal errors, analysis-of-variance model. Define

$$n = n_1 + \cdots + n_k, \quad v_{ij} = n_1 + \cdots + n_{i-1} + j - i - 1$$

$$\bar{x}_{ij} = (x_{i1} + \cdots + x_{ij})/j, \quad SS_{ij} = \sum_{\alpha=1}^{j} (x_{i\alpha} - \bar{x}_{ij})^2$$

and for $v_{ij} > 0$,

$$A_{ij} = \frac{[(j - 1)v_{ij}/j]^{\frac{1}{2}}(x_{ij} - \bar{x}_{i(j-1)})}{\Big[ \sum_{\alpha=1}^{i-1} SS_{\alpha n_\alpha} + SS_{i(j-1)} \Big]^{\frac{1}{2}}} \qquad (6.46)$$

(Remark: Readers familiar with the QGB paper will note that the updating formulas in (6.34) above have been used to rewrite the formula given for $A_{ij}$ in that paper in a more convenient form.)

Case I, $\sigma^2 = \sigma_0^2$ known.

$$u_{ij} = \Phi\{[x_{ij} - \bar{x}_{ij}][j/(j - 1)]^{\frac{1}{2}}/\sigma_0\} = \Phi\{[x_{ij} - \bar{x}_{i(j-1)}][(j - 1)/j]^{\frac{1}{2}}/\sigma_0\} \qquad (6.47)$$

Case II, $\sigma^2$ unknown.

$$u_{ij} = G_{v_{ij}} (A_{ij}) \qquad \qquad (6.48)$$

for $i = 1$ and $j = 3, \ldots, n_1$ and $i = 2, \ldots, k, j = 2, \ldots, n_i$.

A case can arise which is, in a sense, intermediate between cases I and II above. Suppose there is an external mean square estimator $S_v^2$ of $\sigma^2$ available such that $vS_v^2/\sigma^2$ is a $\chi^2(v)$ rv. Then put

$$v_{ij}^* = v_{ij} + v, \; S^* = \left\{ \left( \sum_{\alpha=1}^{i-1} SS_{\alpha n_\alpha} + SS_{i(j-1)} + vS_v^2 \right) \bigg/ v_{ij}^* \right\}^{\frac{1}{2}}$$

and

$$u_{ij} = G_{v_{ij}^*} \{ [(j-1)/j]^{\frac{1}{2}} (x_{ij} - \bar{x}_{i(j-1)})/S^* \} \qquad (6.49)$$

The question can be raised as to what is to be gained by using (6.49) rather than (4.48) since, under the null hypothesis model, we obtain the same number of i.i.d. $U(0,1)$ rv's. The answer is that when some of the model assumptions are incorrect the anomalous data patterns resulting should be more distinct and analyses based on them more sensitive. Another point is that even under the null hypothesis the dependence of the u's upon the ordering of the data is weakened. This point can be seen by observing in (6.49) that as $v \to \infty$ the distribution function in (6.49) converges to $\Phi$ and thus (6.47) is the limiting case of (6.49).

## 6.6  NUMERICAL EXAMPLES

### 6.6.1  Introduction

In this section we use some of the transformations of the last section in numerical examples. In all of these examples the computations were performed using programs written by the author for this purpose.

### 6.6.2  Salinity Data

A large scale program by North Carolina State University to study environmental impact in the Cape Fear Estuary includes a sampling over time by weeks of the larval density in the estuary. Accompanying data includes salinity measured in parts per thousand (ppt) at the time of collection of the larval specimen. The salinity data consists of several samples, and each

TABLE 6.3

| Obs. | Sample Number | | | | | |
|------|------|------|------|------|------|------|
|      | 1    | 2    | 3    | 4    | 5    | 6    |
| 1    | 18.3 | 19.8 | 11.1 | 27.0 | 13.1 | 9.1  |
| 2    | 16.9 | 22.0 | 12.7 | 26.4 | 12.6 | 13.6 |
| 3    | 15.9 | 22.5 | 10.2 | 28.5 | 18.7 | 16.4 |
| 4    | 16.0 | 21.3 | 8.6  | 25.5 | 19.4 | 14.3 |
| 5    | 15.8 | 20.0 | 7.9  | 28.1 | 17.4 | 14.0 |
| 6    | 14.5 | 19.1 | 10.9 | 26.3 | 16.9 | 10.2 |
| 7    | 18.6 | 18.8 | 13.7 | 26.6 | 15.5 | 11.7 |
| 8    | 21.8 | 22.4 | 12.4 | 24.5 | 16.7 | 12.6 |
| 9    | 18.3 | 22.9 | 12.9 | 27.8 | 17.7 | 12.0 |
| 10   | 18.5 | 21.5 | 10.1 | 28.7 |      | 11.1 |
| 11   | 17.1 | 21.4 | 9.9  | 28.7 |      |      |
| 12   | 16.6 | 20.9 |      | 26.8 |      |      |

sample consists of measurements made over a relatively short time period (approximately 24 hours), and the samples are taken at intervals of at least one month apart. It appears reasonable to consider each sample to be an independent random sample from a common parent distribution, and to assume that the functional form of the parent density is the same for all samples. However, the means and variances can be expected to vary widely from sample to sample.

There are available six samples, given in Table 6.3, for studying the functional form of the parent distributions.

## Normal Analysis

We consider the null hypothesis $H_0$ of (6.5) with $F_{\theta_j}$ the df of a $N(\mu_j, \sigma_j^2)$ distribution, i.e., we consider the null hypothesis that each sample is from a normal parent, but allow each of these parents to have different means and variances (cf. Quesenberry, Whitaker, and Dickens (1976)).

We have transformed each of the samples of Table 6.3 using the transformations of equation (6.33). Since a sample of size $n_i$ gives $n_i - 2$ transformed values, there are a total of $N = n_1 + \cdots + n_6 - 12 = 54$ transformed u-values. These 54 u-values are given in Table 6.4, and are graphed against the expected values $(i/(N + 1)) = (i/55)$ in Figure 6.1.

The values of the modified Watson statistic $U_{MOD}^2$ and the Neyman smooth statistic $p_4^2$ are both much too small for significance at the 10 percent

TABLE 6.4  Pooled and Ranked u-Values, Normal Analysis

| | | | | | | | | |
|---|---|---|---|---|---|---|---|---|
| .0380 | .1431 | .2625 | .3487 | .4572 | .5908 | .6861 | .7918 | .9080 |
| .0734 | .1630 | .2676 | .3884 | .4716 | .6030 | .6989 | .8066 | .9106 |
| .0988 | .1823 | .2892 | .3894 | .4779 | .6196 | .7223 | .8197 | .9274 |
| .1012 | .1972 | .2918 | .3971 | .5638 | .6288 | .7607 | .8680 | .9366 |
| .1159 | .2177 | .3139 | .4201 | .5690 | .6380 | .7797 | .8770 | .9765 |
| .1233 | .2450 | .3473 | .4364 | .5705 | .6641 | .7908 | .8926 | .9922 |

$$U^2_{MOD} = 0.013 \qquad\qquad p^2_4 = 1.631$$

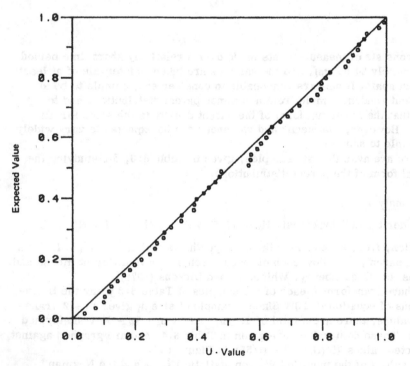

FIGURE 6.1  Salinity data. Normal analysis.

level. The $p_4^2$ statistic is approximately a $\chi^2(4)$ rv and the observed significance level is

$$P(p_4^2 > 1.631) \doteq P(\chi^2(4) > 1.631) = 0.80$$

Consider Figure 6.1 and recall that the points should approximate the line if the underlying parent family is normal. The fit is excellent, and from this and the small values of $U_{MOD}^2$ and $p_4$ we conclude that normal distributions fit these data very well, indeed.

Even though we have concluded that normal distributions fit these data well, we shall in the remainder of this subsection consider fitting two other two-parameter families to the data in order to illustrate the use of the transformations.

## Uniform Analysis

We have transformed each of the samples of Table 6.3 using the transformations of (6.23) for a two-parameter uniform family. The 54 u-values obtained are given in Table 6.5 and are graphed against the null hypothesis expected values in Figure 6.2. This uniform probability plot is typical of the pattern we obtain when normal data are transformed using the transformations for a uniform family. This S-shaped pattern can be anticipated from the discussion of data patterns in subsection 6.4.2, since both distributions are symmetric and on the range of the sample the normal density has thinner tails near the extremes, and is higher than the uniform density in the center. The value of $U_{MOD}^2$ in Table 6.5 falls between the upper 10 and 5 percent points, and the observed significance level of $p_4^2$ is

$$P(p_4^2 > 6.610) \doteq P(\chi^2(4) > 6.610) = .158$$

From these statistics and from Figure 6.2 it is clear that the uniform distributions do not fit the data well, and certainly not as well as normal distributions. It should be borne in mind here that the uniform distribution as an alternative to normality is one that is rather difficult to detect.

## Exponential Analysis

We have transformed each of the samples of Table 6.3 using the transformations of (6.27) for two-parameter exponential distributions; see Case III in subsection 6.5.3. The 54 pooled and ranked values obtained from these transformations are given in Table 6-6. These values are plotted in Figure 6.3 against the expected values. This probability plot is typical of the pattern obtained when normal data are transformed using exponential family transformations—recall the discussion in subsection 6.4.2.

This value of $U_{MOD}^2$ is highly significant since the upper 1 percent point for $U_{MOD}^2$ is .267. Moreover, $P(p_4^2 > 28.260) \doteq P(\chi^2(4) > 28.260) \doteq 0.00001$.

TABLE 6.5  Pooled and Ranked u-Values, Uniform Analysis

| .0732 | .2055 | .3288 | .3973 | .5122 | .5517 | .6324 | .7500 | .8621 |
| .0735 | .2381 | .3448 | .4265 | .5172 | .5616 | .6341 | .7759 | .8780 |
| .1207 | .2439 | .3562 | .4286 | .5205 | .5952 | .6585 | .7805 | .8971 |
| .1507 | .2740 | .3562 | .4524 | .5205 | .6029 | .6712 | .7857 | .9024 |
| .1781 | .2877 | .3793 | .4795 | .5476 | .6098 | .7059 | .8276 | .9525 |
| .1918 | .2927 | .3966 | .5000 | .5479 | .6164 | .7123 | .8571 | 1.0000 |

$$U^2_{MOD} = 0.169 \qquad\qquad p^2_4 = 6.610$$

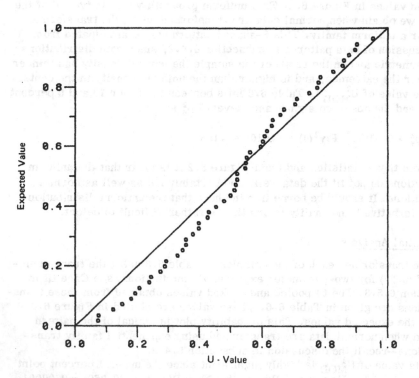

FIGURE 6.2  Salinity data. Uniform analysis.

TABLE 6.6  Pooled and Ranked u-Values, Exponential Analysis

| | | | | | | | | |
|---|---|---|---|---|---|---|---|---|
| .1199 | .3843 | .4254 | .4857 | .5489 | .6009 | .6753 | .7266 | .7952 |
| .1780 | .3871 | .4318 | .5182 | .5550 | .6186 | .6870 | .7301 | .8066 |
| .1790 | .3967 | .4397 | .5313 | .5617 | .6189 | .7073 | .7574 | .8102 |
| .2864 | .4138 | .4546 | .5322 | .5719 | .6294 | .7131 | .7606 | .8233 |
| .3345 | .4146 | .4684 | .5365 | .5926 | .6439 | .7174 | .7619 | .9134 |
| .3539 | .4175 | .4771 | .5386 | .6000 | .6725 | .7256 | .7794 | .9242 |

$$U^2_{MOD} = .747 \qquad\qquad p^2_4 = 28.260$$

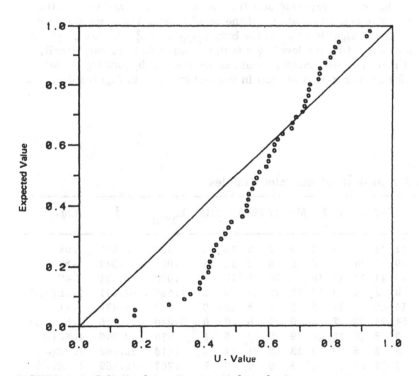

FIGURE 6.3  Salinity data. Exponential analysis.

From these values and Figure 6.3 it is clear that the two-parameter exponential distributions fit these data very poorly.

## 6.6.3 Simulated Data

The appendix gives three samples of 100 observations each that have been drawn from normal, exponential, and uniform distributions. These samples are named NOR, NEX, and UNI, which we shall abbreviate further to N, E, and U in this section. For each of these three samples we have computed 98 u-values using the appropriate transformations for two-parameter normal, exponential, and uniform families. This gives nine sets of u-values. We have partitioned the $(0, 1)$ interval into ten subintervals of equal lengths, and denote by Mi the number of u's that fall in the subinterval $((i - 1)/10, i/10]$ for $i = 1, ..,, 10$ for each set of u-values. These values and those for $U^2_{MOD}$, $p^2_4$ and the observed significance level of $p^2_4$, called Sig., are given in Table 6.7. The label (N, E) on row 4 means the normal sample was subjected to exponential transformations, for example.

The first three rows of Table 6.7 give results when the sample from a distribution is subjected to the transformations for a class to which that distribution belongs. The expected cell frequencies are 9.8, and none of the test statistics is significant at any of the usual levels. The other six rows all give level .05 significant tests for both $U^2_{MOD}$ and $p^2_4$. All except (U, N) have observed significance levels for both statistics that are very small, indeed. A more detailed analysis could be carried out by plotting the 98 u-values for each case as was done in the last example in Figures 6.1, 6.2, and 6.3.

TABLE 6.7  Analysis of Simulated Samples

|        | M1 | M2 | M3 | M4 | M5 | M6 | M7 | M8 | M9 | M10 | $U^2_{MOD}$ | $p^2_4$ | Sig. |
|--------|----|----|----|----|----|----|----|----|----|-----|-------------|---------|------|
| (N, N) | 12 | 12 | 5  | 10 | 7  | 8  | 17 | 3  | 13 | 11  | .068        | 0.818   | .94  |
| (U, U) | 10 | 10 | 14 | 10 | 5  | 9  | 8  | 8  | 11 | 13  | .040        | 1.547   | .82  |
| (E, E) | 13 | 11 | 10 | 10 | 10 | 8  | 10 | 5  | 13 | 8   | .093        | 3.521   | .47  |
| (N, E) | 0  | 3  | 4  | 9  | 12 | 15 | 27 | 20 | 8  | 0   | 1.587       | 63.041  | 2.4E(−12) |
| (U, N) | 14 | 15 | 9  | 11 | 9  | 7  | 1  | 8  | 18 | 6   | .221        | 13.174  | .01  |
| (E, U) | 54 | 21 | 12 | 7  | 3  | 0  | 1  | 0  | 0  | 0   | 3.340       | 281.416 | 4.5E(−12) |
| (N, U) | 3  | 8  | 13 | 12 | 18 | 19 | 9  | 10 | 5  | 1   | .770        | 30.046  | 4.8E(−6) |
| (U, E) | 8  | 5  | 8  | 10 | 9  | 13 | 13 | 21 | 11 | 0   | .416        | 22.083  | 1.9E(−4) |
| (E, N) | 0  | 12 | 27 | 17 | 7  | 8  | 5  | 7  | 6  | 9   | .767        | 42.259  | 1.5E(−8) |

TABLE 6.8

| | Sample A | | | | Sample B | | |
| | Ranked u-values | | | | Ranked u-values | | |
| Data | Normal | Uniform | Exponential | Data | Normal | Uniform | Exponential |
|------|--------|---------|-------------|------|--------|---------|-------------|
| 210  |        |         |             | 196  |        |         |             |
| 190  |        |         |             | 236  |        |         |             |
| 182  | .060   | .053    | .111        | 246  | .022   | .147    | .184        |
| 230  | .067   | .105    | .191        | 187  | .024   | .216    | .281        |
| 236  | .077   | .228    | .356        | 193  | .083   | .224    | .318        |
| 214  | .145   | .263    | .386        | 231  | .122   | .319    | .411        |
| 246  | .244   | .298    | .404        | 199  | .161   | .405    | .418        |
| 186  | .365   | .298    | .500        | 147  | .165   | .422    | .456        |
| 168  | .419   | .351    | .563        | 177  | .225   | .457    | .489        |
| 162  | .524   | .404    | .604        | 232  | .263   | .491    | .496        |
| 196  | .563   | .474    | .625        | 155  | .273   | .491    | .531        |
| 226  | .731   | .509    | .715        | 195  | .296   | .543    | .534        |
| 156  | .789   | .561    | .758        | 179  | .372   | .560    | .562        |
| 202  | .825   | .614    | .758        | 208  | .422   | .569    | .590        |
| 190  | .868   | .649    | .761        | 167  | .445   | .595    | .616        |
| 236  | .891   | .649    | .768        | 130  | .616   | .629    | .643        |
| 220  | .901   | .702    | .777        | 225  | .621   | .672    | .654        |
| 320  | .909   | .702    | .798        | 183  | .724   | .819    | .667        |
| 242  | .918   | .754    | .801        | 187  | .727   | .871    | .669        |
| 270  | .980   | .789    | .868        | 203  | .809   | .879    | .735        |
|      |        |         |             | 156  | .822   | .914    | .744        |

## 6.6.4 The Bliss Data

The appendix gives data from Bliss (1967) on the body weight in grams of 21-day-old white leghorn chicks at two dosage levels of vitamin D. We have randomized each of the samples labeled series A and series B. Table 6.8 gives the data in the order in which it was analyzed here. Table 6.8 gives also the transformed u-values when the samples are subjected to two-parameter normal, uniform, and exponential analyses. The test statistics $U^2_{MOD}$ and $p^2_4$, as well as the observed significance level, Sig., of $p^2_4$, have been computed for each sample as well as for the pooled samples and are given in Table 6.9. Graphs of the pooled u-values are given in Figures 6.4, 6.5, and 6.6 for the normal, uniform, and exponential classes, respectively.

TABLE 6.9  Test Statistics for Bliss Data

| Transformation | Sample | $U^2_{MOD}$ | $p^2_4$ | Sig. |
|---|---|---|---|---|
| Normal | A | .128 | 4.707 | .32 |
|  | B | .060 | 4.329 | .36 |
|  | Pooled | .068 | 3.257 | .52 |
| Uniform | A | .110 | 4.347 | .36 |
|  | B | .135 | 4.269 | .37 |
|  | Pooled | .200 | 7.634 | .11 |
| Exponential | A | .224 | 9.852 | .04 |
|  | B | .423 | 15.550 | .004 |
|  | Pooled | .470 | 18.333 | .001 |

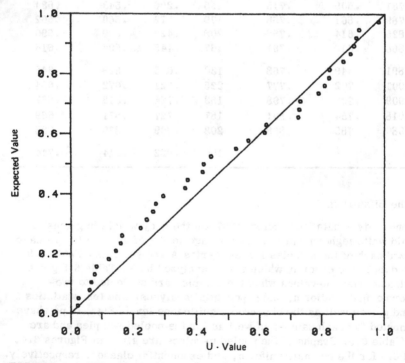

FIGURE 6.4  Bliss data. Normal analysis.

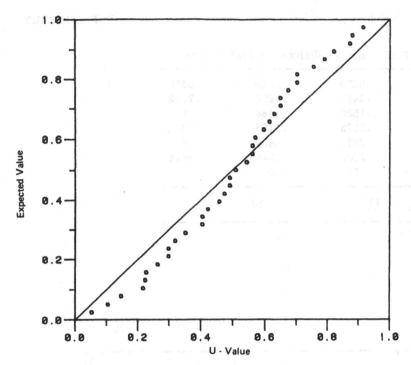

FIGURE 6.5 Bliss data. Uniform analysis.

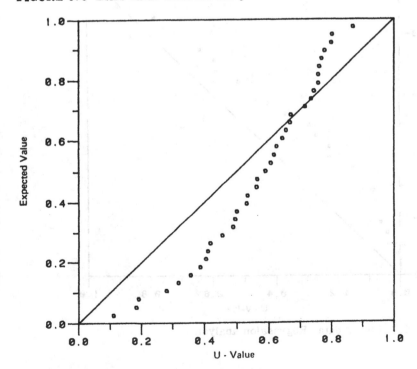

FIGURE 6.6 Bliss data. Exponential analysis.

TABLE 6.10 Ranked u-Values for Fisher Data

| | | | |
|---|---|---|---|
| .0001 | .0870 | .3685 | .6630 |
| .0107 | .1419 | .4275 | .7599 |
| .0163 | .1532 | .4384 | .7848 |
| .0414 | .2273 | .5188 | .9046 |
| .0635 | .2275 | .5899 | .9640 |
| .0636 | .2702 | .5946 | .9842 |
| .0768 | .2940 | .6403 | |

$$U^2_{MOD} = 0.117 \qquad\qquad p^2_4 = 12.16$$

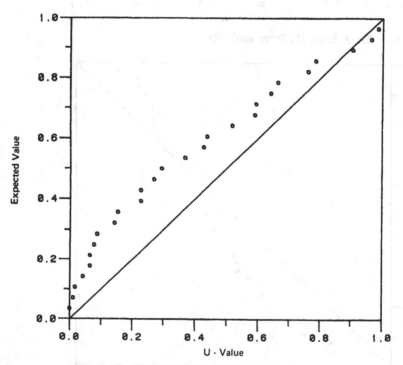

FIGURE 6.7  Fisher data. Regression analysis.

The probability plot in Figure 6.6 as well as the test statistics for the exponential case in Table 6.9 easily eliminate the exponential class from consideration for fitting these data. Comparison of the graphs in Figures 6.4 and 6.5 suggests that the normal class fits better than the uniform class, and this conclusion is supported by the values of $U^2_{MOD}$ and $p^2_4$ computed for the pooled u-values in Table 6.9. The statistic $U^2_{MOD}$ is significant at the .05 level for the uniform class but is much too small for significance at even the .10 level for the normal class. The probability plot of Figure 6.4 suggests that the true distribution (assuming that the i.i.d. assumptions hold, of course) has a density with slightly thicker tails than a normal density, but it does look pretty symmetric. Figure 6.5 suggests that the true density has thinner tails on its interval of support (where the density is positive) than the tails of the uniform density on its interval of support, but again the density appears to be symmetric.

## 6.6.5 Regression Data

In this section we give three examples to illustrate the use of the transformations of (6.37) to test the regression model assumptions of (6.36). It should be borne in mind that we are testing all of the assumptions of the normal linear regression model.

### Fisher's Data

In this example we consider data given by Fisher (1958, p. 137) on the relative effects of two nitrogenous fertilizers in maintaining yields in bushels of wheat over a period of 30 years. Fisher uses a simple linear normal regression model for these data and suggests that a more complex (curved) regression line might give a better fit.

We have applied the regression transformations of (6.43) to these data in the order that they are given by Fisher. Here $n = 30$ and $p = 2$, and the number of u-values is the number of observations less the number of parameters in the model (two regression coefficients and the variance), i.e., $N = n - p - 1 = 30 - 3 = 27$. The ranked u-values and test statistics $U^2_{MOD}$ and $p^2_4$ are given in Table 6.10, and the ranked u-values are plotted against expected values in Figure 6.7.

The value $U^2_{MOD} = 0.117$ is much less than the upper 10 percent point of 0.152 given in Table 6.1, and is not significant at this level. Using the chi-squared approximation to the distribution of $p^2_4$ we obtain $P(p^2_4 > 12.16) \doteq P(\chi^2(4) > 12.16) \doteq 0.016$, and this value is significant at the 2 percent level but not at 1 percent. The graph of Figure 6.7 also raises doubts about the appropriateness of the normal linear model. This pattern suggests that the error distribution is not symmetric and is skewed to the right.

TABLE 6.11  u-Values for Wallace-Snedecor Data

| | | | |
|---|---|---|---|
| .029 | .266 | .687 | .933 |
| .034 | .350 | .717 | .961 |
| .106 | .534 | .819 | .992 |
| .200 | .603 | .826 | |
| .262 | .641 | .886 | |

$$U^2_{MOD} = .063 \qquad\qquad p^2_4 = 2.389$$

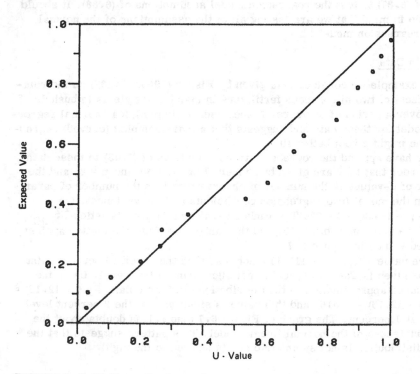

FIGURE 6.8  Wallace-Snedecor data. Regression analysis.

TABLE 6.12  u-Values for Snedecor-Cochran Data

| .057 | .147 | .211 | .453 |
|------|------|------|------|
| .101 | .179 | .258 | .498 |
| .138 | .201 | .429 | .631 |

$$U^2_{MOD} = 0.225 \qquad\qquad p^2_4 = 9.464$$

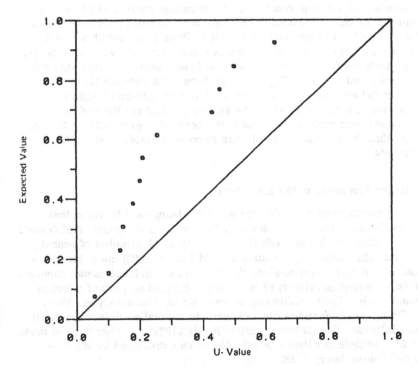

FIGURE 6.9  Snedecor-Cochran data. Regression analysis.

Wallace and Snedecor Data

Wallace and Snedecor (1931) give 25 observations on a dependent variable Y, and five independent variables $(X_1, X_2, X_3, X_4, X_5)$. Ostle (1954, p. 220) also gives these data and uses them in an example to illustrate a standard least squares analysis of a normal multiple linear regression model. We have applied the transformations of (6.43) to these data in the order given by Ostle. Here n = 25 and the number of independent variables is 5, so the number of u-values is $25 - 7 = 18$.

The analysis is given in Table 6.11 and Figure 6.8. The values of $U_{MOD}^2$ and $p_4^2$ are much less than their upper 10 percent points, and the pattern in Figure 6.8 does not give cause to suspect the model.

Snedecor-Cochran Data

Snedecor and Cochran (1967, p. 140) give the initial weight x and the gain in weight y of 15 female rats on a high protein diet, from the 24th to 84th day of age and suggest considering a normal linear regression model. The ranked u-values for these data are given in Table 6.12 and are plotted against expected values in Figure 6.9.

The number of observations here, 15, is rather small and gives only 12 transformed u-values for testing the model. With so little data only rather large violations of the model will be detected. Recall that for this small number of u's that many tests are biased for some alternatives. The $U_{MOD}^2$ statistic is unbiased for all known cases and is our preferred test statistic here. The observed value of $U_{MOD}^2 = .225$ here falls between the upper 5 and 1 percent values, and $P(p_4^2 > 9.464) \doteq .05$. The pattern of points in Figure 6.9 also raises doubts about the adequacy of fitting these data with a normal simple linear regression model. We conclude, even with so few observations, that the normal simple linear regression model assumptions are unwarranted.

6.6.6 Further Examples in the Literature

Some of the transformations of Section 6.5 have been used in numerical examples in the literature. Quesenberry, Whitaker, and Dickens (1976) used the transformations (6.33) to study the normality of 11 samples of peanut aflatoxin data. Quesenberry, Giesbrecht, and Burns (1983) used the transformations (6.33) and (6.48) to study the analysis of variance model assumptions for small samples (size 4) of tall fescue obtained as part of a forage management study. The variable considered was neutral detergent fiber.

The CPIT transformations for multivariate normal models were given by Rincon-Gallardo, Quesenberry, and O'Reilly (1979). Applications of these formulas for multiple multivariate samples were considered by Rincon-Gallardo and Quesenberry (1982).

(Author's note: This chapter was written in 1977 and revised in 1984.)

## REFERENCES

Bartlett, M. S. (1951). An inverse matrix adjustment arising in discriminant analysis. Ann. Math. Stat. 22, 107-111.

Basu, D. (1955, 1960). On statistics independent of a complete sufficient statistic. Sankhyā 15, 337-380; 20, 223-226.

Bliss, C. J. (1946). Collaborative comparison of three ratios for the chick assay of vitamin D. J. Assoc. Off. Agr. Chemists 29, 396-408.

Brown, R. L., Durbin, J., and Evans, J. M. (1975). Techniques for testing the constancy of the regression relationship over time. J. R. Statist. Soc. B37, 149-192.

Chan, T. F., Golub, G. H., and LeVeque, R. J. (1983). Algorithms for computing the sample variance: analysis and recommendations. American Statist. 37(3), 242-247.

Cox, D. R. (1961). Tests of separate families of hypotheses. Proceedings of the 4th Berkeley Symposium on Mathematical Statistics and Probability, Vol. 1. Berkeley: University of California Press, 105-123.

Csörgö, M. and Seshadri, V. (1970). On the problem of replacing composite hypotheses by equivalent simple ones. Rev. Int. Statist. Inst. 38, 351-368.

Csörgö, M. and Seshadri, V. (1971). Characterizing the Gaussian and exponential laws via mappings onto the unit interval. Z. Wahrscheinlichkeitstheorie und Verw. Gebiete 18, 333-339.

Csörgö, M., Seshadri, V. and Yalovsky, M. (1973). Some exact tests for normality in the presence of unknown parameters. J. Roy. Statist. Soc. B35, 507-522.

David, F. N. and Johnson, N. L. (1948). The probability integral transformation when parameters are estimated from the sample. Biometrika 35, 182-192.

Durbin, J. (1961). Some methods of constructing exact tests. Biometrika 40, 41-55.

Dyer, A. R. (1974). Hypothesis testing procedures for separate families of hypotheses. Jour. Am. Stat. Assn. 69, 140-145.

Fisher, R. A. (1930). Inverse probability. Proceedings of the Cambridge Philosophical Society 36(4), 528-535.

Fisher, R. A. (1932). Statistical Methods for Research Workers. 4th Ed., Oliver and Boyd, Ltd.

Galpin, J. S. and Hawkins, D. M. (1984). The use of recursive residuals in checking model fit in linear regression. American Statist. 38(2), 94–105.

Hedayat, A. and Robson, D. S. (1970). Independent stepwise residuals for testing homoscedasticity. Jour. Am. Stat. Assn. 65, 1573–1581.

Hester, R. A., Jr. and Quesenberry, C. P. (1984). Analyzing uniform residuals for heteroscedasticity. Institute of Statistics Mimeo Series No. 1639, North Carolina State University.

Miller, F. L., Jr. and Quesenberry, C. P. (1979). Power studies of some tests for uniformity, II. Commun. Statist. B8(3), 271–290.

Neyman, Jerzy (1937). 'Smooth' test for goodness of fit. Skandinavisk Aktu-arietidskrift 20, 149–199.

O'Reilly, F. and Quesenberry, C. P. (1973). The conditional probability integral transformation and applications to obtain composite chi-square goodness of fit tests. Ann. Statist. 1, 74–83.

O'Reilly, F. J. and Stephens, M. A. (1984). Characterizations of goodness of fit tests. J. R. Statist. Soc. B44(3), 353–360.

Ostle, B. (1954). Statistics in Research. Ames: Iowa State University Press.

Placket, R. L. (1950). Some theorems in least squares. Biometrika 37, 149–157.

Quesenberry, C. P. (1975). Transforming samples from truncation parameter distributions to uniformity. Comm. in Statist. 4, 1149–1155.

Quesenberry, C. P. and Dietz, E. J. (1983). Agreement probabilities for some CPIT-Neyman smooth tests. J. Statist. Comput. Simul. 17, 125–131.

Quesenberry, C. P., Giesbrecht, F. G. and Burns, J. C. (1983). Some methods for studying the validity of model assumptions for multiple samples. Biometrics 39, 735–739.

Quesenberry, C. P. and Hales, C. (1980). Concentration bands for uniformity plots. J. Statist. Comput. Simul. 11, 41–53.

Quesenberry, C. P. and Miller, F. L., Jr. (1977). Power studies of some tests for uniformity. J. Statist. Comput. Simul. 5, 169–191.

Quesenberry, C. P. and Starbuck, R. R. (1976). On optimal tests for separate hypotheses and conditional probability integral transformations. Comm. in Statist. A5(6), 507–524.

Quesenberry, C. P., Whitaker, T. B. and Dickens, J. W. (1976). On testing normality using several samples: an analysis of peanut aflatoxin data. Biometrics 32(4), 753–759.

Rincon-Gallardo, S. and Quesenberry, C. P. (1982). Testing multivariate normality using several samples: applications techniques. Commun. Statist.—Theor. Meth. 11(4), 343-358.

Rincon-Gallardo, S., Quesenberry, C. P. and O'Reilly, F. J. (1979). Conditional probability integral transformations and goodness-of-fit tests for multivariate normal distributions. Ann. Statist. 7(5), 1052-1057.

Rosenblatt, M. (1952). Remarks on a multivariate transformation. Ann. Math. Stat. 23, 470-472.

Sarkadi, K. (1960). On testing for normality. Publications of the Mathematical Institute of the Hungarian Academy of Sciences. Vol. 3, 269-275.

Sarkadi, K. (1965). On testing for normality. Proceedings of the 5th Berkeley Symposium on Mathematical Statistics and Probability. Berkeley: University of California Press, 373-387.

Seheult, A. H. and Quesenberry, C. P. (1971). On unbiased estimation of density functions. Ann. Math. Stat. 42, 1434-1438.

Seshadri, V., Csörgö, M. and Stephens, M. A. (1969). Tests for the exponential distribution using Kolmogorov-type statistics. J. Roy. Statist. Soc. B31, 499-509.

Snedecor, G. W. and Cochran, W. G. (1967). Statistical Methods. 6th Ed., Ames: Iowa State University Press.

Stephens, M. A. (1970). Use of the Kolmogorov-Smirnov, Cramer-von Mises and related statistics without extensive tables. J. R. Statist. Soc. B32, 115-122.

Stephens, M. A. (1974). EDF statistics for goodness of fit and some comparisons. Jour. Am. Stat. Assn. 69, 730-737.

Störmer, Horand (1964). Ein test zum erkennen von Normalverteilungen. Zeitschrift Wahrscheinlichkeitstheorie und Verwande Gebiete 2, 420-428.

Wallace, H. A. and Snedecor, G. W. (1930). Correlation and machine calculation. (Revised edition) Iowa State College Official Publication, Vol. 30, No. 4.

Watson, G. S. (1961). Goodness-of-fit tests on a circle. Biometrika 48, 109-114.

Youngs, E. A. and Cramer, E. M. (1971). Some results relevant to choice of sum and sum-of-product algorithms. Technometrics 13, 657-665.

Zacks, S. (1971). The Theory of Statistical Inference. New York: Wiley.

# 7

# Moment ($\sqrt{b_1}$, $b_2$) Techniques

K. O. Bowman   Oak Ridge National Laboratory, Oak Ridge, Tennessee

L. R. Shenton   University of Georgia, Athens, Georgia

## 7.1 INTRODUCTION

For the random sample $X_1, \ldots, X_n$, with mean

$$m_1 = \sum_{j=1}^{n} X_j/n$$

we define the central moments

$$m_i = \sum_{j=1}^{n} (X_j - m_1)^i/n, \quad i = 2, 3, 4 \tag{7.1}$$

The sample skewness ($\sqrt{b_1}$) and kurtosis ($b_2$) are now defined as

$$\sqrt{b_1} = m_3/m_2^{3/2}$$

$$b_2 = m_4/m_2^2 \tag{7.2}$$

and it is readily seen that they are invariant under origin and scale changes. The corresponding measures for a specified density are denoted by $\sqrt{\beta_1}$ and $\beta_2$. For a normal distribution $\sqrt{\beta_1} = 0$, $\beta_2 = 3$, and in random samples from it there may be wide variations from these values, especially for small samples ($n < 25$). Moreover, the sample may arise from some nonnormal distribution, such as a uniform, negative exponential, or Weibull, etc. Symmetric distributions (or those with non-zero densities extending over negative and positive variate values) are likely to produce samples with small skewness, whereas distributions corresponding to positive valued random variables (such as the negative exponential) are likely to produce samples with large skewness. In sampling from fairly symmetric distributions, one might expect the kurtosis to reflect the nonnormality. Thus a combination of the test statistics $\sqrt{b_1}$ and $b_2$ might provide a more comprehensive test than either taken by itself.

This chapter will be mainly concerned with tests of goodness of fit based on $\sqrt{b_1}$, $b_2$ in sampling from the normal or other distributions such as members of the Pearson system.

The reader may be reminded that of the classical test statistics, Student's t, F-ratio, correlation coefficient (and perhaps mean and variance), the skewness and kurtosis statistics are the only ones whose distributions in normal sampling are still now known exactly. However, Mulholland (1977) has arrived at an approximation to the null distribution of $\sqrt{b_1}$ for samples of at most 25; this is undoubtedly a breakthrough, although the mathematical expressions are very complicated, and it seems unlikely that the method can be applied to sampling from more general populations. In nonnormal sampling very few exact results are known for the distributions of $\sqrt{b_1}$ and $b_2$ or their joint distribution.

## 7.2 NORMAL DISTRIBUTION

Early work goes back to Karl Pearson (1902) who gave in general sampling expressions for the dominant terms ($n^{-1}$ asymptotics) in the variance of $b_1$ and $b_2$, and also the correlation $\rho(b_1, b_2)$. The idea was to use, for example, $\sqrt{\mathrm{Var}\,(b_1)}$ and $\sqrt{\mathrm{Var}\,(b_2)}$ as error assessments, but it was far too early in the development of the subject to consider questions of the validity of the of the asymptotics or their uses.

A quarter of a century later an important development came from E. S. Pearson (1930) who used the work of Fisher (1928) and Wishart (1930) on k-statistics to develop a Taylor series expansion (in terms of the k-statistic discrepancies $k_i - \kappa_i$, $i = 2, 3, 4$) for $\sqrt{b_1}$ and $b_2$. For example, defining

$$y = (n - 1)\sqrt{\{b_1 / (n - 2)\}}$$

Pearson showed for the second and fourth central moments of y (in normal sampling) that

$$\mu_2(y) = 1 - \frac{6}{n} + \frac{22}{n^2} - \frac{70}{n^3} + \cdots$$

and

$$\mu_4(y) = 3 - \frac{1056}{n^2} + \frac{24132}{n^3} - \cdots$$

the odd moments being zero. He developed similar expressions for the 2nd, 3rd, and 4th central moments of $b_2$, along with $E(b_2) = 3(n - 1)/(n + 1)$. To damp-out higher order terms, Pearson used samples of $n \geq 50$ for $\sqrt{b_1}$, $n \geq 100$ for $b_2$ so as to assess the lower and upper 1% and 5% of the distributions in normal sampling. Thirty or so years later (Pearson, 1965), he gave a set of "accepted" percentage points; for a sample of 50 there is no change in the third d.p. entries for $\sqrt{b_1}$ at the 1%, 5% levels; for $b_2$ and $n = 100$ there is no change in the second d.p. entries; in all, quite a remarkable achievement.

The next step forward came from Fisher (1930) who showed that in normal sampling the standardized moments $m_r/m_2^{r/2}$, $r \geq 3$, are distributed independently of the second moment $m_2$ (Fisher used k-statistic notation). Thus, for example, $E(b_1) = Em_3^2/Em_2^3$ follows from the independence of $m_3^2/m_2^3$ and $m_2^3$; here E means the mathematical expectation operator. In this manner, the exact moments of $\sqrt{b_1}$ and $b_2$ can be found. In fact, Fisher derived the first six cumulants of $\sqrt{b_1}$, Hsu and Lawley (1939) the fifth and sixth moments of $b_2$. Later, Geary and Worlledge (1947) gave the seventh noncentral moment of $b_2$; actually they give $E(m_4/m_2^2)^7$. Some of the coefficients are quite large, that of $n^3$ being a 13-digit integer multiplied by 25515 (the whole expression has had scant usage to date).

Knowing exact moments, it was a natural development to search for approximating distributions, reaching out toward percentage points of the distributions. Four-moment fits were studied by Pearson (1963) and at the time he had the choice of the Pearson system, Gram-Charlier series system based on the normal, and the Johnson $S_U$ translation system (Johnson and Kotz, 1970). For $n \geq 30$, the Student-t density gave an acceptable approximation for $\sqrt{b_1}$, the criterion being the closeness of agreement between the standardized sixth and eighth moments for the model and the true values. Johnson's $S_U$, although troublesome to fit, seemed to be equally acceptable to Pearson (1963, p. 106). Recently, D'Agostino (1970) has shown that Johnson's $S_U$ for $n \geq 8$ gives a very acceptable and simple approximation; in fact

$$Z = \delta \ln \left\{ Y/\alpha + \sqrt{(1 + (Y/\alpha)^2)} \right\} \tag{7.3}$$

is approximately a standard normal variate with zero mean and unit standard deviation, where

$$\beta_2 = 3(n^2 + 27n - 70)(n + 1)(n + 3)/\{(n - 2)(n + 5)(n + 7)(n + 9)\}$$
$$W^2 = -1 + \sqrt{(2\beta_2 - 2)}$$
$$\delta = 1/\sqrt{\ln W}, \quad \alpha = \sqrt{\{2/(W^2 - 1)\}}$$
$$Y = \sqrt{\{(n + 1)(n + 3)b_1/(6n - 12)\}}$$

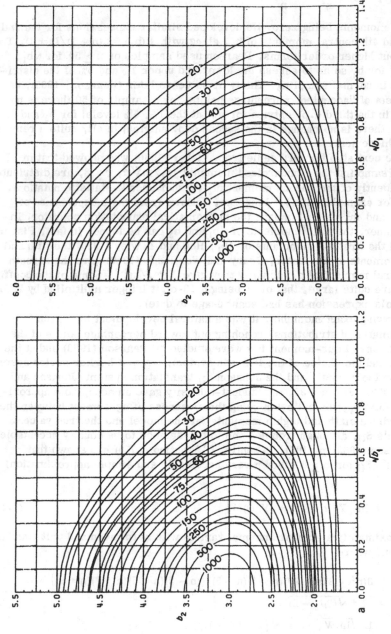

FIGURE 7.1 Contours for $K_S^2$ test; n = 30 (5) 65, 85, 100, 120, 150, 200, 250, 300, 500, 1000; n = 20, 25 based on $K_e^2$ (normal sampling).

For the kurtosis the problem is more difficult because the statistic $b_2$ is one-sided (its range being from 0 to $\infty$) and very skew in general; thus for $n = 25$, $\sqrt{\beta_1}(b_2) = 1.75$, $\beta_2(b_2) = 9.90$ in comparison to $\sqrt{\beta_1}(\sqrt{b_1}) = 0$, $\beta_2(\sqrt{b_1}) = 3.58$. Briefly, Pearson Types VI and IV (or Johnson's $S_U$) can be regarded as acceptable approximations to the distributions of $b_2$ for $n \geq 40$ according to unpublished work of C. T. Hsu (quoted by Pearson). Hsu pointed out that for $n \geq 30$, the $(\beta_1, \beta_2)$ points for $b_2$ were close to the Type V line (Pearson, 1963, p. 106). Noting this, Anscombe and Glynn (1975) suggest a linear function of a reciprocal of a $\chi^2$-variate as an approximation to $b_2$ for $n > 30$ or so; they do not make any comparisons with Johnson's $S_U$ approximation. A brief description of the Pearson system is given in Appendix 1.

### 7.2.1 Omnibus Tests

It is fairly obvious that the behavior of $\sqrt{b_1}$, $b_2$ springs from the values of $\sqrt{\beta_1}, \beta_2$ (in the population) and the sample size. For example, sampling from a uniform population ($\sqrt{\beta_1} = 0$, $\beta_2 = 1.8$) is not likely to produce large values of $\sqrt{b_1}$, whereas sampling from a negative exponential ($\sqrt{\beta_1} = 2$, $\beta_2 = 9$) could result in large $\sqrt{b_1}$ and $b_2$. Thus the skewness and kurtosis statistics are in general correlated (see 7.5), and although for normal sampling the correlation is zero, they are still dependent variables ($E(\sqrt{b_1} b_2) = 0$, but $E(b_1 b_2) \neq E(b_1) E(b_2)$). Put otherwise, there will be situations in which $\sqrt{b_1}$ will dominate the test decision about normality, $b_2$ playing a minor role, and vice versa. For example, monthly rainfall amounts in certain climates are well fitted by a negative exponential distribution so that one might expect the skewness to play a major role in testing for non-normality.

The use of both skewness and kurtosis as a test statistic arose from a study of D'Agostino and Pearson (1973); background theory is given in 7.6.

### 7.2.2 The $K_S^2$ Test Statistic

Calculate $\sqrt{b_1}$ and $b_2$ as defined in (7.2). If a pocket computer is available, it is a simple matter to input $X_1, \ldots, X_n$ and evaluate the first four sample moments along with $\sqrt{b_1}$ and $b_2$. We assume for the omnibus test that $n \geq 20$. Plot the couplet ($\sqrt{b_1}$, $b_2$) on the 90% and 95% contour charts (Figures 7.1(a), 7.1(b)). If the point is internal to the appropriate contour (approximate interpolation for $n$ should not result in much loss of accuracy in the decision process), then accept the hypothesis of normality; the reader should be reminded that this procedure involves only one test for normality and is not necessarily error free. Note also that the contours of acceptance are symmetric with respect to $\sqrt{b_1}$, the negative half of the diagrams being omitted. Obviously, the sign of $\sqrt{b_1}$ ($m_3$ may be negative) plays no part in normality.

### 7.2.3 Numerical Examples

#### 7.2.3.1 Example

Sample values: $X_i = i$, $i = \pm 1, \pm 2, \ldots, \pm m$

(A) $\underline{\quad m = 10 \quad}$  (B) $\underline{\quad m = 25 \quad}$

| (A) | (B) |
|---|---|
| $n = 20$ | $n = 50$ |
| $m_1 = 0$ | $m_1 = 0$ |
| $m_2 = 38.5$ | $m_2 = 221$ |
| $m_3 = 0$ | $m_3 = 0$ |
| $m_4 = 2533.3$ | $m_4 = 86145.8$ |
| $\sqrt{b_1} = 0$ | $\sqrt{b_1} = 0$ |
| $b_2 = 1.71$ | $b_2 = 1.76$ |

#### Conclusion

(A) is borderline at the 90% level whereas (B) is significant.

Comment: The samples nearly follow discrete uniform distributions, so the nearness of $b_2$ to 1.8 is not surprising.

#### 7.2.3.2 Example

Sample values are the first 20 of each of the first four data sets in the appendix.

| | NOR | UNI | EXP | LOG |
|---|---|---|---|---|
| $m_1$ | 100.37 | 6.13 | 4.29 | 100.90 |
| $m_2$ | 88.8095 | 8.1844 | 16.8497 | 306.6382 |
| $m_3$ | 316.32 | -8.4149 | 92.1552 | 487.247 |
| $m_4$ | 16803.04 | 108.81 | 1317.94 | 323979.6 |
| $\sqrt{b_1}$ | 0.378 | -0.359 | 1.33 | 0.09 |
| $b_2$ | 2.130 | 1.624 | 4.642 | 3.45 |
| $K_e^2$ $\begin{cases} 90\% \\ 95\% \end{cases}$ | NS[b] NS[b] | S[a] S[a] | S[a] S[a] | NS[b] NS[b] |

[a]S ≡ significant.
[b]NS ≡ not significant.

## Comment

It is interesting to see how the single tests using $\sqrt{b_1}$, $b_2$ separately would perform. Using the D'Agostino approximation for $\sqrt{b_1}$ under normality, we have for $n = 20$,

$$\beta_2 = 3.5778, \quad W^2 = 1.2706, \quad \alpha = 2.7187, \quad \delta = 2.8899$$

and

$$\sqrt{b_1} = 1.2856 \sinh (0.3460 \, Z) \quad (Z \in N(0,1))$$

The .90 and .95 levels of $\sqrt{b_1}$ are 0.589 and 0.772, respectively. Thus EXP is significant, and the other three not significant. As for $b_2$, using D'Agostino and Pearson (1973, 1974), we find the approximate levels ($n = 20$):

| P | .01 | .025 | .10 | .15 | .20 | .85 | .95 |
|---|-----|------|-----|-----|-----|-----|-----|
| $b_2$ | 1.64 | 1.73 | 1.94 | 2.04 | 2.12 | 3.40 | 4.18 |

We now see that for NOR and UNI the interest is in values of $b_2$ significantly lower than 3, whereas for EXP and LOG, the interest lies in values of $b_2$ larger than 3; thus the tests become directional, a concept introduced by Pearson. Clearly $Pr(b_2 < 2.13)$ is about .20, $Pr(b_2 < 1.54) < 0.01$, $Pr(b_2 > 3.40) = 0.15$, and $Pr(b_2 > 4.15) = 0.05$ approx. For the four cases, the single test summaries are:

|  | NOR | UNI | EXP | LOG |
|---|-----|-----|-----|-----|
| $\sqrt{b_1}$ | NS | NS | S | NS |
| $b_2$ | NS | S | S | NS |

in good agreement with the omnibus test.

## 7.2.3.3 Example

Sample values, series A from BLIS (appendix)

$$n = 20$$
$$m_1 = 209.6$$
$$m_2 = 892.24$$
$$m_3 = -1201.25$$
$$m_4 = 1758724$$
$$\sqrt{b_1} = -0.045$$
$$b_2 = 2.21$$

## Conclusion

Not significant at 90% level for the three tests. Note that under normal sampling, $\sqrt{b_1}$ has $\sigma = 0.473$, $\beta_2 = 3.58$, and $b_2$ has $\mu_1' = 2.714$, $\sigma = 0.761$, $\sqrt{\beta_1} = 1.738$, and $\beta_2 = 8.54$. Evidently, there is a very good chance that $-0.05 < b_1 < 0.05$. For the kurtosis, using a 4-moment Pearson curve as an approximant $p_r$ $(2.18 < b_2 < 3.05) = 0.5$ approx., so the observed value is quite acceptable.

### 7.2.3.4  Example

Sample values, first 50 of $S_U(0,3)$ (appendix)

$$n = 50$$
$$m_1 = -0.024$$
$$m_2 = 0.1084$$
$$m_3 = -0.00118$$
$$m_4 = 0.0401$$
$$\sqrt{b_1} = -0.03$$
$$b_2 = 3.42$$

## Conclusion

Not significant for the single tests nor the omnibus. Note that under normality, $\sqrt{b_1}$ has $\sigma = 0.326$, $\beta_2 = 3.45$, and $b_2$ has $\mu_1' = 2.882$, $\sigma = 0.598$, $\sqrt{\beta_1} = 1.582$, and $\beta_2 = 8.42$. Clearly, the observed $\sqrt{b_1}$ is not significant. A 4-moment Pearson curve for $b_2$ gives $P_r(b_2 > 3.63) = 0.1$ approx., so that the observed value is quite acceptable.

### 7.2.3.5  Example

Sample values are rainfall amounts at Tifton, Georgia for the month of January over 30 years (1928-1957). Data from U.S. Department of Commerce, Weather Bureau.

| Rainfall: | 1.36 | 4.25 | 2.52 | 4.03 | 3.50 | 3.82 |
|-----------|------|------|------|------|------|------|
| (inches)  | 5.11 | 2.09 | 3.18 | 4.90 | 1.44 | 0.72 |
|           | 5.41 | 2.41 | 2.77 | 3.94 | 1.20 | 5.42 |
|           | 2.54 | 5.27 | 1.74 | 4.84 | 1.27 | 3.35 |
|           | 6.06 | 5.55 | 5.02 | 5.94 | 2.57 | 0.81 |

$$n = 30$$
$$m_1 = 3.4343$$
$$m_2 = 2.6857$$
$$\sqrt{b_1} = -0.05$$
$$b_2 = 1.71$$

## Conclusion

The omnibus test rejects at 90%. Also $P_r(b_2 < 1.73) = 0.025$, so the kurtosis test rejects.

### 7.2.4 Further Comments

There is a computerized version of the $K_S^2$ test in Appendix 2. This could be included as a subroutine in a statistical data package and does not depend on graphical displays.

Also, we describe in 7.6 the construction of another set of test contours for normality based on a model for the joint distribution of $\sqrt{b_1}$ and $b_2$. Overall, this new set of contours leads to about the same decision as the $K_S^2$ set.

Given the mean, s.d., skewness, and kurtosis of a distribution, approximate percentage points may be quickly evaluated using a Pearson curve fit (Bowman and Shenton, 1979a, 1979b). The standardized percentiles are given by rational fraction approximants and are suitable for use on a portable calculator.

## 7.3 NONNORMAL SAMPLING

The state of the art for this case is far less developed than for normal sampling. Contours of acceptance under various hypotheses are shown below in Figures 7.3, 7.4, 7.5, 7.7, and 7.8; for each of these the sample values $\sqrt{b_1}$, $b_2$ are plotted and judged against the appropriate percent and sample size. In particular, Figure 7.8 shows 90, 95, and 99 percent contours for samples of 75 from a skew Type I distribution.

An overview of 90% contours (n = 200) for several Pearson populations is shown in Figure 7.9. Note the significant changes in area enclosed for a normal ($\sqrt{\beta_1} = 0$, $\beta_2 = 3$), uniform ($\sqrt{\beta_1} = 0$, $\beta_2 = 1.8$) and a population close to the Type III line (circled as 8). It is evident that discrimination between populations using an omnibus test based on $\sqrt{b_1}$ and $b_2$ raises many unsolved problems.

For a discussion of various aspects of this situation, the reader should turn to 7.6. The remainder of this chapter deals with theoretical aspects of the omnibus contours, including evaluation of the moments of $\sqrt{b_1}$ and $b_2$, their correlation, and the construction of equivalent normal deviates (based on Johnson's $S_U$ system) for $\sqrt{b_1}$ and $b_2$.

## 7.4  MOMENTS OF SAMPLE MOMENTS

### 7.4.1  Series Developments

The independence property of the skewness (and kurtosis) and variance breaks down in nonnormal sampling so that it is no longer possible to express, for example, $E(\sqrt{b_1})$ as the ratio of $E(m_3)$ to $E(m_2^{3/2})$. Thus we have to resort to multivariate Taylor-expansions. The general problem of evaluating moments of functions of moments has been structured on a recursive basis by the present authors and D. Sheehan (1971). The flavor of the results is contained in the following cases:

(a) Univariate; Moments of the mean $m_1'$

If

$$A_s = E(m_1' - \mu_1')^s$$
$$a_s = E(X - \mu_1')^s \tag{7.4}$$

then

$$A_s^{(k)} = \sum_{r=1}^{s} \binom{s}{r} \{ a_{r+1} A_{s-r}^{(k-r)} - a_r A_{s-r+1}^{(k-r)} \} \tag{7.5}$$

where $A_s^{(k)}$ refers to the coefficient of $n^{-k}$ in $A_s$, and $A_s^{(k)} = 0$ for $k \geq s$, $k < 0$. In particular

$$A_2 = \mu_2/n, \quad A_3 = \mu_3/n^2$$
$$A_4 = 3a_2^2/n^2 + (a_4 - 3a_2^2)/n^3$$

(b) Bivariate

$$A_{s+1,t}^{(k)} = \sum_{\lambda=0}^{s} \sum_{\mu=0}^{t} \binom{s}{\lambda}\binom{t}{\mu} a_{s+1-\lambda, t-\mu} A_{\lambda,\mu}^{(k+\lambda+\mu-s-t)}$$

$$- \sum_{\lambda=0}^{s} \sum_{\mu=0}^{t} \delta_{\lambda,\mu} \binom{s}{\lambda}\binom{t}{\mu} a_{\lambda,\mu} A_{s+1-\lambda, t-\mu}^{(k-\lambda-\mu)} \tag{7.6a}$$

$$A^{(k)}_{s,t+1} = \sum_{\lambda=0}^{s} \sum_{\mu=0}^{t} \binom{s}{\lambda}\binom{t}{\mu} a_{s-\lambda,t+1-\mu} A^{(k+\lambda+\mu-s-t)}_{\lambda,\mu}$$

$$- \sum_{\lambda=0}^{s} \sum_{\mu=0}^{t} \delta_{\lambda,\mu} \binom{s}{\lambda}\binom{t}{\mu} a_{\lambda,\mu} A^{(k-\lambda-\mu)}_{s-\lambda,t+1-\mu} \qquad (7.6b)$$

where $[(s+t+1)/2] \leq 1 \leq s+t$, $A_{s,t} = 0$ for $k < 0$, or $k \geq s+t$, $\delta_{\lambda,\mu} = 1$ unless $\lambda = \mu = 0$ when $\delta_{0,0} = 0$ and $[x]$ indicates the largest integer $\leq x$.

$\underline{A}$ and $\underline{a}$ have similar meanings as in (a); for example,

$$A_{r,s} = E(m'_1 - \mu'_1)^r (m'_2 - \mu'_2)^s$$

$$a_{r,s} = E(X - \mu'_1)^r (X^2 - \mu'_2)^s$$

would be a possibility, or similar expression involving linear sums of non-central sample moments and the corresponding expressions for samples $n = 1$.

Using this approach, we have set up, using a computer, the first eight moments of $\sqrt{b_1}$ and the first six moments of $b_2$ up to and including the term in $n^{-8}$ in the sample size. Moments of $\sqrt{b_1}$ involve four-dimensional arrays (corresponding to the first three sample moments and the sample size); similarly those of $b_2$ involve five-dimensional arrays. As for the moments of the population sampled, the first 40 are needed for the eight moments of $\sqrt{b_1}$, and the first 44 for the six moments of $b_2$; it is preferable to set up recursive schemes for these.

### 7.4.2 Illustrations (Bowman and Shenton, 1975a)

(a) Population: Uniform Distribution $(\sqrt{\beta_1} = 0, \beta_2 = 1.8)$
$(\mu_{2s+1}(\sqrt{b_1}) = 0$ by symmetry)

$$\mu_2(\sqrt{b_1}) \sim \frac{2.0571}{n} + \frac{2.2629}{n^2} + \frac{1.8042}{n^3} + \cdots - \frac{5.2943\,E\,05}{n^8}$$

$$\mu_4(\sqrt{b_1}) \sim \frac{1.2696\,E\,01}{n} + \frac{4.7687\,E\,01}{n^2} + \frac{1.5174\,E\,02}{n^3} + \cdots - \frac{4.6669\,E\,05}{n^8}$$

$$\mu_6(\sqrt{b_1}) \sim \frac{1.3058\,E\,02}{n^3} + \frac{1.0406\,E\,04}{n^4} + \cdots + \frac{1.0269\,E\,06}{n^8}$$

$$\mu_8(\sqrt{b_1}) \sim \frac{1.8804\,E\,03}{n^4} + \frac{2.5831\,E\,04}{n^5} + \cdots + \frac{1.8986\,E\,07}{n^8}$$

(b)  Population: Pearson Type 1 ($\sqrt{\beta_1} = 0.2$, $\beta_2 = 3.1$)

$$E(\sqrt{b_1}) \sim 0.2 - \frac{1.4211}{n} + \frac{6.5992}{n^2} - \frac{3.9728\,E\,01}{n^3} + \frac{3.9971\,E\,02}{n^4}$$

$$- \frac{5.9343\,E\,03}{n^5} + \frac{1.14875\,E\,05}{n^6} - \frac{2.7672\,E\,06}{n^7} + \frac{8.0796\,E\,07}{n^8}$$

$$\mathrm{Var}\,(\sqrt{b_1}) \sim \frac{6.7573}{n} - \frac{6.1810\,E\,01}{n^2} + \frac{6.9918\,E\,02}{n^3} + \cdots - \frac{8.5309\,E\,09}{n^8}$$

$$\mu_3(\sqrt{b_1}) \sim \frac{5.9716\,E\,01}{n^2} - \frac{3.0573\,E\,03}{n^3} + \frac{1.0889\,E\,05}{n^4} + \cdots + \frac{3.1041\,E\,11}{n^8}$$

$$\mu_4(\sqrt{b_1}) \sim \frac{1.3698\,E\,02}{n^2} + \frac{1.5310\,E\,03}{n^3} - \frac{3.1549\,E\,05}{n^4} + \cdots - \frac{5.5277\,E\,12}{n^8}$$

(c)  Population: Normal Mixture $pN(v,\sigma^2) + (1-p)N(v',\sigma^2)$
($\sqrt{\beta_1} = 1.0$, $\beta_2 = 4.0$), $p = 0.8706$, $v = -.2961$, $v' = 1.9917$, $\sigma^2 = 0.4103$)

$$E(b_2) \sim 4 - \frac{2.3142}{n} - \frac{9.7524\,E\,01}{n^2} - \frac{9.6665\,E\,02}{n^3} + \cdots - \frac{4.0862\,E\,08}{n^8}$$

$$\mu_4(b_2) \sim \frac{8.5535\,E\,03}{n^2} + \frac{5.2878\,E\,05}{n^3} + \frac{6.0489\,E\,06}{n^4} + \cdots - \frac{5.3930\,E\,12}{n^8}$$

$$\mu_6(b_2) \sim \frac{2.2836\,E\,06}{n^3} + \frac{5.6775\,E\,08}{n^4} + \cdots - \frac{2.5635\,E\,15}{n^8}$$

(d)  Population: Pearson Type I, $\sqrt{\beta_1} = 1.0$, $\beta_2 = 4.0$

$$E(b_2\sqrt{b_1}) \sim 4 - \frac{9.7966}{n} - \frac{4.9612\,E\,02}{n^2} + \frac{6.1251\,E\,03}{n^3} + \cdots + \frac{5.7413\,E\,07}{n^6}$$

$$\mathrm{Var}\,(b_2\sqrt{b_1}) \sim \frac{5.6987\,E\,02}{n} + \frac{3.9918\,E\,03}{n^2} - \frac{4.9240\,E\,05}{n^3} + \cdots + \frac{3.4572\,E\,10}{n^6}$$

$$\mu_3(b_2\sqrt{b_1}) \sim \frac{2.9913\,E\,05}{n^2} + \frac{1.0895\,E\,07}{n^3} + \cdots - \frac{8.0078\,E\,11}{n^6}$$

$$\mu_4(b_2\sqrt{b_1}) \sim \frac{9.7426\,E\,05}{n^2} + \frac{3.1439\,E\,08}{n^3} + \cdots - \frac{9.9575\,E\,13}{n^6}$$

### 7.4.3  Safe Sample Sizes

Our early work in using these expansions to approximate the distributions of,
for example, $\sqrt{b_1}$ and $b_2$, relied on inflating the sample size to damp-out
higher order terms. Thus in the tables of Bowman and Shenton (1975a), safe
sample sizes are indicated for each moment, using the rather arbitrary rule
that the critical sample size is one which adjusts the size of the highest order

to the lowest order terms to be approximately one-tenth. For example, in sampling from Type I with $\sqrt{\beta_1} = 0.6$, $\beta_2 = 3.2$, the safe sample size for $E(\sqrt{b_1})$ is $n = 10$, and

$$E(\sqrt{b_1}) \sim 0.6 - 0.3082 + 0.0444 + 0.0227 - 0.0302$$
$$+ 0.0175 + 0.0186 - 0.0606 + 0.0414 \qquad (7.7)$$

Similarly, in sampling from Pearson Type I with $\sqrt{\beta_1} = 1.4$, $\beta_2 = 3.4$, the critical size is $n = 100$ for $\mu_5(b_2)$, and

$$\mu_5(b_2) \sim 3.4419 + 3.7942 + 2.5590 + 1.4111 + 0.7088 + 0.3417 \qquad (7.8)$$

Clearly, in both cases the sample sizes are only just adequate to damp-out the $n^{-8}$ terms. Pearson (1930) used rather similar damping factors; for example, for normal samples he gave

$$\sigma(\sqrt{b_1}) \sim 0.3464 (1 - .0600 + .0024 - .0001) \qquad (7.9)$$

when $n = 50$, and

$$\beta_2(b_2) \sim 3 + 5.4000 - 2.0196 + 0.4704 \qquad (7.10)$$

when $n = 100$. In the case of $\sigma$ it looks as if a smaller sample size could have been used.

For our illustrations in 7.4.2 the safe sample sizes are for

(a) 8, 8, 10, 18
(b) 18, 28, 28, 61, 86
(c) 21, 43, 102
(d) 36, 57, 72, 179

Evidently using the series indiscriminately would be disastrous.

## 7.4.4 Rational Fraction and Other Approximations

Asymptotic or slowly convergent series may be approximated by the ratio of polynomials in the variable (in our case, n the sample size), and there has been a resurgence of interest in the last decade in the subject, basically initiated by Padé (his thesis was published in 1892). Briefly, the domain of convergence of Padé approximants (which include Stieltjes continued fractions as a special case) is generally more extensive than is the case for series developments; for series in $1/n$, this suggests the possibility that smaller values of n may hold for Padé approximants. Genuinely divergent series (or what appear to be so from the pattern of the first few terms) seem to be quite common in statistics; at least that is our experience, but from a knowledge of a few terms (8, 15, or perhaps 30), one must not expect to arrive at a precise answer; rather, one looks for an optimum assessment.

For fuller accounts of the general Padé approach the reader is referred to Baker (1965, 1970, 1975). An extensive bibliography on Padé approximation and related matters is given by Brezinski (1977, 1978, 1980, 1981). General comments and cautionary remarks on summing divergent series are given by Van Dyke (1974), and problems of error analysis for convergent and divergent series are discussed by Oliver (1974).

An interesting account of the properties of continued fractions (special case of the Padé table) is given by Henrici (1976) who, among other things, pays much attention to the rate of convergence; continued fraction as development for rapidly divergent series fairly frequently converge slowly—remarkable property at worst. A brief account of Padé methods with special reference to statistical series is given in Bowman and Shenton (1984).

Discussion on divergent series for moments of statistics is to be found in Shenton and Bowman (1977a), and Bowman and Shenton (1978, 1983a, 1983b, 1984). A summation algorithm due to Levin (1973) has turned out to be successful (Bowman et al. 1978b, and Bowman and Shenton 1983c) with series of alternating sign and moderately divergent (as for example with the factorial series $1 - x1! + x^2 2! - x^3 3! \ldots$). Cases considered include the standard deviation from exponential, logistic, rectangular, and half Gaussian populations.

Padé algorithms have been used to find low order moments of $\sqrt{b_1}$ and $b_2$ required in the $S_U$ approximations.

## 7.5 THE CORRELATION BETWEEN $b_1$ AND $b_2$

### 7.5.1 Asymptotic Correlation

Early work goes back to Pearson (1905) who gave, for general sampling, the first-order asymptotics for $\mu_2(b_1)$, $\mu_2(b_2)$, and the correlation

$$R(b_1, b_2) = \frac{(2\beta_6 - 3\beta_1\beta_4 - 4\beta_2\beta_3 + 6\beta_1\beta_2^2 + 3\beta_1\beta_2 - 6\beta_3 + 12\beta_1^2 + 24\beta_1)}{[\mu_2(b_1)\mu_2(b_2)]^{\frac{1}{2}}}$$

(7.11)

where

$$\beta_3 = \mu_3\mu_5/\mu_2^4, \quad \beta_4 = \mu_6/\mu_2^3, \quad \beta_5 = \mu_3\mu_7/\mu_2^5, \quad \beta_6 = \mu_8/\mu_2^4$$

Note, as far as first-order terms go, this is the same as the correlation between $\sqrt{b_1}$ and $b_2$.

### 7.5.2  More Exact Results

Further coefficients of higher powers of $n^{-1}$ can be used in

$$\rho(\sqrt{b_1}, b_2) = \frac{E(b_2 \sqrt{b_1}) - E(b_2) E(\sqrt{b_1})}{\sigma(\sqrt{b_1}) \sigma(b_2)} \qquad (7.12)$$

by the method of moments of sample moments. Each of the four terms in (7.12) is developed as a series in descending powers of n, and a summation technique applied in each case (7.4.2(d) gives examples of moment series for the statistic $b_2 \sqrt{b_1}$). Due to digital programming complexities the series for $E(b_2 \sqrt{b_1})$ could only be taken as far as $n^{-6}$; those for $\sqrt{b_1}$ and $b_2$ were taken as far as $n^{-8}$. A selection of results is given in Table 7.1.

The surprising feature is the largeness of the correlation especially at parameter points $(\sqrt{\beta_1}, \beta_2)$ not in the neighborhood of the normal point $(0, 3.0)$. Even at $(0.2, 3.0)$ there is a correlation of $0.43$. Note that in normal sampling the correlation is zero, but $b_1$ and $b_2$ are correlated. Thus (Shenton and Bowman, 1975) in normal sampling,

$$\text{cov}(b_1, b_2) = \frac{216n(n - 2)(n - 3)}{(n + 1)^2 (n + 3)(n + 5)(n + 7)} \qquad (7.13)$$

and the correlation is

$$\rho(b_1, b_2) = \left\{ \frac{54n(n^2 - 9)}{(n - 2)(n + 5)(n + 7)^2 (\beta_2 (\sqrt{b_1}) - 1)} \right\}^{\frac{1}{2}} \quad (n \geq 3) \qquad (7.14)$$

Returning to Table 7.1, also note that sample size beyond 50 only changes the correlation slightly. More extensive tabulations (Bowman and Shenton, 1975a, Table 5) suggest that in sampling from Pearson Type I distributions with samples $n \geq 50$, the correlation between $\sqrt{b_1}$ and $b_2$ is $0.8$ or more if the skewness $(\sqrt{\beta_1})$ of the sampled population exceeds around $0.7$. This property shows that in sampling from Pearson Type I distributions the dot diagram of $(\sqrt{b_1}, b_2)$ will consist of an elongated narrow band for $\sqrt{b_1} \geq 0.7$ or so (doubtless, if $\beta_2$ is also large and the Pearson curve considered is Type III or Type IV, then the narrow band may broaden, but we have insufficient data to be sure of this). The dot diagrams of the couplets $(\sqrt{b_1}, b_2)$ in Figures 7.3, 7.4, and 7.5, below, support these properties (note that a limited investigation of a similar grid of values $(\sqrt{\beta_1}, \beta_2)$ for normal mixtures showed no significant change in the pattern of behavior of the correlation coefficient).

TABLE 7.1 Covariance and Correlation between $\sqrt{b_1}$ and $b_2$ in Pearson Sampling

| $\sqrt{\beta_1}$ | $\beta_2$ | | n | | | | | |
|---|---|---|---|---|---|---|---|---|
| | | | 50 | 75 | 100 | 250 | 500 | 1000 |
| 0.20 | 1.20 | (a) | 0.0344 | 0.0218 | 0.0159 | 0.0061 | 0.0030 | 0.0015 |
| | | (b) | 0.6963 | 0.7478 | 0.7788 | 0.8467 | 0.8740 | 0.8887 |
| 0.20 | 1.80 | (a) | 0.0229 | 0.0150 | 0.0112 | 0.0044 | 0.0022 | 0.0011 |
| | | (b) | 0.5141 | 0.5274 | 0.5343 | 0.5473 | 0.5518 | 0.5541 |
| 0.20 | 3.00 | (a) | 0.0871 | 0.0665 | 0.0534 | 0.0242 | 0.0126 | 0.0064 |
| | | (b) | 0.4325 | 0.4573 | 0.4699 | 0.4930 | 0.5007 | 0.5046 |
| 0.40 | 1.20 | (a) | 0.0816 | 0.0509 | 0.0370 | 0.0140 | 0.0069 | 0.0034 |
| | | (b) | 0.8876 | 0.9192 | 0.9363 | 0.9698 | 0.9818 | 0.9880 |
| 0.40 | 3.00 | (a) | 0.1459 | 0.1085 | 0.0859 | 0.0379 | 0.0195 | 0.0099 |
| | | (b) | 0.6962 | 0.7142 | 0.7226 | 0.7369 | 0.7414 | 0.7435 |
| 0.60 | 1.40 | (a) | 0.1377 | 0.0851 | 0.0616 | 0.0232 | 0.0114 | 0.0056 |
| | | (b) | 0.9379 | 0.9571 | 0.9669 | 0.9850 | 0.9912 | 0.9943 |
| 0.60 | 3.00 | (a) | 0.1671 | 0.1191 | 0.0921 | 0.0387 | 0.0197 | 0.0099 |
| | | (b) | 0.8176 | 0.8252 | 0.8287 | 0.8345 | 0.8363 | 0.8372 |
| 0.60 | 3.40 | (a) | 0.2697 | 0.2163 | 0.1772 | 0.0837 | 0.0444 | 0.0229 |
| | | (b) | 0.8431 | 0.8317 | 0.8359 | 0.8447 | 0.8473 | 0.8485 |
| 0.80 | 1.80 | (a) | 0.1959 | 0.1210 | 0.0876 | 0.0330 | 0.0162 | 0.0080 |
| | | (b) | 0.9559 | 0.9688 | 0.9750 | 0.9860 | 0.9897 | 0.9915 |
| 0.80 | 2.40 | (a) | 0.1306 | 0.0850 | 0.0630 | 0.0246 | 0.0122 | 0.0061 |
| | | (b) | 0.9188 | 0.9263 | 0.9301 | 0.9367 | 0.9388 | 0.9399 |
| 0.80 | 3.00 | (a) | 0.1674 | 0.1133 | 0.0853 | 0.0342 | 0.0171 | 0.0085 |
| | | (b) | 0.8778 | 0.8813 | 0.8831 | 0.8867 | 0.8879 | 0.8885 |
| 0.80 | 3.80 | (a) | 0.3178 | 0.3219 | 0.2725 | 0.1346 | 0.0726 | 0.0378 |
| | | (b) | — | 0.9063 | 0.8918 | 0.8927 | 0.8936 | 0.8939 |
| 1.00 | 3.00 | (a) | 0.1920 | 0.1257 | 0.0932 | 0.0365 | 0.0181 | 0.0090 |
| | | (b) | 0.9271 | 0.9325 | 0.9352 | 0.9402 | 0.9418 | 0.9426 |
| 1.00 | 4.00 | (a) | 0.4262 | 0.3322 | 0.2693 | 0.1239 | 0.0649 | 0.0332 |
| | | (b) | 0.9117 | 0.9141 | 0.9151 | 0.9166 | 0.9170 | 0.9172 |
| 1.40 | 3.00 | (a) | 0.7534 | 0.4229 | 0.2947 | 0.1049 | 0.0506 | 0.0249 |
| | | (b) | 0.9688 | 0.9813 | 0.9867 | 0.9949 | 0.9973 | 0.9984 |
| 1.40 | 4.00 | (a) | 0.4001 | 0.2602 | 0.1922 | 0.0746 | 0.0369 | 0.0183 |
| | | (b) | 0.9546 | 0.9601 | 0.9628 | 0.9676 | 0.9691 | 0.9699 |

(a) Covariance, (b) Correlation. Moments are based on asymptotic series.
— $Eb_2$ not well approximated at this point for $n = 50$.

## 7.6 SIMULTANEOUS BEHAVIOR OF $\sqrt{b_1}$ AND $b_2$

### 7.6.1 Johnson's $S_U$ Distributions and $\sqrt{b_1}$ and $b_2$

We have given in (7.3) the Johnson $S_U$ transformation for $\sqrt{b_1}$ under normality due to D'Agostino. The transformation works well for n > 8 or 9.

We have tested out the $S_U$ system (Shenton and Bowman, 1975) as an approximant to the distributions of $\sqrt{b_1}$ and $b_2$ in nonnormal sampling, including normal mixtures of two components and other distributions. $S_U$ gives excellent results even for relatively small samples (n > 30 or so) and for distributions (determined by specified values of $\sqrt{\beta_1}$ and $\beta_2$) for which $0 < \sqrt{\beta_1}$ < 1.2 and $1.2 < \beta_2 < 4.0$ approximately. The kurtosis statistic is less well fitted due no doubt to its one-tailed nature. But in general for moderate sized samples, $S_U$ provides an acceptable fit.

### 7.6.2 The $S_U$ Transformation

Following Johnson's notation (1949), $S_U$ has density

$$p(y) = (\delta/(2\pi)^{\frac{1}{2}})(1/(1 + y^2)^{\frac{1}{2}}) \exp(-\tfrac{1}{2}z^2) \qquad (7.15)$$

where

$$z = \gamma + \ln\left\{y + (1 + y^2)^{\frac{1}{2}}\right\}$$

so that

$$y = \sinh\left\{(z - \gamma)/\delta\right\}$$

and $z \in N(0,1)$. The first four moments are, defining $\ln \omega = 1/\delta^2$, $\Omega = \gamma/\delta$,

$$\mu_1'(y) = -\sqrt{\omega}\,\sinh\Omega$$
$$\mu_2(y) = (\omega - 1)(\omega\cosh(2\Omega) + 1)/2$$
$$\mu_3(y) = -(\omega - 1)^2\sqrt{\omega}\left\{\omega(\omega + 2)\sinh(3\Omega) + 3\sinh\Omega\right\}/4 \qquad (7.16)$$
$$\mu_4(y) = (\omega - 1)^2\left\{d_4\cosh(4\Omega) + d_2\cosh(2\Omega) + d_0\right\}$$

where $d_4 = \omega^2(\omega^4 + 2\omega^3 + 3\omega^2 - 3)/8$, $d_2 = \tfrac{1}{2}\omega^2(\omega + 2)$, $d_0 = 3(2\omega + 1)/8$.

Suppose now that X is a statistic (or random variable) with mean $K_1$, and cumulants $K_2$, $K_3$, and $K_4$. From the standard cumulants $(K_s/K_2)^{\frac{1}{2}s}$, s = 3, 4, determine $\gamma$, $\delta$ by matching these with the skewness and kurtosis of $S_U$ (quite accurate rational fraction approximations to $\omega$ and $\Omega$ have been given (Bowman and Shenton, 1980a) in terms of $\sqrt{\beta_1}$, $\beta_2$ and capable of implementation on a portable calculator, preferably programmable). Similarly, solutions are available for the neighboring $S_B$ system (Bowman and Shenton, 1980b); otherwise, see Johnson (1965) or Pearson and Hartley (1972) for facilitating tables). Define $t = (X - \xi)/\lambda$, and determine $\xi$, $\lambda$ setting $\mu_1'(t) = \mu_1'(y)$ and $\mu_2(t) = \mu_2(y)$. Thus

$$\lambda^2 = K_2/\mu_2(y)$$

$$\xi = K_1 - \lambda\mu_1'(y)$$

and the first four moments of t are now those of y given in (7.16).

In particular there are the approximate transformations,

$$\sqrt{b_1} = \xi_1 + \lambda_1 \sinh[(X_S(\sqrt{b_1}) - \gamma_1)/\delta_1]$$

$$b_2 = \xi_2 + \lambda_2 \sinh[(X_S(b_2) - \gamma_2)/\delta_2]$$

where $X_S(\sqrt{b_1})$, $X_S(b_2)$ are equivalent normal deviates based on $S_U$. To determine the eight parameters $\xi_1$, $\lambda_1$, $\cdots$, $\delta_2$ we need the first four moments of $\sqrt{b_1}$ and $b_2$ (mean, variance, $\sqrt{\beta_1}$, $\beta_2$ in standardized moments). These are determined by the Taylor series approach (see 7.4), and to set up the two sets of moments we require 40 moments of the population sampled in order to take the moment series out to terms of order $n^{-8}$ (36 for $\sqrt{b_1}$, 40 for $b_2$). The series are summed either by the "safe sample size" technique or the Padé algorithm. Malfunctions of the summation technique can some-times be detected by lack of smoothness in assessments of a moment for a set of equally spaced sample sizes.

## 7.6.3 Omnibus Test Contours

The tests we shall consider are:

Normal Sampling
$$\begin{cases} \text{A:} \quad K_S^2 = X_S^2(\sqrt{b_1}) + X_S^2(b_2) \\[2mm] \text{B:} \quad K_e^2 = X_S^2(\sqrt{b_1}) + X_e^2(b_2) \\[2mm] \text{C:} \quad K_S^2, \text{ and } K_e^2, \text{ treated as } \chi^2\text{-variates with 2 d.f. will be} \\ \qquad \text{labelled } \chi_S^2 \text{ and } \chi_e^2, \text{ respectively} \end{cases}$$

Nonnormal Sampling
$$\begin{cases} \text{D:} \quad K_{SR}^2 = \dfrac{X_S^2(\sqrt{b_1}) - 2RX_S(\sqrt{b_1})X_S(b_2) + X_S^2(b_2)}{1 - R^2} \\[4mm] \qquad \text{where} \\[2mm] \qquad R = \rho(X_S(\sqrt{b_1}),\ X_S(b_2)) \\[2mm] \text{E:} \quad K_{SR}^2 \text{ treated as } \chi^2(\nu = 2) \text{ will be called } \chi_{SR}^2 \end{cases}$$

(In the sequel we shall also describe a bivariate model for the joint distribution of $\sqrt{b_1}$, $b_2$ from which acceptance contours are constructed.)

Case C is the D'Agostino-Pearson test—a fairly obvious concept since the components are squares of approximate normal variates—so that percentage values can be used to test significance. They used extensive simulations to assess the distribution of $b_2$ (sample sizes n = 20(5) 50, 100, 200) from which $X_e(b_2)$, the equivalent normal deviate for $b_2$, could be estimated at various probability levels. However, the statistics $X_S(\sqrt{b_1})$ and $X_e(b_2)$ are not independent (D'Agostino and Pearson, 1974); and whereas this perhaps has little effect on the test contours for large samples, it does have to be taken into account for most applications.

The test statistics in A, B, and D are to be regarded as mappings, so that probability levels are to be discovered by Monte Carlo simulations. A is a quick and easy statistic to simulate if we want a rough test; all that we require are the Johnson transforms for $\sqrt{b_1}$ and $b_2$ followed by simulating $K_S^2$ (see Appendix). B is an improvement on A and C, since it uses the empirical equivalent normal deviate $X_e(b_2)$. D is intended as an approximate testing statistic but is still fairly complicated to construct. First of all, for a given sample size, we must carry out simulations to assess the correlation between $X_S(\sqrt{b_1})$ and $X_S(b_2)$, since this correlation is intractable mathematically. Next, we simulate $K_S^2$ to determine a set of percentage points, and finally, we map the regions in the ($\sqrt{b_1}$, $b_2$) plane. For present purposes, the $\chi_{SR}^2$ version is adequate.

In the next paragraph we set out some supporting material for the statistics involved in A, B, and C. Readers who prefer to follow the main development may move to 7.6.5.

### 7.6.4  Comments on Moments of Statistics in the Test Statistics in Normal Sampling

Moments of $\sqrt{b_1}$, $b_2$, etc. under normality are given in Tables 7.2 and 7.3 (Bowman and Shenton, 1975b). $X_S(\sqrt{b_1})$ has moments very close to the normal even for n = 20. $X_S(b_2)$ is more discrepant but is satisfactory for $n \geq 50$; the improvement using $X_e(b_2)$ is evident, especially for the smaller samples. The theoretical moments of $\sqrt{b_1}$ and $b_2$ are derived from the Taylor series developments and show gratifying closeness to the simulation results. Those for the $\chi^2$-type statistic (A and B) on the assumption of independence should be $\mu_1' = 2$, $\sigma = 2$, $\sqrt{\beta_1} = 2$, $\beta_2 = 9$. The discrepancies (Table 7.2) are marked for n small and persist for $\sqrt{\beta_1}$ and $\beta_2$ even at n = 100.

The upper percentage points in Table 7.3 of $X_S(\sqrt{b_1})$, $X_S(b_2)$, and $X_e(b_2)$ are close to the normal values. However, for the lower percentage points, whereas the agreement for $X_S(\sqrt{b_1})$ is satisfactory, that for $X_S(b_2)$ is quite discrepant, especially for small samples; by contrast the percentage points for $X_e(b_2)$ are in good agreement for all sample sizes studied. Thus $K_S^2$ will give too much weight to large discrepancies especially for small samples,

TABLE 7.2 Moments of $\sqrt{b_1}$, $b_2$ and Related Variates in Normal Sampling Based on 50,000 Simulations[a]

| Sample size n | Variate | Moment parameters | | | |
|---|---|---|---|---|---|
| | | $\mu_1'$ | $\sigma$ | $\sqrt{\beta_1}$ | $\beta_2$ |
| 20 | $\sqrt{b_1}$ | 0.000 (0.000) | 0.472 (0.473) | -0.001 (0.000) | 3.64 (3.58) |
| | $X_S(\sqrt{b_1})$ | 0.001 | 0.998 | -0.002 | 3.03 |
| | $b_2$ | 2.708 (2.714) | 0.767 (0.761) | 1.840 (1.738) | 9.52 (8.54) |
| | $X_S(b_2)$ | -0.018 | 1.033 | -0.257 | 4.09 |
| | $X_e(b_2)$ | -0.008 | 1.000 | 0.040 | 3.22 |
| | $K_S^2$ | 2.063 | 2.653 | 5.012 | 72.38 |
| | $K_e^2$ | 1.995 | 2.441 | 4.297 | 43.38 |
| 25 | $\sqrt{b_1}$ | -0.003 (0.000) | 0.437 (0.435) | -0.015 (0.000) | 3.59 (3.58) |
| | $X_S(\sqrt{b_1})$ | -0.006 | 1.003 | -0.004 | 3.00 |
| | $b_2$ | 2.765 (2.769) | 0.740 (0.731) | 1.834 (1.747) | 9.41 (8.90) |
| | $X_S(b_2)$ | -0.015 | 1.026 | -0.149 | 3.51 |
| | $X_e(b_2)$ | -0.009 | 1.005 | 0.053 | 3.18 |
| | $K_S^2$ | 2.058 | 2.505 | 3.385 | 21.63 |
| | $K_e^2$ | 2.016 | 2.433 | 4.078 | 37.37 |
| 50 | $\sqrt{b_1}$ | 0.000 (0.000) | 0.327 (0.326) | 0.002 (0.000) | 3.57 (3.45) |
| | $X_S(\sqrt{b_1})$ | 0.001 | 1.001 | 0.001 | 3.06 |

| | | | | |
|---|---|---|---|---|
| $b_2$ | 2.880 (2.882) | 0.609 (0.598) | 1.678 (1.582) | 8.90 (8.42) |
| $X_S(b_2)$ | -0.042 | 1.011 | 0.055 | 2.89 |
| $X_e(b_2)$ | -0.020 | 1.011 | 0.057 | 3.06 |
| $K_S^2$ | 2.026 | 2.291 | 3.300 | 22.36 |
| $K_e^2$ | 2.024 | 2.358 | 3.683 | 29.80 |
| 100 $\sqrt{b_1}$ | 0.000 (0.000) | 0.238 (0.238) | -0.016 (0.000) | 3.30 (3.28) |
| $X_S(\sqrt{b_1})$ | 0.002 | 1.001 | -0.015 | 3.01 |
| $b_2$ | 2.939 (2.941) | 0.461 (0.455) | 1.324 (1.277) | 6.99 (6.77) |
| $X_S(b_2)$ | -0.009 | 1.010 | 0.057 | 2.82 |
| $X_e(b_2)$ | -0.011 | 1.012 | 0.007 | 2.96 |
| $K_S^2$ | 2.023 | 2.198 | 3.075 | 20.51 |
| $K_e^2$ | 2.027 | 2.231 | 3.030 | 19.68 |

[a]Parenthetic entries refer to theoretical values of the moment parameters. Most of the simulation for ($\sqrt{b_1}$, $b_2$) and the $\chi^2$-type statistics was carried out on IBM system 360 model 91 with occasional checks on a CDC 6400 system. The basic uniform variates were generated by a multiplicative congruential method; recommended starting values and multipliers are given in a Computer Technology Center Report, Union Carbide, Oak Ridge, Tennessee, by J. G. Sullivan and B. Coveyou. Pseudo-random normal deviates were derived from the uniform variates, using a rejection method attributed to von Neumann (Kahn, 1956, p. 39). This method sets up $x_i = \log(1/u_i)$ (i = 1, 2) and accepts $x_1$ if $(x_1 - 1)^2 < 2x_2$, where $\mu_1$ and $\mu_2$ are identically and independently distributed uniform variates on (0, 1); a normal variate follows by giving $x_1$ an equal chance of being positive or negative.

TABLE 7.3  Percentage Points of $\sqrt{b_1}$, $b_2$ and Related Variates in Normal Sampling Based on 50,000 Simulations

| Sample size n | Variate | Percentage points | | | | | |
|---|---|---|---|---|---|---|---|
| | | 1% | 5% | 10% | 90% | 95% | 99% |
| 20 | $\sqrt{b_1}$ | -1.151 | -0.772 | -0.587 | 0.583 | 0.767 | 1.151 |
| | $X_S(\sqrt{b_1})$ | -2.326 | -1.644 | -1.278 | 1.271 | 1.636 | 2.328 |
| | $b_2$ | 1.642 | 1.828 | 1.948 | 3.657 | 4.143 | 5.394 |
| | $X_S(b_2)$ | -2.631 | -1.717 | -1.299 | 1.261 | 1.631 | 2.359 |
| | $X_e(b_2)$ | -2.316 | -1.651 | -1.289 | 1.262 | 1.618 | 2.333 |
| | $K_S^2$ | 0.020 | 0.101 | 0.208 | 4.702 | 6.787 | 12.756 |
| | $K_e^2$ | 0.020 | 0.103 | 0.214 | 4.473 | 6.352 | 11.654 |
| 25 | $\sqrt{b_1}$ | -1.065 | -0.713 | -0.543 | 0.544 | 0.710 | 1.052 |
| | $X_S(\sqrt{b_1})$ | -2.338 | -1.650 | -1.282 | 1.284 | 1.643 | 2.313 |
| | $b_2$ | 1.722 | 1.912 | 2.028 | 3.675 | 4.137 | 5.381 |
| | $X_S(b_2)$ | -2.601 | -1.704 | -1.303 | 1.264 | 1.634 | 2.382 |
| | $X_e(b_2)$ | -2.321 | -1.670 | -1.289 | 1.268 | 1.634 | 2.370 |
| | $K_S^2$ | 0.020 | 0.101 | 0.212 | 4.695 | 6.728 | 12.44 |
| | $K_e^2$ | 0.021 | 0.103 | 0.217 | 4.514 | 6.354 | 11.66 |
| 50 | $\sqrt{b_1}$ | -0.789 | -0.530 | -0.408 | 0.408 | 0.535 | 0.795 |
| | $X_S(\sqrt{b_1})$ | -2.329 | -1.636 | -0.128 | 0.128 | 1.647 | 2.345 |
| | $b_2$ | 1.964 | 2.143 | 2.251 | 3.631 | 4.000 | 4.925 |
| | $X_S(b_z)$ | -2.305 | -1.707 | -1.358 | 1.251 | 1.623 | 2.327 |
| | $X_e(b_2)$ | -2.311 | -1.671 | -1.310 | 1.283 | 1.645 | 2.355 |
| | $K_S^2$ | 0.021 | 0.106 | 0.217 | 4.515 | 6.146 | 11.15 |
| | $K_e^2$ | 0.021 | 0.107 | 0.217 | 4.529 | 6.259 | 11.30 |

(continued)

TABLE 7.3 (continued)

| Sample size n | Variate | Percentage points | | | | | |
|---|---|---|---|---|---|---|---|
| | | 1% | 5% | 10% | 90% | 95% | 99% |
| 100 | $\sqrt{b_1}$ | -0.569 | -0.389 | -0.301 | 0.300 | 0.389 | 0.561 |
| | $X_S(\sqrt{b_1})$ | -2.332 | -1.642 | -1.285 | 1.281 | 1.640 | 2.305 |
| | $b_2$ | 2.180 | 2.342 | 2.439 | 3.527 | 3.796 | 4.399 |
| | $X_S(b_2)$ | -2.257 | -1.671 | -1.327 | 1.298 | 1.671 | 2.330 |
| | $X_e(b_2)$ | -2.367 | -1.675 | -1.323 | 1.291 | 1.665 | 2.335 |
| | $K_S^2$ | 0.023 | 0.110 | 0.221 | 4.513 | 6.018 | 10.36 |
| | $K_e^2$ | 0.022 | 0.107 | 0.215 | 4.552 | 6.133 | 10.49 |
| | Standard normal | -2.326 | -1.645 | -1.282 | 1.282 | 1.645 | 2.326 |
| | $\chi^2$ ($\nu = 2$) | 0.020 | 0.103 | 0.211 | 4.605 | 5.992 | 9.210 |

but in general there is little to choose between $K_S^2$ and $K_e^2$ for n greater than about 100.

The changes in the upper percentiles of $K_S^2$ as the sample size increases are shown in Table 7.4. For a sample of 1,000, the 99% value is still somewhat larger than the corresponding value for $\chi^2(\nu = 2)$.

It is thought that the $K_S^2$ and possibly $K_e^2$ approximation deteriorates from the chi-squared for percentage points more extreme than those studied here.

TABLE 7.4 Upper Percentage Points for $K_S^2$ for Large Samples Based on 50,000 Simulations

| Sample size n | 90% | 95% | 99% |
|---|---|---|---|
| 150 | 4.497 | 5.950 | 10.010 |
| 200 | 4.507 | 5.997 | 9.944 |
| 250 | 4.522 | 5.961 | 9.737 |
| 300 | 4.561 | 6.021 | 9.972 |
| 500 | 4.595 | 6.029 | 9.574 |
| 1000 | 4.629 | 6.032 | 9.444 |

### 7.6.5 Omnibus Contours for A, B, C (Normal Sampling)

The form of the joint distribution f $(\sqrt{b_1}, b_2)$ and the test contours may be
gained from a dot diagram of 5,000 runs of samples of 20 (Figure 7.2).
Superimposed on the diagram are the 90, 95, and 99% contours based on $K_S^2$,
$K_e^2$, and $\chi_S^2(\nu = 2)$. The parabolic shape adumbrated in the dot diagram is
striking and (from unpublished graphs) becomes less sharp for larger
samples (this feature will be mentioned in the sequel). It is also to be noted
that there is evidence that the conditional distribution of $\sqrt{b_1}$ for $b_2$ small
is unimodal, whereas as $b_2$ increases it becomes markedly bimodal. Contours at 90% and 95% levels of acceptance for numerous sample sizes are
given in Figure 7.1. To use them, merely compute $\sqrt{b_1}$, $b_2$ and enter and
interpret with respect to the appropriate sample size contour. Slightly improved contours are given below in Figures 7.6b and 7.6c, constructed from
an entirely different model.

### 7.6.6 Omnibus Contours Under Nonnormal Sampling

Omnibus contours D (see 7.6.3) and dot diagrams of 1,000 points as illus-

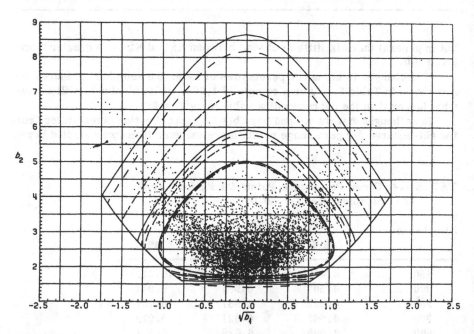

FIGURE 7.2 Dot diagram and 90, 95, and 99 percent contours in normal
sampling for n = 20. $K_S^2$ ——, $K_e^2$ ---, $\chi_S^2$ -·-.

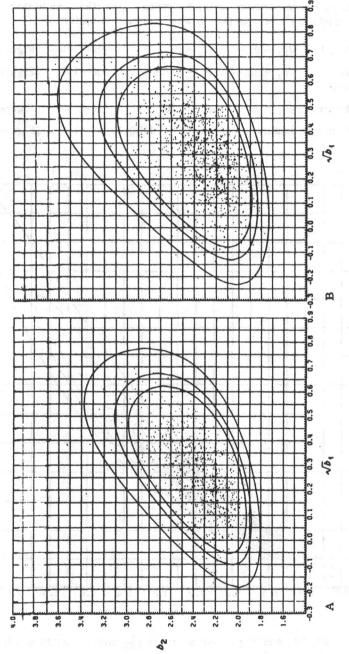

FIGURE 7.3 90, 95, and 99 percent contours ($\chi^2_{SR}$) for two distributions with the same skewness and kurtosis.

trations have been constructed in the following cases (the assumption that $K_{SR}^2$ is approximately a $\chi^2$-variate ($\nu = 2$) was made for simplicity):

Pearson Type I:    $\sqrt{\beta_1} = 2/7$      $\beta_2 = 33/14$        n = 100    (Figure 7.3A)

Normal Mixture:    $\sqrt{\beta_1} = 2/7$      $\beta_2 = 33/14$        n = 100    (Figure 7.3B)
(2 components)

Normal Mixture:    $\sqrt{\beta_1} = 1$        $\beta_2 = 2.6$         n = 50     (Figure 7.4)

Normal Mixture:    density $\rightarrow$ 0.9N(0,1) + 0.1N(3,1)   n = 100    (Figure 7.5)

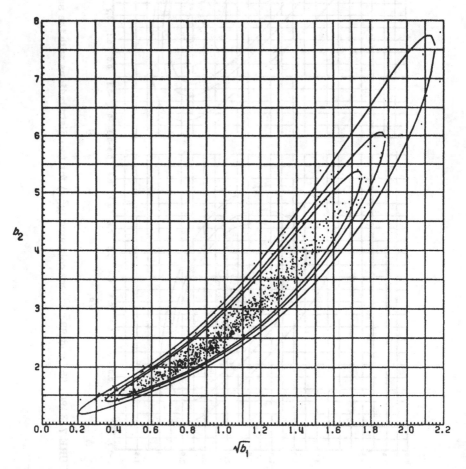

FIGURE 7.4  90, 95, and 99 percent contours $\chi_{SR}^2$ normal mixture $\sqrt{\beta_1} = 1.0$, $\beta_2 = 2.6$, n = 50.

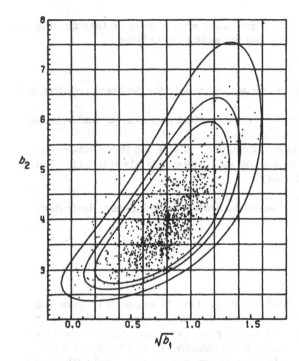

**FIGURE 7.5** 90, 95, and 99 percent contours ($\chi^2_{SR}$) for the normal mixture $f(\cdot) = 0.9N(0,1) + 0.1N(3,1)$; $n = 100$.

**TABLE 7.5** Number of Couplets Outside the 90, 95, 99% $\chi^2_{SR}$-Contours for Five Populations

| | | | | | | Number of ($\sqrt{b_1}$, $b_2$) points in 1000 outside the contour | | |
|---|---|---|---|---|---|---|---|---|
| Figure | Population | $\sqrt{\beta_1}$ | $\beta_2$ | n | R | 90% | 95% | 99% |
| 7.3A | Type I[a] | 2/7 | 33/14 | 100 | 0.571 | 89 | 46 | 9 |
| 7.3B | N.M. | 2/7 | 33/14 | 100 | 0.470 | 85 | 41 | 6 |
| — | N.M. | 1.0 | 2.2 | 70 | 0.998 | 67 | 37 | 16 |
| 7.4 | N.M. | 1.0 | 2.6 | 50 | 0.979 | 97 | 48 | 15 |
| — | Uniform | 0.0 | 1.8 | 50 | 0.000 | 100 | 57 | 16 |

[a]Type I has indices 2 and 3; N.M. = Normal Mixture.

The odd parameter points in the Type I case appear because they correspond to a density $f(n) = kx(1 - x)^2$, chosen for ease of simulation.

Comments

(1) The contours do not seem to change markedly for different populations (Figure 7.3).

(2) The remarks on the effect of the correlation between $\sqrt{b_1}$ and $b_2$ made in 7.5.2 are illustrated in Figures 7.4 and 7.5.

(3) A visual count of the number of couplets outside the three percentage levels for five populations for the $\chi^2_{SR}$ contours is given in Table 7.5. The agreement is satisfactory.

A different set of contours based on another model is now described.

## 7.7 A BIVARIATE MODEL

### 7.7.1 Genesis

We noticed (7.6.5) that in normal sampling there is evidence (Figure 7.2) that the contours of equal probability lie on parabolic arcs and that the omnibus contours have the wrong concavity for $b_2 > 3$ or so. Again there is evidence of a change from unimodality to bimodality of the $\sqrt{b_1}$ arrays for given $b_2$ as $b_2$ increases. So this suggests considering a model with the $S_U$ transformed normal curve for the marginal of $\sqrt{b_1}$. As for the conditional distribution of $b_2$ given $\sqrt{b_1}$, we carried out simulations of 50,000 samples of 10, 20, 30, and 50, from which the $b_2$-arrays appeared to have a gamma-type density bounded by $b_2 = 1 + b_1$ (all distributions for which four moments exist, and samples are such that $b_2 \geq 1 + b_1$). In addition, the means of the $b_2$ arrays ($\sqrt{b_1}$ constant) formed a parabolic arc almost parallel to the bounding parabola. Moreover, the mean and variance of the $b_2$ arrays increased with increasing ($\sqrt{b_1}$) whereas the skewness decreased.

### 7.7.2 The Model for General Sampling

We define a bivariate product density by (Shenton and Bowman, 1977)

$$f(\sqrt{b_1}, b_2) = w(\sqrt{b_1})\, g(b_2 \mid \sqrt{b_1}) \tag{7.17}$$

where $w(\cdot)$ refers to the $S_U$ approximation to the marginal probability density function of $\sqrt{b_1}$, and $g(\cdot \mid \cdot)$ to a conditional gamma distribution. Specifically,

$$w(\sqrt{b_1}) = \frac{\delta}{\sqrt{(2\pi)}\, \{\lambda^2 + (b_1 - \xi)^2\}^{\frac{1}{2}}}\, \exp\left(-\tfrac{1}{2} Z^2\right) \tag{7.18}$$

$$g(b_2 \mid \sqrt{b_1}) = k\{k(b_2 - 1 - b_1)\}^{\theta(\sqrt{b_1})-1}(1/\Gamma\{\theta(\sqrt{b_1})\})\exp\{-k(b_2 - 1 - b_1)\}$$

where

(1)   $Z = \gamma + \delta \sinh^{-1}\{(\sqrt{b_1} - \xi)/\lambda\}$,

(2)   $k > 0$, $\theta(\sqrt{b_1}) > 0$ for all real $\sqrt{b_1}$,

(3)   $\xi$, $\lambda$, $\gamma$, and $\delta$ are the parameters associated with Johnson's $S_U$ for the distribution of $\sqrt{b_1}$.

For our present study we assume a quadratic form for $\theta(\cdot)$, namely,

$$\theta(\sqrt{b_1}) = a + b\sqrt{b_1} + cb_1 \qquad (b^2 < 4ac, \; c > 0)$$

so that the unknowns in the model (bearing in mind (3)) are now a, b, c, and k. The simplest method to determine these (there are several) is to use $E(b_2)$, $\mathrm{var}(b_2)$, $E(b_2\sqrt{b_1})$, and $E(b_2 b_1)$; examples of the latter are given in 7.4. Define

$$\mu_{r,s} = E(b_1 - E(\sqrt{b_1}))^r (b_2 - E(b_2))^s \qquad (7.19)$$

$$\nu_{r,s} = E(\sqrt{b_1})^r b_2^s$$

Then from (7.17) after simplification

$$\nu_{01} = 1 + a/k + (b/k)\nu_{10} + (1 + c/k)\nu_{20}$$

$$\mu_{11} = (b/k)\mu_{20} + (1 + c/k)(\nu_{30} - \nu_{20}\nu_{10})$$

$$\mu_{02} = (1 + c/k)^2(\nu_{40} - \nu_{20}^2) + 2(1 + c/k)(b/k)(\nu_{30} - \nu_{20}\nu_{10}) \qquad (7.20)$$

$$\quad + (b/k)^2\mu_{20} + (a + b\nu_{10} + c\nu_{20})/k^2$$

$$\mu_{21} = (b/k)\mu_{30} + (1 + c/k)(\mu_{40} - \mu_{20}^2 + 2\nu_{10}\mu_{30})$$

For example, in deriving $\mu_{02} = \nu_{02} - \nu_{01}^2$, we use

$$\nu_{02} = \int_{-\infty}^{\infty} w(\sqrt{b_1})\,d(\sqrt{b_1}) \int_{1+b_1}^{\infty} \{b_2 - 1 - b_1 + (1 + b_1)\}^2 g(b_2 \mid \sqrt{b_1})\,db_2$$

The equations (7.20) can be solved explicitly and in sequence

$$r = (\mu_{20}\mu_{21} - \mu_{30}\mu_{11})/(\mu_{20}D - \mu_{30}a_3)$$

$$q = (\mu_{11}D - \mu_{21}a_3)/(\mu_{20}D - \mu_{30}a_3)$$

$$(7.21)$$

$$p = \nu_{01} - q\nu_{10} - r\nu_{20}$$
$$k = (p - 1 + q\nu_{10} + (r - 1)\nu_{20})/(\mu_{02} - r^2 a_4 - rqa_3 - q^2 \mu_{20})$$

where

$$p = 1 + a/k, \quad q = b/k, \quad r = 1 + c/k$$
$$a_3 = \nu_{30} - \nu_{20}\nu_{10}, \quad a_4 = \nu_{40} - \nu_{20}^2$$
$$D = \mu_{40} - \mu_{20}^2 + 2\nu_{10}\mu_{30}$$

In this way having determined k we deduce a, b, and c from p, q, r.

To determine the parameters in the model, we must find the $S_U$ for $\sqrt{b_1}$ and in addition $E(b_2)$, $E(\sqrt{b_1}b_2)$, var $b_2$, and

$$\mu_{21} = E(b_1 b_2) - E(b_1)E(b_2) - 2E(\sqrt{b_1})E(\sqrt{b_1}b_2) + 2E(b_2)(E\sqrt{b_1})^2$$

These in the general case (nonnormal sampling) are found by the series approach described in 7.3.

Note that since $g(b_2 \mid \sqrt{b_1})$ is the conditional distribution of $b_2$, it follows that

$$E(b_2 \mid \sqrt{b_1}) = p + q\sqrt{b_1} + rb_1$$
$$\text{var}(b_2 \mid \sqrt{b_1}) = (a + b\sqrt{b_1} + cb_1)/k^2 \tag{7.22}$$

so that since $c > 0$ (see 7.7.2 (2) and $r = 1 + c/k$, the conditional mean and variance ultimately increase with $|\sqrt{b_1}|$.

Again since the correlation between $b_1$ and $b_2$ is given by

$$\rho(b_1, b_2) = (\mu_{21} + 2\nu_{10}\mu_{11})/\{\sigma(b_1)\sigma(b_2)\} \tag{7.23}$$

and since the moments involved here have been used to determine the model parameters, it follows that the model exactly reproduces $r(\sqrt{b_1}, b_2)$ and $r(b_1, b_2)$; it does not, however, respond directly to the skewness and kurtosis of $b_2$.


7.7.3  The Model Under Normality

In this case it is known (Fisher, 1930)

$$\nu_{20} = 6(n - 2)/((n + 1)(n + 3))$$
$$\nu_{40}/\nu_{20}^2 = 3 + 36(n - 7)(n^2 + 2n - 5)/\{(n - 2)(n + 5)(n + 7)(n + 9)\}$$
$$\nu_{01} = 3(n - 1)/(n + 1)$$
$$\mu_{02} = 24n(n - 2)(n - 3)/\{(n + 1)^2(n + 3)(n + 5)\}$$
$$\mu_{21} = 216n(n - 2)(n - 3)/\{(n + 1)^2(n + 3)(n + 5)(n + 7)\}$$

so that, after simplification from (7.21)

$$a = (n - 2)(n + 5)(n + 7)(n^2 + 27n - 70)/(6\Delta)$$

$$b = 0$$

$$c = (n - 7)(n + 5)(n + 7)(n^2 + 2n - 5)/(6\Delta) \qquad (7.24)$$

$$k = (n + 5)(n + 7)(n^3 + 37n^2 + 11n - 313)/(12\Delta)$$

$$\Delta = (n - 3)(n + 1)(n^2 + 15n - 4)$$

Since $c > 0$ for the existence of the density, we must have $n > 7$, in which case there is always a unique solution. It is also evident that as the sample size increases $a$ and $c \to n/6$, whereas $k \to n/12$, so that $p$ and $r$ approach 3. Parameters of the model are shown in Table 7.6 and comparisons of the conditional means and variances (theory versus simulation) in Table 7.7.

Comparison of the omnibus test contours $K_S^2$ (equivalent to $K_6^2$ for $n > 25$) and those for the bivariate model for $n = 100$ are given in Figure 7.6a. The new contours are more responsive to the bimodality property noticed in the $\sqrt{b_1}$ arrays for $b_2 > 3$. Theoretically, the $\sqrt{b_1}$ arrays for given $b_2$ from the model have a density of the form

$$\phi(x) = ce^{kx^2}(d^2 - x^2)^{a+bx^2-1}/\Gamma(a + bx^2) \qquad (x^2 < d^2,\ a, b > 0;\ x = \sqrt{b_1})$$

and for certain parameter combinations this can show multimodality.

A visual count of the Monte Carlo couplets ($\sqrt{b_1}$, $b_2$) outside the 99, 95, and 90% bivariate contours gave 11, 52, and 94 occurrences. Another run for 1000 samples of $n = 50$ gave 11, 60, and 100 occurrences for the same three levels.

Contours of 90 and 95% in the $\sqrt{b_1}$, $b_2$ plane for samples $n = 20(5)$ 65, 75, 85, 100, 120, 150, 200, 250, 300, 500 and 1000 are shown in Figures 7.6b and 7.6c.

A very interesting study by Tietjen and Low (1975) displays several three dimensional plots of the joint distribution of $\sqrt{b_1}$ and $b_2$, along with a set of contours at the 95% level, sample sizes ranging from 4 to 50, the

TABLE 7.6 Parameters of the Model in Normal Sampling

| n | a | c | k | p | r |
|---|---|---|---|---|---|
| 10 | 5.385 | 0.774 | 5.045 | 2.06 | 1.15 |
| 30 | 8.797 | 4.208 | 5.778 | 2.52 | 1.73 |
| 50 | 12.184 | 7.493 | 7.311 | 2.67 | 2.02 |
| 100 | 20.578 | 15.763 | 11.395 | 2.81 | 2.38 |

TABLE 7.7 Conditional Means and Variances for the Model in Normal Sampling[a]

| Range of $\sqrt{b_1}$ | Sample size $n = 10$ | | | Sample size $n = 50$ | | |
|---|---|---|---|---|---|---|
| | $E(b_2\|\sqrt{b_1})$ | $Var(b_2\|\sqrt{b_1})$ | c | $E(b_2\|\sqrt{b_1})$ | $Var(b_2\|\sqrt{b_1})$ | c |
| $-0.02 < \sqrt{b_1} < 0.02$ | 2.07 (2.11) | 0.21 (0.21) | 2695 | 2.67 (2.68) | 0.23 (0.21) | 5108 |
| $0.18 < \sqrt{b_1} < 0.22$ | 2.11 (2.14) | 0.21 (0.19) | 2581 | 2.75 (2.78) | 0.23 (0.22) | 3747 |
| $0.38 < \sqrt{b_1} < 0.42$ | 2.25 (2.30) | 0.22 (0.21) | 1999 | 2.99 (3.04) | 0.25 (0.26) | 1885 |
| $0.58 < \sqrt{b_1} < 0.62$ | 2.48 (2.58) | 0.22 (0.24) | 1411 | 3.40 (3.48) | 0.28 (0.29) | 645 |
| $0.78 < \sqrt{b_1} < 0.82$ | 2.81 (2.90) | 0.23 (0.23) | 822 | 3.96 (4.04) | 0.32 (0.33) | 188 |
| $0.98 < \sqrt{b_1} < 1.02$ | 3.22 (3.41) | 0.24 (0.24) | 481 | 4.69 (4.46) | 0.37 (0.16) | 32 |

[a]Parenthetic entries refer to Monte Carlo simulations, the number of samples drawn in each range being c.

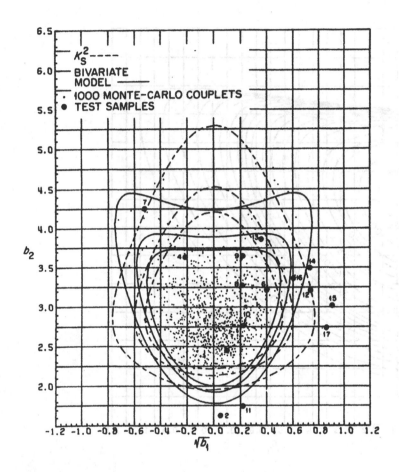

FIGURE 7.6a Normal sampling. Contours of 90, 95, and 99 percent content, $n = 100$. (Remarks on the test samples are given in Sec. 7.8.)

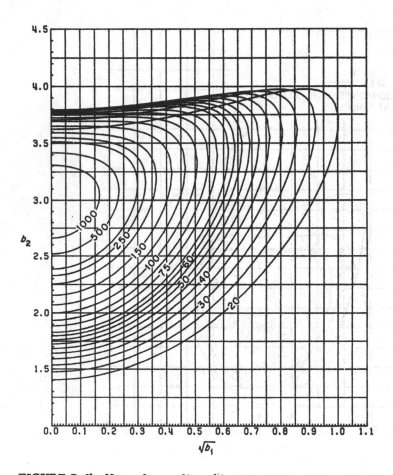

FIGURE 7.6b  Normal sampling, bivariate contours, 90 percent level.
n = 20 (5) 65, 75, 85, 100, 120, 150, 200, 250, 300, 500, and 1000.

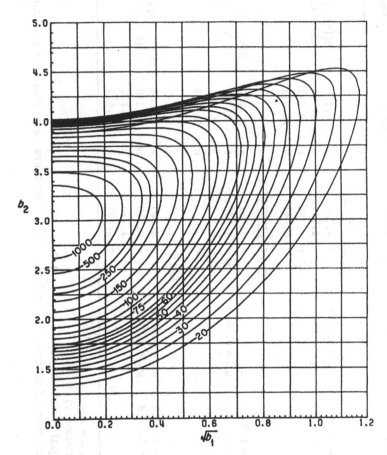

FIGURE 7.6c  Normal sampling, bivariate contours, 95 percent level.

---

**TABLE 7.8 Illustrations of the Bivariate Model for $\sqrt{b_1}$ and $b_2$[d]**

| Moments | | | Coefficients | | |
|---|---|---|---|---|---|
| $\sqrt{b_1}$ | $b_2$ | Joint | $S_U$ for $\sqrt{b_1}$ | Gamma | Regression |
| | | | **Pearson type I[a]** | | |
| $\mu_1' = 0.3660$ | $\mu_1' = 2.7187$ | $\mu_{11} = 0.1095$ | $\delta = 4.766$ | $a = 15.110$ | $E(b_2\|\sqrt{b_1}) =$ |
| $\mu_2 = 0.0752$ | $\mu_2 = 0.3280$ | $\mu_{21} = 0.0280$ | $\gamma = -1.711$ | $b = 1.207$ | $2.32 + 0.11\sqrt{b_1} + 1.70b_1$ |
| $\sqrt{\beta_1} = 0.2260$ | | | $\lambda = 1.198$ | $c = 8.011$ | |
| $\beta_2 = 3.2570$ | | | $\xi = -0.083$ | $k = 11.412$ | |
| | | | **Pearson type I[b]** | | |
| $\mu_1' = 0$ | $\mu_1' = 1.8513$ | $\mu_{11} = 0$ | $\delta = 6.422$ | $a = 22.957$ | $E(b_2\|\sqrt{b_1}) =$ |
| $\mu_2 = 0.0421$ | $\mu_2 = 0.0345$ | $\mu_{21} = 0.0049$ | $\gamma = 1.1301$ | $b = 0$ | $1.80 + 1.31b_1$ |
| $\sqrt{\beta_1} = 0$ | | | $\lambda = 0$ | $c = 9.071$ | |
| $\beta_2 = 3.1006$ | | | $\xi = 0$ | $k = 28.839$ | |
| | | | **Normal mixture[c]** | | |
| $\mu_1' = 0.2037$ | $\mu_1' = 1.2517$ | $\mu_{11} = 0.0232$ | $\delta = 5.590$ | $a = 32.202$ | $E(b_2\|\sqrt{b_1}) =$ |
| $\mu_2 = 0.0503$ | $\mu_2 = 0.0178$ | $\mu_{21} = 0.0063$ | $\gamma = -0.819$ | $b = 0.010$ | $1.15 - 1.08b_1$ |
| $\sqrt{\beta_1} = 0.0806$ | | | $\lambda = 1.220$ | $c = 17.432$ | |
| $\beta_2 = 3.1431$ | | | $\xi = 0.021$ | $k = 211.431$ | |

[a] $\sqrt{\beta_1} = 0.4$, $\beta_2 = 2.8$, n = 50.

[b] $\beta_1 = 0$, $\beta_2 = 1.8$ (rectangular), n = 50.

[c] Two components, equal variances; $\sqrt{\beta_1} = 0.2$, $\beta_2 = 1.2$, n = 75.

[d] Numerical results are rounded to three or four decimal places for convenience from machine output in double precision.

results being mainly based on Monte Carlo simulations. Their contours do have shapes similar to concentric parabolic arcs, although the gradient change is sharp at the intersections.

### 7.7.4  The Model Under Nonnormality

Illustrations of the parameters in a few cases for the bivariate model are given in Table 7.8.

In Figures 7.4 and 7.7 there is a comparison of the $\chi^2_{SR}$ contours and bivariate contours for a normal mixture population. There is little to choose between the two. Again an example of the bivariate contours for another

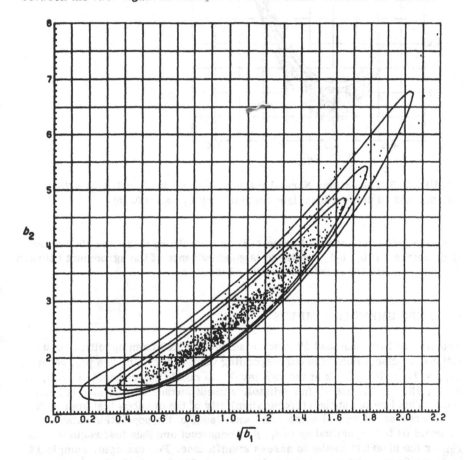

**FIGURE 7.7**  Comparison of bivariate contours and omnibus contours (Figure 7.4) for a normal mixture. $\sqrt{\beta_1} = 1.0$, $\beta_2 = 2.6$, n = 50; 90, 95, and 99 percent levels.

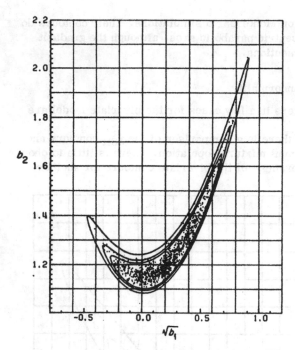

**FIGURE 7.8** Population: Normal mixture $\sqrt{\beta_1} = 0.2$, $\beta_2 = 1.2$, $n = 75$; 90, 95, and 99 percent bivariate contours $(r(\sqrt{b_1}, b_2) = 0.75)$.

normal mixture is given in Figure 7.8. In each figure there are 1000 simulated couplets $(\sqrt{b_1}, b_2)$, and there is good evidence of the agreement between the contours and the trend and density of the dots.

## 7.8 EXPERIMENTAL SAMPLES

The editors have provided 17 random samples of 100 from specified populations for discussion. The corresponding values of $\sqrt{b_1}$, $b_2$ are plotted in Figure 7.6a. If we work at the 99% level, we should reject populations 2, 3, 5, 11, 12, 14, 15, and 17 immediately without further complicated calculations. This brings out the striking simplicity of the omnibus test approach.

Moreover, if we had some prior knowledge of the population sampled, we could (if it is specified by $(\sqrt{\beta_1}, \beta_2)$) construct omnibus test regions from $\chi^2_{SR}$ or the bivariate model to assess significance. For example, sample 12 is drawn from a normal mixture with $\sqrt{\beta_1} = 0.80$, $\beta_2 = 4.02$; the $\chi^2_{SR}$ contours are shown in Figure 7.5. The sample values $\sqrt{b_1} = 0.73$, $b_2 = 3.49$ are well inside the 90% contour.

In Figure 7.9, we give illustrations of the 90% bivariate contours for n = 200 in sampling from Pearson Type I distributions (the moderately large sample of 200 is used merely to contain significant pictures on the scale used). Note that for symmetric populations the size of the acceptance region decreases as $\beta_2$ decreases. Moreover, there is a transition to an elongated shape especially for $\sqrt{\beta_1}$ large and $\beta_2 - \beta_1 - 1$ small; however, this elongated shape broadens considerably as $\beta_2$ increases. The actual area contained in the contours (those for 95, 99% not shown) turns out to be (see Figure 7.9 for the distribution numbers):

| Distribution | 90% | 95% | 99% |
|---|---|---|---|
| 1 | 0.0815 | 0.1064 | 0.1646 |
| 2 | 0.1201 | 0.1566 | 0.2418 |
| 3 | 0.7718 | 1.0161 | 1.6046 |
| 4 | 0.4413 | 0.5771 | 0.8970 |
| 5 | 0.9312 | 1.2492 | 8.8236 |
| 6 | 0.0899 | 0.1179 | 0.1846 |
| 7 | 0.2313 | 0.3024 | 0.4695 |
| 8 | — | — | 16.2689 |

These confirm fairly obvious notions in attempting to discriminate between distributions (Pearson Type I in this case) using $\sqrt{b_1}$ and $b_2$. For

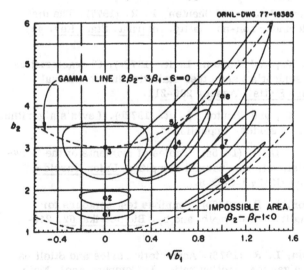

FIGURE 7.9 Examples of different shapes of bivariate contours of 90 percent content for several Type I populations (n = 200); numbers 1-8 (small circles) indicate ($\sqrt{\beta_1}$, $\beta_2$) of population sampled.

example, there should be no problem (with moderate sample sizes) in deciding between a normal distribution and symmetric alternatives for which
$\beta_2 \leq 1.8$ (distributions 3, 2, 1). However, with moderate skewness, the
regions overlap especially for moderate values of $\beta_2$ (i.e., distributions not
close to the "impossible" boundary) and very emphatically in the vicinity of
the Type III boundary.

For comments on the tests described here and other tests for departures
from normality from the power point of view, the reader is referred to
Pearson, D'Agostino, and Bowman (1977).

## REFERENCES

Anscombe, F. J. and Glynn, W. J. (1975). Distribution of the kurtosis
statistic $b_2$ for normal samples. Tech. Rep. 37, Department of Statistics, Yale University.

Baker, G. A., Jr. (1965). The theory and application of the Padé approximation method. In: Brueckner, K. A. (Ed.). Advances in Theoretical
Physics 1, 1-58, Academic Press, New York.

Baker, G. A., Jr. (1975). Essentials of Padé Approximants. Academic
Press, New York.

Baker, G. A., Jr. and Gammel, J. L. (1970). (Editors) The Padé Approximant in Theoretical Physics. Academic Press, New York.

Bowman, K. O., Beauchamp, J. J., and Shenton, L. R. (1977). The distribution of the t-statistic under non-normality. Internat. Stat. Rev. 45,
233-242.

Bowman, K. O., Lam, H. K., and Shenton, L. R. (1978a). Characteristics
of moment series for student's t and other statistics in non-normal
sampling. Proc. Amer. Statist. Assoc. 206-211.

Bowman, K. O., Lam, H. K., and Shenton, L. R. (1978b). Levin's algorithm
applied to some divergent series. Unpublished.

Bowman, K. O. and Shenton, L. R. (1975a). Tables of moments of the skewness and kurtosis statistics in non-normal sampling. Union Carbide
Nuclear Division Report UCCND-CSD-8.

Bowman, K. O. and Shenton, L. R. (1975b). Omnibus test contours for
departures from normality based on $\sqrt{b_1}$ and $b_2$. Biometrika 62, 243-
250.

Bowman, K. O. and Shenton, L. R. (1978). Asymptotic series and Stieltjes
continued fraction for a gamma function ratio. J. Comput. Appl. Math.
4, No. 2, 105-111.

Bowman, K. O. and Shenton, L. R. (1979a). Approximate percentage points
for Pearson distributions. Biometrika 66(1), 145-151.

Bowman, K. O. and Shenton, L. R. (1979b). Further approximate Pearson percentage points and Cornish-Fisher. Comm. Statist. B—Simulation Comput. B8(3), 231-234.

Bowman, K. O. and Shenton, L. R. (1980a). Evaluation of the parameters of $S_U$ by rational fractions. Comm. Statist. B—Simulation Comput. B9(2), 127-132.

Bowman, K. O. and Shenton, L. R. (1980b). Explicit approximate solutions to $S_B$. Comm. Statist. B—Simulation Comput. B10(1), 1-15.

Bowman, K. O. and Shenton, L. R. (1983a). Moment series for estimators of the parameters of a Weibull density. Proceedings of Computer Science and Statistics: 14th Symposium on the Interface. Springer-Verlag, New York, 184-186.

Bowman, K. O. and Shenton, L. R. (1983b). The Distribution of the Standard Deviation and Skewness in Gamma Sampling—A New Look at a Craig-Pearson Study. ORNL/CSD-109.

Bowman, K. O. and Shenton, L. R. (1983c). Levin's summation algorithm. Encyclopedia of Statistical Sciences, Vol. 4 (Samuel Kotz and N. L. Johnson, Eds.). Wiley, New York, 611-617.

Bowman, K. O. and Shenton, L. R. (1986). The dark side of asymptotics and moment series. Sankhyā 48, A. (In press.)

Brezinski, C. (1977). A Bibliography on Padé Approximation and Related Subjects. Université des Science et Techniques de Lille.

Brezinski, C. (1978). A Bibliography on Padé Approximation and Related Subjects, Addendum 1. Université des Science et Techniques de Lille.

Brezinski, C. (1980). A Bibliography on Padé Approximation and Related Subjects, Addendum 2. Université des Science et Techniques de Lille.

Brezinski, C. (1981). A Bibliography on Padé Approximation and Related Subjects, Addendum 3. Université des Science et Techniques de Lille.

D'Agostino, R. B. (1970). Transformations to normality of the null distribution of $g_1$. Biometrika 57, 679-681.

D'Agostino, R. B. and Pearson, E. S. (1973). Tests for departures from normality. Empirical results for the distribution of $b_2$ and $\sqrt{b_1}$. Biometrika 60, 613-622.

D'Agostino, R. B. and Pearson, E. S. (1974). Correction and amendment. Test for departures from normality. Empirical results for the distribution of $b_2$ and $\sqrt{b_1}$. Biometrika 61, 647.

Elderton, W. P. and Johnson, N. L. (1969). Systems of Frequency Curves. Cambridge University Press, London.

Fisher, R. A. (1928). Moments and product moments of sampling distributions. Proc. Lond. Math. Soc. (Series 2) 30, 199-238.

Fisher, R. A. (1930). The moments of the distribution for normal samples of measures of departures from normality. Proc. Roy. Soc., Series A 130, 17-28.

Geary, R. C. and Worlledge, J. P. G. (1947). On the computation of universal moments of tests of statistical normality derived from samples drawn at random from a normal universe. Application to the calculation of the seventh moment of $b_2$. Biometrika 34, 98-110.

Henrici, P. (1976). Applied and Computational Complex Analysis, Vol. 2. Wiley, New York.

Hsu, C. T. and Lawley, D. N. (1939). The derivation of the fifth and sixth moments of $b_2$ in samples from a normal population. Biometrika 31, 238-248.

Johnson, N. L. (1949). Systems of frequency curves generated by methods of translation. Biometrika 36, 149-176.

Johnson, N. L. (1965). Tables to facilitate fitting $S_U$ frequency curves. Biometrika 52, 547-558.

Johnson, N. L. and Kotz, S. (1970). Distributions in Statistics—Continuous Univariate Distributions, Vol. 1. Houghton Mifflin, Boston.

Kahn, H. (1956). Application of Monte Carlo. Rand Corporation, Santa Monica.

Levin, David (1973). Development of non-linear transformations for improving convergence of sequences. Internat. J. Comput. Math. B, 3, 371-388.

Mulholland, H. P. (1977). On the null distribution of $b_1$ for samples of size at most 25, with tables. Biometrika 64, 401-409.

Olver, F. W. J. (1974). Asymptotic and Special Functions. Academic Press, New York.

Pearson, E. S. (1930). A further development of tests for normality. Biometrika 22, 239-249.

Pearson, E. S. (1963). Some problems arising in approximating to probability distributions, using moments. Biometrika 50, 95-111.

Pearson, E. S. (1965). Tables of percentage points of $\sqrt{b_1}$ and $b_2$ in normal samples: a rounding off. Biometrika 52, 282-285.

Pearson, E. S., D'Agostino, R. B., and Bowman, K. O. (1977). Test of departure from normality: comparison of powers. Biometrika 64, 231-246.

Pearson, E. S. and Hartley, H. O. (1972). Biometrika Tables for Statisticians, Vol. 2. Cambridge University Press, London.

Pearson, K. (1905). On the general theory of skew correlation and non-linear regression. Draper's Company Memoirs, Biometric Series II.

Shenton, L. R. and Bowman, K. O. (1975). Johnson's $S_U$ and the kurtosis statistics. J. Amer. Statist. Assoc. 70, 220-228.

Shenton, L. R. and Bowman, K. O. (1977a). A new algorithm for summing divergent series. J. Comput. Appl. Math. 3, 35-51.

Shenton, L. R. and Bowman, K. O. (1977b). A bivariate model for the distribution of $\sqrt{b_1}$ and $b_2$. J. Amer. Statist. Assoc. 72, 206-211.

Shenton, L. R., Bowman, K. O., and Sheehan, D. (1971). Sampling moments of moments associated with univariate distributions. J. Roy. Statist. Soc. B, 33, 444-457.

Tietjen, G. L. and Lowe, V. W., Jr. (1975). Some aspects of the joint distribution of skewness and kurtosis in normal samples. (Private communication).

Wishart, J. (1930). The deviation of certain high-order sampling product-moments from a normal population. Biometrika 22, 224.

Van Dyke, M. (1974). Analysis and improvement of perturbation series. Quar. J. Mech. Appl. Math. 27, 424-450.

## APPENDIX 1

The main Pearson distribution types (Elderton and Johnson, 1969) are:

| Type | Equation | Limits for x |
|------|----------|--------------|
| I | $f(x) = y_0\left(1 + \dfrac{x}{a_1}\right)^{ba_1}\left(1 - \dfrac{x}{a_2}\right)^{ba_2}$ | $-a_1 < x < a_2$ |
| III | $f(x) = y_0\,(x/a)^{\rho-1}\exp(-x/a)$ | $x > 0$ |
| IV | $f(x) = y_0\left(1 + \dfrac{x^2}{a^2}\right)^{-m}\exp\{-b\arctan(x/a)\}$ | $-\infty < x < \infty$ |
| V | $f(x) = y_0 x^{-m}\exp(-a/x)$ | $0 < k \le x < \infty$ |
| VI | $f(x) = y_0 x^{m_2}(x - a)^{-m_1}$ | $a \le x < \infty$ |

Restrictions on the parameters are omitted, but in all cases the probability integrals must exist. As for the structures, Type I can be bell-shaped, J, or U-shaped, the range being finite. The other types have unlimited ranges and generally, at most, one mode.

## APPENDIX 2

### A2.1  A Computer Version of the $K_S^2$ Test Under Normality

Let $X_S(\sqrt{b_1})$ and $X_S(b_2)$ be the normal deviates corresponding to the skewness $(\sqrt{b_1})$ and kurtosis $(b_2)$, respectively. Then $K_S^2$ is

$$K_S^2 = X_S^2(\sqrt{b_1}) + X_S^2(b_2) \tag{A1}$$

We shall need the moments of $\sqrt{b_1}$ and $b_2$.
The first four moments of $\sqrt{b_1}$ are

$$\mu_1'(\sqrt{b_1}) = 0$$

$$\mu_2(\sqrt{b_1}) = 6(n - 2)/\{(n + 1)(n + 3)\}$$

$$\mu_3(\sqrt{b_1}) = 0 \tag{A2}$$

$$\mu_4(\sqrt{b_1}) = \mu_2^2(\sqrt{b_1})\left\{3 + \frac{36(n - 7)(n^2 + 2n - 5)}{(n - 2)(n + 5)(n + 7)(n + 9)}\right\}$$

Similarly those for $b_2$ are

$$\mu_1'(b_2) = 3(n - 1)/(n + 1)$$

$$\mu_2(b_2) = 24n(n - 2)(n - 3)/\{(n + 1)^2(n + 3)(n + 5)\}$$

$$\mu_3(b_2) = \frac{1728n(n - 2)(n - 3)(n^2 - 5n + 2)}{(n + 1)^3(n + 3)(n + 5)(n + 7)(n + 9)} \tag{A3}$$

$$\mu_4(b_2) = \frac{1728n(n - 2)(n - 3)\pi(n)}{(n + 1)^4(n + 3)(n + 5)(n + 7)(n + 9)(n + 11)(n + 13)}$$

where $\pi(n) = n^5 + 207n^4 - 1707n^3 + 4105n^2 - 1902n + 720$.

### A2.2  General Case for $S_U$

To fit a Johnson $S_U$ to a statistic T (or random variable), let its moment parameters be $\mu_1'(T)$, $\mu_2(T)$, $\sqrt{\beta_1}(T)$, and $\beta_2(T)$, where

$$\sqrt{\beta_1}(T) = \mu_3(T)/(\mu_2(T))^{3/2}$$

$$\beta_2(T) = \mu_4(T)/(\mu_2(T))^2$$

For simplicity in the following, we shall write $\sqrt{\beta_1}$, $\beta_2$ for $\sqrt{\beta_1}(T)$, $\beta_2(T)$, respectively.

## A2.2.1 Quick Approximate Solution

We define $w = \exp(1/\delta^2)$, $\Omega = \gamma/\delta$. Using $\beta_1$ and $\beta_2$, compute

$$\Psi(\beta_1, \beta_2) = \Pi_1(\beta_1, \gamma_2)/\Pi_2(\beta_1, \gamma_2) \tag{A4}$$

where for $i = 1$, $2$, with $\gamma_2 = \beta_2 - 3$,

$$\Pi_i(\beta_1, \gamma_2) = \sum\sum_{0 \le r+s \le 3} \alpha_{r,s}^{(i)} \beta_1^r \gamma_2^s \qquad (\alpha_{0,0}^{(2)} = 1)$$

The parameters $\{\alpha_{r,s}^{(i)}\}$ are given in Table A1.

Then an approximation to $w$ is

$$w^* = \sqrt{[\sqrt{\{2\beta_2 - 2\beta_1\Psi(\beta_1, \beta_2) - 2\}} - 1]} \tag{A5}$$

(For $3 < \beta_2 < 75$ and bounded by the lognormal line in the ($\beta_1$, $\beta_2$) plane, the error for $w^*$ is 0.006%.)

TABLE A1  Coefficients $\alpha_{rs}$[a]

| rs | Numerator | | Denominator | |
|----|-----------|---|-------------|---|
| 00 | 1.33384 8465690817 | | 1.0 | |
| 10 | -5.455870858760243 | (1) | -3.621879838877379 | (1) |
| 01 | 4.120727348534858 | (1) | 2.677096382861022 | (1) |
| 20 | 4.557065299738849 | (2) | 2.215014552020006 | (2) |
| 11 | -7.219603313144450 | (2) | -3.727562379795881 | (2) |
| 02 | 2.459166955114775 | (2) | 1.266073347716621 | (2) |
| 30 | -3.989549653042761 | (4) | -9.985226235607946 | (5) |
| 21 | 1.018227677593445 | (3) | 2.517660829639101 | (4) |
| 12 | -8.686256219859072 | (4) | -2.555913695361249 | (4) |
| 03 | 2.000771886697820 | (4) | 6.266097479093280 | (5) |

[a]Parenthetic entries taken negatively give the power of 10 by which the corresponding entry is to be multiplied.

Now compute $\Omega^*$ from

$$\sinh \Omega^* = \sqrt{\left\{ \frac{D(w^*) - 1 - w^*}{2w^*} \right\}} \qquad (A6)$$

where

$$D(w) = -\frac{[B(w) + \sqrt{\{B^2(w) - 4A(w)C(w)\}}]}{2A(w)}$$

and

$$A(w) = 2(\beta_2 - w^4 - 2w^3 - 3w^2 + 3)$$
$$B(w) = 4(w^2 - 1)(w^2 + 2w + 3) \qquad (A7)$$
$$C(w) = (w^2 - 1)^2(w^2 + 2w + 3)$$

$(\sinh^{-1} x = \ln \{x + \sqrt{(x^2 + 1)}\}$, x real). Go to A2.2.6 to determine $\xi$ and $\lambda$ and complete the approximate solution.

### A2.2.2 Exact Solution from Bivariate Iteration

(1) Determine w.
(2) Determine $\Omega$.
(3) Go to A2.2.6.

### Starting value for w

Compute

$$W_1 = \sqrt{[\sqrt{(2\beta_2 - 2)} - 1]}$$

$$u_1 = 1 + \sqrt[3]{\{\underline{P} + \sqrt{Q}\}} - \sqrt[3]{\{-\underline{P} + \sqrt{Q}\}}$$

$$(\underline{P} = 2\beta_2 + 7, \quad Q = (4\beta_2/3 + 3)^3 + \underline{P}^2)$$

$$A = 1 - \sqrt{(u_1 - 2)} \qquad (A8)$$

$$B = \frac{1}{2}u_1 - \sqrt{\left\{ \frac{u_1^2}{4} + \beta_2 + 3 \right\}}$$

$$W_2 = [-A + \sqrt{(A^2 - 4B)}]/2$$

Then a starting value is

$$w_0 = (W_1 + W_2)/2 \qquad (A9)$$

A better starting value in general is that given in (A5). Note that if $(\beta_1, \beta_2)$

is close to the lognormal line (A5), the iterative process may abort (this line is defined by $w = 1 + \beta_1/(w + 2)^2$ and $\beta_2 = w^4 + 2w^3 + 3w^2 - 3$). At this stage check if there is, in any event, a Johnson $S_U$ curve to match $(\beta_1, \beta_2)$.

<u>There is no solution if</u>

$$\beta_1^* > \beta_1$$

where using (A8)

$$\beta_1^* = (W_1 - 1)(W_2 + 2)^2$$

A2.2.3  The Iterates for w and $\Omega$

(1)  From (A6), using $w_0$ for $w_1^*$, compute $\sinh \Omega_0$ and $\Omega_0$, with the restrictions

   (i) $\Omega_0$ is taken to be $< 0$ if $\sqrt{\beta_1} > 0$.

   (ii) $\Omega_0$ is taken to be $> 0$ if $\sqrt{\beta_1} < 0$.

   (iii) $\Omega$ itself is taken to be zero if $\beta_1 = 0$; in this case

$$w = \sqrt{\{\sqrt{(2\beta_2 - 2)} - 1\}}$$

Go to A2.2.6 from which $\xi = \mu_1'(T)$, $\lambda = \sqrt{\{2\mu_2(T)/(w^2 - 1)\}}$

(2)  Using $w_0$, $\Omega_0$ compute $\sqrt{\beta_1^*}(w_0, \Omega_0)$ from

$$\sqrt{\beta_1^*}(w, \Omega) = -\frac{\left[\frac{1}{2}w(w - 1)\right]^{\frac{1}{2}} [w(w + 2)\sinh(3\Omega) + 3\sinh\Omega]}{[w\cosh(2\Omega) + 1]^{3/2}} \tag{A10}$$

Solve for a new w (say $w_1$) the equation

$$\Psi(\beta_1, \beta_2; w) = \Psi(\beta_1^*, \beta_2; w_0)$$

where

$$\Psi(\beta_1, \beta_2; w) = \frac{\beta_2 - \frac{1}{2}(w^4 + 2w^2 + 3)}{\beta_1}$$

Thus

$$w_1 = \sqrt{[\sqrt{\{2\beta_2 - 2\beta_1\Psi(\beta_1^*, \beta_2; w_0) - 2\}} - 1]} \tag{A11}$$

Return this new value to (1) and continue the cycle to meet whatever tolerance is deemed reasonable; for example, at the sth cycle, demand

$$|w_s - w_{s-1}| < c \qquad (c = 10^{-8} \text{ or so})$$

One could also require a tolerance based on the larger of $|w_s - w_{s-1}|$ and $|\Omega_s - \Omega_{s-1}|$.

## A2.2.4 Examples

Check a few cases using the final computed values of w and $\Omega$ in (7.16) to determine $\sqrt{\beta_1}$ and $\beta_2$.

## A2.2.5 Illustrations

Example 1. Computed on HP97 (see ESS, Vol. 4, p. 312)

A random sample of 15 is drawn from a population with density

$$f(x) = \left(\frac{x}{\alpha}\right)^{\rho-1} e^{-(x/\alpha)} / \alpha \Gamma(\rho) \qquad (x > 0, \rho, \alpha > 0)$$

$\hat{\rho}$ is a maximum likelihood estimator of $\rho$ and the solution of

$$\ln \hat{\rho} - \psi(\hat{\rho}) = \ln \left\{ \frac{\text{Arithmetic Mean}}{\text{Geometric Mean}} \right\}$$

Given the moments of $\hat{\rho}$ are

$$\mu_1' = 1.2044$$
$$\mu_2 = 0.2299$$
$$\sqrt{\beta_1} = 2.1680$$
$$\beta_2 = 14.4765$$

Then

$$W_2 = 1.469166402$$
$$\sqrt{\beta_1^*} = 2.376230596 > \sqrt{\beta_1}, \text{ so there is a solution}$$
$$w_0 = 1.758256474, \qquad |w_{11} - w_{10}| < 0.000000001$$

$w_1$  1.579959821  ***
$w_2$  1.597436626  ***
$w_3$  1.595876511  ***
$w_4$  1.596017234  ***

$w_5$   1.596004551   ***
$w_6$   1.596005695   ***   iterates
$w_7$   1.596005591   ***
$w_8$   1.596005602   ***
$w_9$   1.596005601   ***
$w_{10}$ 1.596005601   ***
$w_{11}$ 1.596005601   ***
$\Omega \rightarrow$ -0.905955881   ***


$\mu_1'$   1.204400000   ***
$\mu_2$   0.229900000   ***
$\lambda$   0.358105885   ***
$\xi$   0.736127663   ***

　　　Feed-back

$w$   1.596005601   ***
$\Omega$   -0.905955881   ***
$\mu_1'(y)$ 1.307636524   ***
$\mu_2(y)$ 1.792734830   ***
$\sqrt{\beta_1}$ 2.167999999   ***
$\beta_2$ 14.476500001   ***

### Example 2. HP97 (see ESS, Vol. 4, p. 312)

The density of a random variable X is $P(x) = 1/2 \exp - |x|$, $-\infty < x < \infty$.

with moments: $\mu_1' = 0$, $\mu_2 = 2$, $\beta_1 = 0$, and $\beta_2 = 6$

Then $S_U$ is given by

$\beta_2$   6.000000000   ***
$w_0$   1.315956418   ***
$w$   1.470468517   ***
$\delta$   1.610431098   ***
$\mu_1'$   0.000000000   ***
$\mu_2$   2.000000000   ***
$\lambda$   1.855132999   ***
$\xi$   0.000000000   ***
　　　1.470468517   ***
　　　0.000000000   ***

　　　Feed-back

$w$   1.470468517   ***
$\Omega$   0.000000000   ***
　　　0.000000000   ***

$$\begin{array}{ll}
& 0.581138830 \quad *** \\
\beta_1 & 0.000000000 \quad *** \\
\beta_2 & 5.999999997 \quad ***
\end{array}$$

## A2.2.6 The Scale and Origin Parameters

Use the first two moments of the statistic T, $\mu_1'(T)$ and $\mu_2(T)$ to determine

$$\lambda = \sqrt{\frac{\mu_2(T)}{\frac{1}{2}(w-1)(w \cosh(2\Omega)+1)}}$$

$$\xi = \mu_1'(T) + \lambda\sqrt{w} \sinh \Omega$$

and

$$Z = \gamma + \delta \sinh^{-1}\left(\frac{T-\xi}{\lambda}\right)$$

## A2.3 The Test

Construct $K_S^2 = X_S^2(\sqrt{b_1}) + X_S^2(b_2)$. Bowman and Shenton (1975b) have accumulated extensive simulation results for $K_S^2$ for n = 25 (5) 100 (10) 250,300, 500,1000. A quadratic regression for each of the percentage points 90, 95, and 99 has been worked out to provide acceptance levels for $K_S^2$ under the null hypothesis for n = 30 (1) 5000. The quadratic regressions are, for $30 \le n \le 1000$.

$$90\% \quad F1 = 4.50225 + 5.39716 \times 10^{-5}n + 8.59940 \times 10^{-8}n^2$$
$$95\% \quad F2 = 6.22848 + 1.52485 \times 10^{-3}n + 1.38568 \times 10^{-6}n^2 \qquad (A12)$$
$$99\% \quad F3 = 10.8375 + 2.80482 \times 10^{-3}n + 4.97077 \times 10^{-7}n^2$$

Hence for a given value of n, $\sqrt{b_1}$, $b_2$ proceed as follows:

(1)  Compute $\delta$, $\alpha$ from (7.3), thence $\delta_1$, $\lambda_1$, and finally $X_S(\sqrt{b_1})$.

(2)  Compute $\xi$, $\lambda_2$, $\gamma_2$, $\delta_2$ and thence $X_S(b_2)$.

(3)  Construct $K_S^2 = X_S^2(\sqrt{b_1}) + X_S^2(b_2)$ and compare with the levels F1, F2, F2 for the given n.

It is thought that there may be a use for this computerized version as a sub-routine in a data analysis package, since the omnibus charts are visual aids. There would be an automatic print out of significance at the three levels.

Lastly, note that the $S_U$ transformation for the skewness statistic $\sqrt{b_1} = m_3/m_2^{3/2}$ under a normal population assumption is given explicitly (in terms of the sample size n) from the parameters

$$Y_1 = \sqrt{\{(n + 1)(n + 3)b_1/(6n - 12)\}}$$

$$B_2 = \frac{3(n^2 + 27n - 70)(n + 1)(n + 3)}{(n - 2)(n + 5)(n + 7)(n + 9)}$$

$$W = \sqrt{\{\sqrt{(2B_2 - 2)} - 1\}}$$

$$\delta = 1/\ln W$$

$$\alpha = \sqrt{\{2/(w^2 - 1)\}}$$

Then

$$X_S(\sqrt{b_1}) = \delta \ln \{Y_1/\alpha + \sqrt{(1 + Y_1/\alpha)^2)}\}$$

is an approximate normal variate.

# 8

# Tests for the Uniform Distribution

Michael A. Stephens  Simon Fraser University, Burnaby, B.C., Canada

## 8.1 INTRODUCTION

The uniform distribution plays a special role in goodness-of-fit testing. It sometimes arises, as for most tests, in the natural occurrence of certain types of data; but it appears also because there exist many ways of transforming a given sample of X-values, from a distribution other than the uniform, to produce a set of U-values, which is uniformly distributed between 0 and 1. This distribution is written U(0, 1). A test that the X-sample comes from a certain distribution can then be reduced to a test that the U-sample comes from U(0, 1). For most of this chapter we therefore discuss tests of

$H_o$: a complete random sample of n U-values comes from U(0, 1)  (8.1)

Tests for a uniform distribution with unknown limits are given in Section 8.16, and tests for censored uniforms are summarized in Section 8.17. An enormous literature exists on the properties of a uniform sample, its order statistics, and its spacings, and no attempt will be made to reference this entire literature. Two articles with extensive references are those by Barton and Mallows (1965) and Pyke (1965). References given in subsequent sections will mostly be concerned with a particular test or group of tests, and the provision of tables.

## 8.2 NOTATION

The uniform, or rectangular, distribution for a random variable U' with support over $a \leq u' \leq b$ has the density $f(u') = 1/(b - a)$, $a \leq u' \leq b$, and this density will be denoted by $U(a,b)$. The transformation $U = (U' - a)/(b - a)$ yields a random variable U with density $U(0,1)$. The distribution function of U is

$$F(u) = u, \quad 0 \leq u \leq 1 \tag{8.2}$$

The distribution in (8.2) will be called the standard uniform distribution; U' or U will be described as a uniform random variable, or simply as a uniform and a random sample from $U(0,1)$ will be called a uniform sample, with notation $U_i$, $i = 1, \ldots, n$; when these are placed in ascending order the order statistics $U_{(1)} < U_{(2)} < \cdots < U_{(n)}$ will be called ordered uniforms.

Either $U_i$ or $U_{(i)}$ might be regarded as the components of a vector $U$. It is convenient to define $U_{(0)} \equiv 0$ and $U_{(n+1)} \equiv 1$. The notation will also be extended to sample values which are to be tested to be $U(0,1)$.

Spacings. When a uniform sample is ordered, the spacings $D_i$ are defined by:

$$D_i = U_{(i)} - U_{(i-1)}, \quad i = 1, \ldots, n + 1 \tag{8.3}$$

In particular, $D_1 = U_{(1)}$ and $D_{n+1} = 1 - U_{(n)}$. An important result is that the spacings $D_i$ have the distribution of a sample from the exponential distribution $F(u) = 1 - \exp(-u/\beta)$, $(u > 0)$, with $\beta = 1/(n + 1)$, conditional on the constraint $\Sigma_i D_i = 1$. Spacings are discussed in Section 8.9.

Many of the statistics to follow involve calculations of maxima, or of sums. Unless otherwise stated, the index used (usually i) in these expressions will run from 1 to n.

## 8.3 TRANSFORMATIONS TO UNIFORMS

Some of the most important transformations from a non-uniform distribution to a uniform distribution $U(0,1)$ are:

a)  The Probability Integral Transformation, and the related half-sample method, considered in Chapter 4;
b)  The Conditional Probability Integral Transformation discussed in Chapter 6;
c)  The J and K transformations, discussed in Sections 10.5.4 and 10.5.5, which take exponential random variables to uniform random variables.

There are also methods by which uniform random variables U may themselves be transformed to a new set of values U' or U'' which also has the U(0, 1) distribution. Two of these are transformations G and W below.

## 8.4  TRANSFORMATIONS FROM UNIFORMS TO UNIFORMS

### 8.4.1  The G-Transformation

Suppose $U_{(i)}$, $i = 1, \ldots, n$, is an ordered random sample from U(0, 1), and let $D_k$, $k = 1, 2, \ldots, n + 1$, be the spacings. Suppose further that these spacings are themselves ordered, so that $D_{(1)}, D_{(2)}, \ldots, D_{(n+1)}$ constitutes the ordered spacings, and define $D_{(0)} = 0$. Then construct new variables $D'_r$ from

$$D'_r = (n + 2 - r)(D_{(r)} - D_{(r-1)}), \qquad r = 1, \ldots, n + 1 \qquad (8.4)$$

The set $D'_r$ is another set of <u>unordered</u> uniform spacings (Sukhatme, 1937). From these, we can clearly build up another ordered uniform sample:

$$U'_{(j)} = \sum_{r=1}^{j} D'_r, \qquad j = 1, \ldots, n$$

This transformation, which we call the G transformation, takes a set U of ordered uniforms and produces another set U' of ordered uniforms; we can write U' = GU. A useful purpose of the G transformation is to increase power of a test for uniformity. Suppose the parent population of the U-values is not uniform, but close to uniform. Durbin (1961) showed that, loosely speaking, transformation G makes large spacings larger and small spacings smaller so that, when the transformation is applied to the U-set the resulting set U' will usually appear further removed from uniform. A test for uniformity would then be more powerful applied to U' than to U. There is, of course, a limit to this argument: if the original sample U were far removed from uniform, the transformation will possibly make the resulting U' appear more like uniform, and power would be lost. A more complete discussion of how the spacings might look before G loses power rather than gains it is in Seshadri, Csörgő, and Stephens (1969). In practical terms, G will increase power if the set tested might be thought to be close to the uniform distribution, but not if it is far away.

### 8.4.2  The W Transformation

Another transformation to produce uniforms from uniforms is the W transformation. This is given by

$$U''_i = \left\{ U_{(i)} / U_{(i+1)} \right\}^i, \quad i = 1, \ldots, n$$

with $U_{(n+1)}$ defined to be 1. The resulting $U''$ set is a uniform sample, not ordered. Formally, we may write $U'' = WU$. Wang and Chang (1977) have discussed the use of $W$ in connection with tests for the exponential distribution (see Chapter 10); it arises also in the work of O'Reilly and Stephens (1982).

## 8.5 SUPERUNIFORM OBSERVATIONS

When a test is made for uniformity, the alternative is often that the sample comes from a distribution which gives spacings more irregular than those from a uniform sample. The implication for testing is that for most test statistics a one-tail test is used. There are, however, some occasions when a sample should be tested as too regular to be uniform; such a sample will be called <u>superuniform</u>. Stuart (1954) entertainingly refers to superuniform observations as "too good to be true." One situation in which one wishes to detect superuniformity arises when there is a suspicion that values unhelpful to a certain hypothesis have been deliberately removed from the sample; hopefully this situation will be rare. However, there are other cases where superuniform observations can arise quite naturally, with important implications in test situations. They can occur, for example, when transformations are made (for example, the J transformation of Chapter 10; see Section 10.6).

## 8.6 TESTS BASED ON THE EMPIRICAL DISTRIBUTION FUNCTION (EDF)

Tests based on the EDF of a set of observations are discussed extensively in Chapter 4. Those which apply to a test of uniformity are the Case 0 tests of Section 4.4. In the present context, the values $U_{(i)}$ will replace the $Z_{(i)}$ in equations (4.2) and subsequent equations of Sections 4.5 and 4.7.

If it is desired to guard against superuniform observations, the lower tail of EDF statistics should be used. Significance levels for a given value of a test statistic may be found from Tables 4.2 and 4.3, as described in Section 4.5.1.

Statistics $V$ and $U^2$ were introduced as omnibus tests for observations on a circle (see Section 4.5.3), but they may be used in the same way as the other statistics, for testing $H_0$ in (8.1) above. The various EDF statistics can be expected to possess different powers when the true population is not the uniform distribution. Power properties are discussed further in Sections 8.13 and 8.14.

E 8.6 Example

In Table 8.1 are given 10 values of $U_{(i)}$, the order statistics of a random sample of U-values; these values will be subjected to the various tests for the uniform distribution $U(0,1)$ to be given throughout this chapter. Also included in the table are the EDF statistics calculated from equations (4.2). The modified values of the statistics, for Case 0, are then:

$$D^{+*} = .270, \ D^{-*} = 1.258, \ D^* = 1.258, \ V^* = 1.573,$$

$$W^* = .582, \ U^* = .217, \ A^* = 2.865$$

Reference to Table 4.2.1 shows the significance levels corresponding to these values to be approximately:

$$D^+: \ >.25; \ D^-: \ .045; \ D: \ .09; \ V: \ .13;$$

$$W^2: \ .025; \ U^2: \ .03; \ A^2: \ .03$$

The fact that $W^2$, $U^2$, and $A^2$ all have much the same significance levels is unusual; it reflects the fact that the sample is highly non-uniform in several features. The fact that $D^-$ is much more significant than $D^+$ shows that the drift in the values is toward 1 rather than 0. This is also shown by the mean $\bar{U} = .686$.

Another example of a test for $U(0,1)$ is given in Section 4.4. Here the test is a Case 0 test for normality; because the distribution tested is completely specified, the test reduces to a test that the values of $Z_1$ given in Table 4.1 are uniform. Also, in Chapter 10, tests for exponentiality are

TABLE 8.1 Set of Values $U_{(i)}$ and Derived Statistics

| | |
|---|---|
| Values | U: .004, .304, .612, .748, .771, .806, .850, .885, .906, .977 |

| | | |
|---|---|---|
| 2 | EDF statistics: | D$^+$: .096; D$^-$: .448; D: .448; V: .544 |
| | (Section 8.6) | W$^2$: .554; U$^2$: .207; A$^2$: 2.865 |
| 3 | Mean: $\bar{U}$ = .686 | |
| 4 | Values $v_i$: -.087, .122, .339, .384, .316, .261, .214, .158, .088, .068 | |
| | (Section 8.8.1) | |
| 5 | Statistics C$^+$: .384; C$^-$: .087; C: .384; K: .471 | |
| | (Section 8.8.1) | |
| 6 | Spacings D$_i$: .004, .300, .308, .136, .023, .035, .044, .035, .021, .071, .023 | |
| 7 | Statistics: | G(10) = .214; Q = 0.361 |
| | | M = 14.81   N$_2$ = 6.44 |

converted, by transforming the data by transformations J and K, to tests
for uniformity.

## 8.7 REGRESSION AND CORRELATION TESTS

The expected value of $U_{(i)}$, on $H_0$, is $m_i = i/(n + 1)$. A useful technique for
testing uniformity is to plot the ordered observations against $m_i$, and then
to see how well these fit a straight line. When the limits of the distribution
are not known, the simple correlation coefficient may be used (see Section 5.5). When the limits are known to be $(0, 1)$, the straight line, say L,
will join the origin O to the point $P \equiv (1, 1)$. Quesenberry and Hales (1980),
using the beta distribution of a typical order statistic, have given a useful
pictorial method of testing, by giving bands around L in which the observations can be expected to lie. Test statistics may also be based on the deviations from L, that is, on the $v_i$ in the next section (see Section 5.6).

## 8.8 OTHER TESTS BASED ON ORDER STATISTICS

### 8.8.1 Comparisons of Order Statistics with
####          Expected Values

A measure of the displacement of each variable $U_{(i)}$ from its expected value
$m_i = i/(n + 1)$ is

$$v_i = U_{(i)} - i/(n + 1) \tag{8.5}$$

and statistics for testing $H_0$ can be based on the $v_i$. Some statistics which
have been suggested are

$$C^+ = \max_i v_i; \quad C^- = \max_i (-v_i);$$
$$C = \max(C^+, C^-); \quad K = C^+ + C^-; \tag{8.6}$$
$$T_1 = \Sigma_i v_i^2/n; \quad T_2 = \Sigma_i |v_i|/n$$

The test procedure is to calculate the $v_i$ from (8.5), then the statistic
from (8.6), and refer to the upper tail of the appropriate null distribution.
For the same reasons as for EDF statistics, only the upper tail is usually
used in the test for uniformity; for superuniform observations, the lower
tail would be used. Tables for significance points for $C^+$, $C^-$, and C have
been given by Durbin (1969a), and tables for K by Brunk (1962) and by
Stephens (1969b). From the exact points for each n, Stephens (1970) calcu-

lated modified forms, similar to those for the EDF statistics in Section 8.6, so that the C and K statistics may be modified and used with only the asymptotic points; these points are the same as for the corresponding EDF statistics. The modified forms and the asymptotic points are given in Table 8.2. Johannes and Rasche (1980) have recently given more detailed modifications for $C^+$, $C^-$, and C.

The C-statistics arise in a natural way when testing the periodogram in time-series analysis; see, for example, Durbin (1969a, 1969b). Statistic $T_1$ arises if $U_{(i)}$ are plotted against $m_i$ and a test is based on the sum of squares of residuals (see Section 5.6).

Hegazy and Green (1975) have given moments and percentage points for statistics $T_1$ and $T_2$ above, and also for the statistics $T_1'$ and $T_2'$, calculated using the same formulas as for $T_1$ and $T_2$, but with $v_i = U_{(i)} - (i-1)/(n-1)$; they also gave some power studies for these statistics.

The statistics discussed in this section have much in common with EDF statistics, and overall they have much the same power properties.

## E 8.8.1 Example

The values $v_i$ for the data set in Table 8.1 are included in the table together with the statistics $C^+$, $C^-$, C, and K. When these are modified as in Table 8.2, the significance levels of the test statistics are: $C^+$: .01; $C^-$: > .25; C: .02; K: > .20.

TABLE 8.2 Modifications and Upper and Lower Tail Percentage Points for Statistics $C^+$, $C^-$, C, and K (Section 8.8.1)

| Statistic T | Modified form T* | Significance level $\alpha$ | | | | |
|---|---|---|---|---|---|---|
| | | 0.15 | 0.10 | 0.05 | 0.025 | 0.01 |
| | Upper tail | | | | | |
| $C^+$ | $(C^+ + 0.4/n)(\sqrt{n} + 0.2 + 0.68/\sqrt{n})$ | 0.973 | 1.073 | 1.224 | 1.358 | 1.518 |
| $C^-$ | $(C^- + 0.4/n)(\sqrt{n} + 0.2 + 0.68/\sqrt{n})$ | 0.973 | 1.073 | 1.224 | 1.358 | 1.518 |
| C | $(C + 0.4/n)(\sqrt{n} + 0.2 + 0.68/\sqrt{n})$ | 1.138 | 1.224 | 1.358 | 1.480 | 1.628 |
| K | $\{K - 1/(n+1)\}\{\sqrt{(n+1)} + 0.1555 + 0.24/\sqrt{(n+1)}\}$ | 1.537 | 1.620 | 1.747 | 1.862 | 2.001 |
| | Lower tail | | | | | |
| C | $(C + 0.5/n)(\sqrt{n} + 0.44 - 0.32/\sqrt{n})$ | 0.610 | 0.571 | 0.520 | 0.481 | 0.441 |
| K | $\{K - 1)/(n+1)\}\{\sqrt{(n+1)} + 0.41 - 0.26/\sqrt{(n+1)}\}$ | 0.976 | 0.928 | 0.861 | 0.810 | 0.755 |

Adapted from Stephens (1970), with permission of the Royal Statistical Society.

8.8.2  Statistics Based on One Order Statistic

On $H_0$, the order statistic $U_{(r)}$ has a beta distribution Beta $(x; p, q)$ with density

$$f(x) = [\Gamma(p + q)/\{\Gamma(p)\Gamma(q)\}] x^{p-1}(1 - x)^{q-1} \qquad 0 \le x \le 1$$

where $\Gamma(p)$ is the gamma function; for statistic $U_{(r)}$ the parameters are $p = r$ and $q = n - r + 1$. Thus in a sample of size $n = 2k + 1$, the median $U_{(k+1)}$ has density

$$f(x) = \frac{n!}{(k!)^2} x^k (1 - x)^k, \qquad 0 \le x \le 1$$

with mean $0.5$ and variance $1/\{4(n + 2)\}$.

The following result connects the Beta $(x; p, q)$ and $F_{2p, 2q}$ distributions. If $x$ has the Beta $(x; p, q)$ distribution, then $y$ defined by $y = qx/\{p(1 - x)\}$ has the $F_{2p, 2q}$ distribution; equivalently, $z = 1/y = p(1 - x)/(qx)$ has the $F_{2q, 2p}$ distribution.

It follows that $Z = \{(n - r + 1)U_{(r)}\}/\{r(1 - U_{(r)})\}$ has an $F_{s, t}$ distribution, with degrees of freedom $s = 2r$ and $t = 2(n - r + 1)$. It is therefore possible to base a test for uniformity on one of the order statistics, rejecting if $Z$ is too large or too small compared with the F-distribution percentage points; the median $\tilde{U}$ ($U_{(r)}$ with $r = n/2$ or $r = (n + 1)/2$) has sometimes been suggested. The test is useful if, for some reason, not all of the sample values were known, but only, for example, the smallest values; but in general it appears to be a relatively weak method of testing for uniformity, given a full sample. This is revealed, for example, in Chapter 10, where tests for exponentiality are converted to tests for uniformity and then the mean $\bar{U}$ or the median $\tilde{U}$ used as test statistics: by and large, $\bar{U}$ performs much better than $\tilde{U}$.

## 8.9  STATISTICS BASED ON SPACINGS

Another group of statistics for testing $H_0$ is based on the spacings $D_i$ defined in Section 8.2. These tests have often been introduced in connection with testing for exponentiality rather than as direct tests for uniformity. The reason for this is that an exponential set of $n$ values $X_i$ can be transformed, by means of the J and K transformations of Sections 10.5.4 and 10.5.5, into the $n$ spacings produced by $n - 1$ ordered uniforms $U_{(i)}$. Some tests for exponentiality are tests based on these spacings, using the distribution theory of spacings between uniforms. The treatment of spacings statistics will therefore be shared with Chapter 10. In this section we discuss only the Greenwood statistic and some modifications.

## 8.9.1 The Greenwood Statistic

This statistic was itself introduced by Greenwood (1946) in connection with testing that the intervals between events (the incidence of a contagious disease) were exponential, that is, that the times of the events constituted a Poisson process (see Chapter 10), but it has also received attention specifically as a test for uniformity. The statistic is

$$G(n) = \sum_{i=1}^{n+1} D_i^2$$

Distributional properties of $G(n)$ were investigated by Moran (1947, 1951); recently, percentage points have been given by Burrows (1978), Hill (1979; see the correction 1981), Currie (1981), and Stephens (1981). Percentage points for $nG(n)$ based on these results are given in Table 8.3. Large values of $G(n)$ will indicate highly irregular spacings and small values will indicate superuniform observations; $G(n)$ is well suited to detect superuniforms. For large $n$, $nG(n)$ is approximately normally distributed, with mean $\mu = 2n/(n+2)$ and $\sigma^2 = 4/n$, but this limiting distribution is attained very slowly.

The expected value of $D_i$ is $1/(n+1)$, and a possible test statistic might be the dispersion of the spacings defined by

$$G^*(n) = \Sigma_i \{D_i - 1/(n+1)\}^2$$

However, it is easily shown that $G^*(n)$ is $G(n) - 1/(n+1)$, so that $G^*(n)$ is equivalent to $G(n)$. Several other statistics are also equivalent to $G(n)$; see Section 10.9.3.2.

### 8.9.1.1 The Greenwood Statistic Adapted for Censored Data

Suppose a sample is given, right-censored at value $U_{(r)}$, and define $H_k(n) = \Sigma_{i=1}^{k} D_i^2$. Kimball (1947, 1949) discussed distribution theory and gave moments of $H_k(n)$. Clearly $H_r(n)$ could be used as a test statistic for uniformity; also, because of the exchangeability of uniform spacings, the $D_i$ in $H_k(n)$ could be replaced by any set of available spacings between adjacent order statistics, even if some of the $U_{(i)}$ were missing. A further modification is

$$G_r(n) = H_r(n) + \frac{(1 - U_{(r)})^2}{n - r + 1} \tag{8.7}$$

TABLE 8.3  Upper and Lower Percentage Points for nG(n) (Section 8.9.1)

| Sample size n | Lower tail Significance level $\alpha$ | | | | Upper tail Significance level $\alpha$ | | | |
|---|---|---|---|---|---|---|---|---|
| | .01 | .025 | .05 | .10 | .05 | .025 | .01 |
| 2 | 0.672 | 0.680 | 0.694 | 0.722 | 1.381 | 1.539 | 1.673 | 1.780 |
| 3 | 0.776 | 0.796 | 0.825 | 0.870 | 1.635 | 1.852 | 2.075 | 2.269 |
| 4 | 0.855 | 0.885 | 0.923 | 0.974 | 1.800 | 2.037 | 2.311 | 2.560 |
| 5 | 0.919 | 0.954 | 0.997 | 1.050 | 1.915 | 2.160 | 2.461 | 2.737 |
| 6 | 0.973 | 1.009 | 1.055 | 1.112 | 1.995 | 2.246 | 2.559 | 2.849 |
| 7 | 1.017 | 1.060 | 1.104 | 1.162 | 2.053 | 2.306 | 2.615 | 2.921 |
| 8 | 1.055 | 1.095 | 1.145 | 1.205 | 2.097 | 2.349 | 2.670 | 2.967 |
| 9 | 1.088 | 1.129 | 1.180 | 1.241 | 2.131 | 2.381 | 2.700 | 2.997 |
| 10 | 1.117 | 1.159 | 1.211 | 1.272 | 2.157 | 2.404 | 2.717 | 3.008 |
| 12 | 1.198 | 1.234 | 1.272 | 1.326 | 2.204 | 2.441 | 2.683 | 3.015 |
| 14 | 1.233 | 1.272 | 1.312 | 1.368 | 2.227 | 2.457 | 2.691 | 3.014 |
| 16 | 1.263 | 1.304 | 1.346 | 1.403 | 2.242 | 2.464 | 2.691 | 3.003 |
| 18 | 1.288 | 1.332 | 1.375 | 1.433 | 2.251 | 2.466 | 2.685 | 2.988 |
| 20 | 1.311 | 1.356 | 1.400 | 1.459 | 2.258 | 2.465 | 2.677 | 2.970 |
| 25 | 1.358 | 1.405 | 1.451 | 1.510 | 2.265 | 2.456 | 2.651 | 2.920 |
| 30 | 1.395 | 1.444 | 1.490 | 1.549 | 2.265 | 2.443 | 2.624 | 2.873 |
| 40 | 1.453 | 1.502 | 1.548 | 1.605 | 2.258 | 2.415 | 2.573 | 2.790 |
| 50 | 1.495 | 1.544 | 1.589 | 1.644 | 2.248 | 2.389 | 2.531 | 2.723 |
| 60 | 1.529 | 1.577 | 1.621 | 1.674 | 2.238 | 2.367 | 2.495 | 2.669 |
| 80 | 1.579 | 1.625 | 1.666 | 1.716 | 2.220 | 2.331 | 2.441 | 2.587 |
| 100 | 1.616 | 1.659 | 1.698 | 1.745 | 2.205 | 2.304 | 2.400 | 2.528 |
| 200 | 1.714 | 1.750 | 1.781 | 1.818 | 2.159 | 2.226 | 2.289 | 2.371 |
| 500 | 1.811 | 1.836 | 1.858 | 1.884 | 2.107 | 2.147 | 2.183 | 2.228 |

Adapted from Burrows (1979) and from Stephens (1981), with permission of the first author and of the Royal Statistical Society.

This reduces to G(n) above when r = n. Lurie, Hartley, and Stroud (1974) investigated statistic $S^2 = (n + 2)\{(n + 1)G_r(n) - 1\}$ and gave moments and some null percentage points obtained by curve-fitting. For complete samples $S^2$ had previously been discussed by Hartley and Pfaffenberger (1972). The statistic $S^2$ is clearly equivalent as a test statistic to $G_r(n)$, and the moments of $S^2$ can be used to give moments of $G_r(n)$. The moments of $H_k(n)$ and of $G_r(n)$ have been used by the author to fit Pearson curves to the null distribution to give percentage points for $G_r(n)$. These percentage points are given in Stephens (1986).

TABLE 8.4 Upper Tail Percentage Points
of the Q Statistic (Section 8.9.2)

| n | Significance level $\alpha$ | | | | |
|---|---|---|---|---|---|
| | .50 | .10 | .05 | .01 | .001 |
| 2 | .659 | .811 | .859 | .932 | .977 |
| 3 | .527 | .691 | .736 | .831 | .920 |
| 4 | .447 | .586 | .635 | .727 | .829 |
| 5 | .388 | .505 | .551 | .642 | .739 |
| 6 | .343 | .442 | .483 | .573 | .691 |
| 7 | .307 | .393 | .429 | .512 | .622 |
| 8 | .278 | .355 | .387 | .463 | .551 |
| 9 | .254 | .322 | .350 | .423 | .506 |
| 10 | .234 | .294 | .319 | .378 | .461 |
| 11 | .217 | .272 | .294 | .351 | .434 |
| 12 | .202 | .251 | .272 | .318 | .392 |
| 13 | .189 | .234 | .253 | .298 | .371 |
| 14 | .177 | .220 | .237 | .279 | .348 |
| 15 | .168 | .206 | .222 | .259 | .321 |
| 16 | .159 | .195 | .209 | .245 | .294 |
| 17 | .150 | .184 | .197 | .230 | .278 |
| 18 | .143 | .174 | .187 | .218 | .268 |
| 19 | .137 | .166 | .177 | .206 | .257 |
| 20 | .131 | .158 | .168 | .196 | .246 |
| 21 | .125 | .151 | .162 | .187 | .232 |
| 22 | .120 | .144 | .154 | .178 | .218 |
| 23 | .115 | .138 | .147 | .169 | .206 |
| 24 | .111 | .133 | .141 | .163 | .192 |
| 25 | .107 | .128 | .136 | .156 | .189 |
| 26 | .103 | .123 | .131 | .148 | .182 |
| 27 | .100 | .119 | .126 | .144 | .173 |
| 28 | .097 | .114 | .121 | .138 | .166 |
| 29 | .094 | .111 | .117 | .134 | .158 |
| 30 | .091 | .107 | .114 | .130 | .155 |

(continued)

TABLE 8.4 (continued)

| n | .50 | .10 | .05 | .01 | .001 |
|---|-----|-----|-----|-----|------|
| | Significance level $\alpha$ | | | | |
| 31 | .088 | .104 | .110 | .125 | .148 |
| 32 | .086 | .101 | .106 | .120 | .142 |
| 33 | .083 | .097 | .103 | .116 | .135 |
| 34 | .081 | .095 | .100 | .112 | .131 |
| 35 | .079 | .092 | .097 | .110 | .128 |
| 36 | .077 | .090 | .095 | .107 | .124 |
| 37 | .075 | .087 | .092 | .103 | .122 |
| 38 | .073 | .085 | .089 | .100 | .118 |
| 39 | .071 | .083 | .087 | .097 | .114 |
| 40 | .070 | .081 | .085 | .095 | .112 |
| 41 | .068 | .079 | .083 | .092 | .110 |
| 42 | .067 | .077 | .081 | .090 | .107 |
| 43 | .065 | .075 | .079 | .088 | .103 |
| 44 | .064 | .073 | .077 | .086 | .101 |
| 45 | .062 | .072 | .075 | .084 | .098 |
| 46 | .061 | .070 | .074 | .082 | .094 |
| 47 | .060 | .069 | .072 | .080 | .092 |
| 48 | .059 | .067 | .070 | .078 | .091 |
| 49 | .058 | .066 | .069 | .077 | .090 |
| 50 | .057 | .065 | .068 | .075 | .086 |
| 55 | .052 | .059 | .061 | .068 | .078 |
| 60 | .048 | .054 | .056 | .062 | .070 |
| 65 | .044 | .050 | .052 | .057 | .064 |
| 70 | .041 | .046 | .048 | .052 | .059 |
| 75 | .038 | .043 | .045 | .048 | .054 |
| 80 | .036 | .040 | .042 | .045 | .051 |
| 85 | .034 | .038 | .039 | .042 | .047 |
| 90 | .032 | .036 | .037 | .040 | .044 |
| 95 | .031 | .034 | .035 | .038 | .041 |
| 100 | .029 | .032 | .033 | .036 | .040 |

### 8.9.2 Statistics Related to Greenwood's Statistic

An adaptation of $G(n)$ has been proposed by Quesenberry and Miller (1977). This is the statistic

$$Q = \sum_{i=1}^{n+1} D_i^2 + \sum_{i=1}^{n} D_i D_{i+1}$$

and $H_0$ is rejected if $Q$ is too large. Tables of percentage points for $Q$ are given in Table 8.4, taken from Quesenberry and Miller and based on Monte Carlo studies. $Q$ is designed to take into account the pattern of the spacings (specifically, the autocorrelation) as well as their sizes, and could be a useful statistic in analyzing series of events, where autocorrelation some-times plays a part (see Section 10.6.2).

### E 8.9.2 Example

The values of the spacings $D_i$, for the data set U in Table 8.1 are also given in the table. From these are calculated Greenwood's statistic $G(10) = 0.214$, and $Q = 0.361$. Reference to Table 8.3 shows $G(10)$ to be significant at about the upper 10% level, and reference to Table 8.4 shows $Q$ to be significant at about the upper 6% level.

### 8.9.3 Other Statistics Calculated from Spacings

Other statistics based on spacings are Moran's M and the Kendall-Sherman statistic K (Sections 10.9.3.4 and 10.9.3.5); EDF statistics (Section 10.9.3.6) and $L_n(p)$ (Section 10.9.3.7); and statistics given in Section 10.11.4. Note that these are all defined for n spacings, not n + 1 spacings, and the formulas must be adapted accordingly.

### E 8.9.3 Example

For the U set in Table 8.1, Moran's statistic is 14.81 (Section 10.9.3.4 with n = 11); c = 1.18, so $M(10)/c = 12.57$. This is significant at the 25% level when compared with $\chi_{10}^2$.

### 8.9.4 Higher Order Spacings and Gaps

There has recently been interest in k-spacings, defined by $D_{ki} = U_{(ki)} - U_{(ki-k)}$; these are the spacings between the observations, taken k at a time. This use of spacings suppresses some of the information in the sample, but Hartley and Pfaffenberger (1972) suggested that k-spacings might be useful in tests for large samples. Del Pino (1979) discussed statistics of the form $W = \Sigma_i h(nD_{ki})$ where $h(\cdot)$ is an appropriate function, summed over the range of i for fixed k (for simplicity, suppose k divides into n + 1 and let

$\ell = (n + 1)/k$; then the range of $i$ is $1 \leq i \leq \ell$). Del Pino showed that, by the criteria of asymptotic relative efficiency, $h(x) = x^2$ gives an optimum statistic W; let this be $W_1$. Del Pino also argued for the utility of such statistics for large samples; see also Darling (1953) and Weiss (1957a, 1957b) for more general considerations involving spacings.

Cressie (1976, 1977a, 1978, 1979) and Deken (1980) have considered test statistics which are functions of m-th order gaps $G_i^{(m)} = U_{(i+m)} - U_{(i)}$, for m a fixed integer, and $0 \leq i \leq n + 1 - m$; as before, $U_{(0)} \equiv 0$ and $U_{(n+1)} \equiv 1$. For $m = 1$, $G_i^{(1)} = D_{i+1}$; for higher m, the $G_i^{(m)}$ contain overlapping sets of $D_i$, in contrast with k-spacings above. Deken defined $G_i^{(p+1)}$ as a p-stretch, and gave distribution theory for the maximum p-stretch. Solomon and Stephens (1981) gave percentage points for $n = 5$ and 10 and made a comparison with an approximation given by Deken. Cressie (1977a, 1978) has also studied the minimum p-stretch and the minimum $D_{ki}$; one might suppose the minimum p-stretch to be useful in detecting a "bump" in an otherwise uniform density, and this would be valuable in studying the times of a series of events (see Chapter 10); however, Cressie (1978) found the minimum gap of either type to be less powerful against a specific bump alternative than $L_n^{(m)} = \Sigma_{i=1}^{r} \log G_i^{(m)}$, $r = n + 1 - m$, or its parallel $K_n^{(m)} = \Sigma_i^{r} \log D_{mi}$ $(r = [(n + 1)/m] - 1)$. Cressie (1978) discussed $L_n^{(m)}$, showing asymptotic normality and giving some Monte Carlo power results. For $m = 1$, $L_n^{(1)}$ is essentially Moran's statistic M (Section (10.9.3.4); for $m > 1$, $L_n^{(m)}$ may be useful in overcoming the difficulties of M with very small values (Section 10.10). Tables of the null distributions of $L_n^{(2)}$ and $L_n^{(3)}$ are given by McLaren and Stephens (1985).

An interesting justification for using high-order spacings comes from considerations of entropy, which, under certain conditions, characterizes a distribution. Vasicek (1976) introduced an estimate of entropy and used it to produce a consistent test for normality; the estimate, adapted for the uniform distribution, is

$$H(m,n) = n^{-1} \sum_{i=1}^{n} \log [n \{U_{(i+m)} - U_{(i-m)}\}/(2m)]$$

where now $U_{(r)} \equiv U_{(1)}$ if $r < 1$ and $U_{(r)} \equiv U_{(n)}$ if $r > n$. There are clearly close connections between $L_n^{(m)}$ and $H(m,n)$. Dudewicz and van der Meulen (1981) have proposed $H(m,n)$ as a test statistic for uniformity, and have given tables of percentage points, derived from Monte Carlo methods, for $n = 10, 20, 30, 40, 50, 100$ and for various values of m from $m = 1$ to $m = M$, with M becoming larger with n. They also show asymptotic normality for H(m,n), established by the relationship with $L_n^{(m)}$, but both these statistics attain the asymptotic normality only very slowly; this appears to be a feature of spacings statistics. Dudewicz and van der Meulen also give power results; H(m,n) appears to be particularly good against alternatives with a high density near 0.5 and for m $\approx$ 0.4n. The m to give best power varies with the alternative.

Cressie (1979) discussed statistics of the form $H = \Sigma_i h(nG_i^{(m)})$, and found that, with $h(x) = x^2$ the resulting statistic $S_n^{(m)} = \Sigma_i \{nG_i^{(m)}\}^2$ has higher asymptotic relative efficiency than del Pino's $W_1$. Note that both these statistics can be regarded as extensions of Greenwood's statistics to higher-order spacings and gaps. $S_n^{(2)}$ is closely related, and asymptotically equivalent, to the Quesenberry-Miller statistic Q above. McLaren and Stephens (1985) have given percentage points for $S_n^{(2)}$ and $S_n^{(3)}$. Del Pino (1979) and Cressie (1979) proved asymptotic normality for statistics W, H, and $S_n^{(m)}$, and Holst (1979) gave the mean and variance of H, but tables for finite n are not yet available for these statistics. Greenwood's statistic itself converges only slowly to its asymptotic distribution, and this may be the case also for these related statistics. McLaren and Stephens (1985) have given power studies for statistics $L_n^{(m)}$ and $S_n^{(m)}$, for m = 1, 2, and 3. These included alternatives with spacings derived from Gamma or Weibull variates; the L-class was better than the S-class, and power decreased with m.

Another statistic to detect bumps is the scan statistic; this is S(L), the maximum number of observations (out of n) falling into a window of length L, as the window travels along the interval (0,1). The statistic has been studied by Wallenstein and Naus (1974), who gave the null distribution for finite n, and by Cressie (1977b, 1980) who gave asymptotic theory; see these articles also for references to earlier work by Naus. Ajne (1968) discussed the scan statistic on the circle, with circumference 1 and L = 0.5.

Much interesting work has been done on the gaps and scan statistics (the papers quoted give many earlier references), but more is needed to make them of practical use as test statistics and to compare them with other tests for uniformity.

## 8.10  STATISTICS FOR SPECIAL ALTERNATIVES

The statistics so far considered have been based on various methods of relating the order statistics or their spacings to the pattern expected of them. Many other test statistics for uniformity, usually fairly simple functions of the $U_i$, arise when special distributions are regarded as the alternative if $H_0$ is not true, and likelihood ratio methods are used to find test statistics. Some of these are discussed in this section.

### 8.10.1 The Statistic $\bar{U}$

Suppose the alternative distribution to $H_0$ is density

$$f(u) = \frac{k}{e^k - 1} e^{ku}, \quad 0 < u < 1 \tag{8.8}$$

This is a truncated exponential distribution which reduces to the uniform density when $k = 0$. Thus a test for uniformity becomes a test for $k = 0$, and the likelihood ratio method gives the test statistic $T = \Sigma_i U_i$, or equivalently $\bar{U} = T/n$.

The null distribution of $\bar{U}$ is well known, although its form is quite complicated. Lower tail percentage points are in Table 8.5, adapted from

TABLE 8.5  Lower Tail Percentage Points for $\bar{U}$ (Section 8.10.1)

| n | 0.25 | .15 | .10 | .05 | .025 | .01 | .005 |
|---|---|---|---|---|---|---|---|
| | | | Significance level $\alpha$ | | | | |
| 4 | 0.399 | 0.346 | 0.312 | 0.262 | 0.221 | 0.176 | 0.148 |
| 5 | 0.410 | 0.363 | 0.332 | 0.287 | 0.250 | 0.208 | 0.181 |
| 6 | 0.419 | 0.376 | 0.347 | 0.306 | 0.271 | 0.232 | 0.207 |
| 7 | 0.425 | 0.385 | 0.359 | 0.320 | 0.288 | 0.251 | 0.227 |
| 8 | 0.430 | 0.393 | 0.368 | 0.332 | 0.301 | 0.266 | 0.244 |
| 9 | 0.434 | 0.399 | 0.376 | 0.341 | 0.312 | 0.279 | 0.257 |
| 10 | 0.438 | 0.404 | 0.382 | 0.350 | 0.322 | 0.290 | 0.269 |
| 12 | 0.443 | 0.413 | 0.393 | 0.363 | 0.337 | 0.308 | 0.289 |
| 14 | 0.447 | 0.419 | 0.401 | 0.373 | 0.349 | 0.322 | 0.304 |
| 16 | 0.451 | 0.425 | 0.407 | 0.381 | 0.359 | 0.333 | 0.316 |
| 18 | 0.454 | 0.429 | 0.412 | 0.388 | 0.367 | 0.343 | 0.327 |
| 20 | 0.456 | 0.433 | 0.417 | 0.394 | 0.374 | 0.351 | 0.335 |
| 25 | 0.461 | 0.440 | 0.426 | 0.405 | 0.387 | 0.366 | 0.352 |
| 30 | 0.464 | 0.445 | 0.432 | 0.413 | 0.397 | 0.378 | 0.365 |
| 35 | 0.467 | 0.449 | 0.437 | 0.420 | 0.404 | 0.387 | 0.375 |
| 40 | 0.469 | 0.453 | 0.441 | 0.425 | 0.411 | 0.394 | 0.383 |
| 45 | 0.471 | 0.455 | 0.445 | 0.429 | 0.416 | 0.400 | 0.390 |
| 50 | 0.472 | 0.458 | 0.448 | 0.433 | 0.420 | 0.405 | 0.395 |
| 60 | 0.475 | 0.461 | 0.452 | 0.439 | 0.427 | 0.414 | 0.404 |
| 70 | 0.477 | 0.464 | 0.456 | 0.443 | 0.432 | 0.420 | 0.411 |
| 80 | 0.478 | 0.467 | 0.459 | 0.447 | 0.437 | 0.425 | 0.417 |
| 90 | 0.479 | 0.468 | 0.461 | 0.450 | 0.440 | 0.429 | 0.422 |
| 100 | 0.481 | 0.470 | 0.463 | 0.453 | 0.443 | 0.433 | 0.426 |

Adapted from Stephens (1966), with permission of the Biometrika Trustees.

Stephens (1966); if $Z_\alpha$ is the given point for level $\alpha$, the corresponding upper tail point, that is, for level $1 - \alpha$, is $1 - Z_\alpha$. For large n $(n \geq 20)$ the distribution of $\bar{U}$ is well-approximated by the normal distribution with mean 0.5 and variance $1/\{12n\}$. The distribution (8.8) occurs in connection with points U obtained from a renewal process with a trend (see Section 10.9.1).

### E 8.10.1  Example

The mean of the U-set in Table 8.1 is 0.686, so $1 - \bar{U} = .314$. Reference to Table 8.5 gives a p-level equal to 0.02 (one-tail) or 0.04 (two-tail).

### 8.10.2  The Statistic P

Suppose the alternative distribution to $H_0$ is the density

$$f(u) = (k + 1)u^k, \quad k > -1, \quad 0 < u < 1,$$

which reduces to the uniform density when $k = 0$. This family of densities is sometimes referred to as the Lehmann family. The likelihood ratio test statistic for a test for $k = 0$ against $k \neq 0$ is $P/2$ where

$$P = -2\Sigma_i \log U_i$$

On $H_0$, P has the $\chi^2$ distribution with 2n degrees of freedom. The test of $H_0$ is the test that $k = 0$; against the alternative $k > 0$, low values of P will be regarded as significant, and $H_0$ will be rejected if P is less than the appropriate percentage point in the lower tail of $\chi^2_{2n}$. For the test of $H_0$ against the alternative $-1 < k < 0$, high values of P will be significant, and $H_0$ will be rejected if P exceeds the upper tail percentage point of $\chi^2_{2n}$. For the most general test of $H_0$, that $k = 0$ against the alternative $k \neq 0$, a two-tail test will be used. P has often been used to combine several tests of significance by Fisher's method (see Section 8.15 below).

### 8.10.3  Statistics for the Circle or the Sphere

Suppose points $P_i$, $i = 1, \ldots, n$ are marked on the circumference of a circle of radius 1, and it is desired to test the null hypothesis $H_0$ that the points $P_i$ are uniformly distributed around the circle. Let O be the center of the circle and let N be the north pole; let $\theta$ be the angle between ON and OP, where P is a typical point on the circle. A common distribution used for describing a unimodal population around the circle is the von Mises distribution (Section 4.15) for which the density is

$$f(\theta) = \frac{1}{2\pi I_0(\kappa)} \exp\{\kappa \cos(\theta - \theta_0)\}, \quad 0 < \theta < 2\pi, \quad \kappa > 0$$

This density is symmetric, with a mode along the line OA with coordinate $\theta_0$, and is increasingly clustered around OA as $\kappa$ becomes larger; when $\kappa = 0$ the distribution is uniform around the circle. $I_0(\kappa)$ refers to the Bessel function with imaginary argument, of order zero. For a von Mises alternative, the null hypothesis $H_0$ is equivalent to

$$H_0': \kappa = 0, \text{ against the alternative } H_1: \kappa > 0$$

When the modal vector OA is not known, the likelihood ratio procedure gives a test statistic which is the length R of the resultant, or vector sum, $\underset{\sim}{R}$, of the vectors $OP_i$, $i = 1, \ldots, n$. In the more unlikely event that, on the alternative, the modal vector OA is known, the component of $\underset{\sim}{R}$ along OA, called X, is the test statistic.

The distributions of R and X are very complicated for points on a circle; they have been studied by Greenwood and Durand (1955) and Durand and Greenwood (1957), who have given some percentage points. Stephens (1969a) has given a table of upper tail percentage points for testing $H_0$, for both R and X. For large samples, $2R^2/n$ has the $\chi^2_2$ distribution and X has the normal distribution with mean 0 and variance $n/2$. These statistics arise also in a totally different context, when EDF statistics $W^2$ and $U^2$ are partitioned into components (Section 8.12).

Further applications of the uniform distribution arise in studying randomness of directions on a sphere, against various alternatives. Suppose the sphere has center 0, and radius 1, and let a typical point P on the sphere be located by spherical polar coordinates $(\theta, \phi)$. If P is uniformly distributed on the surface of the sphere, $\cos \theta$ is uniformly distributed between $-1$ and 1. Again, the von Mises distribution, with density per unit area proportional to $\exp(\kappa \cos \gamma)$, is the most important for unimodal data; here $\gamma$ is the angle between OP and the modal vector OA. Likelihood ratio tests for uniformity ($\kappa = 0$) against a von Mises distribution ($\kappa > 0$) (also called the Fisher distribution on the sphere), lead again to the length R of the resultant $\underset{\sim}{R}$ of a sample of n vectors, as test statistic; when the modal vector of the alternative is known, the test statistic is the component X of $\underset{\sim}{R}$ on this vector, as for the circle. Stephens (1964) has given tables of percentage points for R and for X. For large n, $3R^2/n$ is approximately $\chi^2_3$ distributed, and X is normal, with mean 0 and variance $n/3$.

Other alternatives to randomness have been proposed to describe natural data, among them, densities for which the probability per unit area is proportional to

$$
\left.
\begin{aligned}
f_1(\gamma) &= e^{-\kappa |\cos \gamma|} \\
f_2(\gamma) &= e^{\kappa \sin \gamma} \\
\text{or}\quad \\
f_3(\gamma) &= e^{\kappa \cos^2 \gamma}
\end{aligned}
\right\} \quad 0 \le \gamma \le \pi
$$

For $f_1(\gamma)$ and $f_2(\gamma)$, $k \geq 0$; for $f_3(\gamma)$, k is any constant. The densities are all symmetric about the axis OA, are either bimodal or equatorial, and all reduce to the uniform density when $\kappa = 0$. In a test for uniformity of directions against the alternatives given above, when OA is a known axis, the null hypothesis is $H_0$: $\kappa = 0$, against the alternative $\kappa \neq 0$. Likelihood ratio test statistics are, respectively,

$$L_1 = \Sigma_i |V_i|/n; \quad L_2 = \Sigma_i(1 - V_i^2)^{\frac{1}{2}}/n; \quad L_3 = \Sigma_i V_i^2/n$$

where $V_i = \cos \gamma_i$ has, on $H_0$, the uniform distribution between -1 and 1. $L_1$ has the same distribution as $\bar{U}$, considered in Section 8.10.1 above, and $L_2$ and $L_3$ have the same distributions as

$$Q = \Sigma_i(1 - U_i^2)^{\frac{1}{2}}/n$$

and

$$T = \Sigma_i U_i^2/n$$

where $U_i$ are $U(0,1)$. Further, the statistic

$$S^2 = \Sigma_i(U_i - 0.5)^2/n$$

which is a measure of the dispersion of the $U_i$, has the same distribution as $T/4$. Significance points for $\bar{U}$ are in Table 8.5; points for Q and T have been given by Stephens (1966), and the applications to tests for directions are discussed further in that reference. When OA is not known for the distributions above, the tests for uniformity become more complicated. The statistics $\bar{U}$, $S^2$, and T will appear again in the next section in connection with Neyman-Barton tests, and with partitioning the Anderson-Darling statistic into components.

### 8.10.3.1 Ajne's Statistic

Ajne (1968) suggested a test statistic for uniformity on the circumference (of length 1) of a circle, which has optimum properties against the alternative density $f_1(x) = r$ $(0 \leq x \leq 1/2)$, $f_1(x) = s$ $(1/2 \leq x \leq 1)$, where r and s are constants. The test statistic is

$$A = \frac{1}{n}\int_0^1 \left\{ N(x) - \frac{1}{2}n \right\}^2 dx$$

where N(x) is the number of observations falling in the semicircle $(x, x + 1/2)$. Computing formulas have been given by Watson (1967) and by Stephens (1969c).

Suppose the observations are $U_{(1)} < U_{(2)} < \cdots < U_{(n)}$, measured around the circumference. Then

$$A = \frac{1}{2} - \frac{n}{4} + \frac{2}{n} \sum_{j=2}^{n} \sum_{i=1}^{j-1} \left| \frac{1}{2} + U_{(j)} - U_{(i)} \right|$$

Another formula is

$$A = \frac{n}{4} - \frac{2}{n} Z$$

where

$$Z = \sum_{j=2}^{n} \sum_{i=1}^{j-1} m_{ij}$$

with

$$m_{ij} = \begin{cases} U_{(j)} - U_{(i)} & \text{if } U_{(j)} - U_{(i)} \leq \frac{1}{2} \\ 1 - \{U_{(j)} - U_{(i)}\} & \text{if } U_{(j)} - U_{(i)} > \frac{1}{2} \end{cases}$$

Watson (1967) gave the asymptotic null distribution of A.

Stephens (1969c) has given the moments, some exact distribution theory and percentage points for A; also given are some power studies which compare A with $U^2$ and V and which suggest that, in practice, the gain in using A when it is optimal is small compared with the loss when it is not.

## 8.10.3.2 Omnibus Tests

Omnibus tests are not designed for specific alternatives, but it is convenient to mention several of these, especially for the circle, before leaving this section. EDF statistics $U^2$ and V (Chapter 4; here $U_{(i)}$ replaces $Z_{(i)}$ in Equations (4.2)) were designed for the circle because they do not depend on the origin of U; Watson (1976) gave another statistic derived from the EDF and Darling (1982, 1983) has recently given the asymptotic points. Ajne (1968) also gave another statistic for the circle. A review of tests for uniformity on the hypersphere is given by Prentice (1978); see also Beran (1968) and Giné (1975). Such tests are sometimes derived in very general terms, and often give the test statistics in this section when particular cases are taken.

## 8.11 THE NEYMAN-BARTON SMOOTH TESTS

Another application of likelihood ratio methods yields the Neyman-Barton tests. Neyman (1937) considered the problem of testing for uniformity $U(0,1)$, against an alternative density of the form

$$f(u) = c(\underline{\theta}) \exp\left\{1 + \sum_{j=1}^{k} \theta_j \pi_j(u)\right\}, \quad 0 \le u \le 1, \quad k = 1, 2, \ldots \qquad (8.9)$$

where $\pi_j(u)$ are the Legendre polynomials, $\underline{\theta}$ is a vector of parameters with components $\theta_1, \ldots, \theta_k$, and $c(\underline{\theta})$ is the normalizing constant. The Legendre polynomials are orthonormal on the interval $(0,1)$. By varying k, the density may be made to approximate a given density, and it also varies smoothly from the uniform distribution as the $\theta_i$ take increasingly large values. The test for uniformity of U then reduces to testing the null hypothesis

$$H_0: \theta_j = 0 \quad \text{for all } j$$

$H_0$ may be put in the form $\sum_{j=1}^{k} \theta_j^2 = 0$. Neyman found an appropriate statistic, based on likelihood ratio methods, for testing this null hypothesis. For given k, the test statistic is $N_k$, calculated as follows.

(a)  Let

$$v_j = \frac{1}{\sqrt{n}} \sum_{i=1}^{n} \pi_j(U_i) \qquad (8.10)$$

In these calculations, $\pi_j(U)$ is best expressed in terms of $y = U - 0.5$. For the first four polynomials,

$$\pi_1(U) = 2\sqrt{3}y; \qquad \pi_2(U) = \sqrt{5}(6y^2 - 0.5);$$

$$\pi_3(U) = \sqrt{7}(20y^3 - 3y); \qquad \pi_4(U) = 3(70y^4 - 15y^2 + 0.375)$$

(b)  The Neyman statistic of order k is

$$N_k = \sum_{j=1}^{k} v_j^2 \qquad (8.11)$$

The null hypothesis of uniformity will be rejected for large values of $N_k$; for large n, on $H_0$, $N_k$ is asymptotically distributed as $\chi_k^2$. The tests based on $N_k$ are consistent and asymptotically unbiased. Neyman showed that, asymptotically, the $v_j$ are independent $N(0,1)$ variables on $H_O$. David (1939) further showed that the asymptotic $\chi^2$ distributions were very good approximations to the finite-n distributions for $n \geq 20$. Note that $v_1$ and $v_2$ are respectively equivalent to the sample mean $\bar{U}$ and sample variance $\Sigma_{i=1}^{n} (U_i - 0.5)^2/n = S^2$; thus the statistic $N_2$ is a combination of these two basic statistics, and as such has an intuitive appeal for testing uniformity. Furthermore, Locke and Spurrier (1978) and Miller and Quesenberry (1979) have recently shown $N_2$ to be an effective statistic against a wide range of alternatives. In Table 8.6 upper tail percentage points for $N_2$ are given; these were obtained by fitting Pearson curves to the moments, and are taken from Solomon and Stephens (1983). Miller and Quesenberry also recommend $N_4$ against some alternatives; they give tables of $N_1$, $N_2$, $N_3$, and $N_4$ based on Monte Carlo studies. Their tables for $N_3$ and $N_4$ are also given in Table 8.6. The quantities $v_j$ arise again in the next section, in connection with decomposing the EDF statistic $A^2$ into components. Further discussion of the Neyman tests is in Pearson (1938) and David (1939).

Barton (1953) considered a slightly different class of alternatives given by

$$f(u) = 1 + \sum_{j=1}^{k} \theta_j \pi_j(u), \quad 0 \leq u \leq 1, \quad k = 1, 2, \ldots$$

A restriction must now be placed on the $\theta_1$ to ensure that the density is always positive. The same statistic as above, $N_k$, may again be used to test for uniformity against this family of alternatives. Some asymptotic power calculations can also be made. Barton (1955, 1956) has investigated the application of these statistics when the data have been grouped or are discrete, and also the situation when the $U_i$ are not uniform, but have been obtained by the Probability Integral Transformation applied to a distribution with estimated parameters. This situation has also been examined by Thomas and Pierce (1979) and by Bargall and Thomas (1983). For these problems there are some interesting connections with the Pearson $X^2$ test.

## E 8.11 Example

For the data in Table 8.1, Neyman's statistic $N_2 = 6.437$ and is significant at about the 4% level. The individual components $v_1^2$ and $v_2^2$ (equivalent to $\bar{U}$ and $S^2$) have significance levels $p = 0.04$ and $p = 0.12$, respectively.

TABLE 8.6  Upper Tail Percentage Points for $N_2$, $N_3$, and $N_4$ (Section 8.11)

| n | \multicolumn{7}{c}{Significance level $\alpha$} | | | | | | |
|---|---|---|---|---|---|---|---|
|   | 0.5 | 0.25 | 0.1 | 0.05 | 0.025 | 0.01 | 0.005 |
| \multicolumn{8}{c}{Statistic $N_2$} | | | | | | | |
| 2 | 1.587 | 2.244 | 4.023 | 5.903 | 7.771 | 10.012 | 11.530 |
| 3 | 1.589 | 2.565 | 4.013 | 5.682 | 7.372 | 9.717 | 11.526 |
| 4 | 1.530 | 2.712 | 4.116 | 5.566 | 7.287 | 9.643 | 11.472 |
| 5 | 1.491 | 2.763 | 4.227 | 5.573 | 7.226 | 9.517 | 11.340 |
| 6 | 1.464 | 2.776 | 4.316 | 5.618 | 7.148 | 9.384 | 11.214 |
| 7 | 1.445 | 2.774 | 4.382 | 5.640 | 7.096 | 9.326 | 11.100 |
| 8 | 1.438 | 2.777 | 4.421 | 5.683 | 7.110 | 9.276 | 11.030 |
| 9 | 1.434 | 2.777 | 4.453 | 5.735 | 7.142 | 9.208 | 10.940 |
| 10 | 1.432 | 2.772 | 4.476 | 5.774 | 7.167 | 9.265 | 10.870 |
| 11 | 1.429 | 2.779 | 4.489 | 5.790 | 7.174 | 9.173 | 10.820 |
| 12 | 1.420 | 2.766 | 4.486 | 5.822 | 7.198 | 9.170 | 10.770 |
| 14 | 1.406 | 2.736 | 4.517 | 5.897 | 7.311 | 9.235 | 10.735 |
| 16 | 1.403 | 2.740 | 4.527 | 5.908 | 7.319 | 9.233 | 10.720 |
| 18 | 1.402 | 2.744 | 4.536 | 5.918 | 7.327 | 9.235 | 10.716 |
| 20 | 1.400 | 2.746 | 4.542 | 5.925 | 7.332 | 9.234 | 10.706 |
| 25 | 1.398 | 2.751 | 4.554 | 5.937 | 7.341 | 9.230 | 10.684 |
| 30 | 1.396 | 2.755 | 4.562 | 5.947 | 7.348 | 9.230 | 10.677 |
| 35 | 1.395 | 2.757 | 4.568 | 5.962 | 7.352 | 9.226 | 10.662 |
| 40 | 1.394 | 2.759 | 4.573 | 5.958 | 7.357 | 9.230 | 10.666 |
| 45 | 1.393 | 2.760 | 4.576 | 5.961 | 7.357 | 9.221 | 10.645 |
| 50 | 1.392 | 2.762 | 4.579 | 5.964 | 7.360 | 9.223 | 10.646 |
| 60 | 1.391 | 2.763 | 4.584 | 5.969 | 7.364 | 9.224 | 10.644 |
| 80 | 1.390 | 2.766 | 4.589 | 5.974 | 7.367 | 9.218 | 10.627 |
| 100 | 1.390 | 2.768 | 4.592 | 5.979 | 7.370 | 9.220 | 10.626 |
| $\infty$ | 1.386 | 2.773 | 4.605 | 5.991 | 7.378 | 9.210 | 10.597 |
| \multicolumn{8}{c}{Statistic $N_3$} | | | | | | | |
| 2 |  |  | 5.59 | 7.40 |  | 13.50 |  |
| 3 |  |  | 5.75 | 7.48 |  | 12.87 |  |
| 4 |  |  | 5.91 | 7.53 |  | 12.45 |  |
| 5 |  |  | 5.99 | 7.57 |  | 12.15 |  |
| 6 |  |  | 6.04 | 7.60 |  | 11.95 |  |
| 7 |  |  | 6.07 | 7.63 |  | 11.81 |  |
| 8 |  |  | 6.10 | 7.65 |  | 11.71 |  |
| 9 |  |  | 6.11 | 7.67 |  | 11.65 |  |
| 10 |  |  | 6.12 | 7.68 |  | 11.60 |  |

(continued)

**TABLE 8.6** (continued)

| | Significance level $\alpha$ | | |
|---|---|---|---|
| | 0.1 | 0.05 | 0.01 |
| 11 | 6.13 | 7.69 | 11.57 |
| 12 | 6.13 | 7.70 | 11.55 |
| 14 | 6.14 | 7.72 | 11.52 |
| 16 | 6.14 | 7.73 | 11.51 |
| 18 | 6.14 | 7.73 | 11.50 |
| 20 | 6.15 | 7.74 | 11.50 |
| 30 | 6.16 | 7.75 | 11.49 |
| 40 | 6.17 | 7.76 | 11.49 |
| 50 | 6.18 | 7.76 | 11.48 |
| $\infty$ | 6.25 | 7.81 | 11.35 |
| | Statistic $N_4$ | | |
| 2 | 7.19 | 9.52 | 16.14 |
| 3 | 7.34 | 9.51 | 15.80 |
| 4 | 7.46 | 9.50 | 15.43 |
| 5 | 7.53 | 9.49 | 15.12 |
| 6 | 7.57 | 9.48 | 14.86 |
| 7 | 7.60 | 9.47 | 14.65 |
| 8 | 7.62 | 9.47 | 14.47 |
| 9 | 7.63 | 9.46 | 14.32 |
| 10 | | | |
| 11 | 7.65 | 9.45 | 14.09 |
| 12 | 7.65 | 9.45 | 14.00 |
| 14 | 7.66 | 9.44 | 13.87 |
| 16 | 7.66 | 9.43 | 13.78 |
| 18 | 7.67 | 9.42 | 13.71 |
| 20 | 7.67 | 9.42 | 13.67 |
| 30 | 7.68 | 9.40 | 13.58 |
| 40 | 7.68 | 9.40 | 13.52 |
| 50 | 7.69 | 9.40 | 13.48 |
| $\infty$ | 7.78 | 9.49 | 13.28 |

Adapted from Miller and Quesenberry (1979) and from Solomon and Stephens (1983), by courtesy of the authors and of Marcel Dekker, Inc.

## 8.12  COMPONENTS OF TEST STATISTICS

In the expression (8.11) for $N_k$, the individual term $v_j$ may be regarded as a _component_ of the entire statistic $N_k$. Asymptotically these components are independently normally distributed with mean 0 and variance 1. For finite n, their distributions could, in principle, be examined, from the formulas for $v_j$, or from approximations using the moments (see, for example, David, 1939) although for finite n the $v_j$ are not independent. As David (1939) suggests, against certain alternatives, use of one of the individual components will prove more powerful than use of the entire statistic $N_k$. In recent years EDF statistics have also been partitioned into components along similar lines. For example, the EDF statistic $W^2$ can be written

$$W^2 = \sum_{j=0}^{\infty} \frac{z_j^2}{\lambda_j} ,$$

where

$$z_j = \left(\frac{2}{n}\right)^{\frac{1}{2}} \sum_{i=1}^{n} \cos(j\pi U_i) , \qquad\qquad (8.12)$$

and where $\lambda_j$ are weights (Durbin and Knott, 1972; see also Schoenfeld, 1977). Suppose $V_i = \pi j U_i$, $i = 1, 2, \ldots, n$. Starting at the point $(1,0)$ in the usual rectangular coordinates, $V_i$ can be recorded on the circumference of the unit circle, centered at the origin O, and with radius 1. Let point $P_i$ be the point on the circle corresponding to $V_i$, and let $R_j$ be the resultant (vector sum) of the vectors $OP_i$, $i = 1, \ldots, n$. Component $z_j$ is proportional to $X_j$, the length of the projection of $R_j$ on the X-axis. When the $U_j$ are $U(0,1)$, the $V_i$ will be uniform on $(0, j\pi)$ and the distributions of $X_j$ are the same as those discussed in connection with directions in Section 8.10.3. Stephens (1974a) has shown that the components of $U^2$ are proportional to $R_j$, the length of $R_j$, also discussed in Section 8.10.3. For $A^2$, the components are proportional to the $v_j$ in the Neyman-Barton test statistics; thus the sum of the first k components of $A^2$ is related to Neyman's $N_k$ in that they use the same components $v_j$, $j = 1, \ldots, k$, defined in equation (8.10), but with different weights. Against some alternatives one or two components of, say, $W^2$ or $A^2$ may be more powerful, as a test statistic for uniformity of $U_i$, than the entire statistic $W^2$ or $A^2$. Durbin and Knott (1972) have demonstrated this for a test of normality $N(0,1)$, in which the $U_i$ are obtained by the Probability Integral Transformation, against alternatives involving either a shift in mean or a shift in variance. The first component alone, for example, is better than $W^2$ in detecting the shift in mean. However, Stephens (1974a) has shown that the first component is insensitive to an alternative where both mean and

variance have been changed; for such an alternative, at least the first two
components would be needed—this is roughly the same as using the first
component of $U^2$. By expanding an alternative density into a series, using
appropriate orthogonal functions, it should be possible to suggest which
departures are detected by which components, and then perhaps decide how
many to use to get best power, but this will be difficult in the usual situation
where the alternative distribution to the null is not clearly known. Similar
remarks apply to the other statistics partitioned into components.

Durbin, Knott, and Taylor (1975), Stephens (1976), and Schoenfeld (1980)
have discussed partitioning of EDF statistics into components for the case
where the tested distribution contains unknown parameters; here the distribution theory of components is very complicated, and only asymptotic results
are known. Components can be useful in the theoretical examination of test
statistics, especially in calculating asymptotic power properties.

## 8.13 THE EFFECT ON TEST STATISTICS OF
### CERTAIN PATTERNS OF U-VALUES

In the next section we discuss the power of the various test statistics in this
chapter. However, before this, some general observations can be made on
the appearance of the U-set and its effect on different test statistics. If the
U-set is truly uniform, it should be scattered more or less evenly along the
interval $(0, 1)$. If the alternative to uniformity makes the values tend toward 0,
there will be a high value for $D^+$; if they tend toward 1, there will be a high
value for $D^-$. In either case D will be large and perhaps significant. The
statistics $W^2$ and $A^2$ will also detect a shift of values toward 0 or 1. If the
U set has been obtained from the Probability Integral Transformation (Chapter 4), from a Case 0 test that $X_i$ are from a completely specified $F(x)$, a
set of U-values tending toward 0 or toward 1 will suggest that the hypothesized
$F(x)$ has an incorrect mean (it may of course also have other incorrect parameters or be of incorrect shape). If the U-set tends to cluster at some point
in the interval, or to divide into two groups toward 0 and 1, the statistics V
and $U^2$ will be large and will tend to show significance. This indicates that
the variance of the hypothesized $F(x)$ is too large or too small. The statistic
$P = -2 \Sigma_i \log U_i$, like $D^+$ and $D^-$, also indicates which way the points have
moved; if they have moved closer to 0, P will be large, and if closer to 1,
P will be small. The value of P is very much more dependent on low values
of $U_i$ than on high values, because log u, when u is nearly 1, is nearly 0,
while as u approaches 0, log u becomes very large and negative. We shall
see later that this has some importance in methods of combining tests for
several samples. Among the other statistics, clearly $\bar{U}$ or $U_{(n/2)}$ might have
some power against an error in mean, but not against an error in variance of
the tested distribution. The same applies to the first component $v_1$ or $z_1$ in the
decomposition of both the Neyman tests and $W^2$ or $A^2$, and in turn $z_2$ in
equation (8.12) will not be sensitive to an error in mean (Durbin and Knott,

1972; Stephens, 1974a). Greenwood's statistic $G(n) = \Sigma_i D_i^2$ takes its smallest value $1/(n + 1)$ when all spacings are equal, that is, when values $U_i$ are superuniform; large values of $G(n)$ will occur with widely varying patterns of U-values.

## 8.14  POWER OF TEST STATISTICS

A number of studies have been made on tests for uniformity, including those by Stephens (1974b), Quesenberry and Miller (1977), Locke and Spurrier (1978), and Miller and Quesenberry (1979).

In general it can be said that, among the EDF statistics, the quadratic statistics appear to be more powerful than the supremum class; when the discrepancy between the EDF $F_n(u)$ and the theoretical distribution $F(u) = u$ is used all along the interval $0 \leq u \leq 1$ it appears that better use is made of the information in the sample than by using only the maximum discrepancy. When the basic problem is to test an X-set for a distribution $F(x)$, so that the observations $U_i$ have been obtained by the Probability Integral Transformation, $W^2$ and $A^2$ (especially $A^2$) will detect shifts in the mean of the hypothesized distribution from the true mean, and $U^2$ is effective at detecting shifts in variance. $A^2$ is also especially good at detecting irregularity in the tails of the distribution. Among other tests for uniformity, statistics obtained by the likelihood-ratio method are most powerful against their respective families of alternatives, as would be expected. Against a decreasing or increasing density, $\bar{U}$ alone is very efficient, and against distributional alternatives in which the mean is near the uniform mean of 0.5, but the variance is changed either because the distribution is unimodal and symmetric, or U-shaped and symmetric, the quantity $S^2$ above is a powerful statistic. For unimodal nonsymmetric distributions, these statistics lose some of their efficiency. The effect of the good performance of these relatively simple statistics means that the Neyman statistic $N_2$ in Section 8.11, which combines them both, is effective for a wide range of alternatives to uniformity (Locke and Spurrier, 1978; Miller and Quesenberry, 1979). Although the two components in $N_2$ occur again in $A^2$, the presence of further components, and the different weightings, sometimes make $A^2$ less effective than $N_2$; in a similar way, $N_4$ can be less effective than $N_2$, whenever adding further components "dilutes" the power of the first two (Quesenberry and Miller, 1979). This is similar to the situation for EDF statistics (Section 8.12).

## 8.15  STATISTICS FOR COMBINING INDEPENDENT
##       TESTS FOR SEVERAL SAMPLES

The uniform distribution has traditionally played an important role in combining test statistics for several samples. This is probably because of the general use of Fisher's method, which is based on the p-levels of the component tests.

To fix ideas, suppose k tests are to be made of null hypotheses $H_{o1}$, $H_{o2}$, ..., $H_{ok}$. Let $H_o$ be the composite hypothesis that all $H_{ok}$ are true; if any one is not true, $H_o$ should be rejected. Let $T_i$ be the test statistic used for $H_{oi}$, and suppose the test is an upper tail test. When the test is made, let $T_i$ take the value $T_i'$, and suppose $p_i$ is the significance level (often called the p-level) of this value, that is, when $H_{oi}$ is true, $Pr(T_i > T_i') = p_i$. When $H_{oi}$ is true, $p_i$ is $U(0, 1)$, and when k independent tests are made of the k null hypotheses above, we should obtain a random sample of k values of $p_i$ from $U(0, 1)$. Thus all k null hypotheses are tested simultaneously by testing if the $p_i$ appear to be such a uniform sample. This of course can be done by any of the methods described in this chapter. Fisher (1967) suggested the statistic $P = -2 \Sigma_i \log p_i$, already discussed in Section 8.10.2. Effectively the same idea had been put forward before by Karl Pearson, who suggested using the product of the $p_i$. For a summary of early work on some of the problems discussed in this section see Pearson (1938). Note that if $q_i = 1 - p_i$, $q_i$ could replace $p_i$ in P, since clearly $q_i$ is also $U(0, 1)$. Finally, let $r_i$ be the minimum of $p_i$ and $q_i$, formally written $r_i = \min(p_i, q_i)$; it is easily proved that, when $p_i$ is $U(0, 1)$, $r_i$ has the uniform distribution with limits 0 and 0.5, so that P could be calculated using $2r_i$, or $1 - 2r_i = |q_i - p_i|$ instead of $p_i$. Thus we have possible statistics

$$P_1 = -2 \Sigma_i \log p_i; \quad P_2 = -2 \Sigma_i \log q_i$$

$$P_3 = -2 \Sigma_i \log 2r_i; \quad P_4 = -2 \Sigma_i \log (1 - 2r_i)$$

On $H_o$, each of these statistics has the $\chi^2_{2k}$ distribution. An important question in making the overall test is which of these statistics to use. Fisher advocated $P_1$, with significance for large values, and this suggestion appears to have been generally accepted, although Pearson (1938) raised the possibility of using $P_2$. Littell and Folks (1971, 1973) have shown that $P_1$ has desirable properties from the point of view of Bahadur efficiency.

It has already been shown that $P_1$ is the likelihood ratio statistic, for a test of k = 0 against the alternative density $f_1(u) = (k + 1)u^k$, $0 < u < 1$, $k > -1$; for $k > 0$, $P_1$ will be declared significant for small values and for $-1 < k < 0$, $P_1$ will be significant for large values. Similarly, $P_2$ is the likelihood ratio statistic for the alternative density $f_2(u) = (k + 1)(1 - u)^k$, $0 < u < 1$, $k > -1$, with $P_2$ significant for large values when $k > 0$ and for small values when $-1 < k < 0$. Thus Fisher's use of $P_1$, with significance for large values, would imply that the alternative distribution for the $p_i$ values will give small values of $p_i$ large probability, but will allow some values to be close to one (the density $f_1(u)$, with $-1 < k < 0$, gives non-zero density at u = 1).

Thus $P_1$ can be expected to be powerful if some of the component hypotheses $H_{0i}$, $i = 1, \ldots, k$, were true (giving possibly a high p-value) and some were not true. Another possibility is that, when $H_0$ is rejected, all $H_{0i}$ are false together so that small p-values are likely in every test (for example, the tests might all be tests of normality, and it is felt that all the samples are likely to be non-normal if any of them are); then statistic $P_2$, used with significance in the lower tail, might be more effective.

### E 8.15.1 Example

Suppose five independent tests, for example, that five small samples are each from the normal distribution with mean zero and variance one, give p-values .15, .20, .28, .16, .25, so that each test is not significant at the 10% level. Then $P_1 = -2 \Sigma_i \log p_i = 15.99$; this value is exactly significant at the 10% level for $\chi_{10}^2$. This follows the usual procedure as suggested by Fisher. However, if we use the q-values .85, .80, .72, .84, .75 and calculate $P_2 = -2 \Sigma_i \log q_i$ we obtain $P_2 = 2.352$. This is significant at the 1% level in the <u>lower</u> tail of $\chi_{10}^2$; the sample gives greater significance using $P_2$ than using $P_1$.

### E 8.15.2 Example

For this example we take Fisher's first illustration of his test (Fisher, 1967). In this example three tests of fit yielded p-values of .145, .263, .087, and $P_1$ is 11.42. In the upper tail of $\chi_6^2$, this is significant at approximately the 7.5% level. The q-values are .855, .737, .913, and $P_2$ is 1.105. This value is significant at the 2.5% level in the lower tail of $\chi_6^2$. Again the value of $P_2$ is more significant than the value of $P_1$ in its appropriate tail.

### 8.15.3 Use of Fisher's Test with Two-Tail Component Tests

If all the component tests in $H_0$ were two-tail tests, either $P_3$ or $P_4$ above should be used as test statistics. For the same reasons that $P_2$ can be preferred to $P_1$ for one-tail tests, $P_4$ might be better than $P_3$ for two-tail tests. When some of the component hypotheses $H_{0i}$ are tested by one-tail tests and some by two-tail tests, the formula for P could be

$$P_5 = -2 \Sigma_i \log U_i$$

with $U_i = q_i$ for one-tail tests, and $U_i = 1 - 2r_i$ for two-tail tests. Alternatively $P_6 = -2 \Sigma_i \log U_i$ could be used with $U_i = p_i$ for one-tail tests and $U_i = 2r_i$ for two-tail tests. $P_5$ and $P_6$ will again have the $\chi_{2k}^2$ distribution when all $H_{0i}$ are true.

### E 8.15.3 Example

Suppose five independent tests are to be made that five samples from normal distributions have means 0, against the alternative that the means are not 0. Thus five t-tests will be used, with significance in either tail for each test. Suppose the significance levels measured all from the upper tail (for this we use the temporary notation $p_i^*$) are $p_i^* = .15, .04, .75, .92, .07$; thus only the second sample would be declared significant using a two-tail 10% level for $t_4$. The corresponding values of $r_i$ are $.15, .04, .25, .08, .07$ and these give the value for $P_3 = 16.44$. Using $1 - 2r_i$ instead of $2r_i$ we have $P_4 = 2.92$. $P_3$ is significant at the 10% level of $\chi_{10}^2$ (upper tail), and $P_4$ at the 2% level (lower tail).

### 8.15.4 Possible Misuse of Fisher's Test

It is possible to misuse Fisher's test, in the situation where some of the component tests are two-tail tests, by using $r_i$ when $2r_i$ should be used. This is especially easily done when the results of two-tail tests are sometimes reported, using expressions such as "the lower tail p-value equals $0.11$," or "the upper tail p-value equals $0.35$." The statistician might then wrongly use $\log 0.11$ and $\log 0.35$ in the calculation of, say, $P_6$, and obtain false levels of significance for the test of the overall hypothesis $H_0$. As an example, consider Example E 8.15.3 above in which all tests are two-tail tests, and the third and fourth tests, for example, could have been reported as significant at the $.25$ and $.08$ levels in the lower tail. Then if the values of $r_i$ instead of $2r_i$ are used in calculating $P_3$ (which is the same as $P_6$ since all tests are two-tail), we have $P_3 = 23.37$. This is spuriously highly significant; $P_3$ is at the 1% level of $\chi_{10}^2$.

### 8.16  TESTS FOR A UNIFORM DISTRIBUTION WITH UNKNOWN LIMITS

Until now, tests in this chapter have been tests for a U(0,1) distribution; they can be used easily for a test for U(a,b), where a and b are known, by making the transformation given in Section 8.2. If the limits of the distribution are not known, other procedures are available. Two of these are:

(a)  Suppose, given an ordered sample $U'_{(1)}, \ldots, U'_{(n)}$, that the range is $R = U'_{(n)} - U'_{(1)}$, and let $U_{(i)}$ be defined by $U_{(i)} = \{U'_{(i+1)} - U'_{(1)}\}/R$, $i = 1, \ldots, (n-2)$. A test of $H_0$: the U' set is an ordered sample from U(a,b), becomes a test of $H_0$: the U-sample, of size $n-2$, is an ordered sample from U(0,1).

(b)  Use of the correlation test of Section 5.5.1.

## 8.17 TESTS FOR CENSORED UNIFORM SAMPLES

Tests for censored data are described in several other chapters and examples are given in Chapter 11. Tests specifically for censored uniforms are

(a) EDF tests: Sections 4.7.3-4.7.6, and 12.3;
(b) Correlation tests: Sections 5.5, 5.6, and 12.3;
(c) Spacings tests based on Greenwood's statistic: Section 8.9.1.

## REFERENCES

Ajne, B. (1968). A simple test for uniformity of a circular distribution. Biometrika, 55, 343-354.

Bargal, A. I. and Thomas, D. E. (1983). Smooth goodness-of-fit tests for the Weibull distribution with singly censored data. Commun. Statist. Theor. Meth., 12, 1431-1447.

Barton, D. E. (1953). On Neyman's smooth test of goodness-of-fit and its power with respect to a particular system of alternatives. Skand. Aktuartidskr., 36, 24-63.

Barton, D. E. (1955). A form of Neyman's psi-2 test of goodness-of-fit applicable to grouped and discrete data. Skand. Aktuartidskr., 39, 1-17.

Barton, D. E. (1956). Neyman's psi-2 test of goodness-of-fit when the null hypothesis is composite. Skand. Aktuartidskr., 39, 216-245.

Barton, D. E. and Mallows, C. L. (1965). Some aspects of the random sequence. Ann. Math. Statist., 36, 236-260.

Beran, R. J. (1968). Testing for uniformity on a compact homogeneous space. J. App. Prob., 5, 177-195.

Brunk, H. D. (1962). On the range of the difference between hypothetical distribution function and Pyke's modified empirical distribution function. Ann. Math. Statist., 33, 525-532.

Burrows, P. M. (1979). Selected percentage points of Greenwood's Statistic. J. Roy. Statist. Soc., A, 142, 256-258.

Cressie, N. (1976). On the logarithms of high order spacings. Biometrika, 63, 343-355.

Cressie, N. (1977a). The minimum of higher order gaps. Australian J. Statist., 19, 132-143.

Cressie, N. (1977b). On some properties of the scan statistic on the circle and the line. J. Appl. Prob., 14, 272-283.

Cressie, N. (1978). Power results for tests based on high order gaps. Biometrika, 65, 214-218.

Cressie, N. (1979). An optimal statistic based on higher order gaps. Biometrika, 66, 619-627.

Cressie, N. (1980). Asymptotic distribution of the scan statistic under uniformity. Ann. Prob., 9, 828-840.

Currie, I. D. (1981). Further percentage points for Greenwood's Statistic. J. Roy. Statist. Soc., A, 144, 360-363.

Darling, D. A. (1953). On a class of problems related to the random division of an interval. Ann. Math. Statist., 24, 239-253.

Darling, D. A. (1982). On the supremum of a certain Gaussian Process. Ann. Prob., 11, 803-806.

Darling, D. A. (1983). On the asymptotic distribution of Watson's statistic. Ann. Statist., 11, 1263-1266.

David, F. N. (1939). On Neyman's "smooth" test for goodness of fit. Biometrika, 31, 191-199.

Deken, J. G. (1980). Exact distributions for gaps and stretches. Technical Report, Department of Statistics, Stanford University.

Del Pino, G. E. (1979). On the asymptotic distribution of k-spacings with applications to goodness-of-fit tests. Ann. Statist., 7, 1058-1065.

Dudewicz, E. J. and Van Der Meulen, E. C. (1981). Entropy-based tests of uniformity. J. Amer. Statist. Assoc., 76, 967-974.

Durand, D. and Greenwood, J. A. (1957). Random unit vectors II: Usefulness of Gram-Charlier and related series in approximating distributions. Ann. Math. Statist., 28, 978-986.

Durbin, J. (1961). Some methods of constructing exact tests. Biometrika, 48, 41-55.

Durbin, J. (1969a). Test for serial correlation in regression analysis based on the periodogram of least-squares residuals. Biometrika, 56, 1-16.

Durbin, J. (1969b). Tests of serial independence based on the cumulated periodogram. Bull. Inst. Internat. Statist., 42, 1039-1048.

Durbin, J. and Knott, M. (1972). Components of Cramer-Von Mises statistics, I. J. Roy. Statist. Soc., B, 34, 290-307.

Durbin, J., Knott, M., and Taylor, C. C. (1975). Components of Cramer-Von Mises statistics, II. J. Roy. Statist. Soc., B, 37, 216-237.

Fisher, R. A. (1967). Statistical Methods for Research Workers. 4th Edition. New York: Stechert.

Giné, E. (1975). Invariant tests for Uniformity on Compact Riemannian Manifolds Based on Sobolev Norms. Ann. Statist., 3, 1243-1266.

Greenwood, J. A. and Durand, D. (1955). The distribution of length and components of the sum of N random unit vectors. Ann. Math. Statist., 26, 233-246.

Greenwood, M. (1946). The statistical study of infectious disease. J. Roy. Statist. Soc., A, 109, 85-110.

Hartley, H. O. and Pfaffenberger, R. C. (1972). Quadratic forms in order statistics used as goodness-of-fit criteria. Biometrika, 59, 605-611.

Hegazy, Y. A. S. and Green, J. R. (1975). Some new goodness-of-fit tests using order statistics. Appl. Statist., 24, 299-308.

Hill, I. D. (1979). Approximating the distribution of Greenwood's statistic with Johnson distributions. J. Roy. Statist. Soc., A, 142, 378-380. Corrigendum (1981). J. Roy. Statist. Soc., A, 144, 388.

Holst, L. (1979). Asymptotic normalities of sum-functions of spacings. Ann. Prob., 7, 1066-1072.

Johannes, J. M. and Rasche, R. H. (1980). Additional information on significance values for Durbin's $C^{+}$, $C^{-}$ and C statistics. Biometrika, 67, 511-514.

Kimball, B. F. (1947). Some basic theorems for developing tests of fit for the case of the non-parametric probability distribution function, I. Ann. Math. Statist., 18, 540-548.

Kimball, B. F. (1950). On the asymptotic distribution of the sum of powers of unit frequency differences. Ann. Math. Statist., 21, 263-271.

Littell, R. C. and Folks, J. L. (1971). Asymptotic optimality of Fisher's method of combining independent tests. J. Amer. Statist. Assoc., 66, 802-806.

Littell, R. C. and Folks, J. L. (1973). Asymptotic optimality of Fisher's method of combining independent tests, II. J. Amer. Statist. Assoc., 68, 193-194.

Locke, C. and Spurrier, J. D. (1978). On tests of uniformity. Comm. Statist. Theory Methods, A7, 241-258.

Lurie, D., Hartley, H. O., and Stroud, M. R. (1974). A goodness of fit test for censored data. Comm. Statist., 3, 745-753.

McLaren, C. G. and Stephens, M. A. (1985). Percentage points and power for spacings statistics for testing uniformity. Technical Report, Department of Mathematics and Statistics, Simon Fraser University.

Miller, F. L. and Quesenberry, C. P. (1979). Power studies of some tests for uniformity, II. Comm. Statist. Simula. Comput., B, 8, 271-290.

Moran, P. A. P. (1947). The random division of an interval—Part I. J. Roy. Statist. Soc., B, 9, 92-98.

Moran, P. A. P. (1951). The random division of an interval—Part II. J. Roy. Statist. Soc., B, 13, 147-150.

Neyman, J. (1937). "Smooth" tests for goodness-of-fit. Skand. Aktuarie-tidskr., 20, 149-199.

O'Reilly, F. J. and Stephens, M. A. (1982). Characterizations and goodness of fit tests. J. Roy. Statist. Soc.,

Pearson, E. S. (1938). The probability integral transformation for testing goodness-of-fit and combining independent tests of significance. Biometrika, 30, 134-148.

Prentice, M. J. (1978). On invariant tests for uniformity for directions and orientations. Ann. Statist., 6, 169-176.

Pyke, R. (1965). Spacings. J. Roy. Statist. Soc., B, 27, 395-449.

Quesenberry, C. P. and Hales, S. (1980). Concentration bands for Uniformity plots. J. Statist. Comput. Simul., 11, 41-53.

Quesenberry, C. P. and Miller, F. L. Jr. (1977). Power studies of some tests for uniformity. J. Statist. Comput. Simulation, 5, 169-191.

Schoenfeld, D. A. (1977). Asymptotic properties of tests based on linear combinations of the orthogonal components of Cramér-von Mises statistics. Ann. Statist., 5, 1017-1026.

Schoenfeld, D. A. (1980). Tests based on linear combinations of the orthogonal components of the Cramér-von Mises statistic when parameters are estimated. Ann. Statist., 8, 1017-1022.

Seshadri, V., Csorgo, M., and Stephens, M. A. (1969). Tests for the exponential distribution using Kolmogorov-type statistics. J. Roy. Statist. Soc., B, 31, 499-509.

Smirnov, N. V. (1947). Akad. Nauk SSR, C. R. (Dokl.) Akad. Sci. URSS, 56, 11-14.

Solomon, H. and Stephens, M. A. (1981). Tests for uniformity: Greenwood's test and tests based on gaps and stretches. Technical Report 311, Dept. of Statistics, Stanford University.

Solomon, H. and Stephens, M. A. (1983). On Neyman's statistic for testing uniformity. Comm. Statist. Simulation Comput., 12, 127-134.

Stephens, M. A. (1964). The testing of unit vectors for randomness. J. Amer. Statist. Assoc., 59, 160-167.

Stephens, M. A. (1966). Statistics connected with the uniform distribution: percentage points and applications to testing for randomness of directions. Biometrika, 53, 235-239.

Stephens, M. A. (1969a). Tests for randomness of directions against two circular alternatives. J. Amer. Statist. Assoc., 64, 280-289.

Stephens, M. A. (1969b). Results from the relation between two statistics of the Kolmogorov-Smirnov type. Ann. Math. Statist., 40, 1833-1837.

Stephens, M. A. (1969c). A goodness-of-fit statistic for the circle, with some comparisons. Biometrika, 56, 161-168.

Stephens, M. A. (1970). Use of the Kolmogorov-Smirnov, Cramer-von Mises and related statistics without extensive tables. J. Roy. Statist. Soc., B, 32, 115-122.

Stephens, M. A. (1974a). Components of goodness-of-fit statistics. Ann. Inst. H. Poincare, B, 10, 37-54.

Stephens, M. A. (1974b). EDF statistics for goodness-of-fit and some comparisons. J. Amer. Statist. Assoc., 69, 730-737.

Stephens, M. A. (1976). Asymptotic power of EDF statistics for exponentiality against Gamma and Weibull alternatives. Technical Report No. 297: Department of Statistics, Stanford University.

Stephens, M. A. (1981). Further percentage points for Greenwood's statistic. J. Roy. Statist. Soc., A, 144, 364-366.

Stephens, M. A. (1986). Goodness-of-fit for censored data. Technical report, Department of Statistics, Stanford University.

Stuart, A. (1954). Too good to be true. Appl. Statist., 3, 29-32.

Sukhatme, P. V. (1937). Tests of significance for samples of the chi-square population with two degrees of freedom. Annals of Eugenics London, 8, 52-56.

Thomas, D. R. and Pierce, D. A. (1979). Neyman's smooth goodness-of-fit test when the hypothesis is composite. J. Amer. Statist. Assoc., 74, 441-445.

Vasicek, O. (1976). A test for normality based on sample entropy. J. Roy. Statist. Soc., B, 38, 54-59.

Wallenstein, S. R. and Naus, J. I. (1974). Probabilities for the size of largest clusters and smallest intervals. J. Amer. Statist. Soc., 69, 690-695.

Wang, Y. H. and Chang, S. A. (1977). A new approach to the nonparametric tests of exponential distribution with unknown parameters. The Theory and Application of Reliability. 2, 235-258. New York: Academic Press.

Watson, G. S. (1967). Some problems in the statistics of directions. Bulletin of the Int. Statist. Inst. Conference, 36th Session, Sydney, Australia.

Watson, G. S. (1976). Optimal invariant tests for uniformity. Study in Probability and Statistics. Amsterdam: North-Holland, 121-127.

Weiss, L. (1957a). The convergence of certain functions of sample spacings. Ann. Math. Statist., 28, 778-782.

Weiss, L. (1957b). The asymptotic power of certain tests of fit based on sample spacings. Ann. Math. Statist., 28, 783-788.

# 9

# Tests for the Normal Distribution

Ralph B. D'Agostino   Boston University, Boston, Massachusetts

## 9.1 INTRODUCTION

The single most used distribution in statistical analysis is the normal
distribution. Its uses can be classified in two sets. The first relates to the
class of statistics which are taken to be normally distributed due to the
applicability of large sample theorems such as the Central Limit Theorem
(Rao, 1973, Chapter 2), the Delta Theorems (Rao, 1973, Chapter 6), and
theorems related to the asymptotic distribution of linear functions of order
statistics (Chernoff, Gastwirth, and Johns, 1967). The second set relates to
situations where the normal distribution is assumed to be the appropriate
mathematical model for the underlying phenomenon under investigation. The
applied literature is replete with examples of this latter class where, for
examples, the normal distribution or the related lognormal distribution (i.e.
logs of data are normally distributed) are used as models for cadmium and
lead levels in the blood of children (Smith, Temple, and Reading, 1976), the
distribution of hydrologic runoff (Kottegoda and Yevjevich, 1977), body dis-
comfort and transmissibility scores (Griffin and Whitham, 1978), weights of
mammary tumors in rats (Fredholm, Gunnarsson, Jensen, and Muntzing,
1978), levels of toxic gases to which workers are exposed (D'Agostino and
Gillespie, 1978; and Smith, Wagner, and Moore, 1978), nuclear cross sections data (Richert, Simbel, and Weidenmuller, 1975), radio scintillation
data (Rino, Livingston, and Whitney, 1976), earnings and wages (White and
Olson, 1981), and the distributions of air pollutants (Larson, 1971, and
Hunt, 1972). The  chapter deals with this second class of use and discusses
goodness-of-fit tests designed to test formally the appropriateness or adequacy of the normal distribution as a model for the underlying phenomenon

from which data were generated. These tests complement the informal graphical techniques already discussed in Chapter 2 (see Sections 2.4 and 2.5).

This chapter will focus on tests applicable to complete samples. Techniques based on incomplete or censored samples are discussed elsewhere; in Chapter 11, Analysis of Data from Censored Samples, and also in Chapters 3, 4, and 5. We start by discussing tests that assume a complete random sample is available for analysis. These tests occupy the major portion of the chapter and are the primary interest of the chapter. Tests applicable on residuals and tests for multivariate normality will also be discussed.

## 9.2 COMPLETE RANDOM SAMPLES

Until stated otherwise we assume the following. Let $X_1$, $X_2$, ..., $X_n$ be a random sample of size n from a population with probability density function (pdf) f(x) and cumulative distribution function (cdf) F(x). The pdf and cdf of the normal distribution are, respectively, $\phi(x)$ and $\Phi(x)$. Our null hypothesis is

$$Ho: f(x) = \phi(x) \quad \text{or} \quad F(x) = \Phi(x) \tag{9.1}$$

### 9.2.1 Null Hypothesis

The pdf of the normal distribution is given by

$$\phi(x) = \frac{1}{\sqrt{2\pi}\sigma} e^{-\frac{1}{2}((x-\mu)/\sigma)^2} \tag{9.2}$$

$$\begin{pmatrix} -\infty < x < \infty \\ -\infty < \mu < \infty \\ \sigma > 0 \end{pmatrix}$$

In testing for departures from the normal distribution the null hypothesis of (9.1), $H_0$, is that the random variable X under consideration is distributed as a normal variable, or in other words, X has a probability density function given by (9.2). If, further, specific values of both the mean and standard deviation, $\mu$ and $\sigma$, of (9.2) are specified by the null hypothesis (e.g., X is normally distributed with $\mu = 500$ and $\sigma = 100$), then the null hypothesis is a simple hypothesis. This means the null hypothesis concerns itself with only one particular distribution. If either $\mu$ or $\sigma$ are not specified completely, then the null hypothesis under consideration is a composite hypothesis. This chapter deals mainly with the composite null hypothesis with both $\mu$ and $\sigma$ unknown. In most applications prior knowledge of $\mu$ or $\sigma$ is not available. If it is available, it usually is of no help, from a power point of view, in judging goodness-of-fit (see, for example, Chapter 4 on EDF tests).

## 9.2.2 Alternative Hypothesis

The alternative hypothesis, $H_1$, usually employed in these testing situations is the composite hypothesis that X is not normally distributed. Directions of nonnormality or alternative distributions are only rarely considered (e.g., by Uthoff, 1970, 1973). In this chapter alternatives to normality are limited to the following: (1) X is nonnormal and no prior information is available concerning alternative distributions, or (2) X is nonnormal and information is available concerning the deviations from normality in terms of skewness $(\sqrt{\beta_1})$ and/or tail thickness or peakedness as measured, for example, by the kurtosis coefficient $\beta_2$. For a random variable X the skewness and kurtosis coefficients are, respectively,

$$\sqrt{\beta_1} = \frac{E(X - \mu)^3}{\sigma^3} \tag{9.3}$$

and

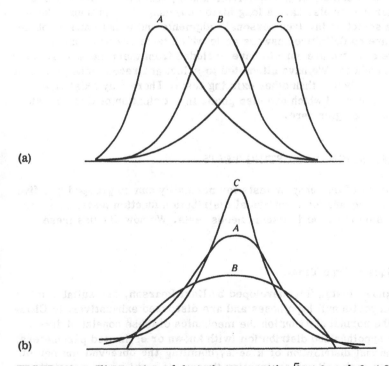

(a)

(b)

FIGURE 9.1 Illustration of distributions with $\sqrt{\beta_1} \neq 0$ and $\beta_2 \neq 3$. (a) Distribution differing in skewness: A. $\sqrt{\beta_1} > 0$, B. $\sqrt{\beta_1} = 0$, C. $\sqrt{\beta_1} < 0$. (b) Distribution differing in kurtosis: A. $\beta_2 = 3$, B. $\beta_2 < 3$, C. $\beta_2 > 3$.

$$\beta_2 = \frac{E(X - \mu)^4}{\sigma^4} \qquad\qquad (9.4)$$

where E represents the expected value operator. For the normal distribution $\sqrt{\beta_1} = 0$ and $\beta_2 = 3$. In the following $\beta_2$ will often be used to refer to "tail thickness" of an alternative distribution with $\beta_2 > 3$ indicating a thick tailed distribution and $\beta_2 < 3$ indicating a thin tailed distribution. The reader is referred to Chapter 2 on graphical analysis and Chapter 7 on moment techniques for further details of these coefficients. Figure 9.1 contains illustrations of distributions with $\sqrt{\beta_1} \neq 0$ and $\beta_2 \neq 3$.

The next two sections consist of classifying and investigating the relative merits of the tests. The reader who is interested in our final recommendations may want to go directly to Section 9.5 and then read sections 9.3 and 9.4 selectively.

The objective of the next three sections is to classify and review a selective number of the various available tests (some of which have already been presented in the above chapters), discuss their relative merits, and then make recommendations concerning which should be used in practice. We have attempted to select, in an objective manner, tests which are serious contenders for use or else have a long historical usage behind them. However, such a selection involves personal judgments and is not entirely objective. There are no definitive answers as to which tests are best, and the tests we have selected are sure to have excluded from them the favorite tests of some researchers. We have attempted to select and recommend tests that are as good as or better than other existing tests. There may exist other tests not recommended which are also good. In anticipation of this omission let us give our apologies here.

## 9.3 CLASSIFICATION OF EXISTING TESTS

For the purposes of this chapter tests for normality can be grouped into five categories, chi-square test, empirical distribution function tests, moment tests, regression tests, and miscellaneous tests. We now discuss these groups.

### 9.3.1 Chi-Square Type Tests

These well-known tests, first developed by Karl Pearson, are suitable for simple or composite null hypotheses and are discussed exhaustively in Chapter 3. For the normal distribution the mechanics of these consist of discretizing the hypothesized distribution (with known or estimated parameters) into a multinormal distribution of k cells, counting the observed number of observations in each cell and contrasting these, via a chi-square statistic or a likelihood ratio statistic, with the expected number of observations for each cell. The latter expected values are computed assuming the data did arise from a normal distribution. Of particular interest to testing for normality are the articles of Chernoff and Lehmann (1954) and Watson (1957). In the former article it is shown that the use of the sample mean and standard deviation based on ungrouped data to obtain the expected values results in the observed chi-square statistic as being asymptotically distributed as

$$X_1^2(k - 3) + \alpha_2 X_2^2(1) + \alpha_3 X_3^2(1) \tag{9.5}$$

In (9.5) $X_i^2(\nu)$ represents a chi square variable with $\nu$ degree of freedom, all these chi square variables are independent and $0 \le \alpha_i < 1$. The often quoted $k - 3$ degrees of freedom is incorrect. The Watson (1957) article describes how switching appropriately to fixed cell probabilities can result in obtaining explicit formulas for the $\alpha_2$ and $\alpha_3$ of (9.5). While the chi square tests are of historical interest and are continuously being modified, we agree with Professor D. S. Moore, the author of Chapter 3, that they should not be recommended for use in testing for departures from normality when the full ungrouped sample of data is available. Other procedures to be discussed below are more powerful. In the cases where the full sample is not available (i.e., data are censored or truncated) or where the data are grouped into classes (see Section 3.2.7, Example 2) these procedures are of use. We refer the reader to Chapter 3 for further details. The remainder of this chapter will not contain any further discussion of these tests.

## 9.3.2 Tests Based on the Empirical Distribution Function (EDF)

Chapter 4 above discussed in detail the concept and applications of the tests based on the empirical distribution function (EDF). Basically for the normal distribution these tests involve measuring the discrepancy between the cumulative distribution function

$$\Phi(x) = \int_{-\infty}^{x} \frac{1}{\sqrt{2\pi}\sigma} e^{-\frac{1}{2}((t-\mu)/\sigma)^2} dt \tag{9.6}$$

of the normal distribution and the empirical distribution function

$$F_n(x) = \frac{\#(X \le x)}{n} \tag{9.7}$$

of the sample. The $\mu$ and $\sigma$ of (9.6) often are not specified and are replaced by the sample mean $\bar{X}$ and standard deviation S, where

$$\bar{X} = \frac{\Sigma X}{n} \quad \text{and} \quad S = \sqrt{\frac{\Sigma(X - \bar{X})^2}{n - 1}} \tag{9.8}$$

## 9.3.2.1 Simple Null Hypothesis

Many tests have been developed for this situation. Some prominent ones are the Kolmogorov (1933)-Smirnov (1939) test, the Kuiper V test (1960), Pyke's C test (1959), Brunk's B test (1962), Durbin's D test (1961), the Cramer-von Mises $W^2$ test (1928), Durbin's M test (1973), Watson's $U^2$ test (1961), the Anderson-Darling $A^2$ test (1954), Fisher's $\pi$ and $\pi'$ tests (1928), and the Hartley-Pfaffenberger $s^2$ test (1972). See Chapter 4 for details on some of these.

## 9.3.2.2 Anderson-Darling Test for the Composite Hypothesis

Some of the above tests have been modified to apply to the composite null hypothesis of normality with $\mu$ and $\sigma$ unknown (Stephens, 1974, and Green and Hegazy, 1976). In Chapter 4, Section 4.2, formulas are given for the Kolmogorov-Smirnov D test, the Kuiper V test, the Cramer-von Mises $W^2$ test, the Watson $U^2$ test, and the Anderson-Darling $A^2$ test, and in Section 4.8 the application of these to the normal distribution is described in detail. For the purposes of this chapter we now present in the notation of this chapter the procedure for performing the Anderson-Darling $A^2$ test, which is the EDF test we recommend for us.

(1) Arrange the sample in ascending order,

$$X_{(1)} \leq \cdots \leq X_{(n)}$$

(2) Calculate standardized values, $Y_{(i)}$, where

$$Y_{(i)} = \frac{X_{(i)} - \bar{X}}{S} \quad \text{for } i = 1, \ldots, n \tag{9.9}$$

(3) Calculate $P_i$ for $i = 1, \ldots, n$ where

$$P_i = \Phi(Y_{(i)}) = \int_{-\infty}^{Y_{(i)}} \frac{e^{-\frac{t^2}{2}}}{\sqrt{2\pi}} \, dt \tag{9.10}$$

$\Phi(y)$ in (9.10) represents the cdf of the standard normal distribution and $P_i$ is the cumulative probability corresponding to the standard score $Y_{(i)}$ of (9.9). $P_i$ can be found from standard normal tables as given in the Appendix or by use of the following approximation due to Hastings (1955). For $Y_{(i)}$ such that $0 \leq Y_{(i)} < \infty$ define $y = Y_{(i)}$ and compute

$$Q_i = 1 - \frac{1}{2}(1 + C_1 y + C_2 y^2 + C_3 y^3 + C_4 y^4)^{-4} \tag{9.11}$$

where

$$C_1 = 0.196854, \quad C_3 = 0.000344$$

$$C_2 = 0.115194, \quad C_4 = 0.019527$$

Here $P_i$ of (9.10) is equal to $Q_i$ of (9.11). For $Y_{(i)}$ such that $-\infty < Y_{(i)} \le 0$ define $y = -Y_{(i)}$ and compute $Q_i$ of (9.11). Here $P_i$ of (9.10) is equal to $1 - Q_i$ of (9.11).

(4) Compute the Anderson-Darling statistic

$$A^2 = -\sum_{i=1}^{n} [(2i - 1)\{\log P_i + \log (1 - P_{n+1-i})\}/n] - n \qquad (9.12)$$

where log is log base e.

(5) Compute the modified statistic

$$A^* = A^2(1.0 + 0.75/n + 2.25/n^2) \qquad (9.13)$$

(6) Reject the null hypothesis of normality if $A^*$ exceeds 0.631, 0.752, 0.873, 1.035, and 1.159 at levels of significance 0.10, 0.05, 0.025, 0.01, and 0.005, respectively.

The above procedure is valid for samples of size $n \ge 8$.

E 9.3.2.2.1 Numerical example of Anderson-Darling test Table 9.1 contains numerical examples employing the first ten observations of the NOR and EXP data sets (of the Appendix). As is to be expected the Anderson-Darling test accepts the hypothesis of normality for the NOR data and rejects it for the EXP data at level of significance 0.05.

The p-level or descriptive level of significance for the Anderson-Darling test can be obtained by use of Table 4.8. The reader is referred to Chapter 4, Section 4.8.2 for details.

## 9.3.2.3 Transformation Methods for the Composite Hypothesis

Csorgo, Seshadri, and Yalovsky (1973) present another approach for applying EDF tests to the composite null hypothesis ($\mu$ and $\sigma$ unknown). In this approach the n ($\ge 4$) observations are first transformed into n - 2 independent observations free of unknown parameters. Then EDF tests (e.g., $A^2$) are applied to these. O'Reilly and Quesenberry (1973) and Quesenberry (1975) present a general theory for obtaining transformations such that the transformed variables are independent uniform random variables. Chapter 6 of this book is devoted to transformation techniques. With use of these it is now possible to apply any EDF test as a test for deviations from normality. However, these transformation procedures require randomization of the data. To many users this is considered an undesirable feature.

**TABLE 9.1 Numerical Examples of Anderson–Darling $A^2$ Test[a,b]**

| i | $X_{(i)}$ | $Y_{(i)}$ | $P_{(i)}$ | $1 - P_{(i)}$ | $\log P_{(i)}$ | $\log(1 - P_i)$ | $(2i - 1)(\log P_{(i)} + \log(1 - P_{n+1-i}))$ |
|---|---|---|---|---|---|---|---|
| NOR data | Normal | $\mu = 100$, $\sigma = 10$ | | $\bar{x} = 98.414$, $s = 8.277$ | | | |
| 1 | 84.27 | -1.71 | .0436 | .9564 | -3.1327 | -0.0446 | -6.5685 |
| 2 | 90.87 | -0.91 | .1814 | .8186 | -1.7071 | -0.2002 | -10.7358 |
| 3 | 92.55 | -0.71 | .2389 | .7611 | -1.4317 | -0.2730 | -12.3790 |
| 4 | 96.20 | -0.27 | .3936 | .6064 | -0.9324 | -0.5002 | -12.8506 |
| 5 | 98.70 | 0.03 | .5120 | .4880 | -0.6694 | -0.7174 | -12.7800 |
| 6 | 98.98 | 0.07 | .5279 | .4721 | -0.6388 | -0.7506 | -14.9182 |
| 7 | 100.42 | 0.24 | .5948 | .4052 | -0.5195 | -0.9034 | -13.2561 |
| 8 | 101.58 | 0.38 | .6480 | .3520 | -0.4339 | -1.0441 | -10.6035 |
| 9 | 106.82 | 1.02 | .8461 | .1539 | -0.1671 | -1.8715 | -6.2441 |
| 10 | 113.75 | 1.85 | .9678 | .0322 | -0.0327 | -3.4358 | -1.4687 |
| | | | | | | | -101.8045 |
| EXP data | (Exponential | $\mu = 5.0$) | | $\bar{x} = 4.257$, $s = 5.169$ | | | |
| 1 | 0.06 | -0.81 | .2090 | .7910 | -1.5654 | -0.2345 | -6.3937 |
| 2 | 0.37 | -0.75 | .2266 | .7734 | -1.4846 | -0.2570 | -8.9076 |
| 3 | 0.44 | -0.74 | .2296 | .7704 | -1.4714 | -0.2608 | -12.8460 |
| 4 | 0.89 | -0.65 | .2578 | .7422 | -1.3556 | -0.2981 | -14.8029 |
| 5 | 2.17 | -0.40 | .3446 | .6554 | -1.0654 | -0.4225 | -13.8663 |
| 6 | 2.63 | -0.31 | .3783 | .6217 | -0.9721 | -0.4753 | -15.3406 |
| 7 | 4.69 | 0.08 | .5319 | .4681 | -0.6313 | -0.7591 | -12.0822 |
| 8 | 6.48 | 0.43 | .6664 | .3336 | -0.4059 | -1.0978 | -10.0005 |
| 9 | 8.15 | 0.75 | .7734 | .2266 | -0.2570 | -1.4846 | -8.7380 |
| 10 | 16.69 | 2.41 | .9920 | .0080 | -0.0080 | -4.8283 | -4.6075 |
| | | | | | | | -107.5853 |

$$a_{A^2} = \frac{101.8045}{10} - 10 = .18045$$

$$A^* = A^2\left(1 + \frac{.75}{10} + \frac{2.25}{100}\right) = .198044 \quad \text{(Accept Normality for NOR data)}$$

$$b_{A^2} = \frac{107.5853}{10} - 10 = .75853$$

$$A^* = A^2\left(1 + \frac{0.75}{10} + \frac{2.25}{100}\right) = .832487 \quad \text{(Reject Normality for EXP data at 0.05 level of significance)}$$

### 9.3.2.4 Components of an EDF Statistic

Durbin, Knott, and Taylor (1975) have employed a procedure from which it is possible to express the test statistic of an EDF test as a weighted linear function of independent chi square variables each with one degree of freedom. This permits computation of asymptotic significance points. These are similar to those obtained by Monte Carlo procedures and presented by Stephens in Chapter 4 of the present volume

The reader is referred to Chapter 4 for further discussion of EDF tests.

### 9.3.3 Moment Tests and Related Tests

Chapter 7, Section 7.2 discusses moment tests as they apply to the normal distribution. The modern theory of tests for normality can be regarded as having been initiated by Karl Pearson (1895), who recognized that deviations from normality could be characterized by the standard third and fourth moments of a distribution. To be more explicit, as previously discussed in Section 9.2.2, the normal distribution with density given by (9.2) has as its standardized third and fourth moments, respectively,

$$\sqrt{\beta_1} = \frac{E(X - \mu)^3}{\sigma^3} = 0 \tag{9.14}$$

and

$$\beta_2 = \frac{E(X - \mu)^4}{\sigma^4} = 3$$

The third standardized moment $\sqrt{\beta_1}$ characterizes the skewness of a distribution. If a distribution is symmetric about its mean $\mu$, as is the normal distribution, $\sqrt{\beta_1} = 0$. Values of $\sqrt{\beta_1} \neq 0$ indicate skewness and so nonnormality. The fourth standardized moments $\beta_2$ characterize the kurtosis or peakedness of a distribution. For the normal distribution, $\beta_2 = 3$. Values of $\beta_2 \neq 3$ indicate nonnormality. $\beta_2$ is also useful as an indicator of tail thickness. For the normal $\beta_2 = 3$. Values of $\beta_2 > 3$ indicate distributions with "thicker" than normal tails, and values of $\beta_2 < 3$ indicate distributions with "thinner" than normal tails.

Pearson suggested that in the sample, the standardized third and fourth moments given by

$$\sqrt{b_1} = m_3/m_2^{\frac{3}{2}} \tag{9.16}$$

and

$$b_2 = m_4/m_2^2 \tag{9.17}$$

where

$$m_k = \Sigma(X - \bar{X})^k/n, \quad k > 1 \tag{9.18}$$

and

$$\bar{X} = \Sigma X/n \tag{9.19}$$

could be used to judge departures from normality. He found the first approximation (i.e., to $n^{-1}$) to the variances and covariances of $\sqrt{b_1}$ and $b_2$ for samples drawn at random from any population, and assuming that $\sqrt{b_1}$ and $b_2$ were distributed jointly with bivariate normal probability, constructed equal probability ellipses. From these approximate assessments could be made if the sample deviated too greatly from normality. For situations where $\sqrt{b_1}$ and $b_2$ deviated substantially from expectation under normality, K. Pearson developed his elaborate system of Pearson curves. These could be used as possible alternative distributions for the populations under investigation. (See Elderton and Johnson (1969) for a full discussion of Pearson curves. Chapter 7 also makes extensive use of them.)

A number of investigators have concerned themselves with obtaining correct significance points for $\sqrt{b_1}$ and $b_2$. The reader is referred to Chapter 7, Section 7.2 for a brief review of the history. In the following we present the present state of the field.

### 9.3.3.1 Third Standardized Moment $\sqrt{b_1}$

The $\sqrt{b_1}$ test can be applied for all sample sizes $n \geq 5$.

#### 9.3.3.1.1 Monte Carlo points for n = 5 to 35

D'Agostino and Tietjen (1973) presented simulation probability points applicable for $n = 5$ to 35 and valid for two sided tests (i.e., $H_1$: Nonnormality with $\sqrt{\beta_1} \neq 0$) for levels of significant $\alpha = 0.002, 0.01, 0.02, 0.05, 0.10,$ and 0.20 and for one sided test (i.e., $H_1 : \sqrt{\beta_1} > 0$ or $H_1 : \sqrt{\beta_1} < 0$) for levels of

**TABLE 9.2** Probability Points of $\sqrt{b_1}$ for n = 5 to 35 (Monte Carlo Points)

| | Two-sided significance levels | | | | | |
|---|---|---|---|---|---|---|
| | 0.20 | 0.10 | 0.05 | 0.02 | 0.01 | 0.002 |
| | One-sided significance levels | | | | | |
| n | 0.10 | 0.05 | 0.025 | 0.01 | 0.005 | 0.001 |
| 5 | 0.819 | 1.058 | 1.212 | 1.342 | 1.396 | 1.466 |
| 6 | 0.805 | 1.034 | 1.238 | 1.415 | 1.498 | 1.642 |
| 7 | 0.787 | 1.008 | 1.215 | 1.432 | 1.576 | 1.800 |
| 8 | 0.760 | 0.991 | 1.202 | 1.455 | 1.601 | 1.873 |
| 9 | 0.752 | 0.977 | 1.189 | 1.408 | 1.577 | 1.866 |
| 10 | 0.722 | 0.950 | 1.157 | 1.397 | 1.565 | 1.887 |
| 11 | 0.715 | 0.929 | 1.129 | 1.376 | 1.540 | 1.924 |
| 13 | 0.688 | 0.902 | 1.099 | 1.312 | 1.441 | 1.783 |
| 15 | 0.648 | 0.862 | 1.048 | 1.275 | 1.462 | 1.778 |
| 17 | 0.629 | 0.820 | 1.009 | 1.188 | 1.358 | 1.705 |
| 20 | 0.593 | 0.777 | 0.951 | 1.152 | 1.303 | 1.614 |
| 23 | 0.562 | 0.743 | 0.900 | 1.119 | 1.276 | 1.555 |
| 25 | 0.543 | 0.714 | 0.876 | 1.073 | 1.218 | 1.468 |
| 30 | 0.510 | 0.664 | 0.804 | 0.985 | 1.114 | 1.410 |
| 35 | 0.474 | 0.624 | 0.762 | 0.932 | 1.043 | 1.332 |

significant $\alpha = 0.001, 0.005, 0.01, 0.025, 0.05$, and $0.10$. These points are given here in Table 9.2. Mulholland (1977) gives good approximations for n = 4 to 25.

### 9.3.3.1.2 $S_U$ approximation

D'Agostino (1970) further showed that the null distribution of $\sqrt{b_1}$ can be well approximated by a Johnson $S_U$ curve. The approximation is given as follows:

(1)  Compute $\sqrt{b_1}$ from the sample data.
(2)  Compute

$$Y = \sqrt{b_1} \left\{ \frac{(n + 1)(n + 3)}{6(n - 2)} \right\}^{\frac{1}{2}} \qquad (9.20)$$

$$\beta_2 = \frac{3(n^2 + 27n - 70)(n + 1)(n + 3)}{(n - 2)(n + 5)(n + 7)(n + 9)} \qquad (9.21)$$

$$W^2 = -1 + \left\{ 2(\beta_2 - 1) \right\}^{\frac{1}{2}} \qquad (9.22)$$

$$\delta = 1/\sqrt{\log W} \qquad (9.23)$$

$$\alpha = \left\{ 2/(W^2 - 1) \right\}^{\frac{1}{2}} \qquad (9.24)$$

(3)  Compute

$$Z = \delta \log[Y/\alpha + \left\{ (Y/\alpha)^2 + 1 \right\}^{\frac{1}{2}}] \qquad (9.25)$$

Z of (9.25) is approximately a standard normal variable with mean zero and variance unity. Once Z of (9.25) is computed, rejection or acceptance is decided by reference to any table of the standard normal distribution (such as given in the Appendix). This transformation is applicable for any sample size $n \geq 8$. Further with it both one-sided and two-sided tests with any desired levels of significance can be performed. For example, for a two-sided test with a $0.05$ level of significance reject if $|Z| \geq 1.96$.

Table 9.3 contains critical values of $\sqrt{b_1}$ computed from this $S_U$ approximation for $n \geq 36$.

### 9.3.3.1.3 t approximation

D'Agostino and Tietjen (1973) investigated a t approximation to the null distribution $\sqrt{b_1}$. It also requires $n \geq 8$ and appears to be as good as the $S_U$ approximation. It is given as follows:
Compute

$$T = \sqrt{b_1} \left( \frac{\nu}{\nu - 2} \right)^{\frac{1}{2}} \Big/ \sigma(\sqrt{b_1}) \qquad (9.26)$$

where

**TABLE 9.3** Probability Points of $\sqrt{b_1}$ for $n \geq 36$ (S$_U$ Approximation Points)

| | Two-sided significance levels | | | | |
|---|---|---|---|---|---|
| | 0.20 | 0.10 | 0.05 | 0.02 | 0.01 |
| | One-sided significance levels | | | | |
| n | 0.10 | 0.05 | 0.025 | 0.01 | 0.005 |
| 36 | 0.469 | 0.614 | 0.747 | 0.912 | 1.032 |
| 37 | 0.464 | 0.607 | 0.738 | 0.901 | 1.019 |
| 38 | 0.459 | 0.600 | 0.730 | 0.891 | 1.007 |
| 39 | 0.454 | 0.594 | 0.722 | 0.881 | 0.996 |
| 40 | 0.449 | 0.588 | 0.714 | 0.871 | 0.985 |
| 41 | 0.445 | 0.581 | 0.707 | 0.861 | 0.974 |
| 42 | 0.440 | 0.576 | 0.699 | 0.852 | 0.963 |
| 43 | 0.436 | 0.570 | 0.692 | 0.843 | 0.953 |
| 44 | 0.432 | 0.564 | 0.685 | 0.835 | 0.943 |
| 45 | 0.428 | 0.559 | 0.678 | 0.826 | 0.934 |
| 46 | 0.424 | 0.553 | 0.672 | 0.818 | 0.924 |
| 47 | 0.420 | 0.548 | 0.666 | 0.810 | 0.915 |
| 48 | 0.416 | 0.543 | 0.659 | 0.803 | 0.906 |
| 49 | 0.412 | 0.538 | 0.653 | 0.795 | 0.898 |
| 50 | 0.409 | 0.534 | 0.648 | 0.788 | 0.889 |
| 51 | 0.405 | 0.529 | 0.642 | 0.781 | 0.881 |
| 52 | 0.402 | 0.525 | 0.636 | 0.774 | 0.873 |
| 53 | 0.399 | 0.520 | 0.631 | 0.767 | 0.865 |
| 54 | 0.395 | 0.516 | 0.626 | 0.760 | 0.858 |
| 55 | 0.392 | 0.512 | 0.620 | 0.754 | 0.850 |
| 56 | 0.389 | 0.508 | 0.615 | 0.748 | 0.843 |
| 57 | 0.386 | 0.504 | 0.610 | 0.742 | 0.836 |
| 58 | 0.383 | 0.500 | 0.606 | 0.736 | 0.829 |
| 59 | 0.380 | 0.496 | 0.601 | 0.730 | 0.822 |
| 60 | 0.378 | 0.492 | 0.596 | 0.724 | 0.816 |
| 61 | 0.375 | 0.489 | 0.592 | 0.718 | 0.809 |
| 62 | 0.372 | 0.485 | 0.588 | 0.713 | 0.803 |
| 63 | 0.370 | 0.482 | 0.583 | 0.708 | 0.797 |
| 64 | 0.367 | 0.478 | 0.579 | 0.702 | 0.791 |
| 65 | 0.365 | 0.475 | 0.575 | 0.697 | 0.785 |
| 66 | 0.362 | 0.472 | 0.571 | 0.692 | 0.779 |
| 67 | 0.360 | 0.468 | 0.567 | 0.687 | 0.774 |
| 68 | 0.357 | 0.465 | 0.563 | 0.683 | 0.768 |
| 69 | 0.355 | 0.462 | 0.559 | 0.678 | 0.763 |
| 70 | 0.353 | 0.459 | 0.556 | 0.673 | 0.758 |

TABLE 9.3 (continued)

| | Two-sided significance levels | | | | |
|---|---|---|---|---|---|
| | 0.20 | 0.10 | 0.05 | 0.02 | 0.01 |
| | One-sided significance levels | | | | |
| n | 0.10 | 0.05 | 0.025 | 0.01 | 0.005 |
| 71 | 0.351 | 0.456 | 0.552 | 0.669 | 0.752 |
| 72 | 0.348 | 0.453 | 0.548 | 0.664 | 0.747 |
| 73 | 0.346 | 0.451 | 0.545 | 0.660 | 0.742 |
| 74 | 0.344 | 0.448 | 0.541 | 0.656 | 0.737 |
| 75 | 0.342 | 0.445 | 0.538 | 0.651 | 0.733 |
| 76 | 0.340 | 0.442 | 0.535 | 0.647 | 0.728 |
| 77 | 0.338 | 0.440 | 0.532 | 0.643 | 0.723 |
| 78 | 0.336 | 0.437 | 0.528 | 0.639 | 0.719 |
| 79 | 0.334 | 0.435 | 0.525 | 0.635 | 0.714 |
| 80 | 0.332 | 0.432 | 0.522 | 0.632 | 0.710 |
| 81 | 0.330 | 0.430 | 0.519 | 0.628 | 0.706 |
| 82 | 0.329 | 0.427 | 0.516 | 0.624 | 0.701 |
| 83 | 0.327 | 0.425 | 0.513 | 0.621 | 0.697 |
| 84 | 0.325 | 0.422 | 0.510 | 0.617 | 0.693 |
| 85 | 0.323 | 0.420 | 0.507 | 0.613 | 0.689 |
| 86 | 0.322 | 0.418 | 0.505 | 0.610 | 0.685 |
| 87 | 0.320 | 0.416 | 0.502 | 0.607 | 0.681 |
| 88 | 0.318 | 0.413 | 0.499 | 0.603 | 0.677 |
| 89 | 0.317 | 0.411 | 0.497 | 0.600 | 0.674 |
| 90 | 0.315 | 0.409 | 0.494 | 0.597 | 0.670 |
| 91 | 0.313 | 0.407 | 0.491 | 0.594 | 0.666 |
| 92 | 0.312 | 0.405 | 0.489 | 0.590 | 0.663 |
| 93 | 0.310 | 0.403 | 0.486 | 0.587 | 0.659 |
| 94 | 0.309 | 0.401 | 0.484 | 0.584 | 0.656 |
| 95 | 0.307 | 0.399 | 0.481 | 0.581 | 0.652 |
| 96 | 0.306 | 0.397 | 0.479 | 0.578 | 0.649 |
| 97 | 0.304 | 0.395 | 0.477 | 0.575 | 0.646 |
| 98 | 0.303 | 0.393 | 0.474 | 0.573 | 0.642 |
| 99 | 0.302 | 0.391 | 0.472 | 0.570 | 0.639 |
| 100 | 0.300 | 0.390 | 0.470 | 0.567 | 0.636 |
| 102 | 0.297 | 0.386 | 0.465 | 0.562 | 0.630 |
| 104 | 0.295 | 0.383 | 0.461 | 0.556 | 0.624 |
| 106 | 0.292 | 0.379 | 0.457 | 0.551 | 0.618 |
| 108 | 0.290 | 0.376 | 0.453 | 0.546 | 0.612 |
| 110 | 0.287 | 0.373 | 0.449 | 0.541 | 0.607 |

(continued)

TABLE 9.3 (continued)

|     | Two-sided significance levels | | | | |
| --- | --- | --- | --- | --- | --- |
|     | 0.20 | 0.10 | 0.05 | 0.02 | 0.01 |
|     | One-sided significance levels | | | | |
| n | 0.10 | 0.05 | 0.025 | 0.01 | 0.005 |
| 112 | 0.285 | 0.369 | 0.445 | 0.536 | 0.601 |
| 114 | 0.283 | 0.366 | 0.441 | 0.532 | 0.596 |
| 116 | 0.280 | 0.363 | 0.438 | 0.527 | 0.591 |
| 118 | 0.278 | 0.360 | 0.434 | 0.523 | 0.586 |
| 120 | 0.276 | 0.358 | 0.431 | 0.519 | 0.581 |
| 122 | 0.274 | 0.355 | 0.427 | 0.514 | 0.576 |
| 124 | 0.272 | 0.352 | 0.424 | 0.510 | 0.571 |
| 126 | 0.270 | 0.349 | 0.421 | 0.506 | 0.567 |
| 128 | 0.268 | 0.347 | 0.417 | 0.502 | 0.562 |
| 130 | 0.266 | 0.344 | 0.414 | 0.499 | 0.558 |
| 132 | 0.264 | 0.342 | 0.411 | 0.495 | 0.554 |
| 134 | 0.262 | 0.339 | 0.408 | 0.491 | 0.550 |
| 136 | 0.260 | 0.337 | 0.405 | 0.488 | 0.546 |
| 138 | 0.258 | 0.335 | 0.403 | 0.484 | 0.542 |
| 140 | 0.257 | 0.332 | 0.400 | 0.481 | 0.538 |
| 142 | 0.255 | 0.330 | 0.397 | 0.477 | 0.534 |
| 144 | 0.253 | 0.328 | 0.394 | 0.474 | 0.530 |
| 146 | 0.252 | 0.326 | 0.392 | 0.471 | 0.526 |
| 148 | 0.250 | 0.324 | 0.389 | 0.468 | 0.523 |
| 150 | 0.249 | 0.322 | 0.387 | 0.465 | 0.519 |
| 155 | 0.245 | 0.317 | 0.381 | 0.457 | 0.511 |
| 160 | 0.241 | 0.312 | 0.375 | 0.450 | 0.503 |
| 165 | 0.238 | 0.307 | 0.369 | 0.443 | 0.495 |
| 170 | 0.234 | 0.303 | 0.364 | 0.437 | 0.488 |
| 175 | 0.231 | 0.299 | 0.359 | 0.430 | 0.481 |
| 180 | 0.228 | 0.295 | 0.354 | 0.425 | 0.474 |
| 185 | 0.225 | 0.291 | 0.349 | 0.419 | 0.467 |
| 190 | 0.222 | 0.287 | 0.345 | 0.413 | 0.461 |
| 195 | 0.219 | 0.284 | 0.340 | 0.408 | 0.455 |
| 200 | 0.217 | 0.280 | 0.336 | 0.403 | 0.449 |
| 210 | 0.212 | 0.274 | 0.328 | 0.393 | 0.439 |
| 220 | 0.207 | 0.267 | 0.321 | 0.384 | 0.428 |
| 230 | 0.203 | 0.262 | 0.314 | 0.376 | 0.419 |
| 240 | 0.199 | 0.256 | 0.307 | 0.368 | 0.410 |
| 250 | 0.195 | 0.251 | 0.301 | 0.361 | 0.402 |

Table 9.3 (continued)

| n | Two-sided significance levels | | | | |
| | 0.20 | 0.10 | 0.05 | 0.02 | 0.01 |
| | One-sided significance levels | | | | |
| | 0.10 | 0.05 | 0.025 | 0.01 | 0.005 |
|---|---|---|---|---|---|
| 275 | 0.186 | 0.240 | 0.287 | 0.344 | 0.383 |
| 300 | 0.178 | 0.230 | 0.275 | 0.329 | 0.366 |
| 325 | 0.172 | 0.221 | 0.265 | 0.316 | 0.352 |
| 350 | 0.165 | 0.213 | 0.255 | 0.305 | 0.339 |
| 375 | 0.160 | 0.206 | 0.247 | 0.294 | 0.327 |
| 400 | 0.155 | 0.200 | 0.239 | 0.285 | 0.317 |
| 425 | 0.151 | 0.194 | 0.232 | 0.277 | 0.307 |
| 450 | 0.146 | 0.188 | 0.225 | 0.269 | 0.299 |
| 475 | 0.143 | 0.184 | 0.219 | 0.262 | 0.291 |
| 500 | 0.139 | 0.179 | 0.214 | 0.255 | 0.283 |
| 550 | 0.133 | 0.171 | 0.204 | 0.243 | 0.270 |
| 600 | 0.127 | 0.164 | 0.195 | 0.233 | 0.258 |
| 650 | 0.122 | 0.157 | 0.188 | 0.224 | 0.248 |
| 700 | 0.118 | 0.152 | 0.181 | 0.216 | 0.239 |
| 750 | 0.114 | 0.146 | 0.175 | 0.208 | 0.231 |
| 800 | 0.110 | 0.142 | 0.169 | 0.202 | 0.224 |
| 850 | 0.107 | 0.138 | 0.164 | 0.196 | 0.217 |
| 900 | 0.104 | 0.134 | 0.160 | 0.190 | 0.211 |
| 950 | 0.101 | 0.130 | 0.155 | 0.185 | 0.205 |
| 1000 | 0.099 | 0.127 | 0.152 | 0.180 | 0.200 |
| 1200 | 0.090 | 0.116 | 0.138 | 0.165 | 0.182 |
| 1400 | 0.084 | 0.107 | 0.128 | 0.152 | 0.169 |
| 1600 | 0.078 | 0.101 | 0.120 | 0.143 | 0.158 |
| 1800 | 0.074 | 0.095 | 0.113 | 0.134 | 0.149 |
| 2000 | 0.070 | 0.090 | 0.107 | 0.127 | 0.141 |
| 2500 | 0.063 | 0.080 | 0.096 | 0.114 | 0.126 |
| 3000 | 0.057 | 0.073 | 0.088 | 0.104 | 0.115 |
| 3500 | 0.053 | 0.068 | 0.081 | 0.096 | 0.107 |
| 4000 | 0.050 | 0.064 | 0.076 | 0.090 | 0.100 |
| 4500 | 0.047 | 0.060 | 0.072 | 0.085 | 0.094 |
| 5000 | 0.044 | 0.057 | 0.068 | 0.081 | 0.089 |
| 10000 | 0.031 | 0.040 | 0.048 | 0.057 | 0.063 |

$$\nu = \frac{4\beta_2 - 6}{\beta_2 - 3} \tag{9.27}$$

$\beta_2$ is given by (9.21)

and

$$\sigma(\sqrt{b_1}) = \left[ \frac{6(n - 2)}{(n + 1)(n + 3)} \right]^{\frac{1}{2}} \tag{9.28}$$

Under the null hypothesis T of (9.26) is approximately a t variable with $\nu$ given by (9.27) degree of freedom. Interpolation in standard t appears adequate for judging significance of test results.

### 9.3.3.1.4 Normal approximation

The normal approximation given by

$$\sqrt{b_1} \left[ \frac{(n + 1)(n + 3)}{6(n - 2)} \right]^{\frac{1}{2}} \tag{9.29}$$

appears to be valid for $n \geq 150$.

E 9.3.3.1.5 Numerical examples of $\sqrt{b_1}$ test  Table 9.4 contains the first ten observations from five data sets given in the Appendix. These are the uniform (UNI), two Johnson Unbounded distributions ((SU(0,2) and SU(1,2)), the negative exponential (EXP), and the normal (NOR). The population values of $\sqrt{\beta_1}$ and $\beta_2$ are included in the table. The $\sqrt{b_1}$ statistic has been computed for all five data sets. Employing a two tailed test the 0.05 critical $\sqrt{b_1}$ obtained from Table 9.2 is 1.157. Rejection of the null hypothesis only occurs for the negative exponential data set (EXP).

While we anticipate computation of $\sqrt{b_1}$ will mainly be done via a computer, a computational formula may still be useful. One such formula is

$$\sqrt{b} = \frac{\sqrt{n} \Sigma (X - \bar{X})^3}{[\Sigma (X - \bar{X})^2]^{\frac{3}{2}}}$$

$$= \frac{n^2 \Sigma X^3 - 3n \Sigma X \Sigma X^2 + 2(\Sigma X)^3}{[n \Sigma X^2 - (\Sigma X)^2]^{\frac{3}{2}}} \tag{9.30}$$

### 9.3.3.1.6 Recommendations for use of $\sqrt{b_1}$

The $S_U$ approximation given by (9.20) to (9.25) is adequate for $n \geq 8$. For computerization of $\sqrt{b_1}$, we recommend its use. For $n = 5, 6,$ and $7$, Table 9.2 must be used. For table look-ups we recommend Table 9.2 for $n = 5$ to 35 and Table 9.3 for $n \geq 36$. Also for $n \geq 150$ the simple normal approximation of (9.29) can be used.

TABLE 9.4 Numerical Examples for Moment and Related Tests

| | Uniform | Johnson Unbounded (0,2) | Johnson Unbounded (1,2) | Negative Exponential, Mean 5 | Normal: $\mu = 100$, $\sigma = 10$ |
|---|---|---|---|---|---|
| | UNI | SU(0,2) | SU(1,2) | EXP | NOR |
| Data[a] | 8.10 | 0.10 | −0.41 | 8.15 | 92.55 |
| | 2.06 | −0.31 | −0.91 | 4.69 | 96.20 |
| | 1.60 | −0.09 | −0.63 | 2.17 | 84.27 |
| | 8.87 | −0.58 | −1.25 | 0.37 | 90.87 |
| | 9.90 | 1.15 | 0.50 | 16.69 | 101.58 |
| | 6.58 | 0.17 | −0.34 | 0.06 | 106.82 |
| | 8.68 | −1.39 | −2.46 | 6.48 | 98.70 |
| | 7.31 | −0.14 | −0.68 | 2.63 | 113.75 |
| | 2.85 | −0.31 | −0.97 | 0.44 | 98.98 |
| | 6.09 | 0.68 | 0.13 | 0.89 | 100.42 |
| $\sqrt{\beta_1}$ | 0 | 0 | −0.87 | 2 | 0 |
| $\beta_2$ | 1.80 | 4.51 | 5.59 | 9 | 3 |
| $\bar{x}$ | 6.204 | −0.072 | −0.702 | 4.257 | 98.414 |
| $s$ | 3.009 | 0.688 | 0.807 | 5.169 | 8.277 |
| $\sqrt{b_1}$ | −0.49 | −0.06 | −0.72 | 1.49[c] | 0.16 |
| $b_2$ | 1.75 | 3.11 | 3.58 | 4.33[b] | 2.76 |
| $a$ | 0.86 | 0.74 | 0.73 | 0.77 | 0.76 |
| $u$ | 2.76 | 3.69 | 3.67 | 3.22 | 3.56 |

[†]Data sets are first ten observations for UNI, SU(0,2), SU(1,2), EXP and NOR data sets of Appendix.

[a]Note that

$$\sqrt{b_1} = \frac{m_3}{m_2^{3/2}}; \quad b_2 = \frac{m_4}{m_2^2}; \quad m_k = \frac{\Sigma (X - \bar{X})^k}{n}$$

$$a = \text{Geary's statistic} = \frac{\Sigma |X - \bar{X}|/n}{m_2^{\frac{1}{2}}}$$

$$u = \text{David et al. (1954) statistic} = (\text{sample range})/s$$

$$s = [n/(n - 1)]^{\frac{1}{2}} m_2^{\frac{1}{2}}$$

[b]Reject null hypothesis of normality at 0.02 level of significance.
[c]Reject null hypothesis of normality at 0.10 level of significance.

TABLE 9.5 Probability Points of $b_2$ for n = 7 to 200

Percentiles

| Sample size n | .5 | 1 | 2 | 2.5 | 5 | 10 | 15 | 20 | 80 | 85 | 90 | 95 | 97.5 | 98 | 99 | 99.5 |
|---|---|---|---|---|---|---|---|---|---|---|---|---|---|---|---|---|
| **Part 1 (n = 7 to 20)** | | | | | | | | | | | | | | | | |
| 7 | | 1.25 | 1.30 | 1.34 | 1.41 | 1.53 | | 1.70 | 2.78 | | 3.20 | 3.55 | 3.85 | 3.93 | 4.23 | |
| 8 | | 1.31 | 1.37 | 1.40 | 1.46 | 1.58 | | 1.75 | 2.84 | | 3.31 | 3.70 | 4.09 | 4.20 | 4.53 | |
| 9 | | 1.35 | 1.42 | 1.45 | 1.53 | 1.63 | | 1.80 | 2.98 | | 3.43 | 3.86 | 4.28 | 4.41 | 4.82 | |
| 10 | | 1.39 | 1.45 | 1.49 | 1.56 | 1.68 | | 1.85 | 3.01 | | 3.53 | 3.95 | 4.40 | 4.55 | 5.00 | |
| 12 | | 1.46 | 1.52 | 1.56 | 1.64 | 1.76 | | 1.93 | 3.06 | | 3.55 | 4.05 | 4.56 | 4.73 | 5.20 | |
| 15 | | 1.55 | 1.61 | 1.64 | 1.72 | 1.84 | | 2.01 | 3.13 | | 3.62 | 4.13 | 4.66 | 4.85 | 5.30 | |
| 20 | | 1.64 | 1.71 | 1.73 | 1.83 | 1.95 | | 2.12 | 3.20 | | 3.68 | 4.18 | 4.68 | 4.87 | 5.38 | |
| **Part 2 (n = 20 to 100)** | | | | | | | | | | | | | | | | |
| 20 | 1.58 | 1.64 | | 1.73 | 1.83 | 1.95 | 2.04 | 2.12 | 3.20 | 3.40 | 3.68 | 4.18 | 4.68 | | 5.38 | 5.91 |
| 25 | 1.66 | 1.72 | | 1.82 | 1.92 | 2.03 | 2.12 | 2.20 | 3.24 | 3.43 | 3.69 | 4.15 | 4.63 | | 5.29 | 5.81 |
| 30 | 1.73 | 1.79 | | 1.89 | 1.98 | 2.10 | 2.19 | 2.26 | 3.26 | 3.44 | 3.69 | 4.12 | 4.57 | | 5.20 | 5.69 |
| 35 | 1.78 | 1.84 | | 1.94 | 2.03 | 2.15 | 2.24 | 2.31 | 3.28 | 3.45 | 3.68 | 4.09 | 4.51 | | 5.12 | 5.58 |
| 40 | 1.83 | 1.89 | | 1.99 | 2.07 | 2.19 | 2.28 | 2.35 | 3.29 | 3.45 | 3.66 | 4.06 | 4.46 | | 5.04 | 5.48 |
| 45 | 1.87 | 1.93 | | 2.03 | 2.11 | 2.23 | 2.31 | 2.38 | 3.29 | 3.44 | 3.65 | 4.02 | 4.41 | | 4.96 | 5.38 |
| 50 | 1.91 | 1.96 | | 2.06 | 2.15 | 2.26 | 2.34 | 2.41 | 3.29 | 3.44 | 3.63 | 4.00 | 4.36 | | 4.88 | 5.28 |
| 55 | 1.94 | 2.00 | | 2.09 | 2.18 | 2.29 | 2.37 | 2.44 | 3.29 | 3.43 | 3.62 | 3.97 | 4.32 | | 4.81 | 5.19 |

| n | | | | | | | | | | | | | | |
|---|---|---|---|---|---|---|---|---|---|---|---|---|---|---|
| 60 | 1.97 | 2.03 | 2.12 | 2.21 | 2.32 | 2.39 | 2.46 | 3.29 | 3.43 | 3.60 | 3.94 | 4.28 | 4.75 | 5.11 |
| 65 | 2.00 | 2.05 | 2.15 | 2.23 | 2.34 | 2.41 | 2.48 | 3.28 | 3.42 | 3.59 | 3.91 | 4.24 | 4.69 | 5.03 |
| 70 | 2.02 | 2.07 | 2.17 | 2.25 | 2.36 | 2.43 | 2.50 | 3.28 | 3.41 | 3.58 | 3.89 | 4.20 | 4.64 | 4.97 |
| 75 | 2.05 | 2.10 | 2.19 | 2.27 | 2.38 | 2.45 | 2.51 | 3.28 | 3.41 | 3.57 | 3.87 | 4.17 | 4.59 | 4.90 |
| 80 | 2.07 | 2.12 | 2.21 | 2.29 | 2.39 | 2.46 | 2.53 | 3.27 | 3.40 | 3.56 | 3.85 | 4.14 | 4.54 | 4.84 |
| 85 | 2.08 | 2.14 | 2.22 | 2.31 | 2.41 | 2.48 | 2.54 | 3.27 | 3.39 | 3.55 | 3.83 | 4.11 | 4.50 | 4.79 |
| 90 | 2.10 | 2.16 | 2.24 | 2.32 | 2.43 | 2.49 | 2.55 | 3.27 | 3.39 | 3.54 | 3.81 | 4.08 | 4.46 | 4.74 |
| 95 | 2.11 | 2.17 | 2.26 | 2.34 | 2.44 | 2.50 | 2.56 | 3.27 | 3.38 | 3.53 | 3.80 | 4.05 | 4.43 | 4.70 |
| 100 | 2.13 | 2.19 | 2.27 | 2.35 | 2.45 | 2.52 | 2.57 | 3.26 | 3.37 | 3.52 | 3.78 | 4.03 | 4.39 | 4.66 |

Part 3 (n = 100 to 200)

| n | | | | | | | | | | | | | | |
|---|---|---|---|---|---|---|---|---|---|---|---|---|---|---|
| 100 | 2.13 | 2.19 | 2.27 | 2.35 | 2.45 | 2.52 | 2.57 | 3.26 | 3.37 | 3.52 | 3.78 | 4.03 | 4.39 | 4.66 |
| 110 | 2.15 | 2.22 | 2.30 | 2.37 | 2.47 | 2.53 | 2.59 | 3.26 | 3.37 | 3.51 | 3.75 | 3.99 | 4.32 | 4.58 |
| 120 | 2.18 | 2.24 | 2.32 | 2.39 | 2.49 | 2.55 | 2.61 | 3.25 | 3.35 | 3.49 | 3.72 | 3.95 | 4.26 | 4.52 |
| 130 | 2.20 | 2.26 | 2.34 | 2.41 | 2.51 | 2.57 | 2.63 | 3.25 | 3.34 | 3.47 | 3.70 | 3.92 | 4.21 | 4.46 |
| 140 | 2.22 | 2.28 | 2.36 | 2.43 | 2.52 | 2.58 | 2.64 | 3.25 | 3.33 | 3.46 | 3.67 | 3.89 | 4.17 | 4.41 |
| 150 | 2.24 | 2.30 | 2.37 | 2.45 | 2.54 | 2.60 | 2.65 | 3.24 | 3.33 | 3.45 | 3.65 | 3.86 | 4.13 | 4.36 |
| 160 | 2.26 | 2.32 | 2.39 | 2.46 | 2.55 | 2.61 | 2.66 | 3.24 | 3.32 | 3.44 | 3.63 | 3.83 | 4.09 | 4.31 |
| 170 | 2.28 | 2.33 | 2.40 | 2.48 | 2.56 | 2.62 | 2.67 | 3.23 | 3.32 | 3.43 | 3.62 | 3.81 | 4.06 | 4.27 |
| 180 | 2.29 | 2.35 | 2.41 | 2.49 | 2.57 | 2.63 | 2.68 | 3.23 | 3.31 | 3.42 | 3.60 | 3.79 | 4.03 | 4.23 |
| 190 | 2.31 | 2.36 | 2.43 | 2.50 | 2.58 | 2.64 | 2.69 | 3.22 | 3.30 | 3.41 | 3.58 | 3.77 | 4.00 | 4.19 |
| 200 | 2.32 | 2.37 | 2.44 | 2.51 | 2.59 | 2.65 | 2.70 | 3.22 | 3.30 | 3.40 | 3.57 | 3.75 | 3.98 | 4.16 |

Adapted from D'Agostino and Tietjen (1971) and D'Agostino and Pearson (1973), with permission of the Biometrika Trustees.

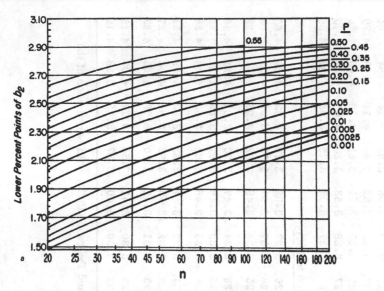

FIGURE 9.2a  Empirical cumulative distribution $b_2$ ($P \leq 0.55$).

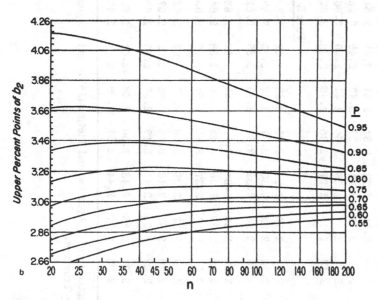

FIGURE 9.2b  Empirical cumulative distribution of $b_2$ ($0.55 \leq P \leq 0.95$).

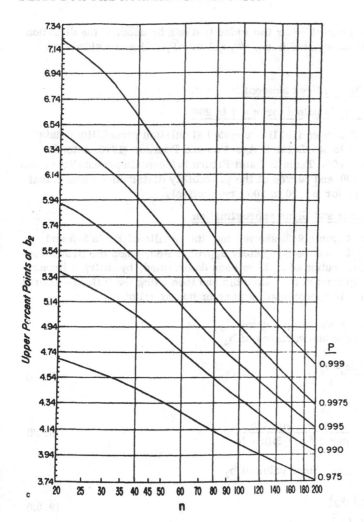

**FIGURE 9.2c** Empirical cumulative distribution of $b_2$ ($P \geq 0.975$).

Finally, both a one sided or two sided test can be used. If the direction of the skewness is anticipated (i.e., $\sqrt{\beta_1} > 0$ or $\sqrt{\beta_1} < 0$) a one sided test should be used.

### 9.3.3.2 Fourth Standardized Moment $b_2$

### 9.3.3.2.1 Monte Carlo points for n = 7 to 200

D'Agostino and Tietjen (1971) presented simulation probability points applicable for n = 7 to 50. Later D'Agostino and Pearson (1973) extended these results to n = 200. Table 9.5 and Figure 9.2 contain probability points valid for n = 7 to 200 and curves of the probability distributions (empirical probability integral) for n = 20 to 200, respectively.

### 9.3.3.2.2 Anscombe and Glynn approximation

Anscombe and Glynn (1983) showed that the results of Table 9.5 and Figure 9.2 for $n \geq 20$ can be adequately approximated, when the first three moments of the distribution of $b_2$ have been determined, by fitting a linear function of the reciprocal of a $\chi^2$ variable and then using the Wilson-Hilferty transformation. Their approximation is computed as follows:

(1) Compute $b_2$ from the sample data.
(2) Compute the mean and variance of $b_2$

$$E(b_2) = \frac{3(n - 1)}{n + 1} \tag{9.31}$$

and

$$\text{var}(b_2) = \frac{24n(n - 2)(n - 3)}{(n + 1)^2(n + 3)(n + 5)} \tag{9.32}$$

(3) Compute the standardized value of $b_2$

$$x = \frac{b_2 - E(b_2)}{\sqrt{\text{var}(b_2)}} \tag{9.33}$$

(4) Compute the third standardized moment of $b_2$

$$\sqrt{\beta_1(b_2)} = \frac{6(n^2 - 5n + 2)}{(n + 7)(n + 9)} \sqrt{\frac{6(n + 3)(n + 5)}{n(n - 2)(n - 3)}} \tag{9.34}$$

(5) Compute

$$A = 6 + \frac{8}{\sqrt{\beta_1(b_2)}} \left[ \frac{2}{\sqrt{\beta_1(b_2)}} + \sqrt{\left\{ 1 + \frac{4}{\beta_1(b_2)} \right\}} \right] \tag{9.35}$$

(6)  Compute

$$Z = \left( \left(1 - \frac{2}{9A}\right) - \left[ \frac{1 - (2/A)}{1 + x\sqrt{\{2/(A-4)\}}} \right]^{\frac{1}{3}} \right) \Big/ \sqrt{\{2/(9A)\}} \qquad (9.36)$$

Z of (9.36) is approximately a standard normal variable with mean zero and variance unity.

The approximation given by (9.31) to (9.36) can be used to test directly null hypotheses concerning $b_2$ for two sided or one sided alternatives. For example, for testing at level of significance 0.05

$H_0$: Normality

versus the one sided composite alternative

$H_1$: Nonnormality with $\beta_2 > 3$

one would reject $H_0$ if Z of (9.36) exceeded 1.645. For

$H_1$: Nonnormality with $\beta_2 < 3$

one would reject $H_0$ if Z of (9.36) was smaller than -1.645.

### 9.3.3.2.3  Normal approximation

The normal approximation given by

$$\frac{b_2 - E(b_2)}{\sqrt{var(b_2)}} \qquad (9.37)$$

is valid only for extremely large n values (i.e., well over 1000). It should not be used.

### 9.3.3.2.4  Bowman and Shenton's $S_U$ approximation

Appendix 2 of Chapter 7 gives details of a computer program for finding an $S_U$ approximation to $b_2$. This can be used to yield tests for $n \geq 40$.

### E 9.3.3.2.5  Numerical examples of $b_2$ test  Table 9.4 contains numerical examples for the five data sets already used for $\sqrt{b_1}$. As before only the EXP data lead to rejection of a two tailed test. Here the level of significance is 0.10.

### 9.3.3.2.6  Recommendation for use of $b_2$

Table 9.5 or Figure 9.2 can be used for $7 \leq n \leq 200$. The Anscombe and Glynn or Bowman and Shenton approximations can be used for $n \geq 20$. Both require computations. The former requires less computations giving an explicit solution.

Again, as with $\sqrt{b_1}$, when knowledge is available concerning the alternative (i.e., $\beta_2 > 3$ or $\beta_2 < 3$) a one sided test should be used.

### 9.3.3.3  Omnibus Tests Based on Moments

The $\sqrt{b_1}$ test is excellent for detecting nonnormality due to skewness ($\sqrt{\beta_1} \neq 0$). The $b_2$ test is primarily directed to detecting nonnormality due to nonnormal kurtosis or nonnormal tail thickness ($\beta_2 \neq 3$). A number of investigators have worked on combining these tests to produce an omnibus test of normality.

#### 9.3.3.3.1  The R-test

The simplest omnibus test consists of performing the $\sqrt{b_1}$ test at level $\alpha_1$ and the $b_2$ test at level $\alpha_2$ and reject normality if either test leads to rejection. The overall level of significance $\alpha$ for these two tests combined would then be, by Bonferroni's inequality,

$$\alpha \leq \alpha_1 + \alpha_2 \tag{9.38}$$

Pearson, D'Agostino, and Bowman (1977) showed that if $\alpha_1 = \alpha_2 = 2\alpha^*$ a good approximation to the overall level of significance is

$$\alpha = 4(\alpha^* - (\alpha^*)^2) \tag{9.39}$$

(9.39) would hold exactly if $\sqrt{b_1}$ and $b_2$ were independent. They are uncorrelated but not independent and use of (9.39) to determine the overall level of significance produces a conservative test. Tables of corrected values are given in Pearson et al. (1966) for $n = 20$, 50, and $\alpha = 0.05$ and $0.10$.

In order to use this test one can determine $\alpha^*$ as

$$2\alpha^* = (1 - (1 - \alpha)^{\frac{1}{2}}) \tag{9.40}$$

where $\alpha$ is the desired overall level. Note $2\alpha^*$ is the level of the individual tests.

The term R test was given to the above omnibus procedure because it can be viewed as employing rectangular coordinates for rejection of normality.

#### 9.3.3.3.2  D'Agostino-Pearson chi square test

D'Agostino and Pearson (1973) suggested the statistic

$$K^2 = X^2(\sqrt{b_1}) + X^2(b_2) \tag{9.41}$$

as an omnibus test where $X(\sqrt{b_1})$ and $X(b_2)$ are standardized normal equivalent deviates and $K^2$ can be viewed as a chi square variable with two degrees of freedom. This test was developed assuming $\sqrt{b_1}$ and $b_2$ were independent. They are not. However, as Bowman and Shenton point out in Chapter 7, they are uncorrelated and nearly independent. So $K^2$ is approximately a chi square variable with two degrees of freedom. For $n \geq 100$ the chi square distribution approximation presents no problem.

The test statistic of (9.41) is trivial to employ given the above material. $X(\sqrt{b_1})$, the normal deviate for $\sqrt{b_1}$, can be found using the $S_U$ approximation of (9.20) to (9.25) and $X(b_2)$, the normal deviate for $b_2$, can be found using the Anscombe and Glynn approximations of (9.31) to (9.36).

### 9.3.3.3.3 Bowman-Shenton chi square test

In Chapter 7 Bowman and Shenton review their $K_S^2$ test which has the same format as (9.41) except their approximation to $X(b_2)$ involved use of Johnson $S_U$ curves (see also Bowman and Shenton (1975)). They also present in Figure 7.1 contours which allow for exact level of significance tests of $\alpha = 0.05$ and $0.10$ for the $K_S^2$ test. These contours are for $n = 25$ to $1000$.

### 9.3.3.3.4 Other omnibus tests

Bowman and Shenton (1975) also suggested

$$(\sqrt{b_1})^2/\sigma_1^2 + (b_2 - 3)^2/\sigma_2^2 \qquad (9.42)$$

where $\sigma_1^2 = 6/n$ and $\sigma_2^2 = 24/n$. Asymptotically (9.42) would be distributed as a chi square variable with two degrees of freedom if the null hypothesis of normality was true. Due to the slow convergence of $b_2$ to normality this test is not useful.

Cox and Hinkley (1974) suggested

$$\max \left( |\sqrt{b_1}| /\sigma_1, \ |b_2 - 3| /\sigma_2 \right) \qquad (9.43)$$

Another possibility is

$$- \log P(\sqrt{b_1}) - \log P(b_2) \qquad (9.44)$$

where $P(\sqrt{b_1})$ and $P(b_2)$ are the probability integral transformation of $\sqrt{b_1}$ and $b_2$. This statistic would be approximately a four degree chi square variable.

### 9.3.3.3.5 Recommendations for use of omnibus test

Figure 7.1 gives the omnibus test, $K_S^2$, for $n = 25$ to $1000$ and $\alpha = 0.05$ and $0.10$. It is preferred to other approximations. For other situations there is little to choose between $K^2$ of (9.41) and $K_S^2$ of Chapter 7. (The reader is referred to Chapter 7 for numerical examples.) Both the $K^2$ and $K_S^2$ tests can be programmed easily. See Section 9.3.3.3.2 for $K^2$.

The R test is not as powerful as $K^2$ or $K_s^2$ but requires in many cases only trivial interpolations in Tables 9.2, 9.3, and 9.5. As a quick test it has much to recommend it.

### 9.3.3.4 Related Tests

### 9.3.3.4.1 Geary's tests

A number of tests are related to moment tests and are of historical interest. Most noticeable are Geary's test involving the ratio of the mean deviation to the standard deviation

$$w = \frac{1}{\sqrt{n}} \frac{\Sigma |X|}{(\Sigma X^2)^{\frac{1}{2}}} \qquad (9.45)$$

and

$$a = \frac{\Sigma |X - \bar{X}|/n}{\sqrt{m_2}} \qquad (9.46)$$

(see Geary (1935)). Tables of a are published in Pearson and Hartley (1972). D'Agostino (1970) showed that

$$\frac{\sqrt{n}(a - .7979)}{.2123} \qquad (9.47)$$

can be considered as a standard normal variable with mean zero and variance unity for $n \geq 41$. Recently Gastwirth and Owen (1977) discussed optimal features of a.

Geary (1947) in one of the most distinguished papers on tests of normality considered tests of the form

$$a(c) = \frac{1}{nm_2^{c/2}} \Sigma |X - \bar{X}|^c \quad \text{for } c \geq 1 \qquad (9.48)$$

Note $a(1) = a$ of (9.46), and $a(4) = b_2$. Geary discussed optimal properties of $b_2$ in this framework. Table 9.4 contains numerical examples of Geary's a test.

### 9.3.3.4.2 Sample range test

David, Hartley, and Pearson (1954) presented a test defined as

$$u = \frac{X_{(n)} - X_{(1)}}{S} \qquad (9.49)$$

that is, the ratio of the sample range to the sample standard deviation. Probability points of u are given in Pearson and Hartley (1972). Table 9.4 contains numerical examples of this u test.

### 9.3.4 Regression Tests

Clearly, the recent interest in tests of normality is due mainly to the exciting work of S. S. Shapiro and M. B. Wilk (1965). Their test for normality and the tests that have resulted as modifications and extensions of their test are called regression and correlation tests (see Chapter 5). These terms are used in that these tests can be viewed as originally arising from considering a linear model

$$X_{(i)} = \mu + \sigma EZ_{(i)} + \epsilon_i \tag{9.50}$$

and estimating, in particular, the parameter $\sigma$ by a regression technique. In (9.50) $X_{(i)}$ is the ith order statistic from the observed sample of size n, $EZ_{(i)}$ is the expected value of the ith order statistic from a sample of size n drawn from the standard normal distribution (mean zero and variance unity), and $\epsilon_i$ is a random error term. Sections 5.7, 5.9, and 5.10 contain a discussion of these tests. It is suggested that the reader read these sections in addition to the following.

#### 9.3.4.1 Shapiro-Wilk Test—An Omnibus Test

In (9.50) the best linear unbiased estimate of $\sigma$ from the Gauss Markov theorem is

$$\hat{\sigma}^2 = \underline{c}'V^{-1}\underline{X}/\underline{c}'V^{-1}\underline{c} \tag{9.51}$$

where $\underline{c}$ is the vector of expected values of the n order statistics from the standard normal distribution and V is the covariance matrix of $\epsilon_i$ in (9.50). (Note $\hat{\sigma}^2$ is a regression estimate.) The Shapiro-Wilk W statistic is basically the ratio of $\hat{\sigma}^2$ to $S^2$, the sample variance. In particular,

$$W = \frac{(\underline{a}'\underline{X})^2}{(n-1)S^2} = \frac{(\Sigma a_i X_{(i)})^2}{\Sigma(X - \bar{X})^2} \tag{9.52}$$

where

$$\underline{a}' = \frac{\underline{c}'V^{-1}}{(\underline{c}'V^{-2}\underline{c})^{\frac{1}{2}}} \tag{9.53}$$

The $a_i$ of (9.52) are the optimal weights for the weighted least squares esti-

TABLE 9.6 Numerical Examples for Shapiro-Wilk W Test and D'Agostino D Test[†]

| $a_i$ | NOR Data | | EXP Data | |
|---|---|---|---|---|
| | $X_{(i)}$ | $a_i X_{(i)}$ | $X_{(i)}$ | $a_i X_{(i)}$ |

**Shapiro-Wilk**

| $a_i$ | $X_{(i)}$ | $a_i X_{(i)}$ | $X_{(i)}$ | $a_i X_{(i)}$ |
|---|---|---|---|---|
| -0.5739 | 84.27 | -48.363 | 0.06 | -0.034 |
| -0.3291 | 90.87 | -29.905 | 0.37 | -0.122 |
| -0.2141 | 92.55 | -19.815 | 0.44 | -0.094 |
| -0.1224 | 96.20 | -11.775 | 0.89 | -0.109 |
| -0.0399 | 98.70 | -3.938 | 2.17 | -0.087 |
| 0.0399 | 98.98 | 3.949 | 2.63 | 0.105 |
| 0.1224 | 100.42 | 12.291 | 4.69 | 0.574 |
| 0.2141 | 101.58 | 21.748 | 6.48 | 1.387 |
| 0.3291 | 106.82 | 35.154 | 8.15 | 2.682 |
| 0.5739 | 113.75 | 65.281 | 16.69 | 9.578 |
| | | 24.627 | | 13.880 |
| $\Sigma (X - \bar{X})^2$ | | 616.554 | | 240.498 |
| $W = \dfrac{(\Sigma aX)^2}{\Sigma (X - \bar{X})^2}$ | | 0.984 | | $0.801^a$ |

**D'Agostino**

| | $X_{(i)}$ | $a_i X_{(i)}$ | $X_{(i)}$ | $a_i X_{(i)}$ |
|---|---|---|---|---|
| -4.5 | 84.27 | -379.215 | 0.06 | -0.270 |
| -3.5 | 90.87 | -318.045 | 0.37 | -1.295 |
| -2.5 | 92.55 | -231.375 | 0.44 | -1.100 |
| -1.5 | 96.20 | -144.300 | 0.89 | -1.335 |
| -0.5 | 98.70 | -49.350 | 2.17 | -1.085 |
| 0.5 | 98.98 | 49.490 | 2.63 | 1.315 |
| 1.5 | 100.42 | 150.630 | 4.69 | 7.035 |
| 2.5 | 101.58 | 253.950 | 6.48 | 16.200 |
| 3.5 | 106.82 | 373.870 | 8.15 | 28.525 |
| 4.5 | 113.75 | 511.875 | 16.69 | 75.105 |
| | | 217.53 | | 123.095 |
| $\Sigma (X - \bar{X})^2$ | | 616.554 | | 240.498 |
| $D = \dfrac{\Sigma c_i X_{(i)}}{n\sqrt{n \Sigma (X - \bar{X})^2}}$ | | 0.27703 | | $0.25101^b$ |

[†]Data sets are first ten observations for NOR and EXP data sets of Appendix.

[a]For the Shapiro-Wilk W test reject the null hypothesis at 0.02 level of significance if $W \le 0.806$. So reject for negative exponential at 0.02 level.

[b]Reject the null hypothesis at 0.05 level of significance if observed $D < 0.2513$, the lower tail critical value of D (see text). So reject at 0.05 level.

mator of $\sigma$ given that the population is normally distributed. W can also be viewed as the $R^2$ (square of the correlation coefficient) obtained from a normal probability plot (see Section 2.4) and thus the notion of a correlation test.

The $a_i$ values for n = 3 to 50 were given by Shapiro and Wilk (1965) and are presented in this book in Table 5.4. Because W is similar to an $R^2$ value, large values (i.e., values close to one) indicate normality and values smaller than unity indicate nonnormality. Thus values in the lower tail of the null distribution of W are used for rejection. Table 5.5 gives the critical values of W for n = 3 to 50.

E 9.3.4.1.1 Numerical example of W test  Table 9.6 contains two numerical examples of the Shapiro-Wilk test. The first ten observations of the NOR data and the EXP data are used. The EXP leads to rejection at the 0.02 level of significance. The weights $a_i$ come from Table 5.4. Chapter 5 contains other numerical examples.

## 9.3.4.2 D'Agostino's D Test

The W test requires a different set of <u>a</u> weights for each sample size n. A modification was presented by D'Agostino (1971) which does not require any tables of weights. It is given as follows:

$$D = \frac{T}{n^2\sqrt{m_2}}$$

$$= \frac{T}{n^{\frac{3}{2}}\{\Sigma(X - \bar{X})^2\}^{\frac{1}{2}}} \tag{9.54}$$

where

$$T = \sum_{i=1}^{n}\left(i - \frac{1}{2}(n + 1)\right)X_{(i)} \tag{9.55}$$

The statistic D is equal, up to a constant, to the ratio of Downton's (1966) linear estimator of the standard deviation to the sample standard deviation.

The expected value of D is approximately $1/(2\sqrt{\pi}) = 0.28209479$ and the standard deviation is asymptotically

$$\left[\frac{12\sqrt{3} - 27 + 2\pi}{24\,m}\right]^{\frac{1}{2}} = \frac{0.02998598}{\sqrt{n}} \tag{9.56}$$

An approximate standardized variable is thus

$$Y = \frac{\sqrt{n}(D - 0.28209479)}{0.02998598} \tag{9.57}$$

TABLE 9.7  Probability Points of D'Agostino's D Test for n = 10 to 2000 (Y statistic of (9.57))

|  |  |  |  |  | Percentiles |  |  |  |  |  |
|---|---|---|---|---|---|---|---|---|---|---|
| n | 0.5 | 1.0 | 2.5 | 5 | 10 | 90 | 95 | 97.5 | 99 | 99.5 |
| 10 | -4.66 | -4.06 | -3.25 | -2.62 | -1.99 | 0.149 | 0.235 | 0.299 | 0.356 | 0.385 |
| 12 | -4.63 | -4.02 | -3.20 | -2.58 | -1.94 | 0.237 | 0.329 | 0.381 | 0.440 | 0.479 |
| 14 | -4.57 | -3.97 | -3.16 | -2.53 | -1.90 | 0.308 | 0.399 | 0.460 | 0.515 | 0.555 |
| 16 | -4.52 | -3.92 | -3.12 | -2.50 | -1.87 | 0.367 | 0.459 | 0.526 | 0.587 | 0.613 |
| 18 | -4.47 | -3.87 | -3.08 | -2.47 | -1.85 | 0.417 | 0.515 | 0.574 | 0.636 | 0.667 |
| 20 | -4.41 | -3.83 | -3.04 | -2.44 | -1.82 | 0.460 | 0.565 | 0.628 | 0.690 | 0.720 |
| 22 | -4.36 | -3.78 | -3.01 | -2.41 | -1.81 | 0.497 | 0.609 | 0.677 | 0.744 | 0.775 |
| 24 | -4.32 | -3.75 | -2.98 | -2.39 | -1.79 | 0.530 | 0.648 | 0.720 | 0.783 | 0.822 |
| 26 | -4.27 | -3.71 | -2.96 | -2.37 | -1.77 | 0.559 | 0.682 | 0.760 | 0.827 | 0.867 |
| 28 | -4.23 | -3.68 | -2.93 | -2.35 | -1.76 | 0.586 | 0.714 | 0.797 | 0.868 | 0.910 |
| 30 | -4.19 | -3.64 | -2.91 | -2.33 | -1.75 | 0.610 | 0.743 | 0.830 | 0.906 | 0.941 |
| 32 | -4.16 | -3.61 | -2.88 | -2.32 | -1.73 | 0.631 | 0.770 | 0.862 | 0.942 | 0.983 |
| 34 | -4.12 | -3.59 | -2.86 | -2.30 | -1.72 | 0.651 | 0.794 | 0.891 | 0.975 | 1.02 |
| 36 | -4.09 | -3.56 | -2.85 | -2.29 | -1.71 | 0.669 | 0.816 | 0.917 | 1.00 | 1.05 |
| 38 | -4.06 | -3.54 | -2.83 | -2.28 | -1.70 | 0.686 | 0.837 | 0.941 | 1.03 | 1.08 |
| 40 | -4.03 | -3.51 | -2.81 | -2.26 | -1.70 | 0.702 | 0.857 | 0.964 | 1.06 | 1.11 |
| 42 | -4.00 | -3.49 | -2.80 | -2.25 | -1.69 | 0.716 | 0.875 | 0.986 | 1.09 | 1.14 |
| 44 | -3.98 | -3.47 | -2.78 | -2.24 | -1.68 | 0.730 | 0.892 | 1.01 | 1.11 | 1.17 |
| 46 | -3.95 | -3.45 | -2.77 | -2.23 | -1.67 | 0.742 | 0.908 | 1.02 | 1.13 | 1.19 |
| 48 | -3.93 | -3.43 | -2.75 | -2.22 | -1.67 | 0.754 | 0.923 | 1.04 | 1.15 | 1.22 |
| 50 | -3.91 | -3.41 | -2.74 | -2.21 | -1.66 | 0.765 | 0.937 | 1.06 | 1.18 | 1.24 |

| n | | | | | | | | | | |
|---|---|---|---|---|---|---|---|---|---|---|
| 60 | -3.81 | -3.34 | -2.68 | -2.17 | -1.64 | 0.812 | 0.997 | 1.13 | 1.26 | 1.34 |
| 70 | -3.73 | -3.27 | -2.64 | -2.14 | -1.61 | 0.849 | 1.05 | 1.19 | 1.33 | 1.42 |
| 80 | -3.67 | -3.22 | -2.60 | -2.11 | -1.59 | 0.878 | 1.08 | 1.24 | 1.39 | 1.48 |
| 90 | -3.61 | -3.17 | -2.57 | -2.09 | -1.58 | 0.902 | 1.12 | 1.28 | 1.44 | 1.54 |
| 100 | -3.57 | -3.14 | -2.54 | -2.07 | -1.57 | 0.923 | 1.14 | 1.31 | 1.48 | 1.59 |
| 150 | -3.409 | -3.009 | -2.452 | -2.004 | -1.520 | 0.990 | 1.233 | 1.423 | 1.623 | 1.746 |
| 200 | -3.302 | -2.922 | -2.391 | -1.960 | -1.491 | 1.032 | 1.290 | 1.496 | 1.715 | 1.853 |
| 250 | -3.227 | -2.861 | -2.348 | -1.926 | -1.471 | 1.060 | 1.328 | 1.545 | 1.779 | 1.927 |
| 300 | -3.172 | -2.816 | -2.316 | -1.906 | -1.456 | 1.080 | 1.357 | 1.528 | 1.826 | 1.983 |
| 350 | -3.129 | -2.781 | -2.291 | -1.888 | -1.444 | 1.096 | 1.379 | 1.610 | 1.863 | 2.026 |
| 400 | -3.094 | -2.753 | -2.270 | -1.873 | -1.434 | 1.108 | 1.396 | 1.633 | 1.893 | 2.061 |
| 450 | -3.064 | -2.729 | -2.253 | -1.861 | -1.426 | 1.119 | 1.411 | 1.652 | 1.918 | 2.090 |
| 500 | -3.040 | -2.709 | -2.239 | -1.850 | -1.419 | 1.127 | 1.423 | 1.668 | 1.938 | 2.114 |
| 550 | -3.019 | -2.691 | -2.226 | -1.841 | -1.413 | 1.135 | 1.434 | 1.682 | 1.957 | 2.136 |
| 600 | -3.000 | -2.676 | -2.215 | -1.833 | -1.408 | 1.141 | 1.443 | 1.694 | 1.972 | 2.154 |
| 650 | -2.984 | -2.663 | -2.206 | -1.826 | -1.403 | 1.147 | 1.451 | 1.704 | 1.986 | 2.171 |
| 700 | -2.969 | -2.651 | -2.197 | -1.820 | -1.399 | 1.152 | 1.458 | 1.714 | 1.999 | 2.185 |
| 750 | -2.956 | -2.640 | -2.189 | -1.814 | -1.395 | 1.157 | 1.465 | 1.722 | 2.010 | 2.199 |
| 800 | -2.944 | -2.630 | -2.182 | -1.809 | -1.392 | 1.161 | 1.471 | 1.730 | 2.020 | 2.211 |
| 850 | -2.933 | -2.621 | -2.176 | -1.804 | -1.389 | 1.165 | 1.476 | 1.737 | 2.029 | 2.221 |
| 900 | -2.923 | -2.613 | -2.170 | -1.800 | -1.386 | 1.168 | 1.481 | 1.743 | 2.037 | 2.231 |
| 950 | -2.914 | -2.605 | -2.164 | -1.796 | -1.383 | 1.171 | 1.485 | 1.749 | 2.045 | 2.241 |
| 1000 | -2.906 | -2.599 | -2.159 | -1.792 | -1.381 | 1.174 | 1.489 | 1.754 | 2.052 | 2.249 |
| 1500 | -2.845 | -2.549 | -2.123 | -1.765 | -1.363 | 1.194 | 1.519 | 1.793 | 2.103 | 2.309 |
| 2000 | -2.807 | -2.515 | -2.101 | -1.750 | -1.353 | 1.207 | 1.536 | 1.815 | 2.132 | 2.342 |

Adapted from D'Agostino (1971) and (1972) with permission of the Biometrika Trustees.

If the null hypothesis of normality is false Y will tend to differ from zero. Simulation studies by D'Agostino (1971) indicated that for alternative distributions with kurtosis less than the normal ($\beta_2 < 3$), Y tends to be greater than zero. For alternative distributions with $\beta_2 > 3$, Y tends to be less than zero. So in order to guard against all possibilities a two sided test needs to be employed. This procedure produces an omnibus test. The statistic can also be used for a one sided test for directional alternatives (i.e., either $\beta_2 < 3$ or $\beta_2 > 3$).

D'Agostino (1971) gave a table of percentile points for Y based on Cornish-Fisher expansions for n = 50 to 1000. D'Agostino (1972) later gave improved points for n = 50 to 100 based on Pearson curves and extensive simulations. Table 9.7 contains probability points for n = 10 to 2000 based on these and other work.

For $n \geq 1000$ a Cornish-Fisher expansion should be adequate to obtain critical values of D. The expansion using the first four cumulants is as follows. If $D_p$ and $Z_p$ are the 100P percentile points ($0 < P < 1$) of D and the standard normal distribution, respectively, then the Cornish-Fisher expansion for $D_p$ in terms of $Z_p$ is

$$D_p = E(D) + V_p \sqrt{\mu_2(D)} \tag{9.58}$$

where

$$V_p = Z_p + \frac{\gamma_1(Z_p^2 - 1)}{6} + \frac{\gamma_2(Z_p^3 - 3Z_p)}{24} - \frac{\gamma_1^2(2Z_p^3 - 5Z_p)}{36} \tag{9.59}$$

Here

$$E(D) = \frac{1}{2\sqrt{\pi}} \sqrt{\frac{n-1}{n}} \left( 1 + \frac{1}{4(n-1)} + \frac{1}{32(n-1)^2} - \frac{5}{128(n-1)^3} \right) \tag{9.60}$$

$$\sqrt{\mu_2(D)} = \frac{0.02998598}{\sqrt{n}} \tag{9.61}$$

$$\gamma_1 = \frac{-8.5836542}{\sqrt{n}} \tag{9.62}$$

and

$$\gamma_2 = \frac{114.732}{n} \tag{9.63}$$

Note with the Cornish-Fisher expansion of (9.58) to (9.63) there is no need to transform to Y of (9.57).

Finally because the range of D is small, it should be calculated to five decimal places.

E 9.3.4.2.1 Numerical examples of D'Agostino's test  Table 9.6 contains
two numerical examples of the D'Agostino D test. As with the Shapiro-Wilk
test the first ten observations of the NOR and EXP data sets are used. With
n = 10 and a level of significance of 0.05 one rejects normality if $Y < -3.25$
or $Y > 0.299$. Using

$$D = 0.28209479 + \frac{0.02998598}{\sqrt{n}} Y$$

one rejects if $D < 0.2513$ or $D > 0.2849$. The NOR data set does not lead to
rejection at the 0.05 level. EXP does.

### 9.3.4.3 Shapiro-Francia's W' Test

Shapiro and Francia (1972) addressed the problem of the $\underline{a}$ weights of the
Shapiro-Wilk test by noting that for large samples the ordered observations
may be treated as if they were independent. With this, the $\underline{a}$ weights of
(9.53) can be replaced by

$$\underline{b}' = \frac{\underline{c}'}{(\underline{c}'\underline{c})^{\frac{1}{2}}} \qquad (9.64)$$

and the W statistic of (9.53) can be replaced by

$$W' = \frac{(\underline{b}'\underline{X})^2}{(n-1)S^2} = \frac{(\Sigma b_i X_{(i)})^2}{\Sigma(X - \bar{X})^2} \qquad (9.65)$$

Recall from Section 9.3.4.1 $\underline{c}$ is the vector of the expected values of the n
order statistics from the standard normal distribution. Values of $\underline{c}$ are
readily available (Harter, 1961) for n up to 400.

Shapiro and Francia (1972) supplied the weights $b_i$ for W' and critical
values for n = 35, 50, 51 (2)99. Pearson, D'Agostino, and Bowman (1977)
noted that these critical values were calculated via simulations from only
1000 samples. They reevaluated percentage points for n = 99, 100, and 125
based on 50,000 simulations. A comparison indicated that the Shapiro-
Francia values in the lower tail were higher than what they should be. This
would result in producing actual levels of significance larger than indicated
by the Shapiro-Francia tables. Further it would indicate in power studies
that the test was more powerful than it actually is.

### 9.3.4.4 Weisberg-Bingham's Ŵ' Test
### and Asymptotic Extensions

Weisberg and Bingham (1975) suggested replacing the $\underline{b}$ of (9.64) with

$$\underline{d} = \frac{\tilde{c}}{(\tilde{c}'\tilde{c})^{\frac{1}{2}}}$$ 

(9.66)

where the elements of the vector $\tilde{c}$ are

$$\tilde{c}_i = \Phi^{-1}\left[\frac{i - 3/8}{n + 1/4}\right] \quad \text{for } i = 1, \ldots, n$$

(9.67)

and $\Phi^{-1}(p)$ is the inverse of the standard normal cumulative distribution function. The resulting statistic is denoted by $\tilde{W}'$ and is given as

$$\tilde{W}' = \frac{(\tilde{c}'\underline{X})^2/(\tilde{c}'\tilde{c})}{\Sigma(X - \bar{X})^2} = \frac{(\Sigma d_i X_{(i)})^2}{\Sigma(X - \bar{X})^2}$$

(9.68)

The approximations $\tilde{c}_i$ to $c_i$ was suggested by Blom (1958) and its use in (9.68) results in the null distribution of $\tilde{W}'$ being very close to $W'$, at least for n = 5, 20, 35 where the authors made a comparison. They suggested using $\tilde{W}'$ in place of $W'$. This removes the need to have tables of weights for computations of the test statistic. With use of $\tilde{W}'$ they suggested use of the critical values of $W'$.

Stephens in Chapter 5 of this book discusses use of

$$Z(X, c) = n(1 - \tilde{W}')$$

(9.69)

and supplies a table of critical values for $n \leq 1000$ (see Section 5.7.3 and Table 5.2 where the statistic (9.69) is written as $Z(X, m)$). Royston (1982) gave an extension of the Shapiro-Wilk test similar to (9.69) for n < 2000. He presented a statistic for the form

$$(1 - W)^\lambda$$

(9.70)

$\lambda$ is a function of sample size.

### 9.3.4.5 Other Extensions/Modifications of the Shapiro-Wilk Test

A number of other investigators have considered extending and modifying the Shapiro-Wilk test. Most noticeable are the works of Filliben (1975) and La Brecque (1977). Filliben's test is exactly the correlation coefficient between the ordered observations $X_{(i)}$ and the order statistic medians $M_i$ from the standard normal distribution. It can be viewed in the context of the last section (9.3.4.4) where the weights $\underline{a}$ of the W statistic are replaced with functions of medians of the order statistics. Filliben gives weights and significance levels for $n \leq 100$.

La Brecque (1977) extended the W test by augmenting (9.50) to detect nonlinearity in normal probability plots. The reader is referred to Section 5.10.1 for a further comment on this text.

Also of interest here is the work of Puri and Rao (1976). They wrote the expected value of the ith order statistic as

$$E(X_{(i)}) = \gamma_1 + \gamma_2 c_i + \gamma_3 (c_i^2 - \lambda) + \gamma_4 (c_i^3 - \mu c_i) + \cdots \qquad (9.71)$$

where $\lambda$ and $\mu$ were selected so as to provide orthogonal polynomials. When the underlying distribution is normal $\gamma_1 = \mu$, $\gamma_2 = \sigma$, $\gamma_3 = \gamma_4 = \cdots = 0$. The Shapiro-Wilk test is basically the test $\gamma_2 = \sigma$. Puri and Rao investigated if a better test can be developed by incorporating $\gamma_3$ and $\gamma_4$ into the test. They ultimately concluded tests using jointly W and a skewness test would be more efficient for testing normality versus skewed distributions (i.e., $\sqrt{\beta_1} \neq 0$) and a test using jointly W and a kurtosis type test would be more efficient for testing normality versus distributions with nonnormal kurtosis (i.e., $\beta_2 \neq 3$).

More on other extensions of the W test are given in Chapter 5.

## 9.3.5 Miscellaneous Tests

There is a plethora of other tests of normality too numerous to mention in detail. Some selected ones we now discuss briefly.

### 9.3.5.1 Locke and Spurrier's U Statistic Test

Locke and Spurrier (1976, 1977) used the theory of U-statistics to develop tests of normality. They showed that both the $\sqrt{b_1}$ test and D'Agostino's D test can be generated from this theory. They also developed other new tests.

### 9.3.5.2 The Gap Test

Andrews, Gnanadesikan, and Warner (1971, 1972) developed gap test for normality. Gaps $g_i$ are defined as

$$g_i = \frac{X_{(i+1)} - X_{(i)}}{c_{(i+1)} - c_{(i)}} \qquad (9.72)$$

where $c_{(i)}$ are the expected values of the ith order statistics from the standard normal distribution. If the null hypothesis of normality is true, the $g_i$ of (9.72) are independent exponential variables. Specific types of deviations from normality reflect themselves in deviations from exponentiality of the $g_i$. Andrews et al. gave an omnibus test for normality that is distributed under the null hypothesis approximately as a chi square variable with two degrees of freedom.

### 9.3.5.3  Likelihood Ratio Tests/Specific Alternatives

Dumonceaux, Antle, and Haas (1973) looked at the theory of likelihood ratio tests to develop tests of normality versus some specific alternative distributions (Cauchy, Exponential, and Double Exponential). Critical values of the tests are given. For the double exponential the likelihood ratio test is similar to Geary's a of Section 9.3.3.4.1. Hogg (1972) presented the family of distributions of the form

$$f(x; \theta) = \frac{e^{-|x|^{\theta}}}{2\Gamma\left(1 + \frac{1}{\theta}\right)} \tag{9.73}$$

For $\theta = 2$, x is normal, for $\theta = 1$, x is double exponential and as $\theta \to \infty$, x tends to the uniform distribution. For testing $\theta = \theta_1$ versus $\theta = \theta_2$, Geary's a test is shown to have optimal properties for $\theta_1 = 1$ versus $\theta_2 = 2$ and for testing $\theta_1 = 2$ versus $\theta_2 = 4$, the $b_2$ test has optimal properties.

### 9.3.5.4  Tiku's Tests

Tiku (1974) presented tests of normality based on the ratio of estimates of $\sigma$ from trimmed samples to the sample standard deviation. These tests are suitable for specific alternative hypotheses in terms of skewness of the alternative. Percentage points are given.

### 9.3.5.5  Spiegelhalter's Combination of Test Statistics

Spiegelhalter (1977) used the theory of most powerful location and scale invariant tests to develop tests of normality against the uniform and the double exponential distributions. He then suggested the sum of these two as a combined test statistic. A Bayesian argument was presented to justify the combination. Asymptotically the two components are equal to Geary's a test of (9.46) and the David, Hartley, and Pearson u test of (9.49). Critical points were given in the article for $n \leq 100$ for level of significance 0.05 and 0.10.

### 9.3.5.6  Other Tests

Other tests of interest are the tests based on the independence of the sample mean and standard deviation (Line and Mudholkar, 1980), the test based on the empirical characteristic function (Hall and Welsh, 1983) and the squeeze test (Burch and Parsons, 1976). The term squeeze was derived from the method used to perform the test, whereby data points plotted on the appropriate probability paper are squeezed between parallel rules.

## 9.4 COMPARISONS OF TESTS

### 9.4.1 Power Studies

There are a large number of tests for judging normality or departures from normality. There is no one test that is optimal for all possible deviations from normality. The procedure usually adopted to investigate the sensitivity of these tests is to perform power studies where the tests are applied to a wide range of nonnormal populations for a variety of sample sizes. A number of such studies have been undertaken. The major ones, in order of completeness and importance, are Pearson, D'Agostino, and Bowman (1977), Shapiro, Wilk, and Chen (1968), Saniga and Miles (1979), Stephens (1974), D'Agostino (1971), Filliben (1975), and D'Agostino and Rosman (1974). Other useful, but less major, studies are Dyer (1974), Prescott (1976), Prescott (1978), Tiku (1974), and Locke and Spurrier (1976).

Presentation of the results of the above power studies produces a number of difficulties. In order to be most informative we will first present results indexed by skewness ($\sqrt{\beta_1}$) and kurtosis ($\beta_2$) and then indexed by specific tests. The former comparisons are mainly from Pearson et al. (1977). The latter are summarized from all of the above articles.

### 9.4.1.1 Power Results for Skewed Alternatives ($\sqrt{\beta_1} \neq 0$)

The Shapiro-Wilk test (Section 9.3.4.1) and the Shapiro-Francia extensions (Sections 9.3.4.3 and 9.3.4.4) are very sensitive omnibus tests against skewed alternatives ($H_1$: Nonnormality with $\sqrt{\beta_1} \neq 0$). For many skewed alternatives they are clearly the most powerful. When we have prior grounds for believing that, if the population is not normal, it will be positively skewed ($\sqrt{\beta_1} > 0$), directional tests are very powerful. Directional tests refer here to the $\sqrt{b_1}$ test (Section 9.3.3.1) based on the upper tail of its distribution and the R test (Section 9.3.3.3.1), employing a one-sided $\sqrt{b_1}$ test and a two-sided $b_2$ test. For negatively skewed alternatives ($\sqrt{\beta_1} < 0$) the lower tail of $\sqrt{b_1}$ should be employed.

### 9.4.1.2 Power Results for Symmetric Distributions with Nonnormal Kurtosis ($\beta_2 \neq 3$)

### 9.4.1.2.1 Platykurtic alternatives ($\beta_2 < 3$)

When the omnibus tests are applied without directional knowledge of the alternative distribution, there is very little to choose between the powers of $K^2$ (Section 9.3.3.3.2) and the R test (Section 9.3.3.3.1) when applied to platykurtic populations. The Shapiro-Wilk W test is on the whole more powerful than these. The D'Agostino D test (Section 9.3.4.2) does not fit consistently in the comparison. In general there is usually some other test more powerful than it.

When knowledge of the direction of $\beta_2$ is known ($\beta_2 < 3$), the lower tail $b_2$ is more powerful than the $K^2$, R, W, and D tests.

### 9.4.1.2.2 Leptokurtic alternatives ($\beta_2 > 3$)

The powers of the omnibus tests are broadly in the following order of descending power: $K^2$, R, D, and W. For very long tailed populations (e.g., $\beta_2 > 36$), D'Agostino's D test is best.

When $\sqrt{\beta_1} = 0$ and $\beta_2 > 3$ is the known direction of the alternative, there is no clear preponderance of the upper tail $b_2$ over the lower tail D'Agostino's D test. However, these tests are more powerful than the omnibus test $K^2$, R, D, and W.

### 9.4.1.3 Power Results for Specific Tests

1. The Shapiro-Wilk W test and the Shapiro Francia extension are very sensitive omnibus tests. For many skewed populations they are clearly the most powerful. When $\sqrt{\beta_1} = 0$ and $\beta_2 > 3$ a number of other tests are more powerful.

2. $\sqrt{b_1}$ and $b_2$ have excellent sensitivity over a wide range of alternative distributions which deviate from normality with respect to skewness and kurtosis, respectively. As n gets large the $\sqrt{b_1}$ test has no power for symmetric alternatives. When directional information is available (e.g., $\sqrt{\beta_1} > 0$ or $\beta_2 < 3$) appropriate one sided versions of these tests are very powerful. In most cases studied in the literature they are usually most powerful.

3. The D'Agostino-Pearson $K^2$ of Section 9.3.3.3.2 or because of its equivalency to $K^2$, the Bowman-Shenton $K_s^2$ of Section 9.3.3.3.3 are sensitive to a wide range of nonnormal populations. They can be considered omnibus tests. For skewed alternatives the Shapiro-Wilk W test is usually more powerful. Also for symmetric alternatives with $\beta_2 < 3$, the W test is often more powerful. For symmetric alternatives with $\beta_2 > 3$, $K^2$ is often most powerful.

4. The R test of Section 9.3.3.3.1 is also an omnibus test. Its power usually does not exceed that of $K^2$.

5. The most powerful EDF test appears to be the Anderson-Darling $A^2$ (Section 9.3.2.2). It is at times presented as being similar in power to the Shapiro-Wilk W test. However, it has not been studied as extensively as either the moments tests or the regression tests. More power studies are required to compare it more fully to the W, $K^2$, R, and D'Agostino's D tests.

6. While D'Agostino's D test is an omnibus test it has best power for distributions with $\beta_2 > 3$. Other tests are better than it for skewed alternatives.

7. Geary's a test (Section 9.3.3.4.1) has good power for symmetric alternatives with $\beta_2 \neq 3$. However, $b_2$ is usually better. For skewed alternatives W is generally superior.

8. Kolmogorov-Smirnov test has poor power in comparison to the many tests described in detail in this chapter.

9. Chi-Square test is in general not a powerful test of normality.

### 9.4.2 Effects of Ties Due to Grouping

Results of power studies are not the only means for judging or comparing the normality tests. In practice the data may often involve ties, either because available figures have been rounded for grouping purposes or because measurements cannot be carried out beyond a certain degree of accuracy. It may not be desirable to reject the null hypothesis just because the data contain these ties or are grouped. Use of the data as if the underlying population were normal may not present any problems if the true population is approximately normal and the resulting data contain ties. (Research is needed on this point.) Pearson et al. (1977) investigated the effect which ties and the grouping of data have on four tests or normality, $\sqrt{b_1}$, D, W, and W'.

For judging the effect what matters is the ratio, say $\ell$, of the standard deviation of the distribution to the rounding interval, i.e., the interval between the nearest possible readings or observations left after rounding. Pearson et al. (1977) considered the effect of grouping on $\sqrt{b_1}$, and W for $\ell = 3$, 5, and 10 and n = 20 and 50, and for D and W' for $\ell = 3$, 5, 8, and 10 and n = 100. The present author also considered $\sqrt{b_1}$ for n = 100, D for n = 20 and 50, and $b_2$ for n = 20, 50, and 100.

The effect of grouping on $\sqrt{b_1}$ and $b_2$ was not significant. That is, grouping did not produce differences between the actual declared or nominal level of significance. The effect on D was to make the test slightly conservative. That is, the actual level of significance was slightly smaller than the declared or nominal level. For the W test the effect was significant for $\ell = 3$ and 5. Here the actual level of significance exceeded significantly the normal level. For $\ell = 10$, the effect was minimal. The statistic W' was extremely unsatisfactory. Usually the actual level of significance exceeded by substantial amounts (e.g., 30%) the nominal level, even for $\ell = 10$.

The above results suggested that W' or its derivatives as given in Sections 9.3.4.4 and 9.3.4.5 and Chapter 5 must be used with caution if there are multiple ties.

Pearson et al. (1977) did not consider the effects of ties on the EDF tests. Until the effects are investigated they should be used with caution on data containing ties.


## 9.5 RECOMMENDATIONS

Attempting to make final recommendations is an unwelcome and near impossible task involving the imposition of personal judgments. Still a set of well justified recommendations can be made at this time. We make the following.

1. A detailed graphical analysis involving normal probability plotting should always accompany a formal test of normality. Section 2.4 gives the necessary steps for this. It is not clear that standard statistical software packages give useful probability plots. However, Chapter 2 explains in detail

how to employ the computer for a good graphical analysis. A detailed examination of the probability plot should be undertaken.

2. The omnibus tests, the Shapiro-Wilks W test and its extensions
(Sections 9.3.4.1, 9.3.4.3, and 9.3.4.4), the D'Agostino-Pearson $K^2$ test
or the Bowman-Shenton version $K_S^2$ (Sections 9.3.3.3.2 and 9.3.3.3.3), and
the Anderson-Darling edf test $A^2$ (Section 9.3.2.2) appear to be the best
omnibus tests available. The Shapiro-Wilk type tests are probably overall
most powerful. However, due to the problem with ties (Section 9.4.2) and
the fact that it gives as a by-product no numerical indications of the nonnormality, the $K^2$ test based jointly on the very informative $\sqrt{b_1}$ and $b_2$ statistics
may be preferred by many. The $\sqrt{b_1}$ and $b_2$ statistics can be very useful for
indicating the type of nonnormality. Also they can be useful for judging if
nonnormality will affect any inferences to be made with the data (e.g., if
a t test is to be applied to the data or a prediction is to be made).

3. The R test (Section 9.3.3.3.1) and D'Agostino's D test (Section
9.3.4.2) can be used as omnibus test. They are convenient and easy to use.
The tests of point 2 are probably more powerful.

4. If the direction of the alternative to normality is known (e.g.,
$\sqrt{\beta_1} > 0$ or $\beta_2 > 3$), then the directional versions of the $\sqrt{b_1}$, $b_2$, and D'Agostino D test should be used.

5. For testing for normality, the Kolmogorov-Smirnov test is only a
historical curiosity. It should never be used. It has poor power in comparison
to the above procedures.

6. For testing for normality, when a complete sample is available the
chi-square test should not be used. It does not have good power when compared to the above tests.

## 9.6  TESTS OF NORMALITY ON RESIDUALS

Chapters 4 and 5 discussed the application of edf and regression tests on
residuals. Much of that discussion involved tests of normality. The reader
is referred to those chapters. Anscombe and Glynn (1983) also discussed the
application of $b_2$ on residuals. These attempts only represent the beginnings.
There is much research that needs to be done.

By residuals we mean the following. A mathematical model of the form

$$Y = f(\vec{\beta}, \vec{x}) + \epsilon \tag{9.74}$$

is under investigation. $\vec{X}$ represents a vector of variables, $\vec{\beta}$ a vector of
unknown, to be estimated, coefficients, Y represents the unknown dependent
variable, and $\epsilon$ represents a random error. Based on a sample size n, the
$\vec{\beta}$ are estimated by $\hat{\beta}$ and the residual for each observation is defined as

$$\hat{\epsilon}_i = Y_i - \hat{Y}_i \tag{9.75}$$

where

$$\hat{Y}_i = f(\hat{\beta}, \vec{X}_i) \tag{9.76}$$

for $i = 1, \ldots, n$. Intuitively if the sample employed to estimate the $\hat{\beta}$ consists of a large number of observations in comparison to the dimension of the $\hat{\beta}$, then the tests of normality given in Section 9.3 above applied directly to the residuals of (9.75) should be approximately correct. This being the case even if no adjustments are made for the statistical dependencies among the residuals and the unequal variances that usually exist with residuals. Unfortunately the correct implementation of this intuition needs substantial work. We now discuss two attempts at its implementation.

### 9.6.1  Residuals from a Linear Regression

White and MacDonald (1980) considered the linear regression model

$$Y = \vec{\beta}^{\mathsf{T}} \vec{X} + \epsilon \tag{9.77}$$

and the residuals

$$\hat{\epsilon} = Y - \hat{Y} \tag{9.78}$$

Here n independent observations are drawn and the resulting design matrix is of full rank. The vector $\vec{\beta}$ is of dimension k.

White and MacDonald showed that under general conditions the $\sqrt{b_1}$, $b_2$, D'Agostino's D, and Shapiro-Francia W' tests computed on the residuals $\hat{\epsilon}_1, \hat{\epsilon}_2, \ldots, \hat{\epsilon}_n$ have asymptotically the same distributions as if computed on n independent identically distributed normal errors $\epsilon$ in (9.77). Further for n = 20, 35, 50, and 100 they performed simulations to judge the effects of using the residuals on the null distributions of the test statistics $\sqrt{b_1}$, $b_2$, D, R, W, and W'. They also examined how the statistics behaved for nonnormal errors $\epsilon$ in model (9.77). For all this simulation the dimension k of the vector $\vec{\beta}$ was 4.

In general they showed that for the cases examined computation of the tests on the residuals did not invalidate them. Overall, D'Agostino's D test produced the best agreement between the test based on the correlated residuals of (9.78) and the one on the independent errors of (9.77), $\sqrt{b_1}$ and $b_2$ exhibited the next best behavior, followed by W' and W.

Weisberg (1980) in a comment article to the White and MacDonald (1980) article emphasized that their results were limited and that n, the sample size, k, the dimension of $\vec{\beta}$ and V, the design matrix, can all influence the validity of the tests of normality on residuals from a linear regression analysis. He demonstrated this with examples for n = 20 using the W test. Unfortunately n = 20 and the W test comprised the weakest combination in the White and MacDonald work. It would have been better had he examined the R or D tests.

Appropriate practical advice here seems to be that n should be large (say $n \geq 50$) and k reasonably small (e.g., 5 or less) before the formal significance levels of the tests on residuals can be taken as appropriate. Further work is needed to clarify the real impact of varying k, n, and the design matrix.

In applying the results of White and MacDonald it should be emphasized that the computations of the test statistics and the corresponding table lookups to determine statistical significance are carried out directly on the n residuals as if they constituted an independent sample of size n. No adjustments are made for k. Also note because the mean of the residuals is zero, in $\sqrt{b_1}$ and $b_2$ the sample mean need not be explicitly computed for them.

### 9.6.2 Residuals from an Autoregressive Model

Alexa Beiser (1985) in an unpublished work has considered first order autoregressive models of the form

$$Y_t = \mu + \rho(Y_{t-1} - \mu) + \epsilon_t \tag{9.79}$$

where $\epsilon_t$ are independent normal variables. She has shown that the $\sqrt{b_1}$ and $b_2$ computed on the residuals produce valid levels of significance for $n \geq 50$ and $\rho \leq 0.9$. In fact, this procedure appeared to be more appropriate than computing $\sqrt{b_1}$ and $b_2$ on $Y_t$ directly incorporating adjustments for the dependencies of the observations.

In her work the $\rho$ of (9.79) is estimated as

$$\hat{\rho} = \frac{\sum\limits_{t=2}^{n} (Y_t - \bar{Y})(Y_{t-1} - \bar{Y})}{\sum\limits_{t=1}^{n} (Y_t - \bar{Y})}$$

where

$$\bar{Y} = \frac{\sum\limits_{t=1}^{n} Y_t}{n} = \hat{\mu}$$

The statistics $\sqrt{b_1}$ and $b_2$ are computed on the $n - 1$ residuals

$$\hat{\epsilon}_t = (Y_t - \bar{Y}) - \hat{\rho}(Y_{t-1} - \bar{Y}) \tag{9.80}$$

for $t = 2, \ldots, n$. Note the tests are employed as if a sample of $n - 1$ is available, not n.

## 9.7  MULTIVARIATE NORMALITY

An excellent review article exists for tests of multivariate normality (Gnanadesikan, 1977). We will not attempt to discuss these tests in the detail given in that treatment. Rather this section will only be a brief overview.

### 9.7.1  Univariate Tests for Marginal Normality

In practice one rarely performs solely a multivariate analysis. Rather, a multivariate analysis is usually one stage in an analysis to be supplemented by univariate analyses considering each variable separately. These univariate analyses often can detect sufficiently what the multivariate analysis contains. At times they are more informative due to their specific attention to each variable. In that view, although normality of each marginal variable does not imply joint normality, the presence of many types of nonnormality is often shown in the marginal distribution. Also if there is multivariate normality then necessarily each marginal distribution will be normal. Detection of one marginal that is nonnormal indicates the multivariate distribution is nonnormal.

Given the above, it is reasonable to test each marginal distribution using the univariate tests discussed in Section 9.3 and recommended in Section 9.5. In order to guard against an inflation of the Type I error it is probably sensible to use Bonferroni's inequality for determining the overall level of significance. Thus if we had a p dimensional distribution under consideration, each marginal should be tested at the

$$\alpha/p$$

level of significance. Here $\alpha$ is the desired overall level of significance. For example, if $p = 5$ and we desired to have an overall $\alpha = 0.05$, then each marginal should be tested at the $0.05/5 = 0.01$ level of significance.

In these assessments of marginal normality normal probability plotting should be employed.

### 9.7.2  Generalization of Univariate Procedures

#### 9.7.2.1  Mardia's Tests

Mardia (1970) proposed tests of multivariate normality via multivariate measure of skewness and kurtosis. These moments have the maximum effect on the distribution of Hotelling's $T^2$ under nonnormality (Mardia, 1975).

Let $X_1, \ldots, X_n$ be a random sample of vectors of p components from a population with mean vector $\vec{\mu}$ and covariance matrix $\Sigma$. Suppose $\bar{X}$ and S denote the sample mean vector and covariance matrix, respectively. Mardia based his test on the following measures of multivariate skewness and kurtosis,

$$b_{1,p} = \frac{1}{n^2} \Sigma_i \Sigma_j r_{ij}^3 \tag{9.81}$$

and

$$b_{2,p} = \frac{1}{n} \Sigma_i r_{ii}^2 \tag{9.82}$$

where

$$r_{ij} = (X_i - \bar{X})'S^{-1}(X_j - \bar{X}) \tag{9.83}$$

Here $\bar{X}$ is the mean vector and S, the sample covariance matrix. Significant points obtained from simulations are given in Mardia (1970) and Mardia (1975). Mardia and Foster (1983) discussed omnibus multivariate tests based on $b_{1,p}$ and $b_{2,p}$.

### 9.7.2.2  Malkovich and Afifi's Tests

Malkovich and Afifi (1973) proposed multivariate skewness and kurtosis tests using Roy's union-intersection principle. Multivariate skewness was given by them as

$$\beta_1(c) = \frac{(E(c'X - c'\mu)^3)^2}{(var\,(c'X))^3} \tag{9.84}$$

and multivariate kurtosis by

$$[\beta_2(c)]^2 = \frac{E(c'X - c'\mu)^4}{(var\,(c'X))^2} \tag{9.85}$$

for some vector c. Roy's principle could lead naturally to appropriate rejection rules. These were in the forms reject for skewness if

$$max\,\{\beta_1(c)\} > k_1 \tag{9.86}$$

where $k_1$ produced an $\alpha$ level test and reject for kurtosis if

$$max\,(\beta_2(c))  \quad or \quad  min\,(\beta_2(c)) \tag{9.87}$$

fall outside the interval $(k_2, k_3)$ where these k's produce an $\alpha$ level test.
Machado (1983) found the asymptotic distributions of the statistics in (9.86) and (9.87) for p = 2, 3, and 4. As with the univariate case, the statistics approach their asymptotic behavior very slowly. Machado used the D'Agostino $\sqrt{b_1}$ suggested Johnson $S_U$ approximation for (9.86) and the Anscombe and Glynn $b_2$ approximation for (9.87) to obtain null distributions

valid for $n \geq 25$ and $p = 2$, 3, and 4. However, he mentioned overestimation for $n < 100$ and underestimation for $n > 100$. He did not give any corrections. More details or work are needed here.

Malkovich and Afifi (1973) also suggested a multivariate generalization of the Shapiro-Wilk W test.

The tests of Malkovich and Afifi appear to require considerable computation.

### 9.7.3 Solely Multivariate Procedures

#### 9.7.3.1 Directional Normality

Andrews, Gnanadesikan, and Warner (1971) defined the scaled residuals as

$$Z_i = S^{-\frac{1}{2}}(X_i - \bar{X}) \tag{9.88}$$

for $i = 1, \ldots, n$ where $S^{-\frac{1}{2}}$ is the symmetric square root of the covariance matrix S. They also defined a normalized weighted sum of the $Z_i$

$$\vec{d}_a = \frac{\Sigma_i w_i Z_i}{\| \Sigma_i w_i Z_i \|} \tag{9.89}$$

Here $\vec{d}_a$ is a vector,

$$w_i = \| Z_i \|^a \tag{9.90}$$

$\| Z \|$ denotes the Euclidean norm, or length of the vector X, and a is a constant to be chosen.

For $a = -1$, $\vec{d}_a$ is a function only of the orientation of the $Z_i$'s, while for $a = 1$, $\vec{d}_a$ becomes sensitive to the observations distant from the mean. More generally, for $a > 0$ the vector $\vec{d}_a$ will tend to point toward any clustering of observations far from the mean, while for $a < 0$, the vector $\vec{d}_a$ will point in the direction of any abnormal clustering near the center of gravity of the data.

Therefore, $\vec{d}_a^* = S^{\frac{1}{2}} \vec{d}_a$ for a given value of a, can be regarded as a univariate sample. Any univariate test of normality can now be employed. The value of a can be selected to be sensitive to certain types of nonnormality. Because of the data-dependence of the approach, the procedure can only be used as a guide. The formal significance levels do not apply.

#### 9.7.3.2 Radius and Angles Decompositions

Consider again the scaled residuals

$$Z_i = S^{-\frac{1}{2}}(X_i - \bar{X}) \tag{9.91}$$

for $i = 1, \ldots, n$. Under the null hypothesis, the scaled residuals are approximately spherically symmetrically distributed. The squared radii, or squared lengths of the $Z_i$

$$r_i^2 = Z_i'Z_i = (X_i - \bar{X})'S^{-1}(X_i - \bar{X}) \tag{9.92}$$

will have approximately a chi-squared distribution with p degrees of freedom. Here p is the dimension of Z. For the bivariate case, define $\theta_i$ to be the angle $Z_i$ makes with a prescribed line. The $\theta_i$ are then approximately uniformly distributed on $(0, 2\pi)$ under the null hypothesis and the $r_i$ and $\theta_i$ are approximately independent. For moderate sample sizes the dependence should be negligible. Probability plots of the $r_i$ and $\theta_i$ can be used to evaluate multivariate normality.

#### 9.7.3.2.1 Bivariate case

Order the n squared radii $r_{(1)}^2 < \cdots < r_{(n)}^2$ and plot these against the corresponding expected values for the cdf from a chi-square distribution with two degree of freedom. Similarly for $\theta_i/2\pi$ plot the ordered values against the expected values of the cdf of a uniform distribution. Both of these plots should be linear under the null hypothesis of normality.

#### 9.7.3.2.2 Higher dimensional data (p > 2)

For higher dimensional data the appropriate chi-square distribution for the squared radius plot is the chi-square with p degrees of freedom. For the angles there are p - 1 plots. See Andrews et al. (1974) for details of these plots.

#### 9.7.3.3 Other Procedures

There are still other procedures for testing multivariate normality. Some of these are:
   1. The nearest distance test of Andrews et al. (1974),
   2. The maximum curvature test of Cox and Small (1978), and
   3. The Dahiya and Gurland (1973) generalized minimum chi-square technique applicable for bivariate normality.

#### 9.7.4 Power of Multivariate Normality Tests

Very little has been done by way of power studies for multivariate normality tests. Malkovich and Afifi (1973) have undertaken a small study. More is needed.

## 9.7.5 Recommendations

It is inappropriate at this stage to give detailed recommendations. Much more research is needed. All the procedures reviewed above have merit. Personally we have found the univariate tests for marginal normality (Section 9.7.1) in conjunction with Mardia's test (Section 9.7.2.1) to be very useful. For bivariate normality the radius and angle procedure of Section 9.7.3.2.1 has good merit.

## REFERENCES

This reference list contains two sets of references. The first set is a brief set of references in which the normal distribution is used as a mathematical model to real world data. The second set is the major set of references and it is on the statistical methods of tests of normality.

Applications of the Normal Distribution

D'Agostino, R. B. and Gillespie, J. C. (1978). Comments on the OSHA accuracy of measurement requirements for monitoring employee exposure to benzene. Amer. Industrial Hygiene Assoc. Jour. 39, 510-513.

Fredholm, B., Gunnarson, K., Jensen, G., and Muntzing, J. (1978). Mammary tumor inhibition and subacute toxicity in rats of predrimostine and of its molecular components chlorambuic and prednisolone. Acta Pharmacol. et Toxicol. 42, 159-163.

Griffin, M. J. and Whitham, E. M. (1978). Individual variability and its effect on subjective and biodynamic response to whole body vibration. Jour. of Sound and Vibration 58, 239-250.

Hunt, W. F. J. (1972). The precision associated with the frequency of lognormally distributed air pollutant measurements. Jour. Air Pollution Cont. Ass. 22(9), 687.

Kollegoda, N. T. and Yevjevich, V. (1977). Preservation of correlation in generated hydrologic samples through two-station models. Jour. of Hydrology 33, 99-121.

Larson, R. I. (1971). A mathematical model for relating air quality measurement to air quality standards. Publication No. AP-89, U.S. EPA, Research Triangle Park, N.C.

Rino, C. L., Livingston, R. C., and Whitney, H. E. (1976). Some new results on the statistics of radio wave scintillation: 1. Empirical evidence for gaussian statistics. Jour. of Geophysical Research 81, 2051-2064.

Richert, J., Simbel, M. H., and Weidenmuller, H. A. (1975). Statistical theory of nuclear cross section fluctuation. Z. Physik 273, 195-203.

Smith, T. J., Temple, A. R., and Reading, J. C. (1976). Cadmium, lead and copper blood levels in normal children. Clinical Toxicology 9, 75-87.

Smith, T. J., Wagner, W. L., and Moore, D. E. (1978). Distribution of $SO_2$ for people with chronic exposure. Jour. of Occupational Medicine 2(2), 83-87.

White, H. and Olson, L. (1981). Conditional distributions of earnings, wages, and hours for blacks and whites. Jour. of Econometrics 17, 263-285.

## Statistical Methods of Tests of Normality

Andrews, D. F., Gnanadesikan, R., and Warner, J. L. (1971). Transformation of multivariate data. Biometrika 27, 825-840.

Andrews, D. F., Gnanadesikan, R., and Warner, J. L. (1972). Methods for assessing multivariate normality. Unpublished memorandum.

Anscombe, F. J. and Glynn, W. J. (1983). Distribution of the kurtosis statistic $b_2$ for normal statistics. Biometrika 70, 227-234.

Beiser, A. (1985). Distributions of $\sqrt{b_1}$ and $b_2$ for autoregressive errors. Unpublished thesis, Boston University.

Blom, G. (1958). Statistical Estimates and Transformal Beta-Variables. Wiley, New York.

Bowman, K. O. and Shenton, B. R. (1975). Omnibus test contours for departures from normality based on $\sqrt{b_1}$ and $b_2$. Biometrika 62, 243-250.

Brunk, H. D. (1962). On the range of the difference between hypothetical distribution function and Pyke's modified empirical distribution function. Am. Math. Statist. 33, 525-532.

Burch, C. R. and Parsons, I. T. (1976). "Squeeze" significance tests. Appl. Statist. 25, 287-291.

Chernoff, H., Gastwirth, J. L., and Johns, M. V. (1967). Asymptotic distribution of linear combinations of functions of order statistics with applications to estimation. Am. Math. Statist. 38, 52-72.

Chernoff, H. and Lehmann, E. L. (1954). The use of the maximum likelihood estimates on $X^2$ tests for goodness of fit. Am. Math. Statist. 25, 579-586.

Cox, D. R. and Hinkley, D. V. (1974). Theoretical Statistics. Chapman and Hall, London.

Cox, D. R. and Small, N. J. H. (1978). Testing multivariate normality. Biometrika 65, 263-272.

Cramer, H. (1928). On the composition of elementary errors. Second paper: statistical applications. Skand. Aktvarietidskr. 11, 141-180.

Csorgo, M., Seshadri, V., and Yalovsky, M. (1973). Some exact tests for normality in the presence of unknown parameters. J. Roy. Statist. Soc. B 35, 507-522.

D'Agostino, R. B. (1971). An omnibus test of normality for moderate and large sample size. Biometrika 58, 341-348.

D'Agostino, R. B. (1972). Small sample probability points for the D test of normality. Biometrika 59, 219-221.

D'Agostino, R. B. and Pearson, E. S. (1973). Testing for departures from normality. I. Fuller empirical results for the distribution of $b_2$ and $\sqrt{b_1}$. Biometrika 60, 613-622.

D'Agostino, R. B. and Rosman, B. (1974). The power of Geary's test of normality. Biometrika 61, 181-184.

D'Agostino, R. B. and Tietjen, G. L. (1971). Simulation probability points for $b_2$ for small samples. Biometrika 58, 669-672.

D'Agostino, R. B. and Tietjen, G. L. (1973). Approaches to the null distribution of $\sqrt{b_1}$. Biometrika 60, 169-173.

Dahiya, R. C. and Gurland, J. (1973). A test of fit for bivariate distribution. J. Roy. Statist. Soc. B. 35, 452-465.

David, H. A., Hartley, H. O., and Pearson, E. S. (1954). The distribution of the ratio, in a single normal sample of range to standard deviation. Biometrika 41, 482-493.

Downton, F. (1966). Linear estimates with polynomial coefficients. Biometrika 53, 129-141.

Dumonceaux, R., Antle, C. E., and Haas, G. (1973). Likelihood ratio test for discrimination between two models with unknown location and scale parameters. Technometrics 15, 19-31.

Durbin, J. (1961). Some methods for constructing exact tests. Biometrika 48, 41-55.

Durbin, J. (1973). Weak convergence of the sample distribution function when parameters are estimated. Ann. Stat. 1, 279.

Durbin, J., Knott, M., and Taylor, C. C. (1975). Components of Cramer-von Mises statistics. II. J. Roy. Statist. Soc. B 37, 216-237.

Dyer, A. R. (1974). Comparisons of tests for normality with a cautionary note. Biometrika 61, 185-189.

Elderton, W. P. and Johnson, N. L. (1969). Systems of Frequency Curves. Cambridge University Press, Cambridge.

Epps, T. W. and Pulley, L. B. (1983). A test for normality based on the empirical characteristic function. Biometrika 70, 723-726.

Filliben, J. J. (1975). The probability plot correlation coefficient test for normality. Technometrics 17, 111-117.

Fisher, R. A. (1928). On a property connecting the X² measure of discrepancy with the method of maximum likelihood. Reproduced in Contributions to Mathematical Statistics (1950). Wiley, New York.

Gastwirth, J. L. and Owens, M. G. B. (1977). On classical tests of normality. Biometrika 64, 135-139.

Geary, R. C. (1935). The ratio of the mean deviation to the standard deviation as a test of normality. Biometrika 27, 310-332.

Geary, R. C. (1947). Testing for normality. Biometrika 34, 209-242.

Gnanadesikan, R. (1977). Methods for Statistical Data Analysis of Multivariate Observations. Wiley, New York.

Green, J. R. and Hegazy, Y. A. S. (1976). Powerful modified EDF goodness-of-fit tests. J. Amer. Statist. Ass. 71, 204-209.

Hall, P. and Welsh, A. H. (1983). A test of normality based on the empirical characteristic function. Biometrika 70, 485-489.

Harter, H. L. (1961). Expected values of normal order statistics. Biometrika 48, 151-165.

Hartley, H. O. and Pfaffenberger, R. C. (1972). Quadratic forms in order statistics used as goodness-of-fit criteria. Biometrika 59, 605-611.

Hastings, C. (1955). Approximations for Digital Computers. Princeton University Press, Princeton, N.J.

Hegazy, Y. A. S. and Green, J. R. (1975). Some new goodness-of-fit tests using order statistics. Appl. Statist. 24, 297-308.

Hogg, R. V. (1972). More light on the kurtosis and related statistics. J. Amer. Statist. Ass. 67, 422-424.

Kolmogorov, A. (1933). Sulla determinazione empirica di una legge di distribuzione. Gior. Ist. Ital. Attuari 4, 83-91.

Kuiper, N. H. (1960). Tests concerning random points on a circle. Proc. Koninkl. Neder. Akad. van Wetenschappen, A 63, 38-47.

La Brecque, J. (1977). Goodness-of-fit tests based on nonlinearity in probability plots. Technometrics 19, 293-306.

Lin, C. and Mudholkar, G. S. (1980). A simple test for normality against asymmetric alternatives. Biometrika 67, 455-461.

Locke, C. and Spurrier, J. D. (1976). The use of U-statistics for testing normality against non-symmetric alternatives. Biometrika 63, 143-147.

Locke, C. and Spurrier, J. D. (1977). The use of U-statistics for testing
normality against alternatives with both tails heavy or both tails light.
Biometrika 64, 638-640.

Machado, S. G. (1983). Two statistics for testing for multivariate normality.
Biometrika 70, 713-718.

Malkovich, J. F. and Afifi, A. A. (1973). On tests for multivariate normality. J. Amer. Statist. Ass. 68, 176-179.

Mardia, K. V. (1970). Measures of multivariate skewness and kurtosis with
applications. Biometrika 57, 519-530.

Mardia, K. V. (1974). Applications of some measures of multivariate skewness and kurtosis for testing normality and robustness studies. Sankhya
A 36, 115-128.

Mardia, K. V. (1975). Assessment of multivariate normality and the robustness of Hotelling's $T^2$ test. Appl. Statist. 24, 163-171.

Mardia, K. V. and Foster, K. (1983). Omnibus test of multinormality based
on skewness and kurtosis. Comm. in Statist. 12, 207-221.

Mulholland, H. P. (1977). On the null distribution of $\sqrt{b_1}$ for samples of size
at most 25, with tables. Biometrika 64, 401-409.

O'Reilly, F. and Quesenberry, C. P. (1973). The conditional probability
integral transformation and application to composite chi-square goodness
of fit test. Am. Statist. 1, 74-83.

Pearson, E. S. (1930). A further development of tests for normality. Biometrika 22, 239.

Pearson, E. S., D'Agostino, R. B., and Bowman, K. O. (1977). Tests for
departure from normality. Comparison of powers. Biometrika 64,
231-246.

Pearson, E. S. and Hartley, H. O. (1972). Biometrika Tables for Statisticians, Vols. I and II. Cambridge University Press, Cambridge.

Pearson, K. (1895). Contributions to the mathematical theory of evolution.
Philosophical Transactions of the Royal Society, London, 91, 343.

Pearson, K. (1900). On a criterion that a given system of deviations from
the probable in the case of a correlated system of variables is such that
it can be reasonably supposed to have arisen in random sampling. Phil.
Mag. 5th Ser. 50, 157-175

Prescott, P. (1976). Comparison of tests for normality using stylized surfaces.
Biometrika 63, 285-289.

Puri, M. L. and Rao, C. R. (1976). Augmenting Shapiro-Wilk test for normality. Contributions to Applied Statistics. Birkhauser (Grossehaus),
Berlin, 129-139.

Pyke, R. (1959). The supremum and infimum of the Poisson process. J. Amer. Math. Soc. 30, 568-576.

Quesenberry, C. P. (1975). Transforming samples from truncation parameter distributions to uniformity. Comm. in Statist. 4, 1149-1155.

Rao, C. R. (1973). Linear Statistical Inference and Its Applications (2nd edition). Wiley, New York.

Royston, J. P. (1982). An extension of Shapiro and Wilk's W Test for normality to large samples. Appl. Statist. 31, 115-124.

Saniga, E. M. and Miles, J. A. (1979). Power of some standard goodness-of-fit tests of normality against asymmetric stable alternatives. J. Amer. Statist. Asn. 74, 861-865.

Shapiro, S. S. and Francia, R. S. (1972). An analysis of variance test for normality. J. Amer. Statist. Ass. 67, 215-216.

Shapiro, S. S. and Wilk, M. B. (1965). An analysis of variance test for normality (complete samples). Biometrika 52, 591-611.

Shapiro, S. S., Wilk, M. B., and Chen, H. J. (1968). A comparative study of various tests for normality. J. Amer. Statist. Ass. 63, 1343-1372.

Spiegelhalter, D. J. (1977). A test for normality against symmetric alternatives. Biometrika 64, 415-418.

Smirnov, N. V. (1939). Sur les ecarts de la courbe de distribution empirique (Russian/French summary). Rec. Math. 6, 3-26.

Stephens, M. A. (1974). EDF statistics for goodness-of-fit and some comparisons. J. Amer. Statist. Ass. 69, 730-737.

Tiku, M. L. (1974). A new statistic for testing for normality. Comm. Statist. 3, 223-232.

Uthoff, V. (1970). An optimum test property of two well-known statistics. J. Amer. Statist. Ass. 65, 1597-1600.

Uthoff, V. (1973). The most powerful scale and location invariant test of the normal versus the double exponential. Ann. Statist. 1, 170-174.

Watson, G. S. (1957). The $X^2$ goodness-of-fit test for normal distributions. Biometrika 44, 336-348.

Watson, G. S. (1961). Goodness-of-fit tests on a circle. Biometrika 48, 109-114.

Weisberg, S. (1980). Comment to some large sample tests for nonnormality in the linear regression model. J. Amer. Statist. Ass. 75, 28-31.

Weisberg, S. and Binham, C. (1975). An approximate analysis of variance test for non-normality suitable for machine calculation. Technometrics 17, 133-134.

White, H. and MacDonald, G. M. (1980). Some large-sample tests for non-normality in the regression model. J. Amer. Statist. Ass. 75, 16-28.

# 10

# Tests for the Exponential Distribution

Michael A. Stephens   Simon Fraser University, Burnaby, B.C., Canada

## 10.1 INTRODUCTION AND CONTENTS

The exponential distribution is probably the one most used in statistical work after the normal distribution. It has important connections with life testing, reliability theory, and the theory of stochastic processes, and is closely related to several other well-known distributions with statistical applications, for example, the gamma and the Weibull distributions.

The general form of the exponential distribution is

$$F(x; \alpha, \beta) = 1 - \exp\{-(x - \alpha)/\beta\}, \quad x \geq \alpha, \tag{10.1}$$

where $\alpha$ and $\beta$ are constants, with $\beta$ positive. The notation $\text{Exp}(\alpha, \beta)$ will be used to refer to $F(x; \alpha, \beta)$, or to a sample from it, and $\alpha$ will be called the origin of the distribution; the mean of $\text{Exp}(\alpha, \beta)$ is $\alpha + \beta$ and the variance is $\beta^2$. If a random sample $X_1, \ldots, X_n$ is $\text{Exp}(\alpha, \beta)$, the ordered sample $X_{(1)} < X_{(2)} < \cdots < X_{(n)}$ will be said to be ordered $\text{Exp}(\alpha, \beta)$. As in other chapters, the notation $U(0, 1)$ will refer to a uniform distribution from 0 to 1, and a sample from $U(0, 1)$ will be called a uniform sample or, if placed in ascending order, an ordered uniform sample.

In this chapter we discuss tests of the null hypothesis that a random sample $X_1, \ldots, X_n$ is $\text{Exp}(\alpha, \beta)$, with possibly $\alpha$, or $\beta$, or both, unknown. This gives four possible cases:

Case 0: where $\alpha$ and $\beta$ are both known;
Case 1: where $\alpha$ is unknown and $\beta$ is known;

Case 2: where $\alpha$ is known and $\beta$ is unknown;
Case 3: where both $\alpha$ and $\beta$ are unknown.

The null hypotheses corresponding to these four cases will be called $H_{00}$, $H_{01}$, $H_{02}$, and $H_{03}$.

Case 0 can be easily handled: the Probability Integral Transform (PIT, Section 4.2.3) gives n values $Z_i = F(X_i; \alpha, \beta)$, which, on $H_{00}$, are uniform $U(0,1)$ and these can be tested by any of the methods of Section 4.4 or Chapter 8. An example of a Case 0 test for exponentiality is given in Section 4.9. The data are the 15 values of X, given in Table 10.1, and are times to failure of airconditioning equipment in aircraft, given by Proschan (1963). We shall use these data throughout the chapter to illustrate test procedures.

Mathematical properties of the exponential distribution can be used to change Case 1 to Case 0, and to change Case 2 and Case 3 to the special test of Case 2 with $\alpha = 0$. Most of the tests in the literature have been proposed for this situation, so we shall reserve the notation $H_0$ for the hypothesis:

$H_0$: a random sample of size n comes from $Exp(0, \beta)$, with $\beta$ unknown

Tests for this hypothesis are those discussed throughout most of this chapter, although we return to Case 3 in Section 10.14.

A large number of test procedures have been given for $H_0$. One reason for this is that, again because of mathematical properties of the exponential distribution, it is possible to transform a sample of n X-values from $Exp(0, \beta)$ in several useful ways; one transformation (N below) takes the sample to a new n-sample X' which is also $Exp(0, \beta)$, and another transformation (J) takes X to a set of n - 1 ordered uniforms U. Further, J can be applied to the X' set, to give a set of n - 1 uniforms U'; we call the conversion of X to U' the K-transformation on X. Thus tests of $H_0$ on X can become tests of $H_0$ on X', or tests of uniformity on U or on U'. Furthermore, the different transformations have useful interpretations, depending on the original motivation for testing the X sample, and on the alternative distributions to $Exp(0, \beta)$ that the X might have. Two of the most important applications of $Exp(0, \beta)$ variables are to modelling time intervals between events, or to modelling lifetimes, or times to failure, in survival analysis and reliability theory. A particular application will tend to lead to a particular group of tests. In a general way, the J transformation, and various tests on U, will arise naturally in connection with a series of events, and the N and K transformations, with tests on X' or on U', will arise in tests on lifetimes. This is partly because the properties of X' and U' are influenced by whether or not the true distribution of X, if not exponential, has a decreasing or increasing failure rate.

The overall plan of this chapter is therefore as follows. After a section on notation, we show how other cases are reduced to Case 0 or to a test of $H_0$. The applications of the exponential distribution are discussed, and fol-

TABLE 10.1  Set of Observations X and Derived Values

| $X^a$ | $T^b$ | $U^c$ | $D^d$ | $X'^e$ | $U'^f$ |
|-------|-------|-------|-------|--------|--------|
| 74    | 74    | .041  | .041  | 180    | .099   |
| 57    | 131   | .072  | .031  | 126    | .17    |
| 48    | 179   | .098  | .026  | 65     | .20    |
| 29    | 208   | .114  | .016  | 12     | .21    |
| 502   | 710   | .390  | .276  | 22     | .22    |
| 12    | 722   | .397  | .007  | 0      | .22    |
| 70    | 792   | .435  | .038  | 171    | .32    |
| 21    | 813   | .447  | .012  | 72     | .36    |
| 29    | 842   | .463  | .016  | 14     | .36    |
| 386   | 1228  | .675  | .212  | 66     | .40    |
| 59    | 1287  | .708  | .033  | 20     | .41    |
| 27    | 1314  | .722  | .014  | 316    | .58    |
| 153   | 1467  | .806  | .084  | 519    | .87    |
| 26    | 1493  | .821  | .015  | 120    | .94    |
| 326   | 1819  | 1.000 | .179  | 116    | 1.00   |
| 1819  |       |       |       | 1819   |        |

$^a$Times to failure of airconditioning equipment for an aircraft.
$^b$Partial sums of $X_i$ (Section 10.4).
$^c$Values of T divided by the largest value (1819), i.e., values U derived from X by the J-transformation (Section 10.5).
$^d$Spacings between the U-values.
$^e$Normalized spacings given by the N transformation (Section 10.5).
$^f$Values U' obtained from X', by the J transformation, or from X by the K transformation (Section 10.5).
Data taken from Proschan (1963), by permission of the author and of the American Statistical Association.

lowed by the details of the N, J, and K transformations, and some of their properties. Then we turn to tests of $H_0$. With the potential applications in mind, these are roughly grouped into three groups, as follows: Group 1, those applied to X; Group 2, those applied to U; Group 3, those applied to X' or to U'. The tests for the three groups occupy Sections 10.7 to 10.11. The test statistics discussed are almost always presented in the context in which they were first suggested, although, obviously, any test first suggested for the X set could equally be applied to set X', and vice versa. There will be some inevitable overlap with other chapters, particularly Chapter 8, containing tests for uniformity. A few statistics are repeated in Chapter 8 (they differ slightly because in Chapter 8 it is natural to calculate the statistics from n uniforms, but in this chapter they are found from m = n - 1 uniforms).

One group which is treated mainly in this chapter are tests based on spacings between uniforms since such spacings are exponentials and these tests have arisen mostly in connection with tests of $H_0$.

If a general random sample is given, with no details as to context, and a test of $H_0$ is required, the question of making a transformation or not becomes one of obtaining the best power for a test procedure against a class of alternatives. Some studies on power of tests are reported in Section 10.13, for both omnibus tests and one-sided tests. One feature of these studies is that they reveal much similarity in power of many of the test procedures, when applied to a general random sample. The user will therefore often be guided by personal preference.

The value of a test statistic often gives information on the set from which it is calculated, and this in turn may sometimes be interpreted in terms of the X-sample. Thus, in modern statistics, it is common practice to calculate several statistics, and to use them to analyze features of the given data, rather than rigorously to apply significance tests. This approach is taken to illustrate the tests, applied to the data set in Table 10.1, in Section 10.12. In Section 10.14 we return to Case 3 tests.

## 10.2  NOTATION

The notation used in this chapter, apart from that already described, is listed in this section.

X: the original data set, $X_1, X_2, \ldots, X_n$. i is the index of $X_i$.

n = size of set X; m = n - 1.

DFR, IFR: decreasing failure rate, increasing failure rate (Section 10.4.5).

DFR (IFR) Sample: a random sample from a DFR (IFR) distribution (Section 10.4.5).

$C_V$: coefficient of variation (Section 10.4.5)

Transformations J, K, N: See Section 10.5.

X': a set of size n, derived from X by transformation N.

U: a set of size m, derived from X by transformation J.

U': a set of size m, derived from X by transformation K.

E: spacings between exponentials X (Section 10.5.2).

D: spacings between set U (Section 10.9.3).

D': spacings between set U' (Section 10.11.4).

Group 1: tests using X (Section 10.8).

Group 2: tests using U (Section 10.9).

Group 3: tests using X' or U' (Section 10.11).

log X means $\log_e$ X.

Significant tail of a test statistic: See Section 10.13.1.

## 10.3 TESTS FOR EXPONENTIALITY: THE FOUR CASES

### 10.3.1 Four Results

In this section we show how Case 1 can be reduced to Case 0, and Cases 2 and 3 to a test of $H_0$. These employ the following properties of a sample from $Exp(\alpha, \beta)$.

Result 1. If $X_i$, $i = 1, \ldots, n$, is a random sample from $Exp(\alpha, \beta)$, the set $Y_i$ given by $Y_i = X_i - \alpha$, $i = 1, \ldots, n$, is a random sample from $Exp(0, \beta)$.

Result 2. If $X_i$, $i = 1, \ldots, n$, is an ordered sample from $Exp(\alpha, \beta)$, the Y-sample obtained from $Y_{(i)} = X_{(i+1)} - X_{(i)}$, $i = 1, \ldots, n - 1$, is an ordered sample of size $n - 1$ from $Exp(0, \beta)$. This result can be successively applied to give

Result 3. If $X_{(i)}$, $i = 1, \ldots, n$, is an ordered sample from $Exp(\alpha, \beta)$, the Z-sample obtained from

$$Z_{(i)} = X_{(i+r)} - X_{(r)}, \quad i = 1, \ldots, n - r, \quad (10.2)$$

where r is fixed, $1 \leq r \leq n - 1$, is an ordered sample of size $n - r$ from $Exp(0, \beta)$.

Result 4. If X is $Exp(0, \beta)$, $Y \equiv 2X/\beta$ has the $\chi_2^2$ distribution.

### 10.3.2 Tests for Cases 1, 2, and 3

#### 10.3.2.1 Tests for Case 1: $\alpha$ is not known but $\beta$ is known

For this case, Result 2 above can be used to change the original X-sample to a Y-sample of size $n - 1$ which, on $H_{01}$, will be $Exp(0, \beta)$, with $\beta$ known, and the test becomes a Case 0 test on the Y set.

#### 10.3.2.2 Tests for Case 2: $\alpha$ is known, and $\beta$ is unknown

For this case, when the known value of $\alpha$ is not zero, Result 1 can be used to produce a Y-sample which, on $H_{02}$, will be $Exp(0, \beta)$. Thus the Case 2 test with $\alpha$ known is reduced to a test of $H_0$, applied to the Y-set.

10.3.2.3  Tests for Case 3: both parameters unknown

This test, of $H_{03}$: that a given set X is from $Exp(\alpha, \beta)$ with both parameters unknown, can be transformed by use of Results 2 or 3 above to a test that the Y-set or Z-set is $Exp(0, \beta)$, that is, a test of $H_0$ applied to set Y or to set Z. Result 2 will be used if a complete X-sample of size n is available; the Y-sample is then of size n - 1. Result 3 will be useful if the first r - 1 ordered observations of the X-set are not available, or if, for some reason, they are suspected to be outliers.

This use of Results 2 and 3 has been a generally accepted way to handle tests with unknown $\alpha$, but it may not be the best way, and we return to tests of Case 3 in Section 10.14.

## 10.4  APPLICATIONS OF THE EXPONENTIAL DISTRIBUTION

### 10.4.1  The Poisson Process

Suppose a series of events is recorded, starting at time $T_0 = 0$; the events occur at times $T_1, T_2, \ldots$, with $T_0 < T_1 < T_2 < \cdots < T_n$. Consider the intervals between events, $X_i$, defined by

$$X_i = T_i - T_{i-1}, \quad i = 1, \ldots, n$$

If the events at times $T_i$ are from a Poisson process the variables $X_i$ will be independently distributed $Exp(0, \beta)$, where $\beta$ is a positive constant. A test that the process generating the events is Poisson can therefore be based on a test that the intervals $X_i$ are $Exp(0, \beta)$. The Poisson process is discussed in many textbooks, see, for example, Cox and Lewis (1966, Chapter 2). If the $T_i$ are recorded on a horizontal time axis, the $X_i$ are the spacings between the $T_i$.

### 10.4.2  Models for Time to Failure

The second application of $Exp(0, \beta)$ is to model the lifetime, or time to failure, of, say, a piece of apparatus, such as the airconditioning equipment for which 15 values of X are given in Table 10.1.

Suppose the item is immediately replaced whenever it fails and let $X_i$ be the lifetime of successive items. If times $T_i$ are calculated, given by

$$T_1 = X_1, \quad T_2 = X_1 + X_2, \quad \ldots, \quad T_n = X_1 + X_2 + \cdots + X_n,$$

these times will be the times of failure for the overall equipment (here an aircraft).

By comparison with the preceding paragraph it may be seen that, whenever the $X_i$ are $\text{Exp}(0, \beta)$, the times $T_i$ can be regarded as a realization of a Poisson process.

### E 10.4.2.1 Example

Values $T_i$, derived from the $X_i$, are shown in Table 10.1.

### 10.4.3 A Lifetesting Experiment

The lifetimes $X_i$ of equipment in the example above were obtained by using the units **successively** in an aircraft as required. If it is desired to test that lifetimes are exponential, the test will be accelerated, if the units are available and expendable, by making a laboratory experiment in which they are all put into use at the same time $T_0 = 0$, and times to failure are recorded either until all units fail, or until a fixed time $T_f$ is reached. Suppose the units are numbered 1 to n, and let $X_i$ be the lifetime of unit number i; notice that in general the labelling of the $X_i$ will be quite arbitrary. Times will be recorded as failures occur, and these times give the order statistics, $T_0 < X_{(1)} < X_{(2)} < \cdots < T_f$ of the X sample. If only r items fail in time $T_f$, only the first r order statistics of the sample will be known. The sample is then said to be right-censored (see Section 4.7 and Chapter 12).

Note that times $T_i^*$ calculated from the <u>order statistics</u> by $T_1^* = X_{(1)}$, $T_2^* = X_{(1)} + X_{(2)}$, etc., will not be times in a Poisson process as described in Section 10.4.2 above, because the order statistics of an exponential random sample are not themselves exponentially distributed. However, transformation N below changes ordered $\text{Exp}(0, \beta)$ variables into random $\text{Exp}(0, \beta)$ variables, and the above construction can then be used to create times in a Poisson process.

### 10.4.4 Alternative Distributions Used
###      in Reliability Theory

In order to compare various test procedures, it will be useful first to discuss distributions alternative to the exponential, which are used in reliability analysis as models for lifetime data. Two of the most important of these are the gamma and Weibull distributions. The most general forms of these distributions are given in Sections 4.11 and 4.12. Here we are interested only in the distributions with origin zero. The gamma distribution then has density

$$f_G(x) = x^{m-1} e^{-x/\beta} / \{ \beta^m \Gamma(m) \}, \quad x > 0, \tag{10.3}$$

where m and $\beta$ are positive constants. When the constant m = 1, the distribution reduces to the exponential. The density is infinite at x = 0 when m < 1, and is zero when m > 1.

The Weibull distribution, with origin zero, has density

$$f_W(x) = (m/\beta)(x/\beta)^{m-1} \exp\{-(x/\beta)^m\}, \quad x > 0, \tag{10.4}$$

with m and $\beta$ positive constants. For $m < 1$, the density $x = 0$ is infinite, and when $m > 1$ it is zero; when $m = 1$, the distribution reduces to the exponential. In shape the gamma and Weibull distributions are somewhat similar.

### 10.4.5 Properties of Distributions: Coefficient of Variation, and Failure or Hazard Rate

Two useful parameters in describing distributions are the coefficient of variation $C_V$, and the failure rate. The $C_V$ is defined as $\sigma/\mu$, where $\mu$ and $\sigma^2$ are the mean and variance of the distribution. A small $C_V$ suggests that the variable X has fairly constant values, but a large $C_V$ suggests they will be widely spread relative to the size of the mean. For the gamma distribution, $\mu = m\beta$ and $\sigma^2 = m\beta^2$; thus $C_V = m^{-\frac{1}{2}}$.

The failure rate, or hazard rate, of X, is defined as

$$h(x) = \frac{f(x)}{1 - F(x)}, \quad x > 0, \tag{10.5}$$

where $f(x)$ and $F(x)$ are the density and distribution functions of X, assumed continuous. If $F(x)$ is a distribution of lifetimes, the quantity $h(x)\,dx$ may be interpreted as the probability of failing in time $dx$ at $x$, given that failure has not occurred up to $x$. For the exponential distribution $\text{Exp}(0,\beta)$, $h(x) = \beta$, a constant; for the gamma and the Weibull distributions, the failure rate increases steadily with x for $m > 1$, and decreases for $0 < m < 1$; for $m = 1$ they both reduce to the exponential distribution with constant failure rate. Abbreviations IFR and DFR are commonly used for increasing or decreasing failure rate. A distribution with IFR (often called an IFR distribution; a sample from such a distribution will be called an IFR sample) will have $C_V < 1$, and a DFR distribution will have $C_V > 1$. We can summarize results for the gamma and Weibull distributions in the following small table.

| Parameter value | $C_V$ | Failure rate |
|---|---|---|
| $m < 1$ | $C_V > 1$ | DFR |
| $m = 1$ (Exp $(0,\beta)$ | $C_V = 1$ | Constant $= \beta$ |
| $m > 1$ | $C_V < 1$ | IFR |

## 10.5  TRANSFORMATIONS FROM EXPONENTIALS
##         TO EXPONENTIALS OR TO UNIFORMS

### 10.5.1  Introduction

We now describe the three important transformations which are often made
to the data set X under test for $Exp(0, \beta)$. These are:

1)  transformation N, which transforms an ordered exponential sample
    $Exp(0, \beta)$, of size n, to a random $Exp(0, \beta)$ sample of size n;
2)  transformation J, which transforms a random $Exp(0, \beta)$ sample of
    size n into an ordered uniform $U(0, 1)$ sample of size $m = n - 1$;
3)  transformation K, which, like J, transforms a random $Exp(0, \beta)$ sample
    of size n into an ordered uniform $U(0, 1)$ sample of size $m = n - 1$.

### 10.5.2  The N Transformation from Exponentials
###          to Exponentials: Normalized Spacings

Suppose $X_{(i)}$, $i = 1, \ldots, n$, is an ordered sample from $Exp(0, \beta)$, and
define the __spacings__ between the exponentials by

$$E_i = X_{(i)} - X_{(i-1)}, \qquad i = 1, \ldots, n$$

with $X_{(0)} \equiv 0$. The new set $X_i'$, defined by

$$X_i' = (n + 1 - i)E_i, \qquad i = 1, \ldots, n$$

will be independently and identically distributed $Exp(0, \beta)$. For more precise
conditions on this result see, for example, Seshadri, Csörgő, and Stephens
(1969). This transformation will be called transformation N and we write
$X' = NX$. The values $X_i'$ are called the __normalized spacings__ of the original
set $X_i$: $E_i$ has expected value $\beta/(n + 1 - i)$, so that $\ell_i = E_i/$(expectation
of $E_i$) $\equiv X_i'/\beta$; the $\ell_i$ are sometimes called __leaps__. For further discussion of
normalized spacings see Section 4.20.

### 10.5.3  The "Total Time on Test" Statistic

Suppose the transformed sample $X_i'$ is used to construct a Poisson process
realization as described in Section 10.4.2:

$$T_1' = X_1', \quad T_2' = X_1' + X_2', \quad \ldots, \quad T_r' = \sum_{i=1}^{r} X_i'$$

It may easily be shown that $T_r'$ is also, in terms of the original X-values,

$$T'_r = X_{(1)} + X_{(2)} + \cdots + X_{(r)} + (n - r)X_{(r)} \qquad (10.6)$$

In the context of the lifetesting experiment discussed in Section 10.4.3, the first r terms on the right side of (10.6) are the times for which the first r failed items were working successfully and the last term is the time so far spent working by those n - r items which have not yet failed. Thus $T'_r$ is interpretable as the <u>total time on test</u> till the r-th failure. At time $X_{(n)}$, all test items have failed, and then

$$T'_n = \sum_{i=1}^{n} X'_i = \sum_{i=1}^{n} X_i = T_n \qquad (10.7)$$

### 10.5.4 The J Transformation from Exponentials to Uniforms

A result in Section 8.2 states that the n + 1 spacings between a sample of size n from U(0, 1) are each Exp$(0, \beta)$ with $\beta = 1/(n + 1)$; the spacings are not a random sample, but are conditional on their sum being fixed. This being so, a sample of ordered uniforms can be produced from an exponential sample as follows:

a)  Let $X_1, X_2, \ldots, X_n$ be a random sample from Exp$(0, \beta)$, and let $T_j = \Sigma_{i=1}^{j} X_i$, $j = 1, \ldots, n$.

b)  Define $U_{(j)} = T_j/T_n$, $j = 1, \ldots, n - 1$.

   <u>Result 1.</u> The $U_{(j)}$ are distributed as the order statistics of a random sample U, of size n - 1, from U(0, 1).
   Note that if the definition were extended to j = n, the value of $U_{(n)}$ would be identically 1. Thus the transformation produces n - 1 ordered uniforms from n original observations; a "degree of freedom" has been lost in eliminating the unknown parameter $\beta$. The above transformation from X to U will be called the transformation J, and we write U = JX. If the $T_i$ are the times in a Poisson process, as described in Section 10.4.2, the $U_{(i)}$ above are the values $T_i/T_n$, $i = 1, \ldots, n - 1$.
   Result 1 was obtained by dividing the times in a Poisson process by the time of the last event observed. If the process is observed to a <u>fixed</u> time, we have a second result:
   <u>Result 2.</u> If a Poisson process is observed from time zero to a fixed time $T_f$, with n events at $0 < T_1 < T_2 < \cdots < T_n < T_f$ occurring in that time, the values $U_{(i)} = T_i/T_f$, $i = 1, \ldots, n$, will give a sample U of n ordered uniforms.

10.5.5  The K Transformation from Exponentials
        to Uniforms

A sample of X-values, all positive, could first be transformed to a new set
X' by transformation N, and then to a set U' by applying transformation J
to X'. On $H_0$, the set U' will be n - 1 ordered uniforms. The combination
of N and J is equivalent to the following transformation, which we call K.

a)  Let $X_{(1)} < X_{(2)} < \cdots < X_{(n)}$ be the order statistics of a sample from
    $Exp(0,\beta)$ and let $T_n = \Sigma_{i=1}^n X_i$, as before.

b)  Write $X_{(0)} = 0$ and let $E_i = X_{(i)} - X_{(i-1)}$; calculate

$$X_i' = (n + 1 - i) E_i, \quad i = 1, \ldots, n$$

c)  Calculate $T_j' = \Sigma_{i=1}^j X_i'$, and $U_{(j)}' = T_j'/T_n'$, $j = 1, \ldots, n - 1$.

Result. The set U' is a sample of n - 1 ordered uniforms from U(0, 1).
Note also that, from (10.7), $T_n' = T_n$. Following earlier notation we write
U' = KX.

## E 10.5.5  Example

In Table 10.1 the values X' are given, obtained by application of transformation N to the set X; also given are the values U and U' obtained from application of transformations J and K.

10.5.6  The N and K Transformations with Censored Data

The N and K transformations can be used with a censored sample. When only
the r smallest values of the original set X are available, the r values
$X_1', X_2', \ldots, X_r'$ of set X' given by the N transformation can still be obtained;
the J transformation can be applied to these to produce r - 1 ordered uniforms. This is equivalent to the following useful result. Suppose we calculate,
using the notation of Section 10.5.5, values

$$U_{(j)}' = T_j'/T_r', \quad j = 1, \ldots, r - 1$$

Result. The set U' is a complete sample of r - 1 ordered uniforms
from U(0, 1).
   Also, since the $X_i'$ are a random sample from $Exp(0,\beta)$, the value of
$(\Sigma_{i=1}^r X_i')/r = T_r'/r$ gives an estimate of $\beta$; this estimate is often used with
censored data.

## 10.6  TEST SITUATIONS AND CHOICE OF PROCEDURES

### 10.6.1  General Comments

The decision whether or not to test $H_0$ directly on the X-set, or to use one of the transformations to X' or to U or U', is related to the three test situations: tests on a general random sample, with no particular context given, tests on intervals between events (where the indexing of the intervals might be important), and tests on lifetimes. We first observe an important difference between transformations J and K: the J transformation preserves the original indexing of the $X_i$, while the K transformation does not—that is to say, in making transformation K (or N) the $X_i$ are first put in ascending order, and the original labelling is irrelevant. Most tests based directly on the $X_i$ (Group 1 below) will also involve putting them in order and losing the original indexing.

In tests on a series of events, the original indexing will probably be important—for example, one wants to know if the intervals are getting longer or shorter as time passes—and then more information will be given by using J followed by tests on U.

On the other hand, when the labelling of the $X_i$ is not important there are some disadvantages to J. Consider, for example, the lifetimes of equipment in the laboratory experiment described in Section 10.4.3. There is no significance (presumably) in the index i attached to lifetime $X_i$, and so different statisticians could label the X-set differently; when this is done the J transformation will produce different values U, and when tests for uniformity are applied to the U-set, different conclusions will be reached.

In contrast, use of the K transformation gives always the same set U', and tests based on U' will, for all statisticians, give the same results. The same holds true of tests based on the X' themselves, and of those tests on the X-set, such as EDF or regression tests, where the observations are first ordered. This invariance is usually considered a desirable property for a test procedure.

Another feature of J is that it can produce superuniform U, that is, a set U which is too evenly spaced to be considered uniform (see Sections 8.5 and 4.5.1). This will occur, sometimes, when the X-set comes from an IFR alternative. Many test procedures are not set up to detect superuniforms; for example, EDF tests or regression tests using the upper tail (as is customary) will not detect them. Several of the test statistics to follow will detect superuniforms, or EDF tests or regression tests may be used with the lower tail, but then power is lost when the tests are used with two tails against DFR alternatives.

The K transformation produces U' values which will never typically be superuniform; in fact, they will drift toward 0 for DFR samples and toward 1 for IFR samples, so the pattern of the U' set gives information about the alternative. Finally, K may be used on a censored sample. For these reasons K is recommended rather than J to apply to a general random sample. These points were discussed further by Seshadri, Csorgo, and Stephens (1969), and we return to them again when we discuss power in Section 10.13.

We now pose two contrasting questions: why transform the data at all or, at the other extreme, why not apply several transformations, one after the other?

In considering these questions, when the indexing is not important, the tester will be most interested in getting good power for a test procedure, with possibly a particular class of alternatives in mind. We shall see in Section 10.13 that, in practice, tests on the original set X, and tests on U', often give much the same power; furthermore, some tests on X also give information on the parent population (IFR or DFR).

On the other hand, if K = JN is a good thing, why not apply, say, N several times to X, to get sets X' = NX, then X'' = NX' = NNX, etc., and test the final set for exponentiality? Equally, one could apply G and W (uniforms-to-uniforms transformations given in Chapter 8) to uniforms U or U', to obtain data sets represented symbolically by, for example, $U_2 = GGJX$, or $U_3 = WGJX$, or $U_4 = GWGJX$, all of which, on $H_0$, should be uniform and can be tested by the methods of Chapter 8. One good reason for not repeating such transformations as G and W ad absurdum must be that they will appear to a practical statistician to be arbitrary and unmotivated, and to produce data sets which are far removed from the original X; then if $H_0$ is rejected, because, say, the final data set has too many values close to zero, it will be difficult or impossible to interpret this phenomenon in terms of properties of the original X set. Other practical reasons exist too; it is pointed out in Chapter 8 that application of G to a nonuniform set may often increase power of a subsequent test for uniformity, but repeated applications may decrease it. No doubt for these reasons tests of $H_0$ in the literature, such as those given below, have been confined, if a transformation of this type is used at all, to one application of N, J, or K. Some tests involving the set $X^W = WX$, that is, one application of W to X, have been discussed by Wang and Chang (1977). Other aspects of transformations have been discussed by O'Reilly and Stephens (1983); there are interesting connections between the CPIT of Chapter 6 and transformations J, K, G, and W.

We now discuss in greater detail tests on events and tests on lifetimes.

## 10.6.2 Tests on a Series of Events

It has been suggested above that in tests on events, the natural index of $X_i$ will play an important role. Consider tests for the Poisson process, against the alternative of trend.

In the Poisson process, events occur at a constant rate as time passes; this leads to the intervals between events being independent and exponential. More precisely, let $\lambda(t)\,dt$ be the probability of an event occurring in the interval dt at time t; for the Poisson process $\lambda(t)$ is constant. An obvious alternative to the Poisson process is the model for which $\lambda(t)$ increases or decreases with t, that is, events occur more quickly or more slowly as time passes. There is then said to be a trend in the rate of occurrence. A possible model for trend considered, for example, by Cox (1955) is to suppose $\lambda(t) = c\,e^{kt}$; if k is positive (negative) there is an increasing (decreasing) rate of events, and if k = 0 the rate is constant. Another model for trend is $\lambda(t) = \mu(\mu t)^a$ (a > 1); this was considered by Bartholomew (1956) who found a sequential test for randomness against this model for trend.

If events occur more quickly, the intervals $X_i$ between events will become shorter on average; thus the $X_i$, as naturally indexed in time, become smaller. If the J transformation is applied to the $X_i$ as naturally indexed, the $U_{(i)}$ values will tend toward 1. If events occur more slowly, the $U_{(i)}$ tend toward zero. Thus trend will be detected by the statistics which detect movements of the U-values toward 0 or 1. Another alternative to a constant arrival rate for events is the possibility that they occur periodically; then the intervals between events are of fairly constant length. The coefficient of variation $C_v$ of the $X_i$ will be small, and the $U_{(i)}$ from the J transformation will be superuniform. Statistics to detect periodicity of events must be well adapted to detect superuniformity of the $U_{(i)}$. In contrast, if there is a wide disparity between the intervals $X_i$, when the longer intervals appear too long compared with the shorter intervals, the $C_v$ of the $X_i$ will be large.

It is possible that the intervals between events, even if exponentially distributed, are not independent. Lack of independence is usually difficult to detect, and the appropriate test statistic will depend on how the intervals are related. Here the indexing of $X_i$ will again be important; as naturally indexed in time, the $X_i$ might, for example, be tested for autocorrelation. An interesting example of events for which the J transform produces superuniform $U_{(i)}$, probably because of lack of independence, are the events recorded by the dates $T_i$ marking the reigns of kings and queens of England (see Pearson (1963)). For further remarks on the problems of correlated intervals see Lewis (1965).

It might be decided to base a test for uniformity of the $U_i$ on the spacings $D_i$ between the $U_i$. The spacing $D_i$ is $X_i/T_n$, so that the remarks above concerning the indexing of the $X_i$ will apply also to the spacings $D_i$. The sizes of the spacings, as naturally indexed by time, will be important both in tests for trend or for independence, whereas the variance of the spacings will be important in measuring periodicity or great disparity between the time intervals.

## 10.6.3  Tests on Lifetime Data

Suppose the original $X_i$ are lifetimes. Application of the N transformation will produce a set $X_i'$ which will be Exp$(0, \beta)$ if the $X_i$ are Exp$(0, \beta)$; however, if the $X_i$ are from an IFR or DFR distribution, the $X_i'$ will not be exponential, and the indexing of the $X_i'$ will become important. Suppose the true lifetime distribution of X has an increasing failure rate. Then a random sample X, when placed in order, should exhibit smaller spacings for large values of X than those given by the exponential distribution with the same mean lifetime. Since the $X_i'$ are normalized values of these spacings, that is, the spacings multiplied by a factor, the $X_i'$ for large i will on the whole be smaller than expected. This is formally stated as follows (see, for example, Barlow, Bartholomew, Bremner, and Brunk, 1972 (Chapter 6).

Result. If the lifetime distribution of X is IFR (DFR) the $X_i'$ are stochastically decreasing (increasing) in i = 1, ..., n.

From this result have come many ideas for testing that the original sample X is Exp $(0, \beta)$, against IFR or DFR alternatives, using tests based on $X_i'$ or $U_i'$. A general discussion, with examples, is in Epstein (1960a, 1960b).

Finally, we should note that powerful tests on lifetimes will be powerful tests on any general random sample, regardless of its source. In the power studies of Section 10.13, it turns out that it is useful to divide alternatives into DFR and IFR classes; this division is naturally meaningful in tests on lifetimes, but it tends also to classify alternatives by $C_V$ value, and by their skewness and kurtosis compared with the exponential.

## 10.7  TESTS WITH ORIGIN KNOWN: GROUPS 1, 2, AND 3

After the preceding discussion we classify test procedures for $H_0$ into three broad groups:

Group 1.  Tests for exponentiality using the basic data set X.

Group 2.  Tests based on the transformation U = JX, with a subsequent test for uniformity of U.

Group 3.  Tests based on the transformations X' = NX, or U' = KX, followed by a test for exponentiality of the X', or a test for uniformity of U'.

It is clear from earlier discussion that J will often be applied to intervals between events, leading to tests of Group 2, and N and K will be applied to lifetime or failure-time data, leading to tests in Group 3. Group 2 tests have been called uniform conditional tests by Lewis (1965) and by Cox and Lewis (1966).

## 10.8  GROUP 1 TESTS

Here the set X will be tested directly for Exp $(0, \beta)$, with no transformation N, J, or K. Tests available include the Pearson chi-square test and modern adaptations, discussed in Chapter 3, EDF tests and regression tests, discussed in Chapters 4 and 5, and tests based on sample moments. No further comments will be offered on chi-square tests, but the other tests will be reviewed below, in the special context of tests for exponentiality.

### 10.8.1 EDF Tests

(a) The direct EDF test, in which $\beta$ is estimated by $\bar{X}$, is described in Section 4.9.

(b) A variation of standard EDF procedures has been suggested by Srinivasan (1970, 1971) (see Section 4.16.3); for the exponential distribution the calculations are easy. The transformation is made to a Z-set by

$$Z_{(i)} = 1 - (1 - X_{(i)}/T_n)^{n-1}, \quad i = 1, \ldots, n$$

where $T_n = \Sigma_{i=1}^n X_i$. The Kolmogorov statistic $\tilde{D}$ is then calculated from the $Z_{(i)}$ using equations (4.2); large values of $\tilde{D}$ are significant. Schafer, Finkelstein, and Collins (1972) have given tables of Monte Carlo significance points. The transformation to $Z_{(i)}$ used in this test and the transformation $Z_{(i)} = 1 - \exp(-X_{(i)}/\bar{X})$, used in the direct EDF test, are very close, and Moore (1973) showed the two tests to be asymptotically equivalent. Power studies show that for small samples they have very similar properties also.

(c) Another test based on the EDF has been proposed by Finkelstein and Schafer (1971). The $Z_{(i)}$ are calculated as for the direct EDF test, that is, from $Z_{(i)} = 1 - \exp\{-X_{(i)}/\bar{X}\}$, $i = 1, \ldots, n$; then $\delta_i$ is defined as $\max\{|Z_{(i)} - (i-1)/n|, |Z_{(i)} - i/n|\}$, $i = 1, \ldots, n$, and the test statistic is $S^* = \Sigma_{i=1}^n \delta_i$. Finkelstein and Schafer have provided tables for $S^*$ based on Monte Carlo studies.

### 10.8.2 Regression Tests

Most regression and correlation tests described in Chapter 5 are devised for Case 3, where both location and scale parameters are unknown. However, two tests are designed specifically for testing $H_0$. These are

a) Stephens' $W_S$, described in Section 5.11.5;
b) Jackson's test, described in Section 5.11.5.

### 10.8.3 Tests Based on s Sample Moments

Gurland and Dahiya (1970) and Dahiya and Gurland (1972) have discussed a general method of deriving a test statistic based on s sample moments. Suppose the r-th sample moment is $m'_r = \Sigma_{i=1}^n (X_i)^r/n$; when the Dahiya-Gurland method is applied to the test for $\text{Exp}(0,\beta)$ we have (Currie and Stephens, 1984)

$$Q_1 = C_1 = n(-1 + m'_2/2m_1'^2)^2 = n\{-1 + m_2/m_1'^2\}^2/4 \quad \text{where } m_2 = m'_2 - (m'_1)^2$$

$$Q_2 = Q_1 + C_2, \text{ with } C_2 = n\{1 - m_2'/m_1'^2 + m_3' /(3m_2'm_1')\}^2$$

$$Q_3 = Q_2 + C_3, \text{ with } C_3 = n\{-1 + 3m_2'/(2m_1'^2) - m_3'/(m_1'm_2') + m_4'/(4m_1'm_3')\}^2$$

Asymptotically $Q_t$ has a $\chi_t^2$ distribution on $H_0$, but the convergence is quite slow. Statistic $Q_1$ is equivalent to Greenwood's statistic, discussed in Section 8.9.1 and in Section 10.9.3 below; percentage points for $Q_2$ and $Q_3$, for finite n, obtained by Monte Carlo sampling, have been given by Currie and Stephens (1984). Tests based on $Q_t$ are upper-tail tests. Currie and Stephens have given power studies for n = 20, against a wide range of alternatives. Dahiya and Gurland (1972) have given power studies, for n = 50 and 100, and for gamma and Weibull alternatives.

TABLE 10.2 Tests for the Set X of Table 10.1

---

## Group 1 tests

---

### EDF and regression statistics applied directly to the X-values

$\bar{X}$ = 121.3, S.D.(X) = 154.3, Coefficient of Variation = 1.272

Direct EDF test statistics (Section 10.8.1): $\hat{\beta} = \bar{X} = 121.2$; $n' = 15$

Statistics followed by approximate significance levels p, when p < 0.10:

$D^+ = 0.277$, $D^- = 0.132$, $D = 0.277$ (p = 0.04), $V = 0.409$

$W^2 = 0.219$ (p = 0.05), $U^2 = 0.170$ (p = 0.04), $A^2 = 1.163$ (p = 0.075)

$\hat{D} = 0.292$ (p = 0.04)   $S^* = 1.9970$ (p = 0.045)

### Regression statistics and p-values for set X (Section 10.8.2)

$W_E = 0.0384$ (p = 0.04, lower tail, so p = 0.080, 2-tail)

$W_S = 0.0397$ (p = 0.075, lower tail, so p = 0.15, 2-tail)

$J = 2.039$

$R(X, m) = .958$, $Z = 15\{1 - R^2(X,m)\} = 1.24$   (p = .25)
$R(X, H) = .950$, $Z = 15\{1 - R^2(X,H)\} = 1.47$   (p = .15)

### Values of Dahiya-Gurland statistics (Section 10.8.3)

$Q_1 = 0.977$   (p = 0.32)
$Q_2 = 3.384$   (p = 0.18)
$Q_3 = 5.081$   (p = 0.16)

---

E 10.8.3  Example

Values of EDF statistics for the X-data of Table 10.1 are given in Table 10.2, Part 1. For these data $\hat{\beta} = 121.2$; thus, following Section 4.9, we calculate

$$Z_{(1)} = 1 - \exp(-X_{(1)}/\bar{X}) = 1 - \exp(-12/121.2) = 0.094,$$

and so on. The values of Z are given in Table 4.13, column 3. Then equations (4.2) give $D^+ = 0.277$ and the other values given in Table 10.2. Some p-levels of the statistics (used with upper tail only) are recorded. Several are significant at the 5% level, suggesting rejection of $H_0$. $\tilde{D}$ and $S^*$ (Section 10.8.1) are also given.

Values of regression statistics are given in Part 2 of Table 10.2. The values of R(X, m) and R(X, H) (Section 5.11) are, respectively, 0.958 and 0.950; then Z(X, m) in Section 5.11.2 is $15(1 - 0.958)^2) = 1.24$ and Z(X, H) is $15\{1 - (0.950)^2\} = 1.47$. These are, respectively, significant at $p = 0.25$ and $p = 0.15$ when referred to Tables 5.6 and 5.7. The value of the Shapiro-Wilk statistic $W_E$, which is designed for use when $\alpha$ is not known, is included for comparison. In Part 3 of Table 10.2 are given the values of the Dahiya and Gurland statistics $Q_1$, $Q_2$, and $Q_3$. The p-values have been found from the Currie-Stephens tables.


10.9  GROUP 2 TESTS, APPLIED TO U = JX

10.9.1  Tests Based Directly on the U-Values

As was stated before, a number of tests for a series of events are based on making transformation J and then testing that the n - 1 values U are U(0, 1). Because these tests have been suggested in this connection they are reviewed here; there will necessarily be some overlap with Chapter 8.

Important Note: In Chapter 8 it is natural to assume that the U set has n values; when tests of Chapter 8 are applied to set U (or to set U' after the K transformation) the value n must be replaced by m (= n - 1) in the formulas for test statistics, and in using the tables.

a) EDF tests. EDF tests (Case 0) can be used on the U-set, as described in Chapter 8. Statistics $D^+$ and $D^-$ are well adapted to detect a shift of U toward 0 or 1, that is, to detect trend in events (Section 10.6.2). $W^2$ and $A^2$ can also be expected to be effective for these alternatives. Notice, however, that as customarily used (employing only the upper tail of test statistics for significance) EDF statistics will not detect superuniform U; thus they will not detect periodicity in events (Section 10.6.2), or the occasions when the J transform can produce superuniformity (Section 10.6.2), unless test statistics are referred to the lower tail of the relevant null distribution.

b) The statistic $\bar{U}$. A simple statistic for testing uniformity is the mean $\bar{U} = \Sigma_{i=1}^{n-1} U_i/(n - 1)$. Percentage points for $\bar{U}$ are given in Table 8.5; the table must be entered at same size $m = n - 1$. For $m \geq 15$, the quantity $P \equiv (\bar{U} - 0.5)(12m)^{\frac{1}{2}}$ will have approximately the standard normal distribution.

c) Statistics based on $U_{(r)}$. The order statistic $U_{(r)}$ has a beta distribution (Section 8.8), and the function of $U_{(r)}$ given by $Z_r = (n - r)U_{(r)}/\{r(1 - U_{(r)})\}$ has an $F_{p,q}$ distribution with $p = 2r$ and $q = 2(n - r)$ degrees of freedom. In particular the median $\tilde{U}$, where $r$ is $n/2$ or $(n + 1)/2$, has been proposed as a test statistic for uniformity.

### 10.9.2 Application to Tests for Trend

When the model for trend in a series of events is $\lambda(t) = c\,e^{kt}$, as discussed in Section 10.6.2, the values $U_{(i)}$, $i = 1, \ldots, n - 1$, instead of being ordered uniforms, will be an ordered sample from density

TABLE 10.3 Tests for the Set X of Table 10.1

---

#### Group 2 tests

---

#### J transformation followed by tests for uniformity on 14 values of U

#### Values of Test Statistics

$D^+ = 0.180$, $D^- = 0.105$, $D = 0.180$, $V = 0.285$

$W^2 = 0.106$, $U^2 = 0.059$, $A^2 = 0.729$, $FS^* = 1.466$

None of the above is significant at the 20% level upper tail, or the 15% level lower tail.

Statistics $\bar{U} = 0.442$, $U_{(7)} = 0.447$, $U_{(8)} = 0.463$, Neyman $N_2^2 = 0.722$

None of the above is significant at the 20% level (1 tail), or the 40% level (2 tails).

$R(U, m) = 0.978$; $Z = 14\{1 - R^2(U, m)\} = 0.61$ $(p > 0.50)$

#### Statistics based on the 15 spacings $D_i$

Moran $M = 19.232$, $M/c = 16.329$ $(p = 0.30$ in the upper tail of $\chi_{14}^2)$

Greenwood $G(14) = 0.167$, $14G(14) = 2.338$ $(p = 0.075$, upper tail)

Kendall-Sherman $K = 0.470$

Quesenberry-Miller $Q = 0.193$ $(p = 0.35$, upper tail)

Lorenz $L_{14}(.5) = 0.106$ $(p = 0.1$, lower tail)

$$f(u) = \frac{k}{e^k - 1} e^{ku}, \quad 0 \le u \le 1$$

Cox (1955) suggested that the test for $k = 0$ (Poisson process) against $k \ne 0$ should be based on $\bar{U}$, which is the likelihood ratio statistic. Large values of $\bar{U}$ will indicate $k > 0$, that is, events are occurring more rapidly with increasing time and the $U_{(i)}$ are tending to drift toward 1 similarly, low $\bar{U}$ indicates that events are happening less often, and the $U_{(i)}$ are moving toward zero. Thus a one-tail test is used if the direction of trend is known, but in general a two-tail test is required. The median $\tilde{U}$ will also detect movements of the U-values toward 0 or 1. Note that neither the mean $\bar{U}$ nor the median $\tilde{U}$ will detect superuniform observations, nor, in general, the case where there is excessive variation among the intervals between events.

### E 10.9.2 Example

Table 10.3, Part 1, shows the values of EDF statistics calculated, following Section 4.4, from the 14 values in the U-set. Also shown are the values of $U_{(7)}$, $U_{(8)}$, and $\bar{U}$. On $H_0$, $U_{(7)}$ has the beta $(x; 7, 8)$ distribution (Section 8.8), and $U_{(8)}$ has the beta $(x; 8, 7)$ distribution. Tables of this distribution, and Table 8.5 for $\bar{U}$, give the approximate p-levels shown. The correlation $R(U, m)$ (Section 5.6) is 0.978, with a p-value greater than 0.5 (Table 5.2); note that $R(U, m)$ has weaknesses as a test statistic (Section 5.6).

### 10.9.3 Statistics Based on the Spacings Between the $U_i$

The spacings between the U-set are defined by

$$D_i = U_{(i)} - U_{(i-1)}, \quad i = 1, \ldots, n, \quad \text{where} \quad U_{(o)} \equiv 0, \quad \text{and} \quad U_{(n)} \equiv 1$$

thus $n - 1$ uniforms give $n$ spacings. The spacings are connected with the original observations $X_i$ by $D_i = X_i/T_n$, $i = 1, \ldots, n$, where $T_n = \Sigma_{i=1}^n X_i$. Basic articles for work on spacings are by Pyke (1965, 1972). Many test statistics for exponentiality have been based on the values $X_i$, divided by $T_n$ to eliminate the scale parameter $\beta$. These statistics are therefore calculated, in effect, from the values $D_i$, and the associated tests can be regarded as tests for uniformity of the set U, based on the spacings. Test statistics of this type are discussed both in this chapter and also in Chapter 8.

### 10.9.3.1 Greenwood's Statistic

The first spacings statistic which we discuss was introduced by Greenwood (1946) in connection with tests on a series of events; specifically, on the incidence of a contagious disease. The statistic here is $G(n - 1) = \Sigma_{i=1}^n D_i^2$, the argument $n - 1$ referring to the fact that $G(n - 1)$ is calculated from n - 1 uniforms, giving n spacings. For use in the present application we have

$$G(n - 1) = \sum_{i=1}^{n} D_i^2$$

$$= \left( \sum_{i=1}^{n} X_i^2 \right) \Big/ T_n^2$$

to make a test, $(n - 1)G(n - 1)$ is referred to Table 8.3 using the percentage points for sample size $m = n - 1$. Small values of $G(n - 1)$ will detect super-uniform values $U_i$, or excessively regular spacings between events, such as would occur if they were periodic. Large values of $G(n - 1)$ will detect if the intervals are too disperse, for example, if the long intervals are too long compared with the short intervals.

### E 10.9.3.1  Example

From the D of Table 10.1, $G(14) = (.041)^2 + (.031)^2 + \cdots + (.179)^2 = .167$. Thus $14G(14) = 2.34$, and reference to Table 8.3 shows this to be significant at $p = 0.075$, upper tail.

### E 10.9.3.2  Equivalence of Greenwood's Statistic and Other Statistics

Greenwood's statistic will detect unusual dispersion of the spacings. The mean value of each spacing is $1/n$, so the dispersion could be measured by

$$V = \sum_{i=1}^{n} (D_i - 1/n)^2$$

a statistic studied by Kimball (1947). It is easily shown that $V = G(n-1) - 1/n$.
   Also, in terms of the original $X_i$, $V$ can be written

$$V = \sum_{i=1}^{n} (X_i - \bar{X})^2 / T_n^2$$

$$= S^2 / (n\bar{X})^2$$

where $S^2 = \Sigma_{i=1}^{n} (X_i - \bar{X})^2$. Moments of the X set are $m_1' = \bar{X}$ and $m_2 = S^2/n$, and the sample coefficient of variation $C_V$ is $(\sqrt{m_2})/m_1'$; thus V is $m_2/(nm_1'^2) = C_V^2/n$. Note also that $Q_1$ of Section 10.8.3 is calculated from $m_2/m_1'^2$. To summarize, the following relations hold:

$$G(n - 1) = V + 1/n = C_v^2/n + 1/n$$

and

$$Q_1 = n\{nG(n - 1) - 2\}^2/4$$

Thus V, $C_v$, and $Q_1$ are all equivalent to $G(n - 1)$ as test statistics. Use of the upper tail of $Q_1$ is equivalent to a two-tail test based on $G(n - 1)$; the two tails contain unequal probabilities for finite n, converging slowly to equal probabilities as n increases.

Furthermore, V is the same as the regression statistic $WE_0$ (Section 5.11.5; Stephens' $W_S$ (Section 5.11.5) is also related to $G(n - 1)$ by

$$(W_S)^{-1} = n(n + 1)\{G(n - 1)\} - n$$

Tests using the upper tail of $G(n - 1)$ or of $WE_0$ are equivalent to tests using the lower tail of $W_S$, and vice versa.

Thus, several statistics which have been derived from very different approaches all turn out to be equivalent to Greenwood's $G(n - 1)$.

### 10.9.3.3 Other Spacings Statistics

A number of statistics have been devised which are directly related to Greenwood's statistic; most of these have been discussed in connection with testing for uniformity and are included in Chapter 8. The Quesenberry-Miller statistic Q (Section 8.9.2) might be useful in detecting autocorrelation in a series of events; so, also, might statistics based on high-order gaps (Section 8.9.4). We now continue with four tests based on spacings which have been developed specifically in connection with tests for exponentiality on X or X'. They are defined in terms of both $D_i$ and the original $X_i$.

### 10.9.3.4 Statistic M (Moran, 1951)

This statistic is

$$M(n - 1) = -2 \sum_{i=1}^{n} \log(n D_i)$$

$$= -2 \sum_{i=1}^{n} \log(X_i/\bar{X})$$

$$= -2 \left\{ \sum_{i=1}^{n} \log X_i \right\} + 2n \log \bar{X}$$

When X is $\text{Exp}(0, \beta)$, the distribution of $2X/\beta$ is $\chi_2^2$ (Result 4 of Section

10.3.1); thus the $X_i$ can be regarded as sample variances from normal samples with true variance $\beta$, based on two degrees of freedom. $M(n - 1)$ is then equivalent to Bartlett's (1934) statistic to test that such samples come from populations with the same variance; on $H_0$, the distribution of $M(n-1)/c$, where $c = 1 + (n + 1)/(6n)$, is approximately $\chi^2$ with $n - 1$ degrees of freedom. As a general test for exponentiality, $M(n - 1)$ is two-tailed.

Moran (1951) showed that $M(n - 1)$ is the asymptotically most powerful test against gamma alternatives (see also Shorack 1972), and Bartholomew (1957) showed it to be a strong test against the Weibull alternative. Cox and Lewis (1966, Chapter 6), Bartholomew (1957), and Jackson (1967), among other authors, have referred to the effect on $M(n - 1)$ of inaccurate measurement of the values $X_i$, particularly for small values; a small inaccuracy in $X_i$ produces a big error in log $X_i$. Difficulties due to small or zero values are discussed in Section 10.10. Bartholomew (1956) has based a sequential test for exponentiality on $M(n - 1)$.

### 10.9.3.5 The Kendall-Sherman Statistic

Kendall (1946) suggested a statistic for testing the randomness of events in time. This is

$$K(n - 1) = \frac{1}{2} \sum_{i=1}^{n} \left| D_i - \frac{1}{n} \right| = \frac{1}{2} \sum_{i=1}^{n} \left| (X_i/T_n) - \frac{1}{n} \right|$$

so that $K(n - 1)$ is based on a comparison of all the $D_i$ with the common expected value $1/n$. Another form of K is

$$K(n - 1) = \frac{\sum_{i=1}^{n} |X_i - \bar{X}|}{2n\bar{X}}$$

This statistic, introduced at about the same time as Greenwood's $G(n - 1)$, has many similar properties. It measures the dispersion of the $D_i$, and small values will detect superuniform $U_i$, or periodicity in events. The statistic $K(n)$, that is, derived from n uniforms, and $n + 1$ spacings, was discussed by Sherman (1950, 1957) who gave its null distribution, moments, and upper tail percentage points for $n < 20$. Bartholomew (1954) fitted an F-approximation to the null distribution of a function of $K(n)$.

### 10.9.3.6 EDF Tests for Spacings

When there are $n - 1$ values $U_i$, giving rise to n spacings $D_i$, the marginal distribution of any one spacing is

$$F_D(x;n) \equiv P(D_1 < x) = 1 - (1 - x)^{n-1}, \quad 0 < x < 1$$

This is a fully specified distribution, and it might be thought that EDF tests, Case 0 (Section 4.4) could be made. The Probability Integral Transformation would be $Z_{(i)} = 1 - (1 - D_{(i)})^{n-1}$, where the $D_{(i)}$ are the ordered spacings, and EDF statistics could be calculated from the $Z_{(i)}$. In particular, suppose $\check{D}$ is the Kolmogorov statistic. Because the spacings are not independent (their sum is 1), the $Z_{(i)}$ are not ordered uniforms, so Case 0 tables cannot be used. However, fortuitously, the $Z_{(i)}$ are exactly those which arise in Srinivasan's test (Section 10.8.1), and $\check{D}$ will be the same as $\check{D}$ of that section, and will be referred to the tables referenced there.

### 10.9.3.7 Test Based on the Lorenz Curve

Let p be a value between 0 and 1, and let $r = [np]$, that is, the greatest integer less than or equal to np. The Lorenz curve statistic, derived from the n ordered spacings $D_{(i)}$, is

$$L_n(p) = \sum_{i=1}^{r} D_{(i)}$$

$$= \left( \sum_{i=1}^{r} X_{(i)} \right) \Big/ T_n$$

Gail and Gastwirth (1978) proposed $L_n(0.5)$ as a test statistic for $H_0$, and they gave tables for a two-tail test, for values $2 \leq n \leq 40$, and a normal approximation for larger values of n. They also gave values of the asymptotic relative efficiency (ARE) of this test, compared with that based on using the maximum likelihood estimate of the shape parameter $\alpha$, for both gamma and Weibull alternatives, and some power studies.

### E 10.9.3.7 Example

Part 2 of Table 10.3 shows the values of some of the above statistics, based on the 15 spacings $D_i$ given in Table 10.1. For example, Moran's statistic is $M(14) = -2[\log(15 \times .041) + \log(15 \times .031) + \cdots + \log(15 \times .179)] = 19.232$; c is then $1 + 16/90 = 1.178$, so $M/c = 16.329$. This must be compared to $\chi^2_{14}$, to give a p-level = 0.30, approximately.

### 10.10 THE EFFECT OF ZERO VALUES, AND OF TIES

It may be that a value of X is recorded as zero; if this is so, the value of $A^2$ in the Group 1 EDF tests, and the value of Moran's statistic $M(n - 1)$ in the previous section, will become infinite and $H_0$ will automatically be rejected. Clearly, if a set X to be tested to be $Exp(0, \beta)$ contains one or more values which are recorded as zero, the reason is that a true small value has been rounded to zero, and a correction can be applied. Suppose the rounding interval is d; for the Moran statistic, Gail and Ware (1978) have shown that an adequate correction is to replace the zero by d/4, for d up to 0.2 times the mean of the exponential distribution. Thus for a mean life of 5 (say hours), the values should be recorded to at least the nearest hour and then a zero would be replaced by 0.25. In many practical situations, measurements will be measured to at least this level of accuracy, and then no correction will be needed.

The problem arises again if there are ties in the X-set (as there are in Table 10.1), and if tests are to be applied to the transformed values X' or U'. This is because two equal values in the X-set gives a zero in the X' set. Pyke (1965) has given a correction to separate two X-values recorded as equal. Nevertheless, even if corrections are used for zero values of X or X', significant values of $A^2$ or M(n - 1) should be examined carefully to see if they are due only to one or two excessively small values in the X or X' set, and, if so, why these are so small.

## 10.11 GROUP 3 TESTS APPLIED TO X' = NX, OR TO U' = KX

### 10.11.1 Introduction

Clearly, after transformations N or K, tests of $H_0$ for X, such as those given above, may equally be applied to X', and tests for uniformity for U can be applied to U'. Good reasons exist for making these transformations, particularly when the original X are lifetimes. These come from the results in Section 10.6.3, namely, that if the distribution of X is not exponential but is IFR, the $X_i'$ are stochastically decreasing with i, while if X is DFR the $X_i'$ are stochastically increasing with i. These properties have led to further tests being proposed in connection with lifetime data X. The new tests are functions of $X_i'/T_n'$, that is, of the spacings $D_i'$ between the U' set, defined as were the $D_i$ from $U_i$ in Section 10.9.3. Tests on U' are discussed in the next section, followed by the new group of tests based on D'.

### 10.11.2 Direct Tests for Exponentiality on the Transformed X'

EDF and regression tests, applied directly to set X' have not been much emphasized, perhaps because the X' would first be ordered, and the information given by the indexing of the X' is then lost. Other tests on the X' themselves have been proposed by Epstein (1960a, 1960b). These make use of Result 4 of Section 10.3.1 but now applied to X'; on $H_0$, $y_i = 2X_i'/\beta$ has the

$\chi_2^2$ distribution, and the $y_i$ are independent. Therefore ratios of independent sums of $y_i$, times a constant, have the F distribution. Epstein has suggested tests based on such partial sums, to test if the value of $\beta$ has changed, or to test if the time to the first failure is significantly longer or smaller than expected if all failure times come from the same exponential distribution. The times $y_i$ can also be divided into groups, and the several groups tested to see if they have a common $\beta$; for example, Bartlett's test that several normal samples have the same population variance, or other well-known tests of this hypothesis, can be adapted to make a test for common $\beta$. When the $y_i$ values are divided into only two groups, $Y_1 = \Sigma_{i=1}^r y_i$ and $Y_2 = \Sigma_{i=r+1}^n y_i$, it is easily seen that tests based on the ratio $Y_1/Y_2$ are equivalent to tests based on $U'_{(r)}$, to be discussed below. As can be seen, much of the emphasis in Epstein (1960a, 1960b) is on tests for $\beta$; however, there is a fine line between such tests for a parameter and tests of fit, and several of Epstein's illustrations may be viewed as tests of fit.

### E 10.11.2  Example

Values of EDF statistics and regression statistics, calculated from the X' given in Table 10.1, are given in Table 10.4.

TABLE 10.4  Tests for the Set X of Table 10.1

| Group 3 tests |
| --- |

Statistics calculated from set X'

Mean = 121.3,  S.D. = 138.6,  coefficient of variation = 1.143

Direct EDF test-statistics: $D^+ = 0.167$,  $D^- = 0.082$,  $D = 0.167$,

$V = 0.250$, $W^2 = 0.050$, $U^2 = 0.041$, $A^2 = \infty$; corrected $A^2 = 0.462$ (see Section 10.10).  $S^* = 1.151$,  $\tilde{D} = 0.177$

None of the above is significant at the 25% level, upper tail; $W^2$ and $U^2$ are significant at approximately the 25% level, lower tail.

Regression statistics

$W_E = 0.058$ (p = 0.20 lower tail).  $R(Z,m) = 0.98$, $Z = 0.54$ (p > 0.50).

$W_s = 0.049$ (p = 0.12 lower tail).  $R(Z,H) = 0.97$, $Z = 0.90$ (p = 0.35)

$J = 1.957$

Greenwood  $G(14) = .148$; $14G(14) = 2.071$ (p > 0.10 upper tail)

Dahiya–Gurland  $Q_2 = 0.569$ (p = 0.48); $Q_3 = 0.680$ (p > 0.50)

## 10.11.3 Tests Based on the U' Values

Since on $H_0$ the $n - 1$ $U_i'$ values should be uniform $U(0, 1)$, tests can be based on testing this hypothesis concerning U'. For IFR alternatives, the $U_i'$ move toward 1, and for DFR alternatives, they move toward zero. EDF statistics (Case 0) or the statistics $\bar{U}'$ or $U_{(r)}'$ for some r, might be expected to be useful in detecting such alternatives. Of the EDF statistics, $D^+$ will be significant when the $U_i'$ move near zero, and $D^-$ when they move near 1. Statistics $W^2$ and $A^2$, and to a lesser extent D, will detect either of these alternatives.

### E 10.11.3  Example

Values of EDF statistics based on the U' derived from the data set X in Table 10.1, are given in Table 10.5. The statistics are found by using $U_{(i)}'$ in equations (4.2), and p-values are found from Tables 4.2.1 and 4.2.2.

### 10.11.3.1  Tests Based on $\bar{U}'$ or $U_{(r)}'$

The simplest test for the uniformity of the U'-set is based on U', the mean of the $n - 1$ values, or equivalently, on their sum $S = (n - 1)\bar{U}'$; this statistic was suggested by Lewis (1965). Some algebra will show that, in terms of the original X-values,

$$S = 2n - 2 \sum_{i=1}^{n} i X_{(i)}/T_n = 2\left\{ \sum_{i=1}^{n} (n - i) X_{(i)}\right\} / T_n$$

$\bar{U}'$ will tend to be large for an original IFR sample, and to be small for a DFR sample. Thus $\bar{U}'$ or S can be used as a one-tail test to guard against alternatives with IFR or DFR, but as a statistic against unknown or general alternatives it will be two-tailed. Percentage points for $\bar{U}' = S/(n - 1)$ are given in Table 8.5; the table must be entered for sample size $m = n - 1$.

For $m \geq 15$, $(\bar{U}' - 0.5)(12m)^{\frac{1}{2}}$ will have approximately the standard normal distribution.

Lewis (1965) also suggested the statistic $U_{(r)}'$ as test statistic, with r a suitable integer. The statistic $Z_r'$ given by $Z_r' = (n - r)U_{(r)}'/\{r(1 - U_{(r)}')\}$ has, on $H_0$, the $F_{p,q}$ distribution with $p = 2r$ and $q = 2(n - r)$ degrees of freedom. A commonly suggested statistic is the median $U_{(r)}'$, with $r = (n + 1)/2$ or $n/2$. For IFR alternatives, $U_{(r)}'$ can be expected to be large, so that $Z_r'$ is significant in the upper tail of $F_{p,q}$; for DFR alternatives $Z_r'$ will be significant in the lower tail. This statistic was again examined by Gnedenko, Belyayev, and Solovyev (1969), by Fercho and Ringer (1972), and Tiku, Rai, and Mead (1974); the statistic y of Tiku, Rai, and Mead (1974, Section 4) designed for testing $H_0$, is equivalent to $U_{(n-r)}'$.

In Table 10.5 are given values of $\bar{U}'$ and $U_{(7)}'$ and $U_{(8)}'$ for the U' set derived from U of Table 10.1; p-values are found from Table 8.5 and the beta (x; 7, 8) and beta (x; 8, 7) distributions (Sections 8.8.2 and 8.10.1).

TABLE 10.5  Tests for the Set X of Table 10.1

Group 3 tests

Statistics for uniformity calculated from the 14 values of U'

Statistics followed by approximate significance levels p in parentheses:

$D^+ = 0.374$ (p = 0.015)            $D^- = 0.099$ (p > 0.25)

$D = 0.374$ (p = 0.03)              $V = 0.473$ (p = 0.025)

$W^2 = 0.417$ (p = 0.07)            $U^2 = 0.227$ (p = 0.02)

$A^2 = 1.894$ (p = 0.10)            $S^* = 2.433$ (p = 0.09)

Statistics $\bar{U}' = 0.383$ (p = 0.07 lower tail, p = 0.14 2-tail)

$\qquad \bar{U}'_{(7)} = 0.317$ (p = 0.12 lower tail)

$\qquad U'_{(8)} = 0.356$ (p = 0.09 lower tail)

Neyman $N_2 = 2.558$ (p = 0.30)

$R(U, m) = 0.90$,  $Z = 2.62$ (p < 0.01)

Statistics based on the 15 spacings $D'_i$

Moran M = 25.579, M/C = 21.714 (p = 0.09 upper tail)
    (a zero spacing corrected, see Section 10.10)

Greenwood G(14) = 0.148,  14G(14) = 2.072 (p > 0.10 upper tail)

Quesenberry-Miller Q = 0.237 (p = 0.05 upper tail)

$S_1^* = -0.60$;  $S_2^* = 1.20$;  $S_3^* = 0.24$ (Section 10.11.4).

10.11.4  Tests Based on the Spacings Between the U' Set

All the tests in Section 10.9.3 for uniformity of U, based on the spacings D, can of course be applied to the new spacings D' calculated from the U':
$D'_i = U'_{(i)} - U'_{(i-1)}$, i = 1, ..., n, with $U'_{(o)} \equiv 0$ and $U'_{(n)} \equiv 1$. Some new statistics have also been proposed for the set X', based essentially on the $D'_i$.

10.11.4.1  The "Cumulative Total Time on Test" Statistic

The total time on test statistic was defined in Section 10.5.3 as $T'_r = \Sigma^r_{i=1} X'_i$.
Suppose, for given k, we define

$$V_k = \frac{\sum\limits_{r=1}^{k-1} T'_r}{T'_n}, \quad k = 2, \ldots, n$$

$V_k$ is called the k-th <u>cumulative total time on test</u> statistic. When $k = n$, $V_n = \sum_{r=1}^{n-1} U'_{(r)}$, since $U'_{(r)} = T'_r / T'_n$. Thus $V_n = (n - 1)\bar{U}'$. Another formula for $V_n$ is

$$V_n = \frac{\sum\limits_{i=1}^{n} (n - 1)X'_i}{T'_n} = n - \frac{\sum\limits_{i=1}^{n} i X'_i}{T'_n}$$

In terms of the spacings $D'_i$, this is

$$V_n = \sum\limits_{i=1}^{n} (n - 1)D'_i = n - \sum\limits_{i=1}^{n} i D'_i$$

Another group of tests, proposed for the set X' by Bickel and Doksum (1969) includes

$$S^*_1 = \sum\limits_{i=1}^{n} [-i X'_i / \{(n + 1)T_n\}] = \left\{ -\sum\limits_{i=1}^{n} i D'_i \right\} / (n + 1)$$

$$S^*_2 = \sum\limits_{i=1}^{n} X'_i H_i / T_n = \sum\limits_{i=1}^{n} D'_i H_i, \quad \text{where } H_i = -\log(1 - i/(n + 1))$$

$$S^*_3 = \sum\limits_{i=1}^{n} X'_i (-\log H_i)/T_n = -\sum\limits_{i=1}^{n} D'_i (\log H_i)$$

Recall, in these formulas, that $T_n = T'_n$ (Equation 10.7). Statistics $V_n$ and $S^*_r$ have a resemblance to the regression statistics of Chapter 5, but there is an important difference; the set $X'_i$ are not ordered in the above formulas, and they would be for regression statistics.

Statistic $S^*_3$ is asymptotically most powerful against Weibull alternatives for X, and statistics $S^*_1$ and $S^*_2$ against two other alternatives (the Makeham and linear failure rate alternatives) discussed by Bickel and Doksum. Other statistics may be derived using the properties of $X'_i$. For example, on average, $X'_i < X'_j$ for $i > j$ if the distribution is IFR and a test could be based on the number of <u>reversals</u>, that is, the number of occasions when this inequality is realized, for all pairwise comparisons. Alternatively, values $X_i'$ could be plotted against $i$, or against $n - i + 1$; the slope of the regression line would, if the original X were $Exp(0, \beta)$, be zero, but if the X were from an IFR distribution, it would be negative for the first plot and positive for the second. (The slope has the same sign as $\Sigma_i i D_i' - n/2$.) Proschan and Pyke (1967) and Bickel and Doksum (1969) have also investigated tests based on the ranks of the $X_i'$.

### E 10.11.4  Example

Values of the Greenwood G(14), Moran M, and Quesenberry-Miller Q, calculated from the $D_i'$, are given in Table 10.5. These tend to have lower p-levels than the corresponding statistics based on the $D_i$, in Table 10.3. The X' set, and hence the D' set, contains a zero, and the Moran statistic (also the Anderson-Darling $A^2$) has been calculated using the correction suggested in Section 10.10: $X_{(1)}$ has been given value 0.25, since the rounding interval is 1.

### 10.11.5  The Equivalence of $\bar{U}'$ and Other Test Statistics

Several of the statistics given in Section 10.11.4 are equivalent to the statistic $\bar{U}'$ discussed in Section 10.11.3. Since $V_n = (n - 1)\bar{U}'$, $V_n$ is the same as S of Section 10.11.3. Also $V_n$ and $S_1^*$ are related by $V_n = n + (n + 1)S_1^*$. Thus, $V_n$, S, and $S_1^*$ are all equivalent to $\bar{U}'$ as test statistics.

Another statistic equivalent to $\bar{U}'$ is the Gini statistic $G_n$, discussed by Gail and Gastwirth (1975). $G_n$ is related to the Lorenz curve discussed in Section 10.9.3.8 above, and, like the Lorenz curve, derives from concepts used in economics. The Gini index for a distribution is twice the area between the population Lorenz curve $y = L_n(p)$ and the line $y = p$. The Gini statistic $G_n$ derived from this index can be calculated in two ways. In terms of the original $X_i$, $G_n$ is

$$G_n = \frac{\sum_{i,j=1}^{n-1} |X_i - X_j|}{2(n-1)T_n}$$

this may be shown to be the same as

$$G_n = \frac{\sum_{i=1}^{n-1} i X_{i+1}'}{(n-1)T_n'} = \frac{\sum_{i=1}^{n-1} i D_{i+1}'}{n-1}$$

Singpurwalla has shown (see Gail and Gastwirth, 1975) that $G_n = 1 - \bar{U}'$, so that $G_n$ too is equivalent to $\bar{U}'$ as a test statistic.

## 10.12 DISCUSSION OF THE DATA SET

The values of all the various statistics can now be used to give an overall assessment of the X-set in Table 10.1.

   a) The direct tests on X in Group 1 (Table 10.2) point towards rejection of $H_0$: that the X are $Exp(0, \beta)$, with the large value of $D^+$ indicating that there are too many small values of X compared with large values. This implies a DFR population for X (see next section and Table 10.6). However, transformation J, from X to U, gives little information, although Greenwood's statistic is quite large (0.167), implying high dispersion among X-values.

   b) The values X are lifetimes, and the discussion in this chapter suggests that tests on X' and on U' will be informative. The high value of $D^+$ for the EDF tests on U' (Table 10.5) means that the U' set tends toward zero, and this is confirmed by the low values of $\bar{U}'$, $U'_{(7)}$ and $U'_{(8)}$. These low values are because the normalized spacings $X'_i$ are, on the whole, increasing with i, implying that the original X are from a DFR distribution (Section 10.6.3). Thus the Group 1 tests and the tests on U' give supporting conclusions.

   c) The indexing of set X' is giving information about the parent population for X. Then Moran's or Greenwood's statistics found from the spacings $D'_i$, which are, in effect, symmetric functions of the $X'_i$ (in which the indexing is lost), are not significant. Neither are direct EDF tests on X', based on first ordering the X'-set, so that the indexing is again lost.

   d) The lack of significance of the statistics where indexing is lost in set X' indicates how the indexing can be important. Here, when the X' are only regarded as a random sample, as in c) above, they are acceptably exponential, and one would then accept that the original X-set is exponential; however, when the indexing of the $X'_i$ gives information, as it does in the tests on U', the indications are that the X-set comes from a DFR distribution.

## 10.13 EVALUATION OF TESTS FOR EXPONENTIALITY

### 10.13.1 Omnibus Tests

The author (Stephens, 1978, 1986) has conducted a large power study on the various test procedures for $H_0$, using Monte Carlo samples of sizes n = 10 and n = 20 from a wide range of alternatives. Tables 10.6, 10.7 give a selection of these results for n = 20. These permit some comparisons between tests, applied to a general random sample, when the special considerations of preserving indexes, etc., are not important.

**TABLE 10.6 Power Studies for Tests of Exponentiality, Origin Known (n = 20)**

| | Group 1 EDF tests on X | | | | | | | | Group 1 | Group 1 | Group 1 moments | | | Group 2 based on U | | | | Group 3 EDF tests on U' | | | | | | | Group 3 | Group 3 |
|---|---|---|---|---|---|---|---|---|---|---|---|---|---|---|---|---|---|---|---|---|---|---|---|---|---|---|
| Alternatives | $D^+$ | $D^-$ | $D$ | $V$ | $W^2$ | $U^2$ | $A^2$ | $\bar{D}$ | $S^*$ | $J$ | $Q_1$ | $Q_2$ | $Q_3$ | $M$ | $G$ | $K$ | $L_n$ | $D^+$ | $D^-$ | $D$ | $V$ | $W^2$ | $U^2$ | $A^2$ | $\bar{U}'$ | $U'_{(n/2)}$ |
| **IFR Alternatives** | | | | | | | | | | | | | | | | | | | | | | | | | | |
| $\chi^2_4$ | 1 | 70 | 55 | 51 | 63 | 55 | 62 | 49 | 64 | 56 | 51 | 35 | 42 | 71 | 57 | 57 | 61 | 1 | 67 | 54 | 34 | 57 | 35 | 60 | 61 | 25 |
| $U(0,1)$ | 16 | 82 | 69 | 79 | 81 | 75 | 79 | 63 | 83 | 93 | 85 | 69 | 54 | 87 | 72 | 71 | 62 | 0 | 80 | 75 | 50 | 80 | 48 | 89 | 83 | 73 |
| Weib(1.5)[a] | 1 | 70 | 55 | 51 | 61 | 55 | 50 | 50 | 65 | 64 | 58 | 45 | 34 | 67 | 63 | 62 | 62 | 0 | 71 | 56 | 30 | 61 | 31 | 61 | 64 | 36 |
| $\frac{1}{2}N$[c] | 1 | 43 | 29 | 26 | 33 | 28 | 28 | 23 | 32 | 37 | 29 | 18 | 11 | 27 | 32 | 33 | 29 | 1 | 38 | 28 | 16 | 30 | 17 | 31 | 33 | 27 |
| **DFR Alternatives** | | | | | | | | | | | | | | | | | | | | | | | | | | |
| $\chi^2_1$ | 69 | 1 | 57 | 69 | 62 | 51 | 75 | 61 | 62 | 53 | 51 | 67 | 72 | 80 | 47 | 6 | | 1 | 70 | 59 | 40 | 65 | 40 | 75 | 65 | 47 |
| Weib(0.8)[a] | 39 | 3 | 21 | 21 | 26 | 22 | 35 | 30 | 27 | 29 | 27 | 36 | 40 | 32 | 25 | 33 | 34 | 1 | 21 | 30 | 19 | 33 | 22 | 34 | 34 | 26 |
| lognor(1)[b] | 18 | 18 | 21 | 22 | 23 | 35 | 22 | 24 | 24 | 23 | 24 | 24 | 23 | 14 | 25 | 19 | 12 | 33 | 6 | 23 | 30 | 23 | 33 | 22 | 18 | 18 |
| $\frac{1}{2}$Cauchy[d] | 73 | 2 | 66 | 59 | 69 | 61 | 70 | 69 | 70 | 75 | 77 | 79 | 79 | 56 | 73 | 72 | 61 | 79 | 1 | 73 | 66 | 74 | 67 | 72 | 72 | 63 |

[a] Weib (m) refers to density (10.4) with $\beta = 1$.

[b] lognor (m) refers to density $f(x) = \text{const} \exp\{-(\log x)^2/(2m^2)\}$, $x > 0$.

[c] $\frac{1}{2}N$ : X is |Y|, where Y = N(0, 1).

[d] $\frac{1}{2}$Cauchy : X is |Y|, where Y has the Cauchy distribution, median 0.

TABLE 10.7  Power Studies for Tests of Exponentiality, Origin Known (n = 20)

5% one-sided tests: Power results

| Alternatives | D⁺ | D⁻ | D | V | W² | U² | A² | D̂ | S* | J | Q1 | Q2 | Q3 | M | G | K | $L_n$ | D⁺ | D⁻ | D | V | W² | U² | A² | Û | U'(n/2) |
|---|---|---|---|---|---|---|---|---|---|---|---|---|---|---|---|---|---|---|---|---|---|---|---|---|---|---|
| | Group 1 EDF tests on X | | | | | | | | | Group 1 moments | | | | Group 2 based on U | | | | Group 3 EDF tests on U' | | | | | | | | Group 3 |
| **IFR** | | | | | | | | | | | | | | | | | | | | | | | | | | |
| Significant tail | L | | | | | | | | | L | | | | L | L | U | | U | U | | | | | | | |
| $\chi^2_4$ | 0 | 53 | 39 | 36 | 47 | 41 | 43 | 32 | 47 | 56 | 24 | 10 | 5 | 71 | 57 | 57 | 61 | 0 | 52 | 37 | 21 | 44 | 20 | 42 | 61 | 25 |
| U(0,1) | 2 | 67 | 54 | 66 | 66 | 61 | 64 | 48 | 73 | 93 | 60 | 27 | 12 | 56 | 87 | 72 | 71 | 0 | 75 | 63 | 35 | 68 | 34 | 82 | 83 | 72 |
| Weib(1.5)[a] | 0 | 53 | 38 | 33 | 46 | 39 | 42 | 33 | 51 | 64 | 27 | 10 | 5 | 67 | 63 | 62 | 62 | 0 | 52 | 39 | 18 | 44 | 19 | 47 | 64 | 34 |
| ½N[c] | 0 | 31 | 17 | 17 | 23 | 20 | 19 | 14 | 23 | 37 | 10 | 3 | 1 | 27 | 32 | 33 | 29 | 0 | 28 | 18 | 8 | 18 | 8 | 22 | 33 | 27 |
| **DFR** | | | | | | | | | | | | | | | | | | | | | | | | | | |
| Significant tail | U | | | | | | | | | U | | | | U | U | L | | L | L | | | | | | | |
| $\chi^2_1$ | 60 | 0 | 45 | 39 | 54 | 44 | 71 | 51 | 52 | 53 | 45 | 56 | 61 | 80 | 47 | 66 | 70 | 62 | 0 | 50 | 31 | 55 | 30 | 69 | 65 | 47 |
| Weib(0.8)[a] | 24 | 0 | 16 | 13 | 20 | 15 | 26 | 22 | 21 | 29 | 23 | 30 | 31 | 32 | 25 | 33 | 34 | 30 | 1 | 22 | 14 | 23 | 13 | 27 | 34 | 26 |
| lognor(1)[b] | 11 | 11 | 13 | 15 | 15 | 16 | 14 | 15 | 15 | 23 | 19 | 18 | 18 | 14 | 25 | 19 | 12 | 14 | 9 | 15 | 21 | 14 | 20 | 13 | 18 | 18 |
| ½Cauchy[d] | 65 | 1 | 58 | 52 | 65 | 53 | 62 | 61 | 61 | 75 | 72 | 74 | 75 | 56 | 73 | 72 | 61 | 73 | 1 | 66 | 62 | 70 | 62 | 68 | 7 | 63 |

[a] Weib (m) refers to density (10.4) with $\beta = 1$.

[b] lognor (m) refers to density $f(x) = \text{const} \exp\{-(\log x)^2/(2m^2)\}$, $x > 0$.

[c] ½N: X is |Y|, where Y = N(0,1).

[d] ½Cauchy: X is |Y|, where Y has the Cauchy distribution, median 0.

Roughly speaking, for a sample as large as $n = 20$, a statistic falls into one or other tail (called the significant tail) according to whether the parent population is IFR or DFR. This is, therefore, a natural way to divide the alternatives; it also coincides with $C_V < 1$ (IFR) and $C_V > 1$ (DFR) for most populations.

We first look for good omnibus tests, that is, tests which will declare significance for the whole range of (IFR and DFR) alternatives. Statistics to be compared are: Group 1, using the upper tail for significance for EDF and Gurland-Dahiya statistics calculated from the X; Group 2, statistics derived from U, using two tails for the statistics Moran M, Greenwood G, Kendall-Sherman S, and Lorenz L because samples from IFR alternatives are likely to be significant in one tail (superuniforms U) and those from DFR alternatives in the other tail; and Group 3, EDF statistics (upper tail) and $\bar{U}'$ and $U'_{(n/2)}$, two-tail. The Jackson statistic J in Group 1 is also two-tail; other correlation statistics treated in Chapter 5 cannot rightfully be compared because they assume unknown origin, except for $W_s$, which is equivalent to Greenwood's. Recall also that several other statistics are equivalent to Greenwood's (Section 10.9.3), several to $\bar{U}'$ (Section 10.11.5), and several to $U'_{(n/2)}$ (Section 10.11.3).

It is obviously impossible to give best procedures against all alternatives, but some salient features emerge from the power results (Stephens, 1978, 1986).

(a) As omnibus tests against both IFR and DFR alternatives (Table 10.6), there is not much to choose between the following sets of statistics: Group 1, $A^2$ or $W^2$ or $S^*$ or J; Group 2, Moran M, Kendall-Sherman K, and Lorenz L; and Group 3, $A^2$ or $W^2$ or $\bar{U}'$. Perhaps the most striking feature of the power results is how similar they are for these statistics. Any of them, at the preference of the user, should do well to provide omnibus tests.

It is noteworthy that statistic $\bar{U}'$ has very high power; after transformation K, $\bar{U}'$ is, of course, easily calculated and has a null distribution which converges quickly to the normal. Note that statistic $U'_{(n/2)}$, by contrast, often gives poor power.

(b) Transformation J gives good results when followed by two-tail statistics M, G, K, or L; recall that $\bar{U}$, $U_{(n/2)}$, and EDF statistics (as usually used) cannot detect the possibility of superuniforms.

(c) For gamma and Weibull alternatives, Moran M is very good (often best, as expected), but loses power against some other alternatives; recall that there can be problems with small values. The Lorenz L and Kendall-Sherman K compare well with M overall. For these reasons M might be regarded as a "risky" statistic compared to others in (a) above (see the discussion in Section 10.10).

## 10.13.2 One-Sided or Directional Tests

Since many statistics have a significant tail for IFR alternatives and the opposite tail significant for DFR alternatives, such statistics can be used with one tail only if it is desired to guard against only one of these types of alternative. The tests must be used with care, since they will be biased against the other alternative family. Table 10.7 gives power comparisons when some statistics are used with one tail only; the size of the test is now 5%. For a fair comparison, statistics <u>always</u> used with one tail only, such as EDF or sample moment statistics, should now be compared for test size 5%. The significant tail of one-sided statistics is indicated in the table.

When the direction of the alternative is known, the Group 3 statistic $\bar{U}'$ is again effective, and now is overall better than $A^2$ in Group 3 or Group 1. The Greenwood, Sherman, and Lorenz statistics compare with $\bar{U}'$ but on occasion are less powerful; $U'_{(n/2)}$ is poor in terms of power. Moran's statistic is again best of all against gamma and Weibull alternatives, but drops behind $\bar{U}'$ for other alternatives. Again, EDF statistics (Group 1) and EDF statistics (Group 3) show remarkably similar results. The power results reported form part of a larger study, and values are available (showing similar trends) for $n = 10$ and $n = 50$.

## 10.14 TESTS WITH ORIGIN AND SCALE UNKNOWN

We now return to tests of exponentiality for which both $\alpha$ and $\beta$ in Exp $(\alpha, \beta)$ are unknown. There are several ways of dealing with two unknown parameters:

(a) In Section 10.3.2 it was shown how such a test situation could be reduced to a test of $H_0$, by making the transformation to $m = n - 1$ new variables $Y_{(i)} = X_{(i+1)} - X_{(1)}$, $i = 1, \ldots, m$; $H_0$ may then be tested on the m values in set Y using any of the methods so far given. Several authors have suggested this as a way of dealing with the unknown $\alpha$; for example, $L_n^*(p)$ of Gail and Gastwirth (1978) is explicitly derived in this way, and statistic y of Tiku, Rai, and Mead (1974, Section 2) can be derived by applying the K transformation to set Y to give $n - 2$ values U', and obtaining y as statistic $U'_{(n-r-1)}$, with $r = [n + 1]/2$.

However, the transformation to Y may not be the best way to handle unknown $\alpha$. It may be better, for example, to use any of the following methods:

(b) EDF tests on X, with $\alpha$ and $\beta$ both estimated from the data, as described in Section 4.9.4.

(c) The Shapiro-Wilk statistic $W_E$ given in Section 5.11.5.

(d) The correlation statistics $R^2(X, m)$ or $R^2(X, H)$ (Section 5.11.2).

Spinelli and Stephens (1987) have compared powers of these several techniques, using samples of size $n = 20$ and 10% tests, and have given a number of tables. Tiku, Rai, and Mead (1974) have also given power studies comparing their statistic with the Shapiro-Wilk $W_E$. Spinelli and Stephens used only Group 1 EDF tests on set Y. From the various comparisons the following points emerge:

(1) EDF statistics on set Y are slightly less powerful than direct EDF statistics where $\alpha$ and $\beta$ are both estimated, as in (b) above, except possibly for alternatives with a very high probability of small values (e.g., $\chi_1^2$ or Weibull (0.5)).

(2) The direct EDF statistics $A^2$ and $W^2$ (method (b) above) and the Shapiro-Wilk $W_E$ give similar power results, but other correlation statistics have lower power.

(3) The results of Tiku, Rai, and Mead suggest that $U'_{(n-r-1)}$, with $r = [(n+1)/2]$ (equivalent to their statistic $T_E$), has better power than has the median $U'_{(n/2)}$ in tests of $H_0$ when $\alpha$ was known to be zero, but is still not as powerful overall as $W^2$ or $A^2$ in direct EDF tests. These two are, therefore, the recommended statistics for omnibus tests, since there is a problem of consistency with $W_E$ (see Section 5.12).

(4) The significant tail for $W_E$ or $T_E$ is the same as that for $W_S$, and opposite to that for G. These statistics, and also $D^+$ or $D^-$ for direct EDF statistics, can be used as one-tail tests on one-sided families of alternatives (DFR or IFR), with a consequent increase in power.

(5) Spinelli and Stephens showed that when $\alpha$ is <u>known</u>, it is best, on the whole, to use this fact, and therefore to apply the tests given earlier in the chapter. Note that the opposite effect has been observed in connection with EDF tests for normality (Section 4.16.2).

## 10.15 SUMMARY

From the plethora of tests and power results which have been given in this chapter, it would be useful to extract a simple strategy for the practical statistician to follow, but this is not easy. Perhaps we should summarize by simply repeating four themes which have surfaced throughout the chapter:

(a) Test statistics should be regarded as giving information about the data and their parent population, rather than as tools for formal testing procedures.

(b) If the data are intervals between events, and if the times of these events are known, the natural questions to ask will be more readily answered by converting these times to the U-set via the J-transformation, and looking at the configuration of the U.

(c) If the data are lifetimes, one must ask what alternative populations are of interest. Information on IFR or DFR parent populations can be deduced from the spacings between the X; this leads naturally to the N-transformation (which gives the normalized spacings) or the K-transformation, with tests to follow on the U'-set. This approach has been much advanced, especially as the T' which lead to the U' have the "total time on test" interpretation: however, very similar information is given by direct EDF tests on the original X, or, for the important gamma or Weibull alternatives, and if the measurements are non-zero, by Moran's statistic.

(d) For data from other sources, referred to above as "a general random sample," it may still be of interest to classify possible alternative populations as IFR or DFR, which is roughly equivalent to shorter-tail or longer-tail, or to low-$C_V$ or to high-$C_V$; then comments in (c) will still apply.

# REFERENCES

Barlow, R. E., Bartholomew, D. J., Bremner, J. M., and Brunk, H. D. (1972). Statistical Inference Under Order Restrictions. New York: Wiley.

Bartholomew, D. J. (1954). Note on the use of Sherman's statistic as a test for randomness. Biometrika 41, 556-558.

Bartholomew, D. J. (1956). Tests for randomness of a series of events when the alternative is a trend. J. Roy. Statist. Soc., B 18, 234-239.

Bartholomew, D. J. (1957). Testing for departure from the exponential distribution. Biometrika 44, 253-257.

Bartlett, M. S. (1934). The problem in statistics of testing several variances. Proc. Camb. Phil. Soc. 30, 164-169.

Bickel, P. J. and Doksum, K. A. (1969). Tests for monotone failure based on normalized spacings. Ann. Math. Statist. 40, 1212-1235.

Cox, D. R. (1955). Some statistical methods connected with series of events. J. Roy. Statist. Soc., B 17, 129-164.

Cox, D. R. and Lewis, P. A. W. (1966). Statistical Analysis of Series of Events. London: Methuen.

Currie, I. and Stephens, M. A. (1984). On sample moments and tests of fit. Technical Report: Department of Statistics, Stanford University.

Dahiya, R. C. and Gurland, J. (1972). Goodness of fit tests for the gamma and exponential distributions. Technometrics 14, 791-801.

Epstein, B. (1960a). Tests for the validity of the assumption that the underlying distribution of life is exponential. Part I. Technometrics 2, 83-101.

Epstein, B. (1960b). Tests for the validity of the assumption that the underlying distribution of life is exponential. Part II. Technometrics 2, 167-183.

Fercho, W. W. and Ringer, L. J. (1972). Small sample power of some tests of the constant failure rate. Technometrics 14, 713-724.

Finkelstein, J. M. and Schafter, R. E. (1971). Improved goodness-of-fit tests. Biometrika 58, 641-645.

Gail, M. H. and Gastwirth, J. L. (1975). A scale free goodness-of-fit for the exponential distribution based on the Gini statistic. J. Roy. Statist. Soc. 40, 350-357.

Gail, M. H. and Gastwirth, J. L. (1978). A scale-free goodness-of-fit test for the exponential distribution based on the Lorenz curve. J. Amer. Statist. Assoc. 73, 787-793.

Gail, M. H. and Ware, J. (1978). On the robustness to measurement error of tests of exponentiality. Biometrika 65, 305-309.

Gnedenko, B. V., Belyayev, Yu. K., and Solovyev, A. D. (1969). Mathematical Methods of Reliability Theory. New York: Academic Press.

Greenwood, M. (1946). The statistical study of infectious disease. J. Roy. Statist. Soc., A 109, 85-110.

Gurland, J. and Dahiya, R. C. (1970). A test of fit for continuous distributions based on generalised minimum chi-squared. Statistical Papers in Honor of G. W. Snedecor, T. A. Bancroft, Editor, 115-127. Iowa State University Press.

Jackson, O. A. Y. (1967). An analysis of departure from the exponential distribution. J. Roy. Statist. Soc., B 29, 540-549.

Kendall, M. G. (1946). Discussion of Professor Greenwood's paper. J. Roy. Statist. Soc. 109, 103-105.

Kimball, B. F. (1947). Some basic theorems for developing tests of fit for the case of the non-parametric probability distribution function, I. Ann. Math. Statist. 18, 540-548.

Lewis, P. A. W. (1965). Some results on tests for Poisson processes. Biometrika 52, 67-77.

Moore, D. S. (1973). A note on Srinivasan's goodness-of-fit test. Biometrika 60, 209-211.

Moran, P. A. P. (1951). The random division of an interval—Part II. J. Roy. Statist. Soc., B 13, 147-150.

O'Reilly, F. J. and Stephens, M. A. (1982). Characterizations and goodness of fit tests. J. Roy. Statist. Soc.

Pearson, E. S. (1963). Comparison of tests for randomness of points on a line. Biometrika 50, 315-325.

Proschan, F. (1963). Theoretical explanation of observed decreasing failure rate. Technometrics 5, 375-383.

Proschan, F. and Pyke, R. (1967). Tests for monotone failure rate. Proceedings Fifth Berkeley Symposium, 3, 293-312.

Pyke, R. (1965). Spacings. J. Roy. Statist. Soc., B 27, 395-449.

Pyke, R. (1972). Spacings revisited. Proceedings Sixth Berkeley Symposium, 1, 417-427.

Schafer, R. E., Finkelstein, J. M., and Collins, J. (1972). On a goodness-of-fit test for the exponential distribution with mean unknown. Biometrika 59, 222-224.

Seshadri, V., Csörgő, M., and Stephens, M. A. (1969). Tests for the exponential distribution using Kolmogorov-type statistics. J. Roy. Statist. Soc., B 31, 499-509.

Sherman, B. (1950). A random variable related to spacings of sample values. Ann. Math. Statist. 21, 339-361.

Sherman, B. (1957). Percentages of the $\omega_n$ statistic. Ann. Math. Statist. 28, 259-261.

Shorack, G. R. (1972). The best test of exponentiality against gamma alternatives. J. Amer. Statist. Assoc., 67, 213-214.

Spinelli, J. J. and Stephens, M. A. (1987). Tests for exponentiality when origin and scale parameters are unknown. Technometrics 29, 471-476.

Srinivasan, R. (1970). An approach to testing the goodness-of-fit of incompletely specified distributions. Biometrika 57, 605-611.

Srinivasan, R. (1971). Tests for exponentiality. Statist. Hefte 12, 157-160.

Stephens, M. A. (1978). Goodness of fit tests with special reference to tests for exponentiality. Technical Report, Department of Statistics, Stanford University.

Stephens, M. A. (1986). Power studies for tests for exponentiality. Technical Report, Department of Statistics, Stanford University.

Tiku, M. L., Rai, K., and Mead, E. (1974). A new statistic for testing exponentiality. Comm. Statist. 3, 485-493.

Wang, Y. H. and Chang, S. A. (1977). A new approach to the nonparametric tests of exponential distribution with unknown parameters. The Theory and Application of Reliability 2, 235-258. New York: Academic Press.

# 11

# Analysis of Data from Censored Samples

John R. Michael  Bell Telephone Laboratories, Holmdel, New Jersey*

William R. Schucany  Southern Methodist University, Dallas, Texas

## 11.1 INTRODUCTION

In this chapter we consider a variety of techniques which are appropriate as tests of fit when only a certain portion of the random sample from a continuous underlying distribution is available. The censoring or deletion of observations can occur in several ways. The type or manner of censoring determines the appropriate method of analysis.

The most common and simple censoring schemes involve a planned limit either to the magnitude of the variables or to the number of order statistics which can be observed. These are called singly Type 1 and Type 2 censored data, respectively. The number of small (or large) order statistics which will be observed in Type 1 censoring is a random variable. In life testing applications it is quite common for an experiment to produce a Type 1 right censored sample by having n items placed on test and recording the values $0 < Y_{(1)} < \cdots < Y_{(r)}$ of the failure times which are observed up to a fixed test time. (In this chapter observations will be referred to as Y, rather than X, since in plotting techniques we shall wish to plot observations on the vertical, or y-axis.) Data arising from such a procedure are occasionally also referred to as being truncated. If the life test is planned to continue until a fixed number, r, of failures occur, then the resulting failure data are Type 2 right censored. As another example, if one records only the 10 largest independent competitive bids on an oil lease, the observed sample is singly Type 2 censored on the left. Types 1 and 2 censoring are sometimes referred to as time censoring and failure censoring, respectively.

In the more complicated situation in which the variables are subject to different censoring limits the sample is said to be multiply censored. If the

---

*Current affiliation: Westat Inc., Rockville, Maryland.

differing censoring limits are preplanned, as would result from placing
items on a life test at different starting times with a single fixed termination
time for the test, the data are progressively censored (Type 1). Samples
which are progressively censored (Type 2) occur less often in practice but
could result, again in life testing, if the units are put on test at the same
time and then selected fixed numbers of (randomly chosen) unfailed items
are removed from test immediately after different preplanned numbers of
failures have occurred.

The unplanned type of censored data which arises most often in practice
is randomly time censored or arbitrary right censored data. The larger
values (again usually in life testing) are not observed due to random censoring times which are statistically independent of the variable of interest (usually failure times). If some of the units are accidentally lost, destroyed, or
removed from the study prior to the measurement of the variable (failure
time) and if these independent censoring times are recorded then the data
can still be analyzed for goodness of fit. In certain situations competing
modes of failure will produce randomly censored data (see Example 11.2.3.2.)
Combinations of multiply right and left censored data can also arise in practice (see Section 11.2.4).

The graphical technique of examining probability plots (Chapter 2) adapts
quite easily to the censored sample situation. Subjective impressions should
be formed with somewhat more caution than in the complete sample case,
but the computational aspects are essentially unchanged. Probability plots
are discussed in Section 11.2.

When the null distribution is completely specified, the probability integral
transformation (see Section 4.2.3) may be employed to reduce the problem
to a test for uniformity. Section 11.3 presents a number of examples of
standard EDF (Chapter 4) goodness-of-fit statistics which have been modified
in a straightforward fashion to accommodate a censored uniform sample.
Adaptations for correlation (Chapter 5) and spacings (Chapter 8) statistics
are also discussed. For Type 2 censored samples a transformation of the
uniform order statistics is described which makes it possible to analyze the
data as if it were a complete random sample.

In testing fit, it is a common situation for the null hypothesis to be
composite; the hypothesized parent population is not completely specified,
but only the form $F(x|\theta)$ of the cumulative distribution function (cdf) is given.
Here $\theta$ is an indexing parameter; it may be a vector of several components,
some known and some unknown. One very natural approach which has been
taken in the complete sample case is to replace the unknown components in $\theta$
by efficient estimators (for example, the m.l.e. $\theta$) and then to calculate a
statistic based on $F(x|\theta)$ as if it were the completely specified distribution
function. This has been done, for example, in many of the tests in Chapters
4 and 5. Censoring presents an extra complication for this approach simply
because of increased complexity of efficient estimators of $\theta$. A variety of
results for the composite hypothesis problem are examined in the final

Section 11.4. Adaptations of the chi-square procedure are not covered in this chapter. For some discussion on this topic, see Section 3.4.2.

## 11.2 PROBABILITY PLOTS

Probability plotting has been described in Chapter 2 as a valuable technique for assessing goodness of fit with complete samples. This extends naturally to incomplete samples for most types of censoring. Even in the case of multiple censoring a probability plot can often be constructed quickly using only ordinary graph paper and a hand calculator.

In Section 11.2.1, the construction of probability plots for complete samples is reviewed. The method is extended to singly-censored samples in Section 11.2.2, to multiply right-censored samples in Section 11.2.3, and to other types of censoring in Sections 11.2.4-11.2.6. An easy-to-use summary of the steps required in constructing a probability plot is given in Section 11.2.7.

### 11.2.1 Complete Samples

Let $Y_{(1)}$, $Y_{(2)}$, ..., $Y_{(n)}$ be a complete ordered random sample of size n and let $F(y|\mu,\sigma)$ be the corresponding cdf where $\mu$ and $\sigma$ are unknown location and scale parameters, respectively. (Note that $\mu$ and $\sigma$ are not necessarily the mean and standard deviation.) When there is no ambiguity $F(y|\mu,\sigma)$ will be shortened to $F(\cdot)$ or $F$.

Since $\mu$ and $\sigma$ are location and scale parameters, we can write (as was done in Formula (2.9))

$$F(y|\mu,\sigma) = G\left(\frac{y-\mu}{\sigma}\right) = G(z) \tag{11.1}$$

where $Z = (Y - \mu)/\sigma$ is referred to as the standardized variable and $G(z)$, also referred to as $G(\cdot)$ or $G$, is the cdf of the standardized random variable. Using obvious notation, it follows that, using E for expectation or mean,

$$E\{Y_{(i)}\} = \mu + \sigma E\{Z_{(i)}\} = \mu + \sigma m_i$$

where $Z_{(i)}$ is the ith order statistic from the standardized distribution, and $m_i$ is $E\{Z_{(i)}\}$. Similarly, for $0 \le p_i \le 1$,

$$p_i\text{-th quantile of } F(y|\mu,\sigma) = \mu + \sigma\{p_i\text{-th quantile of } G(z)\}$$

$$= \mu + \sigma[G^{-1}(p_i)]$$

where $G^{-1}$ is the inverse function of G.

We can regard $Y_{(i)}$ as an estimate of its mean, or of the $p_i$-th quantile of $F(y; \mu, \sigma)$, where $p_i$ is an appropriate probability. In constructing a probability plot we could plot the $Y_{(i)}$ on the y-axis versus $m_i$ on the x-axis. If the sample is in fact from $F(y; \mu, \sigma)$ then the points will tend to fall on a straight line with intercept $\mu$ and slope $\sigma$. We then test our distributional assumption by visually judging the degree of linearity of the plotted points. Methods based on regression and correlation are discussed in Chapter 5.

It should be noted that if the null hypothesis is simple, that is, the values of all distributional parameters are specified beforehand, we can plot the $Y_{(i)}$ against their hypothesized means and then judge whether the plotted points fall near a straight line with intercept 0 and slope 1.

A drawback to using means of order statistics is that they are often difficult to compute. Quantiles, on the other hand, are easy to compute as long as F is easy to invert. A plot of the sample quantiles $Y_{(i)}$ versus theoretical quantiles of G is a probability plot as defined in Chapter 2; it is also called a quantile-quantile or Q-Q plot (Wilk and Gnanadesikan, 1968). However, the plots will be different from those in Chapter 2 where the observations were plotted on the horizontal or x-axis; here they are plotted on the verticle or y-axis. Special probability plotting paper is available for many families of distributions, but as was stated in Chapter 2 no special graph paper is required if F can be inverted in closed form or if standard quantiles are available from tables or approximations. Often a scientific calculator and ordinary graph paper are all that one needs.

Table 11.1 lists the cdf's of some common families of distributions along with the formulas required to construct probability plots. The reader is referred to Chapter 2 for further discussion of these distributions. In this context the $p_i$ will be referred to as quantile probabilities.

There is much discussion in the literature over the best choice of quantile probabilities for Q-Q plots (see Kimball (1960) and Barnett (1975)). A frequently used formula is given by $p_i = (i - c)/(n - 2c + 1)$, where c is some constant satisfying $0 \le c \le 1$. The choices $c = 0$ and $c = 0.5$ (see Chapter 2) are both popular. Here we use $c = 0.3175$ since the resulting probabilities closely approximate medians of uniform $(0, 1)$ order statistics (Filliben, 1975). This choice has the attractive invariance property that if $p_i$ is the median of the ith order statistic from the uniform $(0, 1)$ distribution, then $G^{-1}(p_i)$ is the median of $Z(i)$ and $F^{-1}(p_i)$ is the median of $Y(i)$, for any continuous F. Medians may also be preferred as measures of central tendency since the distributions of most order statistics are skewed. In the examples that follow we will adhere to the convention of choosing $c = 0.3175$ unless stated otherwise. Thus we will plot the points

$$\left\{ G^{-1}(p_i), \ Y_{(i)} \right\} \tag{11.2}$$

where $p_i = (i - 0.3175)/(n + 0.365)$. The particular choice of quantile probabilities is not crucial since for any reasonably large sample different choices

TABLE 11.1 CDFs and Plotting Formulas for Selected Families
of Distributions

| Distribution[a] | $F(y)$ | Abscissa | Ordinate |
|---|---|---|---|
| Uniform | $\dfrac{y - \mu}{\sigma}$ | $p_i$ | $Y_{(i)}$ |
| Normal | $\Phi\left(\dfrac{y - \mu}{\sigma}\right)$ | $\Phi_i^{-1}(p_i)$ | $Y_{(i)}$ |
| Lognormal | $\Phi\left[\dfrac{\log(y) - \mu}{\sigma}\right]$ | $\Phi_i^{-1}(p_i)$ | $\log\{Y_{(i)}\}$ |
| Exponential | $1 - \exp\left[-\left(\dfrac{y - \mu}{\sigma}\right)\right]$ | $\log[1/(1 - p_i)]$ | $Y_{(i)}$ |
| Extreme-value | $1 - \exp\left[-\exp\left(\dfrac{y - \mu}{\sigma}\right)\right]$ | $\log\{\log[1/(1 - p_i)]\}$ | $Y_{(i)}$ |
| Weibull | $1 - \exp\left[\left(\dfrac{y}{\sigma}\right)^m\right]$ | $\log\{\log[1/(1 - p_i)]\}$ | $\log(Y_{(i)})$ |
| Laplace | $\begin{cases} \dfrac{1}{2} \cdot \exp\left(\dfrac{y - \mu}{\sigma}\right), & y \le \mu \\ 1 - \dfrac{1}{2}\exp\left(-\dfrac{y - \mu}{\sigma}\right), & y > \mu \end{cases}$ | $\begin{cases} \log(2p_i), & p_i \le \dfrac{1}{2} \\ \log[1/(2 - 2p_i)], & p_i \ge \dfrac{1}{2} \end{cases}$ | $Y_{(i)}$ |
| Logistic | $1/\left[1 + \exp\left(-\dfrac{y - \mu}{\sigma}\right)\right]$ | $\log[p_i/(1 - p_i)]$ | $Y_{(i)}$ |
| Cauchy | $\dfrac{1}{2} + \dfrac{1}{\pi} \cdot \arctan\left(\dfrac{y - \mu}{\sigma}\right)$ | $\tan\left[\pi \cdot \left(p_i - \dfrac{1}{2}\right)\right]$ | $Y_{(i)}$ |

[a]Support of each distribution is $(-\infty < y < \infty)$ except for the uniform $(\mu < y < \mu + \sigma)$, lognormal $(y > 0)$, exponential $(y > \mu)$, and Weibull $(y > 0)$.

will have little effect on the appearance of the main body of the plot. There may be some noticeable differences, however, for extreme order statistics from long-tailed distributions. (The reader should note that in Chapter 2 the $p_i$ of (11.2) was symbolized by $F_n(y)$, the empirical distribution function.)

## E 11.2.1.1 Uncensored Normal Example

Data for this example consist of the first 40 values from the NOR data set which were simulated from the normal distribution with $\mu = 100$ and $\sigma = 10$. A normal probability plot is shown in Figure 11.1. The normal distribution provides a good fit to the data. Note that the intercept and slope of a straight line drawn through the points provide estimates of the theoretical mean and standard deviation. (The reader should compare Figure 11.1 to Figure 2.15

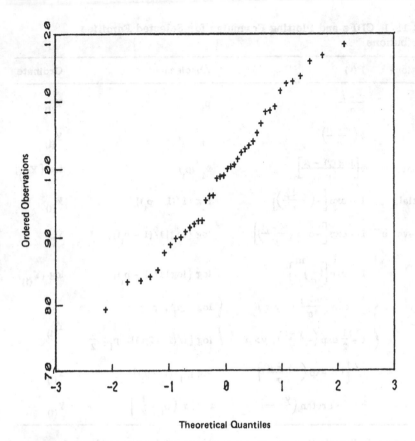

**FIGURE 11.1** Normal probability plot of the first 40 observations from the NOR data set.

where the full NOR data set is plotted with X and Y axes interchanged from Figure 11.1.)

### 11.2.2 Singly-Censored Samples

The method of the previous section can be applied directly in any situation where the data consist of some known subset of order statistics from a random sample. This is because the available $Y_{(i)}$ are still sample quantiles from the complete sample and appropriate quantiles of G can be calculated as before. Although only a portion of the observations from the hypothetical complete sample can be plotted, the plotted positions of the uncensored points are the same as when the complete sample is available. The only difference is that points corresponding to censored observations do not appear. The simple example of this is the case of a singly-censored sample.

### E 11.2.2.1 Right-Censored Normal Example

Data for this example consist of the smallest 20 values among the first 40 values listed in the NOR data set. A normal probability plot is shown in Figure 11.2. This plot is merely an enlargement of the lower portion of the plot shown in Figure 11.1.

The plotting procedure is the same for Type 1 as for Type 2 singly-censored samples; however, with Type 1 censoring there is one additional piece of information, namely the censoring time, that can be represented graphically. Suppose we observe the r smallest observations from a random

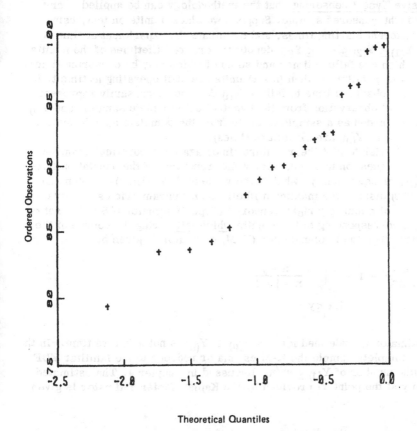

Theoretical Quantiles

FIGURE 11.2 Normal probability plot of the smallest 20 of the first 40 values listed in the NOR data set.

sample of size n. For a location and scale family we plot the points $(G^{-1}(p_i), Y_{(i)})$ for $i = 1, 2, \ldots, r$. Now suppose that the censoring is Type 1 and that the observations are all those that are less than some predetermined value t; thus $Y_{(r+1)}$ must be greater than t. This additional information can be given by plotting the point $\{G^{-1}(p_{r+1}), t\}$ with a symbol such as an arrow pointing up, thus indicating the range of possible values for $Y_{(r+1)}$. Nelson (1973) illustrates this technique.

### 11.2.3 Multiply Right-Censored Samples

The method of probability plotting extends easily to multiply right-censored samples; however, the computation of quantile probabilities is more complicated. For ease of explanation we will first consider the special case of <u>progressive Type 1 censoring</u>, but the methodology can be applied to any multiply right-censored sample. Suppose we place n units on test, using several different starting times, and terminate the experiment at time t. Now let $Y_{(1)} \leq Y_{(2)} \leq \cdots \leq Y_{(n)}$ denote the ordered lifetimes of the n units, some of which are failure times and some of which may be censoring times. If we observe r failures, then $(n - r)$ units are still operating at time t. In this case the observed time to failure $Y_{(i)}$ does not necessarily represent the ith largest observation from the hypothetical complete sample, and $Y_{(i)}$ cannot be regarded as a sample quantile from the complete sample (unless $Y_{(1)}, Y_{(2)}, \ldots, Y_{(n)}$ are all failure times).

We still wish to plot the r failure times against theoretical quantiles from G. The question now becomes, what proportion of the population falls below $Y_{(i)}$, or equivalently, what is the value of $F(Y_{(i)} | \mu, \sigma)$. Kaplan and Meier (1958) discuss the maximum likelihood nonparametric estimator of F for the case of a multiply right-censored (Type 1) sample. If S is the set of subscripts corresponding to those units <u>which fail</u> during the course of the experiment, then the Kaplan-Meier (K-M) estimator is given by

$$\text{K-M} \hat{F}_n(y) = 1 - \prod_{\substack{j \in S \\ j: Y_j \leq y}} \frac{n - j}{n - j + 1}$$

(This estimator is undefined for $y > Y_{(n)}$ if $Y_{(n)}$ is not a failure time.) In the case of a complete sample the K-M estimator reduces to the familiar EDF $F_n(y) =$ (the number of $Y_{(j)} \leq y)/n$, discussed in Chapter 4. The estimated probability at the point $Y_i$ provided by the Kaplan-Meier estimator is given by

$$p_i(\text{K-M}) = 1 - \prod_{\substack{j \in S \\ j \leq i}} \frac{n - j}{n - j + 1} \tag{11.3}$$

for $i \in S$. Herd (1960) and Johnson (1964) propose the similar quantile probabilities

$$p_i(H\text{-}J) = 1 - \prod_{\substack{j \in S \\ j \leq i}} \frac{n - j + 1}{n - j + 2} \tag{11.4}$$

for $i \in S$. Implicit in the work of Nelson (1972) are the quantile probabilities

$$p_i(N) = 1 - \prod_{\substack{j \in S \\ j \leq i}} \exp\left(- \frac{1}{n - j + 1}\right) \tag{11.5}$$

for $i \in S$. Nelson refers to his method as (cumulative) <u>hazard plotting</u>, but it is equivalent to probability plotting with the above special choice of quantile probabilities. An algebraic comparison reveals that $p_i(K\text{-}M) > p_i(N) > p_i(H\text{-}J)$ for all $i \in S$. For a discussion of the properties of the Kaplan-Meier estimator see Peterson (1977). Results by Breslow and Crowley (1974) apply to the Kaplan-Meier estimator and the estimator implicit in the work of Nelson. See Gaver and Miller (1983) for a discussion of the jackknife technique for approximate confidence intervals in this setting. For a complete sample the formulas (11.3), (11.4), and (11.5) for quantile probabilities reduce to $i/n$, $i/(n + 1)$, and $[1 - \exp(-s_i)]$, respectively, where $s_i = \Sigma_{j=1}^{i} (n - j + 1)^{-1}$. The choice of probabilities given by

$$p_i(c) = 1 - \frac{n - c + 1}{n - 2c + 1} \prod_{\substack{j \in S \\ j \leq i}} \frac{n - j - c + 1}{n - j - c + 2} \tag{11.6}$$

reduces to $(i - c)/(n - 2c + 1)$ with a complete sample. As a special case, $p_i(c) = p_i(H\text{-}J)$ when $c = 0$. In the examples that follow we will remain consistent with Section 11.2.1 and use (11.6) with $c = 0.3175$ unless stated otherwise. Again for purposes of assessing goodness of fit the particular formulation for quantile probabilities is of little consequence.

### E 11.2.3.1 Multiply Right-Censored Example

Data for this example consist of the 100 observations from the WE2 data set which were simulated from the Weibull distribution with $\sigma = 1$ and $m = 2$. The data were censored as follows: observations among the first, second, third, and fourth sets of 25 were recorded that were less than 1, 0.75, 0.50, and 0.25, respectively. This type of progressive Type 1 censoring could have occurred if four sets of 25 devices were placed on test at times 0, 0.25, 0.50, and 0.75 with the experiment terminating at time 1. The 100 values

TABLE 11.2  Progressively Censored Data from the WE2 Data Set

| i | Failure time | Quantile Probabilities | | | |
|---|---|---|---|---|---|
| | | K-M | N | H-J | c = 0.3175 |
| 1 | 0.09 | 0.010 | 0.010 | 0.010 | 0 0.007 |
| 2 | 0.14 | 0.020 | 0.020 | 0.020 | 0.017 |
| 3 | 0.16 | 0.030 | 0.030 | 0.030 | 0.027 |
| 4 | 0.18 | 0.040 | 0.040 | 0.040 | 0.037 |
| 5 | 0.18 | 0.050 | 0.050 | 0.050 | 0.047 |
| 6 | 0.20 | 0.060 | 0.060 | 0.059 | 0.057 |
| 30 | 0.27 | 0.073 | 0.073 | 0.072 | 0.070 |
| 31 | 0.30 | 0.086 | 0.086 | 0.086 | 0.083 |
| 32 | 0.32 | 0.100 | 0.099 | 0.099 | 0.096 |
| 33 | 0.33 | 0.113 | 0.112 | 0.112 | 0.109 |
| 34 | 0.33 | 0.126 | 0.125 | 0.125 | 0.122 |
| 35 | 0.34 | 0.139 | 0.139 | 0.138 | 0.136 |
| 36 | 0.34 | 0.153 | 0.152 | 0.151 | 0.149 |
| 37 | 0.36 | 0.166 | 0.165 | 0.164 | 0.162 |
| 38 | 0.38 | 0.179 | 0.178 | 0.177 | 0.175 |
| 39 | 0.40 | 0.192 | 0.191 | 0.190 | 0.188 |
| 40 | 0.42 | 0.206 | 0.204 | 0.203 | 0.201 |
| 41 | 0.43 | 0.219 | 0.218 | 0.216 | 0.215 |
| 42 | 0.47 | 0.232 | 0.231 | 0.229 | 0.228 |
| 43 | 0.49 | 0.245 | 0.244 | 0.242 | 0.241 |
| 61 | 0.51 | 0.264 | 0.262 | 0.261 | 0.260 |
| 62 | 0.56 | 0.283 | 0.281 | 0.279 | 0.278 |
| 63 | 0.62 | 0.302 | 0.300 | 0.298 | 0.297 |
| 64 | 0.65 | 0.321 | 0.318 | 0.316 | 0.316 |
| 65 | 0.68 | 0.340 | 0.337 | 0.335 | 0.334 |
| 66 | 0.71 | 0.359 | 0.356 | 0.353 | 0.353 |
| 67 | 0.74 | 0.377 | 0.375 | 0.372 | 0.371 |
| 85 | 0.76 | 0.416 | 0.412 | 0.409 | 0.409 |
| 86 | 0.78 | 0.455 | 0.450 | 0.446 | 0.447 |
| 87 | 0.92 | 0.494 | 0.488 | 0.483 | 0.485 |
| 88 | 0.93 | 0.533 | 0.526 | 0.520 | 0.522 |
| 89 | 0.95 | 0.572 | 0.564 | 0.556 | 0.560 |
| 90 | 0.97 | 0.611 | 0.602 | 0.593 | 0.598 |

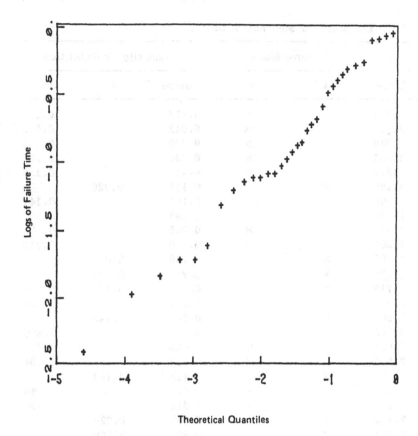

FIGURE 11.3 Weibull probability plot of progressively censored data from the WE2 data set.

(censored and failed) were ranked from smallest to largest. The 33 failure times are listed in Table 11.2 along with four different choices of quantile probabilities. Of the 67 censored devices, 23, 17, 17, and 10 devices had censoring times of 0.25, 0.50, 0.75, and 1.00, respectively. One purpose of this example is to show how close the agreement can be for different choices of quantile probabilities. Note also the relationship $p_i(K-M) > p_i(N) > p_i(H-J)$. A Weibull probability plot for the data using $p_i(c)$ with $c = 0.3175$ is shown in Figure 11.3.

The remarks made in Section 11.2.1 and Chapter 2 concerning the interpretation of probability plots with complete samples hold also for the case of multiple censoring. However, in the case of a multiply right censored sample, the effect of censoring is to increase the variability on the right-hand side of the plot.

TABLE 11.3  Life Data for Mechanical Device

| i | Time | Failure Mode | | Quantile Probabilities | | |
|---|------|---|---|---|---|---|
|   |      | A | B | Device | A | B |
| 1 | 1.151 |   | × | 0.017 |   | 0.017 |
| 2 | 1.170 |   | × | 0.042 |   | 0.042 |
| 3 | 1.248 |   | × | 0.066 |   | 0.066 |
| 4 | 1.331 |   | × | 0.091 |   | 0.091 |
| 5 | 1.381 |   | × | 0.116 |   | 0.116 |
| 6 | 1.499 | × |   | 0.141 | 0.020 |   |
| 7 | 1.508 |   | × | 0.166 |   | 0.141 |
| 8 | 1.534 |   | × | 0.190 |   | 0.167 |
| 9 | 1.577 |   | × | 0.215 |   | 0.192 |
| 10 | 1.584 |   | × | 0.240 |   | 0.218 |
| 11 | 1.667 | × |   | 0.265 | 0.052 |   |
| 12 | 1.695 | × |   | 0.289 | 0.084 |   |
| 13 | 1.710 | × |   | 0.314 | 0.116 |   |
| 14 | 1.955 |   | × | 0.339 |   | 0.246 |
| 15 | 1.965 | × |   | 0.364 | 0.149 |   |
| 16 | 2.013 |   | × | 0.389 |   | 0.276 |
| 17 | 2.051 |   | × | 0.413 |   | 0.305 |
| 18 | 2.076 |   | × | 0.438 |   | 0.334 |
| 19 | 2.109 | × |   | 0.463 | 0.187 |   |
| 20 | 2.116 |   | × | 0.488 |   | 0.365 |
| 21 | 2.119 |   | × | 0.512 |   | 0.396 |
| 22 | 2.135 | × |   | 0.537 | 0.228 |   |
| 23 | 2.197 | × |   | 0.562 | 0.269 |   |
| 24 | 2.199 |   | × | 0.587 |   | 0.430 |
| 25 | 2.227 | × |   | 0.611 | 0.313 |   |
| 26 | 2.250 |   | × | 0.636 |   | 0.466 |
| 27 | 2.254 | × |   | 0.661 | 0.360 |   |
| 28 | 2.261 |   | × | 0.686 |   | 0.505 |
| 29 | 2.349 |   | × | 0.711 |   | 0.544 |
| 30 | 2.369 | × |   | 0.735 | 0.415 |   |
| 31 | 2.547 | × |   | 0.760 | 0.470 |   |
| 32 | 2.548 | × |   | 0.785 | 0.524 |   |
| 33 | 2.738 |   | × | 0.810 |   | 0.597 |
| 34 | 2.794 | × |   | 0.834 | 0.586 |   |
| 35 | 2.883 | (Working) | | | | |
| 36 | 2.883 | (Working) | | | | |
| 37 | 2.910 | × |   | 0.870 | 0.675 |   |
| 38 | 3.015 | × |   | 0.905 | 0.763 |   |
| 39 | 3.017 | × |   | 0.941 | 0.851 |   |
| 40 | 3.793 | (Working) | | | | |

For Type 2 multiple right censoring consider the following simple situation. We place n units on a life test and when the rth unit fails we remove all but a fraction $\phi$ of the remaining working units. We then observe the failure time of those units not removed. In this situation the $p_i(c)$ values can be obtained from (11.6) where $Y_{(1)} < \cdots < Y_{(r)}$ are the first r failure times, $Y_{(r+1)} = \cdots = Y_{(r+(n-r)(1-\phi))}$ are the censoring times of the removed items and $Y_{(r+(n-r)(1-\phi)+1)}, \cdots, Y_{(n)}$ are the failure times of the items that were not removed. The set $S$ in formula (11.6) consists of the indices of the first r failure times and the last $(n-r)\phi$ failure times, and a probability plot can be drawn as described above. More elaborate Type 2 multiply right-censored samples are handled in the obvious manner.

### E 11.2.3.2 Competing Modes Example

Data for this example consist of the lifetimes, measured in millions of operations, of 40 mechanical devices. The devices were placed on test at different times, and three were still working at the end of the experiment. The data are presented in Table 11.3. Only two modes of failure were observed: either component A failed or component B failed. These two components are identical in construction, but they are subject to different stresses when the device is operated. Thus their life distributions need not be identical. Quantile probabilities are given in Table 11.3 for the device as a whole, component A, and component B under the columns headed "device," "A," and "B," respectively. The data for the device are multiply censored since the 35th, 36th, and 40th ordered lifetimes are incomplete. In addition, observations on component A are censored by failures of component B and vice versa. This is an example of random censoring caused by competing modes of failure.

Probability plots for the individual components were constructed using several common life distributions. The lognormal distribution seemed to offer the best fit. Lognormal probability plots are shown for components A and B in Figure 11.4(A). The intercepts and slopes of the two lines suggested by the plots appear to be different. This raises the possibility that, while the life distributions of the two components may be of the same family, the distribution parameters may be different. A distracting feature is the noticeable gap near the center of the plots. The natural tendency is to expect too much orderliness and to declare that something unusual has occurred. But such anomalies frequently arise by chance and should not be taken too seriously. The reader is referred to Hahn and Shapiro (1967), pages 264-265, for an example of a plot in which the same unusual feature has arisen by chance.

If the life distributions for components A and B are independent and lognormal, then the life of the device is distributed as the minimum of two lognormal random variables. For illustration we assume the equality of parameters. The cdf of the device is then given by

$$F(y|\mu,\sigma) = 1 - \left\{1 - \Phi\left[\frac{\log(y) - \mu}{\sigma}\right]\right\}^2$$

A probability plot for this distribution is constructed by plotting the points

$$\left\{\Phi^{-1}(1 - \sqrt{1 - p_i}),\ \log(Y_{(i)})\right\} \qquad (11.7)$$

Such a plot is shown in Figure 11.4(B). If it is desired to fit different sets of parameters to the individual components, we can always estimate them using, say, the method of maximum likelihood. The estimated cdf of the device, however, would then be difficult to invert. One way around this is to estimate the probability integral transformation with $F(\cdot|\hat{\mu}_1,\hat{\mu}_2,\hat{\sigma}_1,\hat{\sigma}_2)$ and plot the $F(y_i|\hat{\mu}_1,\hat{\mu}_2,\hat{\sigma}_1,\hat{\sigma}_2)$ versus the $p_i$. This approach is described more fully in

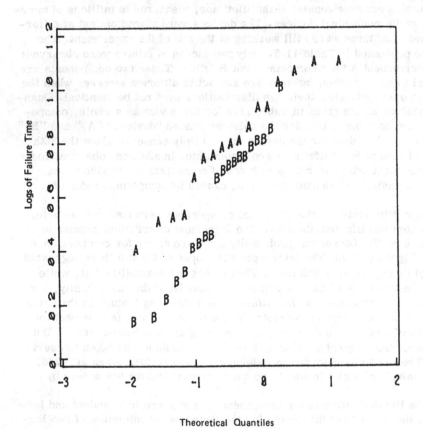

Theoretical Quantiles

FIGURE 11.4(A)  Lognormal probability plots for components A and B.

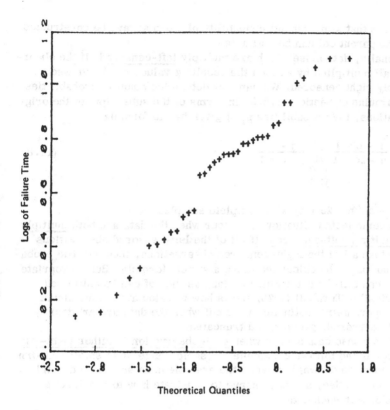

**FIGURE 11.4(B)** Probability plots for the mechanical device where the assumed distribution is the minimum of 2 I.I.D. lognormals.

Section 11.3. Note finally that the derivation of special theoretical quantiles given in (11.7) would not have been necessary if we had modeled the lifetimes of the components as exponential, extreme-value, or Weibull random variables. This is because the minimum of any number of independent identically distributed random variables from one of these families is also of the same family.

Although the development in the last example is somewhat speculative in nature, it does serve to illustrate the versatility and usefulness of probability plotting, as well as its subjective and limited interpretability.

**11.2.4 Other Types of Multiple Censoring**

There are other more complicated types of multiple censoring which can arise in practice. A few of these will be discussed below. The thought to keep in mind is that a meaningful probability plot can always be constructed as long as the parent cdf can be estimated.

Occasionally, data arise which are multiply <u>left-censored</u>. If the observations are all multiplied by -1 then the resulting values can be viewed as being multiply right censored. We can now determine quantile probabilities using the formulas of Section 11.2.3. In terms of the subscripts of the original observations, the probabilities $p_i(c)$ given by the formula

$$p_i(c) = \frac{n - c + 1}{n - 2c + 1} \prod_{\substack{j \in S \\ j \geq i}} \frac{j - c}{j - c + 1} \tag{11.8}$$

reduce to $(i - c)/(n - 2c + 1)$ with complete samples.

A more complicated situation can occur when the data are both <u>multiply right- and multiply left-censored</u>. If all of the left-censored observations are not less than all of the right-censored observations, then quantile probabilities can no longer be calculated using a simple formula. But appropriate probabilities can still be determined as long as the cdf can be estimated nonparametrically. Turnbull (1976) shows how to calculate the maximum likelihood nonparametric estimate of the cdf when the data are arbitrarily right and left-censored, grouped and truncated.

Quantal response data occurs when <u>each observation is either right- or left-censored</u>. In the following example the sample size is so small that firm inferences cannot be drawn; however, the example does show how quantal response data can arise, and does serve to illustrate how to construct a probability plot with such data.

## E 11.2.4.1 Quantal Response Example

It is desired to investigate the nature of the distribution of the shelf life of a certain electronic set. A total of 47 sets are involved in the study. After $Y_{(i)}$ days on the shelf the ith set is tested and is found to be either good or bad. The set is never observed again. Thus a good set constitutes a right censored observation whereas a bad set constitutes a left censored observation. The number of days on the shelf at the times of test are as follows with failures indicated by an asterisk: 20, 22, 23, 25, 26, 27, 28, 29*, 30, 31, 37, 37, 37, 41, 42, 43, 62, 69, 69, 78, 92, 92, 93, 114, 117, 124*, 128*, 130, 136, 151, 211, 226, 231, 242, 244, 244, 244, 244, 245*, 245, 245, 250, 259*, 259, 287, 317, and 340 days. Using the recursive algorithm given by Turnbull (1976), the maximum likelihood nonparametric estimate of the cdf is found to be

$$F_n(y) = \begin{cases} 0 & -\infty < y < 28 \\ \text{undefined} & 28 \leq y \leq 29 \\ .056 & 29 < y < 117 \\ \text{undefined} & 117 \leq y \leq 124 \\ .143 & 124 < y < 244 \\ \text{undefined} & 244 \leq y \leq 245 \\ .222 & 245 < y < 340 \\ \text{undefined} & 340 \leq y < \infty \end{cases}$$

Four values of y were selected for purposes of probability plotting: 28.5, 120.5, 244.5, and 340. The first three are the midpoints of the three closed

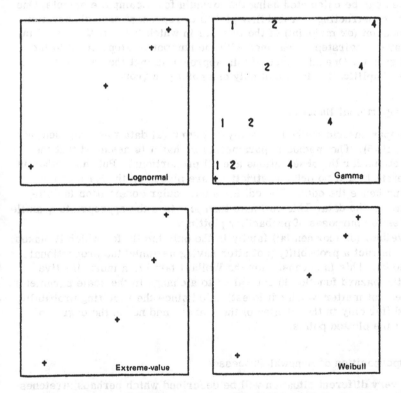

FIGURE 11.5 Probability plots for the shelf life of electronic sets (gamma shape parameter = 1, 2, and 4; origin for gamma plots is shown as "+").

intervals which are assigned probability, and the last is the largest value for which $F_n(y)$ is defined. The four probabilities used are 0.028, 0.099, 0.127, and 0.222. The first three are the midpoints of the jumps and the fourth is equal to $\hat{F}_n(340)$. Probability plots are shown in Figure 11.5 for four families commonly used to model lifetimes. The lognormal, gamma (with origin 0 and shape near 1), and Weibull distributions all appear to fit the data well. These results are not inconsistent since the gamma distribution described (exponential distribution with origin zero) is a member of the Weibull family, and the lower portions of the Weibull and lognormal cdfs are very similar.

Any conclusions, however, are highly tentative because of the small sample size and the severity of the censoring. If we use a jackknife technique or the theory of m.l.e. (see Turnbull, 1976) to estimate the variances of probabilities assigned to each of the four intervals, it then appears that none of the models considered in Figure 11.5 can be soundly rejected.

Grouping is perhaps the most common form of censoring encountered in practice. Each grouped observation is both right- and left-censored. Quantile probabilities can be calculated using the formula for a complete sample. One approach to constructing a probability plot is to represent each observation with the endpoint (or midpoint) of the interval in which it falls. The resulting plot will have a stairstep appearance with the number of steps equal to the number of groups. One advantage of this approach is that the sample size is evident. A simplification is to plot only one point per group.

## 11.2.5 Proportional Hazards

A quasi-nonparametric method for analyzing survival data was proposed by Cox (1972, 1975). The method is parametric in that it is assumed that the hazard functions for the observations are all proportional. But the method is nonparametric in that no prior restrictions are placed on the form of the hazards (and hence the cdfs). The cdf for a particular observation is estimated using all the data. This estimate then provides the appropriate quantile probabilities for purposes of probability plotting.

The Weibull (or exponential) family is the only family for which it makes sense to construct a probability plot after having assumed the proportional hazards model. This is because, for the Weibull family, a multiplicative change in the hazard function is equivalent to a change in the scale parameter. Thus it does not matter which cdf is estimated since the resulting probability plots will differ only in the labeling of their axes, and not in the degree of linearity of the plotted points.

## 11.2.6 Superposition of Renewal Processes

Finally, a very different situation will be described which perhaps stretches the definition of the term "censoring." Suppose we have n units that all begin operation at the same time. If a unit fails, it is instantly replaced with a new unit. It is assumed that the lifetimes of the original and replacement units are independent and identically distributed with cdf F. The exact times of failures are known but not the identities of the failed units. Except for the first failure, then, we cannot be sure for the ages of the failed units. We thus observe a superposition of renewal processes. The failure times are not censored here, but the identities and therefore the ages of the failed units are censored in a sense.

Trindade and Haugh (1979) describe a method for the nonparametric estimation of F in the above situation. The renewal function, M, is estimated using a straightforward nonparametric method. The parent cdf is then estimated by exploiting the relationship of F to M through the fundamental renewal equation. For any particular set of points in time, the estimate of F provides appropriate probabilities for determining corresponding theoretical standard quantiles for purposes of probability plotting. Again we will emphasize that a meaningful probability plot can always be constructed as long as the parent cdf can be estimated using a nonparametric method.

## 11.2.7 Summary of Steps in Constructing a Probability Plot

Below are given the steps required in constructing a probability plot with uncensored, singly-censored, multiply right-censored, and multiply left-censored data. The user must provide a value of the constant c with $0 \leq c \leq 1$. The values $c = 0.3175$ and $c = .5$ are popular.

(1) Let $Y_{(1)}$, $Y_{(2)}$, $\ldots$, $Y_{(n)}$ denote n ordered observations, some of which may be censored, and let S be the set of subscripts corresponding to the observations in the ordered list that are not censored.

(2) Determine quantile probabilities $p_i$ for each $i \in S$ using one of the following formulas:

$$(i - c)/(n - 2c + 1, \text{ for complete or singly-censored samples}$$

$$1 - \frac{n - c + 1}{n - 2c + 1} \prod_{\substack{j \in S \\ j \leq i}} \frac{n - j - c + 1}{n - j - c + 2}, \text{ for multiply right-censored samples}$$

$$\frac{n - c + 1}{n - 2c + 1} \prod_{\substack{j \in S \\ j \geq i}} \frac{j - c}{j - c + 1}, \text{ for multiply left-censored samples}$$

(11.9)

(3) Enter Table 11.1 and find the line corresponding to the hypothesized family of distributions. Plot the entry under "abscissa" versus the entry under "ordinate" for each $i \in S$.

## 11.3 TESTING A SIMPLE NULL HYPOTHESIS

For this section it is assumed that the hypothesis of interest is $H_0$: the sampled population has the completely specified absolutely continuous cdf $F(y)$. As in other chapters, this situation is called Case 0. For most of the discussion the data at hand will consist of a singly right-censored sample (Type 1 or Type 2), that is, the set of r smallest order statistics $Y_{(1)}, \ldots, Y_{(r)}$. The probability integral transformation $U_{(i)} = F(Y_{(i)})$, $i = 1, \ldots, r$, can be applied, and an equivalent test of fit is that the $U_{(1)} \le \cdots \le U_{(r)}$ are the r smallest order statistics of a random sample of size n from the uniform $(0, 1)$ distribution. If the data are Type 1 censored at $y = y*$ and if $t = F(y*)$, then r is a random variable giving the number of order statistics for the uniform random sample which are less than t; if the data are Type 2 censored, then r is fixed in advance.

Many of the methods which have been discussed in earlier chapters have been adapted to accommodate censoring of both types. These include censored versions of EDF statistics (Section 4.7), correlation-type tests (Section 5.5) and tests based on spacings (Section 8.9). Later we examine procedures in which the order statistics $U_{(i)}$, $i = 1, \ldots, r$, are transformed to new values which under $H_0$ are distributed as a complete uniform sample. Then any of the many tests for uniformity for a complete sample (Chapter 8) may be applied to the transformed values.

### 11.3.1 EDF Statistics

In Chapter 4, censored versions of EDF statistics were introduced. We will now illustrate the use of these statistics by applying them directly to a censored sample.

TABLE 11.4 Hypothetical Survival Data
and Transformed Observations

| i | $Y_{(i)}$ | $U_{(i)}$ | $Z_{(i)}$ |
|---|---|---|---|
| 1 | .1 | .00995 | .03979 |
| 2 | .2 | .01980 | .07918 |
| 3 | .3 | .02955 | .11815 |
| 4 | .4 | .03921 | .15681 |
| 5 | .7 | .06761 | .27038 |
| 6 | 1.0 | .09516 | .38056 |
| 7 | 1.4 | .13064 | .52245 |

### E 11.3.1.1 Exponential Example

This is an example is of Type 1 censoring. Barr and Davidson (1973) give the smallest 7 observations in a Type 1 censored sample of size n = 20. The hypothesized null distribution is the exponential distribution $F(y) = 1 - \exp(-y/10)$, $y > 0$, with a censoring value at $y^* = 2.2$. Table 11.4 gives the values $Y_{(i)}$, with the values $U(i)$ given by $U_{(i)} = F\{Y_{(i)}\}$, $i = 1, \ldots, 7$. The Type 1 censoring value for u is then $t = F(2.2) = 0.1975$. The $Z_{(i)}$ values shown here are first discussed in Section 11.3.3.

In Section 4.7.2, the <u>Kolmogorov-Smirnov</u> statistics $_1D_{t,n}$ and $_2D_{r,n}$ for censored data of Types 1 and 2, respectively, are defined; it is also shown how these may be transformed and referred to the asymptotic distribution tabulated in Table 4.4. Alternatively, the exact tables for finite n, given by Barr and Davidson (1973) may be used. Working from the values $U_{(i)}$ of Table 11.4, the statistic $_1D_{t,n}$ is found to be $7/20 - 0.131 = 0.219$, with $t = 0.1975$ and $n = 20$. Direct interpolation in the tables of Barr and Davidson gives the approximate significance level $p = 0.11$. Alternatively, use of the formulas of Dufour and Maag (1978) (Section 4.7) yields the modified statistic $D_t^* = 4.472(0.219) + 0.19/4.472 = 1.022$. Reference to the asymptotic points in Table 4.4 then gives a p-value of approximately 0.10. If the data had been Type 2 censored, the formulas would give a modified statistic $D_r^* = 1.033$ with p-value approximately 0.095. Percentage points of the asymptotic distribution are derived and tabled for the Kuiper statistic, $V_{t,n} = {}_1D_{t,n}^+ + {}_1D_{t,n}^-$, under Type 1 right censoring by Koziol (1980a).

In Section 4.7.3 two types of <u>Cramér-von Mises</u> statistics for censored data are given. The first type is denoted by $_1W_{t,n}^2$ for right censored data of Type 1. The second is for Type 2 censoring. Both were derived by Pettitt and Stephens (1976a,b) by adapting the complete-sample definitions of these statistics. For the $U_{(i)}$ of Table 11.4, we have $_1W_{0.2,20}^2 = 0.104$ and $_1A_{0.2,20}^2 = 1.214$. Referring to Table 4.4 with $t = 0.2$ gives p-values 0.008 and 0.005, respectively. If the data are treated as Type 2 censored, the test statistics become $_2W_{7,20}^2 = 0.057$ and $_2A_{7,20}^2 = 0.863$; referring to Table 4.5 with $p = r/n = .35$ gives approximate p-values of 0.25 and 0.08.

### E 11.3.1.2 Uniform Example

This is an example of Type 2 censoring. Consider the first n = 25 values from the UNI data set, rescaled to the unit interval. Suppose that the smallest r = 15 values from this set are available and are to be tested for fit as uniform order statistics. The observed value of $_2D_{r,n} = 0.216$ and $(\sqrt{n})_2D_{r,n} = 1.08$; then $D_r^*$ (Section 4.7) is $1.08 + .24/5 = 1.128$ and reference to the tabulated asymptotic distribution gives a p-value of 0.1296.

The p-value may also be computed from the formula for the asymptotic distribution given by Schey (1977) and quoted in Section 4.7.2. The value of t = 15/25 = 0.60; this gives $A_t$ = 2.041, $B_t$ = 0.408 in the notation of Section 4.7.2, and then $G_t(1.128)$ is 0.9362. The observed significance level for the two-sided test is then approximately $2(1 - 0.936) = 0.128$.

## Comment A. Use of the censoring information

When the observations $U_{(i)}$ are to be tested for uniformity, the value of t, or of r (whichever is given) is important, in addition to the values $U_{(i)}$. Thus, for example, in Table 11.4, there are 7 observations out of 20 below t = 0.1975, a number larger than expected. If the sample had been Type 2 censored, we could observe that the largest observation $U_{(7)}$ is 0.131, much smaller than the expected value 0.333. These facts are implicitly used in calculating the EDF statistics. Also, the value of $U_{(r)}$, although not the value of t, is used in the censored version of spacings statistics (Section 8.9).

## Comment B. Random censoring

Extensions of EDF statistics to situations involving randomly censored data generally involve a Kaplan-Meier estimator for the true distribution function. For versions of the Kolmogorov-Smirnov, Kuiper, and Cramér-von Mises see Koziol (1980b), Nair (1981), or Fleming et al. (1980), who obtain asymptotic distributions and examine the adequacy of small sample approximations.

## 11.3.2 Correlation Statistics

The statistics in this class, as discussed in Chapter 5, basically focus upon the strength of the pattern of linear association which is present in probability plots (see Chapter 2 and Section 11.2). Suppose all the observations $U_{(i)}$ are known between $U_{(s)}$ and $U_{(r)}$. These may be plotted against $m_i = i/(n + 1)$, $i = s, \ldots, r$, and the coefficient $R(X, m)$ may be calculated as described in Section 5.1.2. Because $R^2$ is scale-free, the value obtained is the same as if the $U_{(i)}$ were correlated with i, from 1 to r - s + 1. The $U_{(i)}$ are a subset of order statistics from (0, 1) and will themselves be uniform between limits which may or may not be known. In either case, the distribution of $R^2$ so calculated is the same under $H_0$, as that of $R^2$ for a full sample of size r - s + 1. Thus Table 5.1 may be used to make a test. The weakness in this test procedure is that it does not make use of any Type 1 censoring values. In Chapter 5 it is shown how this may be overcome, by including the censoring limits in the observed sample. The value of $R^2$ calculated from these values will have the same null distribution as $R^2$ calculated from a complete uniform sample of size r - s + 3, so that again Table 5.1 can be used.

## E 11.3.2.1 Exponential Example Revisited

For the r = 7 values of $U_{(i)}$ in Table 11.4 the correlation coefficient $R(X, m)$ = 0.964, which yields $T = r(1 - R^2(X, m)) = 0.49$. Reference to Table 5.1

shows that this value is not significant even at the 50% level. If the endpoints $s = 0$ and $t = 0.1975$ are included, then $T = 1.071$, which is significant at the .20 level approximately.

## 11.3.3 Transformations to Enable Complete-Sample Tests

### 11.3.3.1 Conditioning on the Censoring Values

When all the values from $U_{(s)}$ to $U_{(r)}$ $(s < r)$ are available, the test of $H_0$: that these are a subset from an ordered uniform sample, can be changed to a test for a complete sample. There are several ways to do this. The simplest method is as follows. For Type 1 censoring suppose the lower censoring value is A and the upper censoring value is B, and let $R = B - A$; then under $H_0$ the values $V_{(i)} = (U_{(i)} - A)/R$, $i = 1, \ldots, r - s + 1$, will be a complete ordered uniform sample on the unit interval and can be so tested.

For Type 2 censoring, under $H_0$ the values $U_{(s+1)}, \ldots, U_{(r-1)}$ will be distributed as a complete ordered sample from the uniform distribution between limits $A = U_{(s)}$ and $B = U_{(r)}$. The transformation $V_{(i)} = (U_{(s+1)} - U_{(s)})/R$, can be made for $i = 1, \ldots, n^*$, where $n^* = r - 1 - s$ and $R = B - A$; the $V_{(i)}$, $i = 1, \ldots, n^*$ can then be tested for uniformity between 0 and 1.

### E 11.3.3.1 Exponential Example Again

Consider, again, the data of Table 11.4. The upper (Type 1) censoring value is $t = 0.1975$; thus we can first transform the $V_{(i)}$ as $V_{(i)} = U_{(i)}/0.1975$ and then test the $V_{(i)}$ for uniformity between 0 and 1. The $V_{(i)}$ are then 0.050, 0.100, 0.150, 0.199, 0.342, 0.482, and 0.661. The EDF statistics are $D^+ = 0.375$, $D^- = 0.050$, $D = 0.375$, $V = 0.426$, $W^2 = 0.413$, $U^2 = 0.085$, and $A^2 = 2.107$. Reference to Case 0 tables (Table 4.2) gives p-values of 0.21 for D, 0.07 for $W^2$, and 0.08 for $A^2$.

### 11.3.3.2 Handling Blocks of Missing Observations

Suppose censoring occurs in a uniform sample other than at the ends; for example, $U_{(r)}$ and $U_{(r+q)}$ might be known, but the $q - 1$ observations in between are not known. A spacing $U_{(r+q)} - U_{(r)}$ is called a q-spacing. Now suppose that S is a q-spacing covering unknown observations, and let its length be d. Keeping in mind the exchangeability of uniform spacings (see Chapter 8) we exchange S with the set of all spacings to the right of S. Under $H_0$ the new sample $U_{(1)}, U_{(2)}, \ldots, U_{(r)}, U^*_{(r+1)}, \ldots, U^*_{(n-q+1)}$, where $U^*_{(j)} = U_{(j+q)} - d$, for $j = r + 1, \ldots, n - q + 1$, will be distributed as an ordered uniform sample which is right-censored at $U^*_{n-q+1}$. The process may be repeated if there is more than one such spacing. The method can be used only if it is known how many values are missing in the spacings. Thus a uniform (0, 1) sample with known blocks of missing observations can be transformed to behave like a right (or left) censored sample. Techniques for this simpler kind of censoring can then be used.

E 11.3.3.2.1 Example from Chapter 4

In Table 4.13, a set of 15 values for Z is given which are distributed uniformly on $(0, 1)$ under the null hypothesis that the original set X (also given there) is exponential. Suppose the four values $0.237, 0.252, 0.252, 0.381$ are lost from the set $Z_1^*$. Then $0.434 - 0.229$ is a 5-spacing of length $d = 0.205$. We subtract d from all the values of Z starting with .446 to obtain $0.113, 0.189, 0.229, 0.241 (= 0.446 - d), 0.298, 0.317, 0.578,$ $0.757, 0.774, 0.778, (= 0.993 - d), 0.795 (= 1.0 - d)$. These 11 values can be analyzed as being right censored (Type 2) at $0.795$, and thus can be tested by any of the methods of Section 11.3.1. Alternatively, they can be transformed to be a complete sample, as in Section 11.3.3.1 above, or by another method to be described after the next example.

E 11.3.3.2.2 Exponential Example Modified

These various techniques may be combined to handle blocks of missing observations within, say, right-censored data. Thus suppose the values in the U-set of Table 11.4, which are Type 1 right censored at $t = 0.1975$, are, in fact, the values $U_{(1)}, U_{(2)}, U_{(3)}, U_{(4)}, U_{(8)}, U_{(9)}, U_{(10)}$ —that is, values $U_{(5)}, U_{(6)}, U_{(7)}$ constitute a block of missing observations. First the set of U's may be transformed to a uniform sample as described in Section 11.3.3.1 above to give a new set $V_{(i)} = U_{(i)}/0.1975$. The values are those given in Example E 11.3.3.1.1, but they now represent the order statistics with indices 1, 2, 3, 4, 8, 9, 10. There is thus a 4-spacing of length $d = 0.342 - 0.199 = 0.143$ between $V_{(4)}$ and $V_{(8)}$. Following the steps of this section, new values V* are found to be $V_{(5)}^* = V_{(9)} - 0.143 = 0.339$, $V_{(6)}^* = 0.518$, $V_{(7)}^* = 1 - d = 0.857$. These 7 values are to be treated as a right-censored sample of size 7, now of Type 2, with $n = 10$ (the 7 given values plus the 3 missing in the 4-spacing). Since the lower end-point of the distribution is known to be zero, the values $0.050, 0.100, 0.150, 0.199, 0.339, 0.518$ can be divided by $0.857$ and then tested for uniformity on the unit interval.

11.3.3.3 More Powerful Transformations

A disadvantage of the above method of transforming to a complete sample for Type 2 censoring is that the resulting test examines the values of the $U_{(i)}$ relative to $U_{(s)}$ and $U_{(r)}$ but takes no account of whether these values themselves are too large or too small. (See Comment A in Section 11.3.1.)

Michael and Schucany (1979) propose a modification of the above technique by which a subset of r uniform $U(0, 1)$ order statistics can be transformed monotonically to behave like a complete $U(0, 1)$ sample of size r from the $U(0, 1)$ distribution. For definiteness the result is presented here in terms of right censorship; however, the technique can be applied to any

kind of Type 2 censoring. For example, a q-spacing representing a block
of $(q - 1)$ missing observations in a sample of size n can be shrunk to a
1-spacing in a "complete" sample of size $n - q + 1$. The relative spacings
between consecutive order statistics are not affected by this transformation.

Let $U_{(1)}$, $U_{(2)}$, $\ldots$, $U_{(r)}$ be the smallest r order statistics from a
random sample of size n from the uniform $(0, 1)$ distribution, and let $B(*)$
denote the cdf of $U_{(r)}$, which is known to have the beta $(r, n - r + 1)$ distribution (see Section 8.8.2). If $Z_{(1)}$, $Z_{(2)}$, $\ldots$, $Z_{(r)}$ are defined by

$$Z_{(i)} = U_{(i)} \cdot h_{r,n}(U_{(r)}) \tag{11.10}$$

where $h_{r,n}(x) = \{B^*(x)\}^{1/r}/x$, then the $Z_{(i)}$, $i = 1, \ldots, r$, are distributed
like a complete uniform $(0, 1)$ sample of size r. The proof is straightforward
by change of variable.

The computations for the transformation are easily performed on a
scientific calculator since the beta cdf can be expressed as the binomial sum

$$B(u) = \sum_{i=r}^{n} \binom{n}{i} u^i (1 - u)^{n-i}$$

Any standard goodness-of-fit test for uniformity (Chapter 8) may now be
applied to the transformed observations. The Anderson-Darling statistic is
recommended because of its sensitivity to departures from uniformity in the
tails of the distribution. The reason why this is important is best presented
by illustration.

### E 11.3.3.3.1  Artificial Uniform Right-Censored Sample

Three artificial but informative examples of the transformation are shown
in Figure 11.6 where the smallest five of nine observations are plotted both
before and after the transformation. In each example the values of the $U_{(i)}$
were artificially chosen to satisfy $U_{(i)}/U_{(5)} = i/5 = E(U_{(i)}/U_{(5)} | U_{(5)})$. The
values for $U_{(5)}$ were chosen to be .500, .103, and .897 which correspond,
respectively, to the .500, .001, and .999 quantiles of the beta $(5, 5)$ distribution, which is the null distribution of $U_{(5)}$ when testing the hypothesis of
uniformity for the $U_{(i)}$. Note the manner in which small and large values of
$U_{(5)}$ affect the appearance of the transformed points. Small values of $U_{(5)}$
lead to small values of $Z_{(5)}$ which, in turn, will inflate most reasonably
formulated goodness-of-fit statistics. But if $Z_{(5)}$ is large, the departure
from uniformity may appear less pronounced; however, $Z_{(5)}$ will be very
close to 1 and this will inflate a statistic like the Anderson-Darling statistic
which is sensitive to such an apparent departure from uniformity.

a. $U_{(5,9)} = .500$

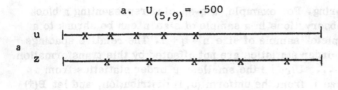

b. $U_{(5,9)} = .103$

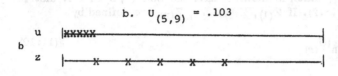

c. $U_{(5,9)} = .897$

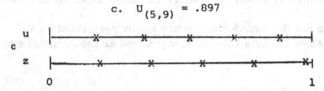

0                                                                      1

FIGURE 11.6 Examples of the transformation with r = 5, n = 9.

## E 11.3.3.3.2 Exponential Example

Consider again the values $U_{(i)}$, i = 1, ..., 7 given in Table 11.4. Using the transformation above, we first compute the scale factor h = $h_{7,20}(U_{(7)})$ as

$$h = [B*(0.13064)]^{1/7}/0.13064 = 3.9991$$

where B*(·) is the beta (7,14) distribution. The values Z(i) = $hU_{(i)}$ are given in Table 11.4. The Cramér-von Mises statistic, $W^2$, calculated from the seven values of $Z_{(i)}$ by the full-sample formulas (Equation 4.2) is 0.673, and the Anderson-Darling statistic if $A^2 = 3.404$. These have approximate p-values of 0.035 and 0.02. The Kolmogorov-Smirnov statistic is D = 0.47755 which has a p-value of approximately 0.056. The p-values using the transformed Z-values are lower than those using the statistics directly adapted for censoring (see Section 11.3.1).

## Comment

This transformation technique for goodness-of-fit analysis of censored samples has some advantages over the other procedures which have been proposed for this problem. No new or additional tables of critical points are required. Any subset of order statistics can be analyzed. The power of the Anderson-Darling statistic based on the transformed sample appears to be generally greater than that of existing methods in the presence of left or right censorship. A minor disadvantage is the slight increase in computation to evaluate the scaling factor, $h_{r,n}(U_{(r)})$. The technique can be extended to all kinds of Type 2 censoring, even progressive censoring. For details and asymptotic results see Michael and Schucany (1979).

## 11.4 TESTING A COMPOSITE HYPOTHESIS

In this section the hypothesis of interest is that the sampled population has an absolutely continuous cdf $F(y|\theta)$, where $\theta$ is a vector of unknown (nuisance) parameters. Typically the censored data at hand must be a singly-censored sample if published tables of critical points are to be used. For more complicated types of censoring, such as multiple right censoring, little work has been done. For a particular set of data, it may be possible to modify a standard statistic and then estimate certain percentiles, or the observed significance level, using simulation techniques. When the censoring is Type 2, test statistics can often be constructed which have parameter-free null distributions. When the censoring is Type 1, statistics with asymptotically parameter-free distributions are a possibility.

### 11.4.1 Omnibus Tests

Turnbull and Weiss (1978) present an omnibus test for a composite null hypothesis based on the generalized likelihood ratio statistic. Their procedure is appropriate for discrete or grouped data and accommodates multiple censoring by employing the Kaplan-Meier estimate to maximize the alternative likelihood. In less complicated cases of Type 1 or 2 censoring several standard goodness-of-fit statistics have been modified to test a composite null hypothesis.

### 11.4.1.1 EDF Statistics for Censored Data with Unknown Parameters

Modifications of EDF statistics which accommodate certain types of censoring when the null hypothesis is simple were discussed in Section 11.3.1.1. Similar modifications for use in testing normality with unknown parameters, or exponentiality with unknown scale, are given in Sections 4.8.4 and 4.9.5.

### E 11.4.1.1.1 Normal example

The data consist of the smallest 20 values among the first 40 values listed in the NOR data set. We wish to test that the underlying distribution is normal. Gupta's estimates (Gupta, 1952) here are $\hat{\mu} = 98.233$ and $\hat{\sigma} = 9.444$. Relevant calculations are given in Table 11.5. The value of the Cramér-von Mises statistic is found to be, using Section 4.7.3,

$$2\hat{W}^2_{20,40} = \frac{20}{12(40)^2} + 0.02512 - \frac{40}{3}(0.5 - 0.53741)^3 = 0.02686$$

TABLE 11.5 Steps in Calculating $\hat{W}^2$ for the Smallest 20 Order Statistics Among the First 40 Observations in the NOR Data Set

| $i$ | $Y_{(i)}$ | $\dfrac{i-0.5}{40}$ | $\hat{U}_{(i)}$ | $\left[\hat{U}_{(i)} - \dfrac{i-0.5}{n}\right]^2$ |
|---|---|---|---|---|
| 1 | 79.43 | 0.0125 | 0.02323 | 0.00012 |
| 2 | 83.53 | 0.0375 | 0.05974 | 0.00049 |
| 3 | 83.67 | 0.0625 | 0.06152 | 0.00000 |
| 4 | 84.27 | 0.0875 | 0.06962 | 0.00032 |
| 5 | 85.29 | 0.1125 | 0.08525 | 0.00074 |
| 6 | 87.83 | 0.1375 | 0.13531 | 0.00000 |
| 7 | 89.00 | 0.1625 | 0.16411 | 0.00000 |
| 8 | 89.90 | 0.1875 | 0.18878 | 0.00000 |
| 9 | 90.03 | 0.2125 | 0.19252 | 0.00040 |
| 10 | 90.87 | 0.2375 | 0.21778 | 0.00039 |
| 11 | 91.46 | 0.2625 | 0.23662 | 0.00067 |
| 12 | 92.02 | 0.2875 | 0.25529 | 0.00104 |
| 13 | 92.45 | 0.3125 | 0.27014 | 0.00179 |
| 14 | 92.55 | 0.3375 | 0.27364 | 0.00408 |
| 15 | 95.45 | 0.3625 | 0.38411 | 0.00047 |
| 16 | 96.13 | 0.3875 | 0.41188 | 0.00059 |
| 17 | 96.20 | 0.4125 | 0.41477 | 0.00001 |
| 18 | 98.70 | 0.4375 | 0.51972 | 0.00676 |
| 19 | 98.98 | 0.4625 | 0.53152 | 0.00476 |
| 20 | 99.12 | 0.4875 | 0.53741 | 0.00249 |
| | | | | 0.02512 |

Referring to Table 4.5, we find that the observed value is smaller than the .15 point which, by interpolation, is approximately 0.03. The value of $_2\hat{A}^2_{20,40}$ is 0.233; this is significant at about the .10 level.

EDF tests for exponentiality with an unknown scale parameter are set out in Section 4.9. Note that use of the N transformation of Chapter 10 (see Section 10.5.6) converts a right-censored exponential sample to a complete sample of exponentials, with the same scale, and then any of the tests of Chapter 10 can be used. This property is explored in a test based on leaps, in Section 11.4.1.3 below.

## 11.4.1.2 Correlation Statistics

Consider again the sample correlation coefficient between the $Y_{(i)}$ and a set of constants $K_i$, denoted, as in Chapter 5, by $R(Y, K)$. Because $R(Y, K)$ is invariant with respect to a linear transformation of the $Y_{(i)}$, it follows that its null distribution does not depend on location or scale parameters of the distribution. This makes it a useful statistic for testing fit. Suppose $F(y)$ is the cdf of Y for location parameter 0 and scale parameter 1, and let $F^{-1}(\cdot)$ be its inverse. Suitable sets of constants are then $K_i = m_i$, where $m_i$ is the expected value of the ith order statistic of a sample of size n from $F(y)$, or $K_i = H_i = F^{-1}(i/(n + 1))$. The statistic $Z = n\{1 - R^2(Y, K)\}$ has been discussed in Chapter 5, and percentage points have been given for censored versions of Z.

Chen (1984) presents a correlation statistic as an omnibus test for the composite hypothesis of exponentiality in the presence of random censoring. Asymptotic distributions are derived under a particular censorship model, which is quite robust provided that less than 40% of the observations are censored.

### E 11.4.1.2.1 Normal example revisited

Consider again the smallest 20 values among the first 40 values in the NOR data set. When testing for normality using $R(Y, K)$, we obtain $Z = 0.035$ which falls just below the .50 point. Thus on the basis of the statistic Z we cannot reject the hypothesis of normality at the usual levels. Another example of a correlation type statistic is given in the next section.

### 11.4.1.3 Statistics Based on Spacings and on Leaps

Spacings between ordered uniforms were defined in Section 10.9.3. Similarly, spacings can be defined between order statistics $Y_{(i)}$ of a sample from any distribution. If the distribution has no lower limit, the first spacing will be $D_1 = Y_{(2)} - Y_{(1)}$, and so on. Similarly leaps $\ell_i$ can be defined by $\ell_i = D_i/E(D_i)$.

An important property of leaps is that, for continuous distributions, they will (under regularity conditions) be asymptotically exponentially distributed with mean 1. Then a test for a given distribution with unknown location and scale parameters is reduced to a test similar to a test for exponentiality of the $\ell_i$. The test will not be exactly the same as a test for exponentiality because the $\ell_i$ do not become an independent sample, even asymptotically. We illustrate the technique with an example given by Mann, Scheuer, and Fertig (1973) who created a test for the extreme-value distribution by using leaps; the test is for right-censored data and can be adapted to a test for the Weibull distribution. Both these features are illustrated in the example.

### E 11.4.1.3.1 Weibull Example

The following values are $t_{(i)}$, $i = 1, \ldots, 15$, the first 15 order statistics of a sample of 22 t-values; the null hypothesis is $H_0$: the t-sample comes from a two-parameter Weibull distribution (Section 10.4.4), against a three-parameter Weibull (with positive origin) as alternative. Values are: 15.5, 15.6, 16.5, 17.5, 19.5, 20.6, 22.8, 23.1, 23.5, 24.5, 26.5, 26.5, 32.7, 33.8, 33.9. The steps in making the test are as follows. First find $X_{(i)} = \log t_{(i)}$, $i = 1, \ldots, 15$; $H_0$ then reduces to a test that $X_{(i)}$ are from an extreme-value distribution (equation 4.7 or 5.22) with unknown location and scale parameters. The denominator $E(D_i)$ of $\ell_i$ depends on the unknown scale, so test statistics are calculated from <u>normalized spacings</u>. These are defined by $y_i = \{X_{(i+1)} - X_{(i)}\}/(m_{i+1} - m_i)$, where $m_i$ are the expected values of the order statistics from an extreme-value distribution with location 0 and scale 1. Tests based on normalized spacings, including the Mann, Scheuer, and Fertig statistic S, are discussed in Section 4.20. From the data above, the 14 normalized spacings are .0063, .1070, .1640, .3912, .2408, .5177, .0752, .1089, .2858, .5724, 0.0, 1.6651, .2676, .0241.

In the notation of Section 4.20, these give 13 values $z_{(i)} = \Sigma_{j=1}^{i} y_i / \Sigma_{j=1}^{14} y_i$; these are .0014, .0256, .0626, .1510, .2054, .3224, .3394, .3640, .4286, .5579, .5579, .9341, .9946. The Mann-Scheuer-Fertig statistic is $S = 1 - z_{(7)} = 0.661$; the authors suggest a one-tail test, and reference to their tables shows S to be significant at about the 11% level. Tiku and Singh (1981) proposed using the mean $\bar{Z}$ of the $z_{(i)}$, and Lockhart, O'Reilly, and Stephens (1985) have suggested $A_S^2$, the Anderson-Darling statistic calculated from the $z_{(i)}$. These statistics also are discussed in Section 4.20. For the data above, $\bar{Z} = 0.380$ and $A_S^2 = 1.878$; these are significant at about the 4% and 5% levels, respectively. These statistics appear to be more sensitive than S. Mann and Fertig (1975) consider ratios of other sums of leaps as well as ratios of weighted sums of leaps, and describe how their approach can be extended to progressively censored samples. For further discussion see Section 4.20.

We can use this example also to illustrate the use of correlation statistics for the extreme value distribution. The test is for distribution (5.22), which has a short tail to the right and a long left tail. The $r = 15$ values of $X_{(i)}$ are tested to correlate with $H_i = \log[-\log\{1 - i/(n + 1)\}]$, $i = 1, \ldots, 15$, $n = 22$; then $R = 0.9446$ and $Z = n\{1 - R^2(X, H)\} = 1.616$. Interpolation in Table 5.10, for $n = 22$ and $p = r/n = 0.68$, shows Z to be significant at approximately the 0.25 level.

It might be useful also to illustrate a danger which may arise in testing for the extreme-value distribution. For a full sample, it does not matter whether one takes $X = \log t$, where t is Weibull data, and tests that X is from (5.22), or takes $X' = -\log t$ and tests that $X'$ is from (5.21); the same value of the correlation coefficient is obtained by both methods, and both recommendations are seen in the literature. However, for a censored sample, it is important to follow the correct procedure: for right-censored Weibull data, take $X = \log t$ and test for right-censored data from (5.22), as in the example above, and for left-censored Weibull data, take $X' = -\log t$ and test for right-censored data from (5.21). This second test for Weibull is probably less likely to occur in practice, and the Mann, Scheuer, and Fertig test is not set up for this case, although it could be adapted.

Two tests for the two-parameter exponential that can be used with doubly censored samples have been presented by Brain and Shapiro (1983). These tests combine the properties of spacings and of the correlation statistic to have good sensitivity to alternatives with monotone and nonmonotone hazard functions, respectively. Still other related work on statistics based on spacings may be found in Mehrotra (1982). Some statistics based on modified leaps have been studied by Tiku (1980, 1981).

## 11.4.2 Alternative Families of Distributions

Typically when testing for goodness of fit we assume only that the underlying cdf is absolutely continuous. Occasionally we may wish to limit our choices to, say, two families of distributions. In particular we may wish to test the composite null hypothesis

$$H_1: \; F(y) = F_1(y|\theta_1)$$

against the composite alternate hypothesis

$$H_2: \; F(y) = F_2(y|\theta_2)$$

where $\theta_1$ and $\theta_2$ are unknown (nuisance) parameters. Because we have narrowed the set of alternate distributions considerably, we should be able to tailor tests to the specific hypothesis of interest which are more powerful than omnibus goodness-of-fit tests. There have been several approaches to this problem.

Let $f_i(y|\theta)$ be the probability density function for family i, i = 1, 2. We will denote by $L_i$ the sample likelihood under $H_i$ after $\theta_i$ has been replaced with its maximum likelihood estimator, $\hat{\theta}_i$. This maximized likelihood is then

$$L_i = \prod_{j=1}^{n} f_i(Y_{(j)}; \hat{\theta}_i)$$

for the complete sample $Y_{(1)}, \ldots, Y_{(n)}$. We will denote the ratio of maximum likelihoods by

$$RML = L_1/L_2$$

Cox (1961, 1962) formulates a test of $H_1$ versus $H_2$ which is based upon the statistic

$$T = \ln(RML) - E[\ln(RML)]$$

where E is the expectation under the null hypothesis, $H_1$. For complete samples the large sample distribution of T is approximated using maximum likelihood theory. Hoadley (1971) extends maximum likelihood theory to situations which include censoring. Thus valid approximations to the distribution of T are also possible with censored samples.

For location-scale families with pdfs $\sigma_i^{-1} f_i(y - \mu_i)/\sigma_i)$, Lehmann (1959) shows that the uniformly most powerful invariant (under linear transformations) test is based upon the (Lehmann) ratio of integrals

$$LRI = I_1/I_2$$

where

$$I_i = \int_{-\infty}^{\infty} \int_0^{\infty} v^{n-2} f_i(vy_1 + u, \ldots, vy_n + u)\, dv\, du$$

The RML statistic and some modified versions are discussed in a series of papers: Antle (1972, 1973, 1975); Dumonceaux, Antle, and Haas (1973); Dumonceaux and Antle (1973); Klimko and Antle (1975); Kotz (1973). Percentage points are given for the null distribution of RML for comparisons involving a number of different families of distributions. In some cases, the LRI and RML tests coincide. In others, the RML test is almost as powerful as the LRI test. The authors make use of the fact that the distribution of the RML statistic is parameter-free whenever the families to be compared are both location-scale families. This result appears to hold for any Type 2 censored sample. The only tables of critical points which have been constructed for use with censored samples appear in Antle (1975) and apply to the situation where one is testing the null hypothesis that the underlying distribution is Weibull (or extreme-value) against the alternate hypothesis of lognormality (or normality), and vice versa.

## E 11.4.2.1 Lognormal vs. Weibull Example

We once more consider the smallest 20 values among the first 40 values listed in the NOR data set. We first exponentiated the data, and then proceeded to test the lognormal against the Weibull family. An interactive procedure was used to determine the value of RML. Entries in Table IX of Antle (1975) must be compared to $RML^{1/20}$ which here was determined to be 1.063. This value is just above the 95 percent point and so we have the surprising result that we can reject the (true) hypothesis of normality in favor of the extreme-value distribution at the 0.05 level of significance.

Finally, a somewhat different approach to this general problem deserves mention. Farewell and Prentice (1977) construct a three-parameter family of generalized gamma distributions which includes the Weibull, lognormal, and gamma families as special cases. Likelihood ratio tests using asymptotic likelihood results are recommended which can accommodate censoring as well as regression variables.

## REFERENCES

Antle, C. E. (1972). Choice of model for reliability studies and related topics. ARL 72-0108, Aerospace Technical Laboratories, Wright-Patterson AFB, Ohio.

Antle, C. E. (1973). Choice of model for reliability studies and related topics, II. ARL 73-0121, Aerospace Technical Laboratories, Wright-Patterson AFB, Ohio.

Antle, C. E. (1975). Choice of model for reliability studies and related topics, III. ARL 75-0133, Aerospace Technical Laboratories, Wright-Patterson AFB, Ohio.

Barnett, V. (1975). Probability plotting methods and order statistics. J. R. Statist. Soc. C 24, 95-108.

Barr, D. R. and Davidson, T. (1973). A Kolmogorov-Smirnov test for censored samples. Technometrics 15, 739-757.

Brain, C. W. and Shapiro, S. S. (1983). A regression test for exponentially: Censored and complete samples. Technometrics 25, 69-76.

Breslow, N. and Crowley, J. (1974). A large sample study of the life table and product limit estimates under random censorship. Ann. Statist. 2, 437-453.

Chen, C. H. (1984). A correlation goodness-of-fit test for randomly censored data. Biometrika 71, 315-322.

Cox, D. R. (1961). Tests of separate families of hypotheses. Proc. 4th Berkeley Symp. 1, 105-123.

Cox, D. R. (1962). Further results on tests of separate families of hypotheses. J. R. Statist. Soc. B 24, 406-424.

Cox, D. R. (1972). Regression models and life tables (with discussion). J. R. Statist. Soc. B 34, 187-220.

Cox, D. R. (1975). Partial likelihood. Biometrika 62, 269-276.

Dufour, R. and Maag, U. R. (1978). Distribution results for modified Kolmogorov-Smirnov statistics for truncated or censored samples. Technometrics 20, 29-32.

Dumonceaux, R. and Antle, C. E. (1973). Discrimination between the lognormal and Weibull distributions. Technometrics 15, 923-926.

Dumonceaux, R., Antle, C. E., and Haas, G. (1973). Likelihood ratio test for discrimination between two models with unknown location and scale parameters. Technometrics 15, 19-27.

Farewell, V. T., and Prentice, R. L. (1977). A study of distributional shape in life testing. Technometrics 19, 69-75.

Filliben, J. J. (1975). The probability plot correlation coefficient test for
normality. Technometrics 17, 111-117.

Fleming, T. R., O'Fallon, J. R., O'Brien, P. C., and Harrington, D. P.
(1980). Modified Kolmogorov-Smirnov test procedures with application
to arbitrarily right-censored data. Biometrics 36, 607-625.

Gaver, D. P. and Miller, Jr., R. G. (1983). Jackknifing the Kaplan-Meier
survival estimator for censored data: Simulation results and asymptotic
analysis. Commun. Statist. 12, 1701-1718.

Gupta, A. K. (1952). Estimation of the mean and standard deviation of a
normal population from a censored sample. Biometrika 39, 260-273.

Hahn, G. J. and Shapiro, S. S. (1967). Statistical Methods in Engineering.
Wiley, New York.

Herd, G. R. (1960). Estimation of reliability from incomplete data. Proceedings of the Sixth National Symposium on Reliability and Quality
Control, 202-217.

Hoadley, B. (1971). Asymptotic properties of maximum likelihood estimators
for the independent not identically distributed case. Ann. Math. Statist.
42, 1977-1991.

Johnson, L. G. (1964). The Statistical Treatment of Fatigue Experiments.
Elsevier, New York.

Kaplan, E. L. and Meier, P. (1958). Nonparametric estimation from incomplete observations. J. Amer. Statist. Assoc. 53, 457-481.

Kimball, B. F. (1960). On the choice of plotting positions on probability
paper. J. Amer. Statist. Assoc. 55, 546-560.

Klimko, L. A. and Antle, C. E. (1975). Tests for normality versus lognormality. Commun. Statist. 4, 1009-1019.

Kotz, S. (1973). Normality versus lognormality with applications. Commun.
Statist. 1, 113-132.

Koziol, J. A. (1980a). Percentage points of the asymptotic distributions of
one and two sample Kuiper statistics for truncated or censored data.
Technometrics 22, 437-442.

Koziol, J. A. (1980b). Goodness-of-fit tests for randomly censored data.
Biometrika 67, 693-696.

Lehmann, E. L. (1959). Testing Statistical Hypotheses. Wiley, New York.

Mann, N. R., Scheuer, E. M., and Fertig, K. W. (1973). A new goodness-
of-fit test for the two-parameter Weibull or extreme-value distribution
with unknown parameters. Commun. Statist. 2, 383-400.

Mann, N. R. and Fertig, K. W. (1975). A goodness-of-fit test for the two parameter vs. three parameter Weibull; confidence bounds for threshold. Technometrics 17, 237-245.

Mehrotra, K. G. (1982). On goodness-of-fit tests based on spacings for type II censored samples. Commun. Statist. 11, 869-878.

Michael, J. R. (1977). Goodness of fit: Type II censoring, influence functions. Unpublished Ph.D. dissertation, Southern Methodist University.

Michael, J. R. and Schucany, W. R. (1979). A new approach to testing goodness of fit for censored samples. Technometrics 21, 435-441.

Nair, V. N. (1981). Plots and tests for goodness of fit with randomly censored data. Biometrika 68, 99-103.

Nelson, W. (1972). Theory and applications of hazard plotting for censored failure data. Technometrics 14, 945-966.

Nelson, W. (1973). Analysis of residuals from censored data. Technometrics 15, 697-715.

Peterson, A. V. (1977). Expressing the Kaplan-Meier estimator as a function of empirical subsurvival functions. J. Amer. Statist. Assoc. 72, 854-858.

Pettitt, A. N. and Stephens, M. A. (1976a). Cramér-von Mises type statistics for the goodness of fit of censored data—simple hypothesis. Technical Report No. 229, Department of Statistics, Stanford University.

Pettitt, A. N. and Stephens, M. A. (1976b). Modified Cramér-von Mises statistics for censored data. Biometrika 63, 291-298.

Schey, H. M. (1977). The asymptotic distribution for the one-sided Kolmogorov-Smirnov statistic for truncated data. Commun. Statist. A6, 1361-1366.

Tiku, M. L. (1980). Goodness-of-fit statistics based on spacings of complete or censored samples. Australian J. of Statist. 22, 260-275.

Tiku, M. L. (1981). A goodness-of-fit statistic based on the sample spacings for testing a symmetric distribution against symmetric alternatives. Australian J. of Statist. 23, 149-158.

Trindade, D. C. and Haugh, L. D. (1979). Nonparametric estimation of a lifetime distribution via the renewal function. Technical Report 19.0463, IBM.

Turnbull, B. W. (1976). The empirical distribution function with arbitrarily grouped, censored and truncated data. J. R. Statist. Soc. B 38, 290-295.

Turnbull, B. W. and Weiss, L. (1978). A likelihood ratio statistic for testing goodness of fit with randomly censored data. Biometrics 34, 367-375.

Wilk, M. B. and Gnanadesikan, R. (1968). Probability plotting methods for the analysis of data. Biometrika 55, 1-17.

# 12

# The Analysis and Detection of Outliers

G. L. Tietjen  Los Alamos National Laboratory, Los Alamos, New Mexico

## 12.1 INTRODUCTION

The term "outlier" (straggler, sport, maverick, flyer or a wild, aberrant, discordant, or anomalous observation) has at best a subjective definition. It is an observation "so far separated . . . from the remainder that [it] gives rise to the question of whether [it is] not from a different population" (Kendall and Buckland, 1957), or "It is one that appears to deviate markedly from other members of the sample" (Grubbs, 1959). We shall adopt in this chapter the definition given by Beckman and Cook (1983): A discordant observation is one that appears surprising or discrepant to the investigator; a contaminant is one that does not come from the target population; an outlier is either a contaminant or a discordant observation.

In order for an observation to "appear surprising" to the investigator, he must have in mind some model of the data (symmetry, normality, upper bounds) which he is applying. We shall discuss here only the underlying assumption of normality since there is very little theory for any other case. The assumption of normality should not be taken lightly: The investigator needs some experience with his data generating process in order to decide whether the assumption holds. Lacking experience, he needs to ask himself whether the assumption is theoretically reasonable; he must bear in mind that there are numerous sets of data with genuinely skewed distributions (in which case the outlier theory is not applicable unless the data can be transformed to normality) and other instances where the data arise from a mixture of distributions. Gumbel (1960) has stated that "The rejection of outliers on a purely statistical basis is and remains a dangerous procedure. Its very existence may be a proof that the underlying population is, in reality, not what it was assumed to be."

Before beginning his search for outliers, the experimenter will need to ask himself <u>why</u> he is looking for outliers. It may be that he wishes only to estimate the mean and variance of his population. In that case, however dangerous the procedure, it may be more dangerous to do nothing. Anscombe (1960) has noted that "No observations are absolutely trustworthy," and that "one sufficiently erroneous reading can wreck the whole of a statistical analysis." A set of bivariate data is especially sensitive to outliers, and one observation can easily change the correlation coefficient from .01 to .99. A second important reason for looking for outliers is that interest may be centered in the outliers themselves. In prospecting for uranium, the prospector is interested only in the discordant observation; he is not at all interested in estimating the average background of a region. Beckman and Cook (1983) cite the search for the Russian satellite which crashed in Canada as a similar instance. Barnett (1978) discusses a court case of doubtful paternity where the mother gave birth to a child 349 days after the father went overseas. Is this gestation period an outlier or is it within the range of variation? The main interest here is not in estimating the mean background or the standard deviation of the human gestation period. A third reason for looking for outliers is for the information they may yield about the data gathering process. Kruskal (1960) pursues this issue: "An apparently wild observation is a signal that says: "Here is something from which we may learn a lesson, perhaps of a kind not anticipated beforehand and perhaps more important than the main object of the study. Examples of such serendipity have been frequently discussed—one of the most popular is Fleming's recognition of the virtue of penicillin . . . . Much depends on what we are after . . . ."

Kruskal cites an example of five determinations on the concentration of a chemical in a certain mixture, one of which is badly out of line. It is determined that the outlier stemmed from a miscalibration affecting only the one observation. If the objective is to estimate the concentration of that particular mixture, the outlier could be forgotten or a correction made. If the goal is to investigate the <u>method</u> of measurement, the presence of the outlier "tells us something about the frequency and magnitude of serious errors in the method. If finding an outlier results in correcting a flaw in the measurement process, its discovery will be worthwhile. When an unusual observation is encountered, we should ask: (1) What was the likelihood, before taking the measurement, that something would go wrong with the experiment and that it would be wild? (2) Is there any evidence, other than its magnitude, that something <u>did</u> go wild? Can we check the notebooks to see whether the procedure was carried out properly or that the results were recorded correctly? What is done with an outlier may depend upon the answers to these questions. We agree with Kruskal (1960a) that "it is of great importance to preach the doctrine that apparent outliers should <u>always</u> be reported, even when one feels that their causes are known or when one rejects them for whatever good rule or reason. The immediate pressures of practical statistical analysis are almost uniformly in the direction of suppressing announcement of observations that do not fit the pattern; we must maintain a strong sea-wall against these pressures."

Anscombe (1960) has identified three sources of error in any measurement process: (1) the inherent variability in the experimental units themselves, (2) the error in the measuring instruments, and (3) execution error, or any discrepancy between what we intend to do and what is actually done. The latter may include measuring a subject not belonging to the population or measuring some characteristic other than the one intended or selecting a biased sample. "If we could be sure," he continues, "that an outlier was caused by a large measurement or execution error which could not be rectified, we should be justified in discarding the observation and all memory of it. The act of observation would have failed; there would be nothing to report. Such an observation could just be described as spurious." [Following the Kruskal doctrine, we would report such a value, tell what caused the error, then forget about it.] In some cases, measurement or execution errors may be giving us measurements which are not extreme. In an interlaboratory experiment, for example, one laboratory may be reporting the mean of several measurements while another is reporting single observations. The means will tend to fall toward the center of the data rather than toward the extremes. Goodness-of-fit techniques, rather than outlier methods should be resorted to in these cases.

Having given some thought to the objectives of an analysis, we need to realize that there are two principal methods of dealing with outliers: identification and accommodation. If the outliers are detected or identified, they may be treated in one of several ways:

1. Omit the outliers and treat the reduced sample as a "new" sample.
2. Omit the outliers and treat the reduced sample as a censored sample.
3. Winsorize the outliers, i.e., replace them with the value of the nearest "good" observation. This at least preserves the direction of measurement.
4. Ask the experimenter to take additional observations to replace the outliers.
5. Present one analysis including the outliers and another excluding them. If the results are very different, view the conclusions cautiously.

Accommodation of the outliers without previously identifying them falls into the area of robust estimation. This may take the form of using trimmed means, Winsorized means, using the median instead of the mean (an extreme form of the other two), or it may involve the use of a weighted estimator (omission and trimming corresponds to zero weights, Winsorization to others, and Huber's M-estimation to still others). Estimation of the variance in these circumstances is a very different matter from estimation of the mean and may be much more difficult.

## 12.2  A SINGLE OUTLIER IN A UNIVARIATE SAMPLE

We begin first with the identification of a single outlier in a univariate sample of size n. Without going into the long history of work in this area (see Beckman and Cook, 1983), the best one can do is to obtain the mean ($\bar{x}$) and the standard deviation (s) of the entire sample, and calculate the extreme studentized residual $T_n = (x_{(n)} - \bar{x})/s$ where $x_{(n)}$ is the single largest suspect observation. If the least observation $x_{(1)}$ is suspect, $T_n = (\bar{x} - x_{(1)})/s$. If $T_n$ is larger than the critical value given by Grubbs (1959) in Table 12.1, the suspect observation is not regarded as being part of the underlying normal population. If the population standard deviation $\sigma$ is considered "known" (from considerable experience), one may use the fourth and fifth columns of Table 12.1 as critical values. If the standard deviation is estimated independently of the present sample, the second part of Table 12.1 should be used for critical values.

The test we have given is a one-sided test. To use it appropriately, we must decide, in advance, whether the outliers will occur only on the high side or only on the low side. Alternately, we may have decided that we were interested only in outliers on the high side or on the low side. If we do not know in advance whether the outlier will occur on the high side or on the low side, we should use a two-sided test. For a two-sided test at the $\alpha$-level of significance, we calculate both test statistics and compare the maximum of the two statistics it to the tabled critical value for $\alpha/2$.

### E 12.2.1  Example

In this and the following examples, sample sizes were chosen partly for convenience in computation; the tests may not be as powerful as one would like. A set of eight mass spectrometer measurements were made on a single sample of a particular isotope of uranium. The data, arranged in order, are as follows: 199.31, 199.53, 200.19, 200.82, 201.92, 201.95, 202.18, 245.57. Experience has shown that outliers usually occur on the high side. Assuming normality, can the largest observation be rejected as an outlier? We calculate $\bar{x} = 206.43$, s = 15.85, and $T_n = (245.57 - 206.43)/15.85 = 2.460$. Since this is greater than the 5% critical value of 2.03 from Table 12.1, we reject the hypothesis (i.e., we decide that 245.57 is an outlier).

Anscombe (1970) saw no particular reason for treating outliers as a hypothesis-testing problem, partly because significance levels can be swamped by the assumptions made. He advocated a data analysis approach in which we treat a rejection rule like a homeowner's fire insurance policy. A fire occurs when an observation is spurious (comes from a different population). Before buying fire insurance, we should ask: (1) What is the premium? (2) How much protection does the policy give when there is a fire? (3) How much danger is there of a fire? Reduced to statistical terms, the premium measures the

TABLE 12.1  Critical Values for Grubbs' One-Outlier Statistic $T_n$

| | Std Dev Calculated from Sample | | | Std Dev Known | |
|---|---|---|---|---|---|
| n | 5% | 2.5% | 1% | 5% | 1% |
| 3 | 1.15 | 1.15 | 1.15 | 1.74 | 2.22 |
| 4 | 1.46 | 1.48 | 1.49 | 1.94 | 2.43 |
| 5 | 1.67 | 1.71 | 1.75 | 2.08 | 2.57 |
| 6 | 1.82 | 1.89 | 1.94 | 2.18 | 2.68 |
| 7 | 1.94 | 2.02 | 2.10 | 2.27 | 2.76 |
| 8 | 2.03 | 2.13 | 2.22 | 2.33 | 2.83 |
| 9 | 2.11 | 2.21 | 2.32 | 2.39 | 2.88 |
| 10 | 2.18 | 2.29 | 2.41 | 2.44 | 2.93 |
| 11 | 2.23 | 2.36 | 2.48 | 2.48 | 2.97 |
| 12 | 2.29 | 2.41 | 2.55 | 2.52 | 3.01 |
| 13 | 2.33 | 2.46 | 2.61 | 2.56 | 3.04 |
| 14 | 2.37 | 2.51 | 2.66 | 2.59 | 3.07 |
| 15 | 2.41 | 2.55 | 2.71 | 2.62 | 3.10 |
| 16 | 2.44 | 2.59 | 2.75 | 2.64 | 3.12 |
| 17 | 2.47 | 2.62 | 2.79 | 2.67 | 3.15 |
| 18 | 2.50 | 2.65 | 2.82 | 2.69 | 3.17 |
| 19 | 2.53 | 2.68 | 2.85 | 2.71 | 3.19 |
| 20 | 2.56 | 2.71 | 2.88 | 2.73 | 3.21 |
| 21 | 2.58 | 2.73 | 2.91 | 2.75 | 3.22 |
| 22 | 2.60 | 2.76 | 2.94 | 2.77 | 3.24 |
| 23 | 2.62 | 2.78 | 2.96 | 2.78 | 3.26 |
| 24 | 2.64 | 2.80 | 2.99 | 2.80 | 3.27 |
| 25 | 2.66 | 2.82 | 3.01 | 2.81 | 3.28 |
| 30 | 2.75 | 2.91 | 3.10 | | |
| 35 | 2.82 | 2.98 | 3.18 | | |
| 40 | 2.87 | 3.04 | 3.24 | | |
| 45 | 2.92 | 3.09 | 3.29 | | |
| 50 | 2.96 | 3.13 | 3.34 | | |

(continued)

**TABLE 12.1** (continued)

| Std Dev Calculated from Sample | | | Std Dev Known | |
|---|---|---|---|---|
| n | 5% | 2.5% | 1% | 5% | 1% |
| 60 | 3.03 | 3.20 | 3.41 | | |
| 70 | 3.09 | 3.26 | 3.47 | | |
| 80 | 3.14 | 3.31 | 3.52 | | |
| 90 | 3.18 | 3.35 | 3.56 | | |
| 100 | 3.21 | 3.38 | 3.60 | | |

Critical Values for $T_n$: Std Dev Independently Estimated[a]

| n | 3 | 4 | 5 | 6 | 7 | 8 | 9 | 10 | 12 |
|---|---|---|---|---|---|---|---|---|---|
| d.f. | | | | | 1% points | | | | |
| 10 | 2.78 | 3.10 | 3.32 | 3.48 | 3.62 | 3.73 | 3.82 | 3.90 | 4.04 |
| 11 | 2.72 | 3.02 | 3.24 | 3.39 | 3.52 | 3.63 | 3.72 | 3.79 | 3.93 |
| 12 | 2.67 | 2.96 | 3.17 | 3.32 | 3.45 | 3.55 | 3.64 | 3.71 | 3.84 |
| 13 | 2.63 | 2.92 | 3.12 | 3.27 | 3.38 | 3.48 | 3.57 | 3.64 | 3.76 |
| 14 | 2.60 | 2.88 | 3.07 | 3.22 | 3.33 | 3.43 | 3.51 | 3.58 | 3.70 |
| 15 | 2.57 | 2.84 | 3.03 | 3.17 | 3.29 | 3.38 | 3.46 | 3.53 | 3.65 |
| 16 | 2.54 | 2.81 | 3.00 | 3.14 | 3.25 | 3.34 | 3.42 | 3.49 | 3.60 |
| 17 | 2.52 | 2.79 | 2.97 | 3.11 | 3.22 | 3.31 | 3.38 | 3.45 | 3.56 |
| 18 | 2.50 | 2.77 | 2.95 | 3.08 | 3.19 | 3.28 | 3.35 | 3.42 | 3.53 |
| 19 | 2.49 | 2.75 | 2.93 | 3.06 | 3.16 | 3.25 | 3.33 | 3.39 | 3.50 |
| 20 | 2.47 | 2.73 | 2.91 | 3.04 | 3.14 | 3.23 | 3.30 | 3.37 | 3.47 |
| 24 | 2.42 | 2.68 | 2.84 | 2.97 | 3.07 | 3.16 | 3.23 | 3.29 | 3.38 |
| 30 | 2.38 | 2.62 | 2.79 | 2.91 | 3.01 | 3.08 | 3.15 | 3.21 | 3.30 |
| 40 | 2.34 | 2.57 | 2.73 | 2.85 | 2.94 | 3.02 | 3.08 | 3.13 | 3.22 |
| 60 | 2.29 | 2.52 | 2.68 | 2.79 | 2.88 | 2.95 | 3.01 | 3.06 | 3.15 |
| 120 | 2.25 | 2.48 | 2.62 | 2.73 | 2.82 | 2.89 | 2.95 | 3.00 | 3.08 |
| ∞ | 2.22 | 2.43 | 2.57 | 2.68 | 2.76 | 2.83 | 2.88 | 2.93 | 3.01 |

(continued)

TABLE 12.1 (continued)

| | \multicolumn{9}{c}{Critical Values for $T_n$: Std Dev Independently Estimated[a]} | | | | | | | | |
|---|---|---|---|---|---|---|---|---|---|
| n | 3 | 4 | 5 | 6 | 7 | 8 | 9 | 10 | 12 |
| d.f. | | | | | 5% points | | | | |
| 10 | 2.01 | 2.27 | 2.46 | 2.60 | 2.72 | 2.81 | 2.89 | 2.96 | 3.08 |
| 11 | 1.98 | 2.24 | 2.42 | 2.56 | 2.67 | 2.76 | 2.84 | 2.91 | 3.03 |
| 12 | 1.96 | 2.21 | 2.39 | 2.52 | 2.63 | 2.72 | 2.80 | 2.87 | 2.98 |
| 13 | 1.94 | 2.19 | 2.36 | 2.50 | 2.60 | 2.69 | 2.76 | 2.83 | 2.94 |
| 14 | 1.93 | 2.17 | 2.34 | 2.47 | 2.57 | 2.66 | 2.74 | 2.80 | 2.91 |
| 15 | 1.91 | 2.15 | 2.32 | 2.45 | 2.55 | 2.64 | 2.71 | 2.77 | 2.88 |
| 16 | 1.90 | 2.14 | 2.31 | 2.43 | 2.53 | 2.62 | 2.69 | 2.75 | 2.86 |
| 17 | 1.89 | 2.13 | 2.29 | 2.42 | 2.52 | 2.60 | 2.67 | 2.73 | 2.84 |
| 18 | 1.88 | 2.11 | 2.28 | 2.40 | 2.50 | 2.58 | 2.65 | 2.71 | 2.82 |
| 19 | 1.87 | 2.11 | 2.27 | 2.39 | 2.49 | 2.57 | 2.64 | 2.70 | 2.80 |
| 20 | 1.87 | 2.10 | 2.26 | 2.38 | 2.47 | 2.56 | 2.63 | 2.68 | 2.78 |
| 24 | 1.84 | 2.07 | 2.23 | 2.34 | 2.44 | 2.52 | 2.58 | 2.64 | 2.74 |
| 30 | 1.82 | 2.04 | 2.20 | 2.31 | 2.40 | 2.48 | 2.54 | 2.60 | 2.69 |
| 40 | 1.80 | 2.02 | 2.17 | 2.28 | 2.37 | 2.44 | 2.50 | 2.56 | 2.65 |
| 60 | 1.78 | 1.99 | 2.14 | 2.25 | 2.33 | 2.41 | 2.47 | 2.52 | 2.61 |
| 120 | 1.76 | 1.96 | 2.11 | 2.22 | 2.30 | 2.37 | 2.43 | 2.48 | 2.57 |
| ∞ | 1.74 | 1.94 | 2.08 | 2.18 | 2.27 | 2.33 | 2.39 | 2.44 | 2.52 |

inflation in the mean square error (MSE) of an estimator of location when in fact all the observations are from the underlying population (by falsely rejecting the hypothesis a fraction of the time, the MSE of our estimator is larger than it would be if we had no rejection rule). Protection measures the reduction in the mean square error of the estimator when there are outliers present, i.e., we get a smaller MSE using the rejection rule than not using it. Guttman has stated that "while the above concepts of premium and protection are relevant and appealing, numerical computation turns out to be quite difficult." Nevertheless, considerable work has been done in measuring and comparing the premium and protection of different rejection rules.

The null hypothesis in identifying outliers is that all the observations come from a normal population; rejection of the hypothesis can mean many things. Since only the extreme observations have been tested, we see that outlier-detection statistics would not be very useful as tests of normality (goodness-of-fit tests would be more appropriate). Two models for generating a population containing outliers are widely used. The first is the mean shift model where n - k observations are from a $N(\mu, \sigma^2)$ population and k from a $N(\mu + \lambda, \sigma^2)$ population. In practice this is done by generating all n from the first population and adding $\lambda$ to the first k of these. If some of the first k are below the mean, adding $\lambda$ will make them close to the mean, so that they will be well "hidden" among the others. This naturally limits the power of any outlier procedure. An erroneous method of contamination is to add $\lambda$ to the largest k of the n observations generated from the $N(\mu, \sigma^2)$ population. This creates a non-normal truncated distribution for the bulk of the data, and makes it easy for any outlier test to perform well because of the large gap in the data. The second model for contamination contains k observations from a $N(\mu, \lambda^2 \sigma^2)$ population, and is called the variance-shift model.

## 12.3 MULTIPLE OUTLIERS IN A UNIVARIATE SAMPLE

When there is a possibility of more than one outlier in the sample, complications quickly arise. Grubbs (1950) derived exact critical values for the two largest (or two smallest) outliers, but did not obtain critical values for the largest and smallest observations. Tietjen and Moore (1972) extended Grubbs' critical values, by simulation, for up to 10 outliers. The statistic used was the ratio of the sum of squares in a sample which omits the outliers to the sum of squares for the complete sample. This statistic was called $L_k$. Tietjen and Moore obtained another statistic, $E_k$, which took the same form as $L_k$ but the numerator was based on omitting the k most extreme observations from the mean (from either or both ends). The test was based on the assumption that k was known and in practice determined by looking at the data. Since one does not anticipate any outliers in a sample, k could not be known in advance, and looking at the sample interfered with the $\alpha$-level in an unknown way. Furthermore, if k were determined automatically, $E_k$ could pick the wrong observations to test. (In a sample of size 12, let 10 values be from a $N(0, 1)$, one be at 10 and one be at 100. Since the mean is close to 10, the two most extreme observations are the smallest one and the one at 100. Clearly the smallest observation is not an outlier.) The last problem is remedied simply by picking out the single observation furthest from the mean as outlier candidate #1, then omitting it from the sample and picking the farthest from the new mean as outlier candidate #2, etc.

Yet another problem arises in trying to choose a value for k, the number of observations to be tested as outliers. If there are two large observations which are nearly equal, and if one uses a one-outlier test on the largest, the test will usually fail to reject because the second outlier masks the presence

TABLE 12.2  Critical Values for Rosner's ESD ($T_n$) Statistic. Sample Estimates of 10, 5, and 1% Points for $ESD_{(\ )}$, for Selected N

| N | .10 | .05 | .01 | | N | .10 | .05 | .01 |
|---|---|---|---|---|---|---|---|---|
| k = 2 (for $ESD_1$, $ESD_2$) | | | | | 40 | 3.01 | 3.17 | 3.52 |
| 10 | | 2.39 | 2.55 | | | 2.64 | 2.77 | 2.98 |
| | | 2.17 | 2.32 | | 45 | | 3.17 | 3.57 |
| 11 | | 2.45 | 2.62 | | | | 2.82 | 3.05 |
| | | 2.23 | 2.41 | | 50 | 3.10 | 3.27 | 3.61 |
| 12 | | 2.50 | 2.71 | | | 2.72 | 2.85 | 3.08 |
| | | 2.27 | 2.49 | | 60 | 3.15 | 3.34 | 3.70 |
| 13 | | 2.57 | 2.84 | | | 2.77 | 2.90 | 3.17 |
| | | 2.31 | 2.56 | | 80 | 3.28 | 3.45 | 3.80 |
| 14 | | 2.62 | 2.86 | | | 2.85 | 2.97 | 3.23 |
| | | 2.39 | 2.61 | | 100 | 3.34 | 3.52 | 3.87 |
| 15 | | 2.65 | 2.91 | | | 2.92 | 3.03 | 3.28 |
| | | 2.42 | 2.66 | | k = 3 (for $ESD_1$, $ESD_2$, $ESD_3$) | | | |
| 16 | | 2.70 | 2.95 | | 20 | 2.76 | 2.88 | 3.13 |
| | | 2.44 | 2.64 | | | 2.47 | 2.60 | 2.83 |
| 17 | | 2.75 | 3.03 | | | 2.34 | 2.45 | 2.68 |
| | | 2.48 | 2.65 | | 30 | 2.97 | 3.12 | 3.41 |
| 18 | | 2.79 | 3.08 | | | 2.61 | 2.73 | 3.01 |
| | | 2.46 | 2.68 | | | 2.44 | 2.56 | 2.75 |
| 19 | | 2.80 | 3.10 | | 40 | 3.07 | 3.22 | 3.58 |
| | | 2.49 | 2.71 | | | 2.69 | 2.81 | 3.03 |
| 20 | 2.69 | 2.83 | 3.09 | | | 2.52 | 2.62 | 2.82 |
| | 2.41 | 2.52 | 2.76 | | 50 | 3.18 | 3.34 | 3.68 |
| 25 | | 2.99 | 3.34 | | | 2.76 | 2.89 | 3.15 |
| | | 2.62 | 2.82 | | | 2.58 | 2.68 | 2.89 |
| 30 | 2.89 | 3.05 | 3.35 | | 60 | 3.26 | 3.42 | 3.75 |
| | 2.55 | 2.67 | 2.92 | | | 2.83 | 2.95 | 3.20 |
| 35 | | 3.09 | 3.41 | | | 2.64 | 2.73 | 2.95 |
| | | 2.74 | 2.96 | | 80 | 3.32 | 3.49 | 3.85 |
| | | | | | | 2.90 | 3.03 | 3.27 |
| | | | | | | 2.71 | 2.81 | 3.01 |

(continued)

TABLE 12.2 (continued)

| N | α | | | N | α | | |
|---|---|---|---|---|---|---|---|
|  | .10 | .05 | .01 |  | .10 | .05 | .01 |
| 100 | 3.44 | 3.60 | 3.97 | k = 5 (for $ESD_1$, $ESD_2$, $ESD_3$, | | | |
|  | 2.97 | 3.10 | 3.34 | $ESD_4$, $ESD_5$) | | | |
|  | 2.77 | 2.86 | 3.06 | 20 | 2.85 | 2.97 | 3.10 |
| k = 4 (for $ESD_1$, $ESD_2$, $ESD_3$, $ESD_4$) | | | |  | 2.55 | 2.65 | 2.89 |
|  | | | |  | 2.40 | 2.51 | 2.69 |
| 20 | 2.81 | 2.95 | 3.20 |  | 2.33 | 2.42 | 2.61 |
|  | 2.51 | 2.63 | 2.83 |  | 2.27 | 2.37 | 2.57 |
|  | 2.38 | 2.49 | 2.68 | 30 | 3.05 | 3.19 | 3.48 |
|  | 2.29 | 2.39 | 2.58 |  | 2.67 | 2.78 | 3.03 |
| 30 | 3.02 | 3.16 | 3.48 |  | 2.51 | 2.60 | 2.80 |
|  | 2.65 | 2.77 | 3.02 |  | 2.42 | 2.51 | 2.74 |
|  | 2.48 | 2.59 | 2.79 |  | 2.35 | 2.45 | 2.62 |
|  | 2.39 | 2.49 | 2.70 | 40 | 3.16 | 3.31 | 3.63 |
| 40 | 3.14 | 3.32 | 3.64 |  | 2.76 | 2.88 | 3.13 |
|  | 2.74 | 2.86 | 3.10 |  | 2.59 | 2.69 | 2.89 |
|  | 2.57 | 2.67 | 2.87 |  | 2.46 | 2.55 | 2.74 |
|  | 2.45 | 2.55 | 2.74 |  | 2.39 | 2.47 | 2.65 |
| 50 | 3.24 | 3.40 | 3.74 | 50 | 3.28 | 3.45 | 3.77 |
|  | 2.81 | 2.93 | 3.18 |  | 2.84 | 2.96 | 3.21 |
|  | 2.62 | 2.72 | 2.92 |  | 2.65 | 2.74 | 2.94 |
|  | 2.50 | 2.59 | 2.78 |  | 2.52 | 2.61 | 2.79 |
| 60 | 3.31 | 3.48 | 3.82 |  | 2.44 | 2.52 | 2.70 |
|  | 2.85 | 2.98 | 3.20 | 60 | 3.34 | 3.51 | 3.81 |
|  | 2.67 | 2.77 | 2.97 |  | 2.88 | 3.01 | 3.24 |
|  | 2.54 | 2.63 | 2.82 |  | 2.68 | 2.77 | 2.96 |
| 80 | 3.40 | 3.57 | 3.91 |  | 2.56 | 2.65 | 2.83 |
|  | 2.94 | 3.05 | 3.31 |  | 2.48 | 2.56 | 2.72 |
|  | 2.74 | 2.84 | 3.04 | 80 | 3.44 | 3.61 | 3.93 |
|  | 2.61 | 2.69 | 2.87 |  | 2.98 | 3.11 | 3.36 |
| 100 | 3.47 | 3.64 | 3.96 |  | 2.77 | 2.86 | 3.08 |
|  | 3.00 | 3.13 | 3.34 |  | 2.63 | 2.72 | 2.89 |
|  | 2.79 | 2.89 | 3.06 |  | 2.54 | 2.62 | 2.76 |
|  | 2.66 | 2.74 | 2.90 | 100 | 3.53 | 3.70 | 4.01 |
|  | | | |  | 3.04 | 3.16 | 3.42 |
|  | | | |  | 2.81 | 2.91 | 3.10 |
|  | | | |  | 2.68 | 2.77 | 2.93 |
|  | | | |  | 2.59 | 2.67 | 2.84 |

of the first; the procedure cannot get started. Thus the repeated application of a single outlier test can easily fail. Furthermore, if there is only one large outlier and one uses a test for two outliers, the test is likely to reject $H_0$ and claim that there are two outliers, a phenomenon known as <u>swamping</u>. Rosner (1975) devised a procedure which successfully overcame masking but is still subject to some swamping. Let $I_0$ be the full data set and $I_{t+1}$ be the set obtained by omitting from $I_t$ the point most extreme from the mean of $I_t$. Let k be an upper bound on the number of outliers in the sample. Apply a one-outlier test in succession to $I_0$, $I_1$, $\ldots$, $I_{k-1}$, and let the last significant result be for $I_{m-1}$. Decide that the m observations omitted from $I_m$ are outliers. The critical values have to hold simultaneously for the several tests, hence are difficult to generate. It should be easier to estimate $k_u$ than k, but the amount of swamping will depend on how badly we estimate $k_u$. Despite the swamping, we recommend Rosner's test for several outliers if the $\alpha$-level is important to maintain. We cannot state how effective $L_k$ or $E_k$ might be because there is no <u>objective</u> way of deciding upon a value of k. The best tables are given by Jain (1981) as Table 12.2.

## E 12.3.1  Example

Twenty laboratories did an analysis on a single blood sample for lead content. Assuming the data are normally distributed, are there any outliers? .000, .015, .016, .022, .022, .023, .026, .027, .027, .028, .028, .031, .032, .033, .035, .037, .038, .041, .056, .058.

Using Rosner's ESD procedure, we set 20% as an upper limit on the number of outliers and check for up to 4 outliers. $I_0$ is the full set of data ($\bar{x}$ = .0298, s = .0131), $I_1$ the set with .000 omitted ($\bar{x}_1$ = .0313, $s_1$ = .0114), $I_2$ is the set with .000 and .058 omitted ($\bar{x}_2$ = .0298, $s_2$ = .0097), $I_3$ is the set with .000, .058, and .056 omitted ($\bar{x}_3$ = .0283, $s_3$ = .0073), $I_4$ is the set with .000, .058, .056, and .015 omitted. We calculate the $T_n$ statistic for each set, obtaining $R_1$ = 2.27 (for set $I_0$), $R_2$ = 2.34, $R_3$ = 2.70, $R_4$ = 1.82. From Table 12.2 we obtain the 5% critical values for $R_1$, $R_2$, $R_3$, and $R_4$ of 2.95, 2.63, 2.49, and 2.39, respectively. $R_3$ is the only one significant, hence we declare .000, .056, and .058 to be outliers.

## 12.4  THE IDENTIFICATION OF A SINGLE OUTLIER
### IN LINEAR MODELS

In the univariate case the residuals are correlated and have a common variance. In the regression case, the residuals are also correlated but each residual has its own variance which depends, to some extent, on the arrangement of the x-values. Let $Y = X\beta + \varepsilon$ be the linear model in which $Y$ is the (n × 1) vector of responses, $X$ an (n × p) matrix of known constants, $\beta$ a (p × 1) vector of unknown parameters, and $\varepsilon$ an (n × 1) vector of normally

TABLE 12.3 Critical Values for a Single Outlier in Linear Models
($\alpha = .10$)

| n | 1 | 2 | 3 | 4 | q 5 | 6 | 8 | 10 | 15 | 25 |
|---|---|---|---|---|---|---|---|---|---|---|
| 5 | 1.87 | | | | | | | | | |
| 6 | 2.00 | 1.89 | | | | | | | | |
| 7 | 2.10 | 2.02 | 1.90 | | | | | | | |
| 8 | 2.18 | 2.12 | 2.03 | 1.91 | | | | | | |
| 9 | 2.24 | 2.20 | 2.13 | 2.05 | 1.92 | | | | | |
| 10 | 2.30 | 2.26 | 2.21 | 2.15 | 2.06 | 1.92 | | | | |
| 12 | 2.39 | 2.37 | 2.33 | 2.29 | 2.24 | 2.17 | 1.93 | | | |
| 14 | 2.47 | 2.45 | 2.42 | 2.39 | 2.36 | 2.32 | 2.19 | 1.94 | | |
| 16 | 2.53 | 2.51 | 2.50 | 2.47 | 2.45 | 2.42 | 2.34 | 2.20 | | |
| 18 | 2.58 | 2.57 | 2.56 | 2.54 | 2.52 | 2.50 | 2.44 | 2.35 | | |
| 20 | 2.63 | 2.62 | 2.61 | 2.59 | 2.58 | 2.56 | 2.52 | 2.46 | 2.11 | |
| 25 | 2.72 | 2.72 | 2.71 | 2.70 | 2.69 | 2.68 | 2.66 | 2.63 | 2.50 | |
| 30 | 2.80 | 2.79 | 2.79 | 2.78 | 2.77 | 2.77 | 2.75 | 2.73 | 2.66 | 2.13 |
| 35 | 2.86 | 2.85 | 2.85 | 2.85 | 2.84 | 2.84 | 2.82 | 2.81 | 2.77 | 2.55 |
| 40 | 2.91 | 2.91 | 2.90 | 2.90 | 2.90 | 2.89 | 2.88 | 2.87 | 2.84 | 2.72 |
| 45 | 2.95 | 2.95 | 2.95 | 2.95 | 2.94 | 2.94 | 2.93 | 2.93 | 2.90 | 2.82 |
| 50 | 2.99 | 2.99 | 2.99 | 2.99 | 2.98 | 2.98 | 2.98 | 2.97 | 2.95 | 2.89 |
| 60 | 3.06 | 3.06 | 3.05 | 3.05 | 3.05 | 3.05 | 3.05 | 3.04 | 3.03 | 3.00 |
| 70 | 3.11 | 3.11 | 3.11 | 3.11 | 3.11 | 3.11 | 3.10 | 3.10 | 3.09 | 3.07 |
| 80 | 3.16 | 3.16 | 3.16 | 3.15 | 3.15 | 3.15 | 3.15 | 3.15 | 3.14 | 3.12 |
| 90 | 3.20 | 3.20 | 3.19 | 3.19 | 3.19 | 3.19 | 3.19 | 3.19 | 3.18 | 3.17 |
| 100 | 3.23 | 3.23 | 3.23 | 3.23 | 3.23 | 3.23 | 3.23 | 3.22 | 3.22 | 3.21 |

(continued)

TABLE 12.3 (continued)

$$\alpha = .05$$

| n | 1 | 2 | 3 | 4 | 5 | 6 | 8 | 10 | 15 | 25 |
|---|---|---|---|---|---|---|---|---|---|---|
| 5 | 1.92 | | | | | | | | | |
| 6 | 2.07 | 1.93 | | | | | | | | |
| 7 | 2.19 | 2.08 | 1.94 | | | | | | | |
| 8 | 2.28 | 2.20 | 2.10 | 1.94 | | | | | | |
| 9 | 2.35 | 2.29 | 2.21 | 2.10 | 1.95 | | | | | |
| 10 | 2.42 | 2.37 | 2.31 | 2.22 | 2.11 | 1.95 | | | | |
| 12 | 2.52 | 2.49 | 2.45 | 2.39 | 2.33 | 2.24 | 1.96 | | | |
| 14 | 2.61 | 2.58 | 2.55 | 2.51 | 2.47 | 2.41 | 2.25 | 1.96 | | |
| 16 | 2.68 | 2.66 | 2.63 | 2.60 | 2.57 | 2.53 | 2.43 | 2.26 | | |
| 18 | 2.73 | 2.72 | 2.70 | 2.68 | 2.65 | 2.62 | 2.55 | 2.44 | | |
| 20 | 2.78 | 2.77 | 2.76 | 2.74 | 2.72 | 2.70 | 2.64 | 2.57 | 2.15 | |
| 25 | 2.89 | 2.88 | 2.87 | 2.86 | 2.84 | 2.83 | 2.80 | 2.76 | 2.60 | |
| 30 | 2.96 | 2.96 | 2.95 | 2.94 | 2.93 | 2.93 | 2.90 | 2.88 | 2.79 | 2.17 |
| 35 | 3.03 | 3.02 | 3.02 | 3.01 | 3.00 | 3.00 | 2.98 | 2.97 | 2.91 | 2.64 |
| 40 | 3.08 | 3.08 | 3.07 | 3.07 | 3.06 | 3.06 | 3.05 | 3.03 | 3.00 | 2.84 |
| 45 | 3.13 | 3.12 | 3.12 | 3.12 | 3.11 | 3.11 | 3.10 | 3.09 | 3.06 | 2.96 |
| 50 | 3.17 | 3.16 | 3.16 | 3.16 | 3.15 | 3.15 | 3.14 | 3.14 | 3.11 | 3.04 |
| 60 | 3.23 | 3.23 | 3.23 | 3.23 | 3.22 | 3.22 | 3.22 | 3.21 | 3.20 | 3.15 |
| 70 | 3.29 | 3.29 | 3.28 | 3.28 | 3.28 | 3.28 | 3.27 | 3.27 | 3.26 | 3.23 |
| 80 | 3.33 | 3.33 | 3.33 | 3.33 | 3.33 | 3.33 | 3.32 | 3.32 | 3.31 | 3.29 |
| 90 | 3.37 | 3.37 | 3.37 | 3.37 | 3.37 | 3.37 | 3.36 | 3.36 | 3.36 | 3.34 |
| 100 | 3.41 | 3.41 | 3.40 | 3.40 | 3.40 | 3.40 | 3.40 | 3.40 | 3.39 | 3.38 |

(continued)

TABLE 12.3 (continued)

$(\alpha = .01)$

| | | | | | q | | | | | |
|---|---|---|---|---|---|---|---|---|---|---|
| n | 1 | 2 | 3 | 4 | 5 | 6 | 8 | 10 | 15 | 25 |
| 5 | 1.98 | | | | | | | | | |
| 6 | 2.17 | 1.98 | | | | | | | | |
| 7 | 2.32 | 2.17 | 1.98 | | | | | | | |
| 8 | 2.44 | 2.32 | 2.18 | 1.98 | | | | | | |
| 9 | 2.54 | 2.44 | 2.33 | 2.18 | 1.99 | | | | | |
| 10 | 2.62 | 2.55 | 2.45 | 2.33 | 2.18 | 1.99 | | | | |
| 12 | 2.76 | 2.70 | 2.64 | 2.56 | 2.46 | 2.34 | 1.99 | | | |
| 14 | 2.86 | 2.82 | 2.78 | 2.72 | 2.65 | 2.57 | 2.35 | 1.99 | | |
| 16 | 2.95 | 2.92 | 2.88 | 2.84 | 2.79 | 2.73 | 2.58 | 2.35 | | |
| 18 | 3.02 | 3.00 | 2.97 | 2.94 | 2.90 | 2.85 | 2.75 | 2.59 | | |
| 20 | 3.08 | 3.06 | 3.04 | 3.01 | 2.98 | 2.95 | 2.87 | 2.76 | 2.20 | |
| 25 | 3.21 | 3.19 | 3.18 | 3.16 | 3.14 | 3.12 | 3.07 | 3.01 | 2.78 | |
| 30 | 3.30 | 3.29 | 3.28 | 3.26 | 3.25 | 3.24 | 3.21 | 3.17 | 3.04 | 2.21 |
| 35 | 3.37 | 3.36 | 3.35 | 3.34 | 3.34 | 3.33 | 3.30 | 3.28 | 3.19 | 2.81 |
| 40 | 3.43 | 3.42 | 3.42 | 3.41 | 3.40 | 3.40 | 3.38 | 3.36 | 3.30 | 3.08 |
| 45 | 3.48 | 3.47 | 3.47 | 3.46 | 3.46 | 3.45 | 3.44 | 3.43 | 3.38 | 3.23 |
| 50 | 3.52 | 3.52 | 3.51 | 3.51 | 3.51 | 3.50 | 3.49 | 3.48 | 3.45 | 3.34 |
| 60 | 3.60 | 3.59 | 3.59 | 3.59 | 3.58 | 3.58 | 3.57 | 3.56 | 3.54 | 3.48 |
| 70 | 3.65 | 3.65 | 3.65 | 3.65 | 3.64 | 3.64 | 3.64 | 3.63 | 3.61 | 3.57 |
| 80 | 3.70 | 3.70 | 3.70 | 3.70 | 3.69 | 3.69 | 3.69 | 3.68 | 3.67 | 3.64 |
| 90 | 3.74 | 3.74 | 3.74 | 3.74 | 3.74 | 3.74 | 3.73 | 3.73 | 3.72 | 3.70 |
| 100 | 3.78 | 3.78 | 3.78 | 3.77 | 3.77 | 3.77 | 3.77 | 3.77 | 3.76 | 3.74 |

n = number of observations
q = number of independent variables (including count for intercept if fitted)

TABLE 12.4  Critical Values for Balanced Two-Way
and Three-Way Layouts

| R C | 3 | 4 | 5 | 6 | 7 | 8 | 9 | 10 |
|---|---|---|---|---|---|---|---|---|

### Critical Values for the Two-Way Layout
(asterisks denote theoretically exact values)

$$\alpha = 0.01$$

| R C | 3 | 4 | 5 | 6 | 7 | 8 | 9 | 10 |
|---|---|---|---|---|---|---|---|---|
| 3 | .66033* | | | | | | | |
| 4 | .67484* | .66511* | | | | | | |
| 5 | .66434* | .63995* | .60797* | | | | | |
| 6 | .64597* | .61302* | .57774* | .54628* | | | | |
| 7 | .62576* | .58767* | .55080* | .51901* | .49193 | | | |
| 8 | .60584* | .56463* | .52707* | .49538 | .46870 | .44599 | | |
| 9 | .58696* | .54386* | .50611* | .47475 | .44857 | .42641 | .40736 | |
| 10 | .56935* | .52516* | .48750 | .45658 | .43094 | .40931 | .39079 | .37469 |

$$\alpha = 0.05$$

| R C | 3 | 4 | 5 | 6 | 7 | 8 | 9 | 10 |
|---|---|---|---|---|---|---|---|---|
| 3 | .64810* | | | | | | | |
| 4 | .64512* | .62066* | | | | | | |
| 5 | .62415* | .58971* | .55513* | | | | | |
| 6 | .60008* | .56079* | .52491* | .49459 | | | | |
| 7 | .57666* | .53513* | .49897 | .46899 | .44396 | | | |
| 8 | .55498* | .51256* | .47660 | .44715 | .42273 | .40213 | | |
| 9 | .53521* | .49265 | .45712 | .42827 | .40447 | .38447 | .36736 | |
| 10 | .51724 | .47498 | .43998 | .41175 | .38856 | .36911 | .35251 | .33812 |

### Critical Values for the Three-Way Layout
(asterisks denote theoretically exact values; differences ($\times 10^{-5}$) between upper
and lower bounds are given in parentheses after the bound when necessary)

$$\alpha = 0.01$$

| R | C | 3 | 4 |
|---|---|---|---|
| 3 | 3 | .50294* | |
| 4 | 3 | .48778* | |
| | 4 | .46011* | .42529 |

(continued)

TABLE 12.4 (continued)

| R | C | 3 | 4 | 5 | 6 | 7 | 8 | 9 | 10 |
|---|---|---|---|---|---|---|---|---|----|
| 5 | 3 | .46503* | | | | | | | |
|   | 4 | .43209 | .39515 | | | | | | |
|   | 5 | .40257 | .36509 | .33582 | | | | | |
| 6 | 3 | .44276 | | | | | | | |
|   | 4 | .40755 | .37024 | | | | | | |
|   | 5 | .37783 | .34087 | .31267 | | | | | |
|   | 6 | .35352 | .31756 | .29060 | .26969 | | | | |
| 7 | 3 | .42260 | | | | | | | |
|   | 4 | .38649 | .34951 | | | | | | |
|   | 5 | .35709 | .32100 | .29387 | | | | | |
|   | 6 | .33341 | .29858 | .27279 | .25290 | | | | |
|   | 7 | .31397 | .28044 | .25585 | .23697 | .22191 | | | |
| 8 | 3 | .40468 | | | | | | | |
|   | 4 | .36836 | .33200 | | | | | | |
|   | 5 | .33949 | .30436 | .27824 | | | | | |
|   | 6 | .31647 | .28278 | .25804 | .23903 | | | | |
|   | 7 | .29770 | .26538 | .24186 | .22386 | .20953 | | | |
|   | 8 | .28203 | .25099 | .22852 | .21138 | .19776 | .18660 | | |
| 9 | 3 | .38877 | | | | | | | |
|   | 4 | .35261 | .31697 | | | | | | |
|   | 5 | .32434 | .29018 | .26498 | | | | | |
|   | 6 | .30198 | .26937 | .24557 | .22734 | | | | |
|   | 7 | .28383 | .25264 | .23005 | .21281 | .19911 | | | |
|   | 8 | .26872 | .23882 | .21728 | .20089 | .18788 | .17723 | | |
|   | 9 | .25591 | .22715 | .20653 | .19086 | .17845 | .16829 | .15977 | |
| 10 | 3 | .37461 | | | | | | | |
|   | 4 | .33878 | .30391 | | | | | | |
|   | 5 | .31115 | .27791 | .25355 | | | | | |

(continued)

TABLE 12.4 (continued)

| R | C | 3 | 4 | 5 | 6 | 7 | 8 | 9 | 10 |
|---|---|---|---|---|---|---|---|---|---|
| | 6 | .28941 | .25779 | .23484 | .21730 | | | | |
| | 7 | .27182 | .24166 | .21991 | .20334 | .19019 | | | |
| | 8 | .25723 | .22835 | .20763 | .19190 | .17942 | .16921 | | |
| | 9 | .24487 | .21714 | .19731 | .18228 | .17038 | .16065 | .15250 | |
| | 10 | .23422 | .20751 | .18847 | .17406 | .16265 | .15333 | .14553 | .13887 |

$$\alpha = 0.05$$

| R | C | 3 | 4 | 5 | 6 | 7 | 8 | 9 | 10 |
|---|---|---|---|---|---|---|---|---|---|
| 3 | 3 | .47790* | | | | | | | |
| 4 | 3 | .45465* | | | | | | | |
| | 4 | .42314 | .38800 | | | | | | |
| 5 | 3 | .42912 | | | | | | | |
| | 4 | .39495 | .35936 | | | | | | |
| | 5 | .36652 | .33144 | .30467 | | | | | |
| 6 | 3 | .40625 | | | | | | | |
| | 4 | .37136 | .33624 | | | | | | |
| | 5 | .34338 | .30930 | .28370 | | | | | |
| | 6 | .32096 | .28814 | .26380 | .24500 | | | | |
| 7 | 3 | .38640 | | | | | | | |
| | 4 | .35157 | .31724 | | | | | | |
| | 5 | .32426 | .29127 | .26675 | | | | | |
| | 6 | .30259 | .27101 | .24779 | .22993 | | | | |
| | 7 | .28496 | .25468 | .23259 | .21565 | .20214 | | | |
| 8 | 3 | .36920 | | | | | | | |
| | 4 | .33476 | .30131 | | | | | | |
| | 5 | .30817 | .27625 | .25270 | | | | | |
| | 6 | .28723 | .25679 | .23455 | .21750 | | | | |
| | 7 | .27025 | .24116 | .22004 | .20389 | .19104 | | | |
| | 8 | .25614 | .22824 | .20808 | .19271 | .18049 | .17046 | | |

(continued)

TABLE 12.4 (continued)

| R | C | 3 | 4 | 5 | 6 | 7 | 8 | 9 | 10 |
|---|---|---|---|---|---|---|---|---|---|
| 9 | 3 | .35418 (1) | | | | | | | |
|   | 4 | .32028 | .28770 | | | | | | |
|   | 5 | .29441 | .26347 | .24079 | | | | | |
|   | 6 | .27414 | .24474 | .22337 | .20702 | | | | |
|   | 7 | .25776 | .22972 | .20945 | .19399 | .18170 | | | |
|   | 8 | .24417 | .21732 | .19800 | .18329 | .17161 | .16204 | | |
|   | 9 | .23266 | .20686 | .18836 | .17430 | .16314 | .15401 | .14634 | |
| 10 | 3 | .34094 (1) | | | | | | | |
|   | 4 | .30765 | .27590 | | | | | | |
|   | 5 | .28246 | .25244 | .23054 | | | | | |
|   | 6 | .26280 | .23435 | .21375 | .19801 | | | | |
|   | 7 | .24696 | .21987 | .20036 | .18550 | .17369 | | | |
|   | 8 | .23384 | .20794 | .18935 | .17522 | .16401 | .15484 | | |
|   | 9 | .22275 | .19788 | .18009 | .16659 | .15589 | .14713 | .13979 | |
|   | 10 | .21320 | .18924 | .17415 | .15920 | .14894 | .14055 | .13351 | .12750 |

distributed errors with mean zero and variance $\sigma^2 I$. If $\hat{\underline{Y}}$ is the vector of fitted values, $\underline{e} = \underline{Y} - \hat{\underline{Y}}$ is the vector of residuals. Letting $\underline{V} = (v_{ij}) = \underline{X}(\underline{X}'\underline{X})^{-1}\underline{X}'$, $\hat{\sigma}^2 = \underline{e}'\underline{e}/(n - p)$ and var $(e_i) \simeq \hat{\sigma}^2(1 - v_{ij})$. The ith studentized residual is $r_i = e_i/\sqrt{\text{var } e_i}$ and the maximum of the absolute values of the n will be denoted by $r_m$. If $r_m$ is greater than some critical value $h_\alpha$, the observation which gave rise to $r_m$ is declared to be an outlier. Much of the early work in this area is due to Srikantan (1961), but the best tables to date are those of Lund (1975), Table 12.3.

In cases where we "know" $\sigma^2$ or have an independent estimate $s_V^2$ of it, we may use this knowledge in calculating $r_m$, but should use the critical values given by Joshi (1972).

For a certain large class of designed experiments the residuals have a common variance. This class of experiments includes all balanced factorial arrangements, balanced incomplete blocks, and Latin squares. For these arrangements we fit the model and find the ith residual. The ith normalized residual is $Z_i = e_i/(\Sigma e_i^2)^{\frac{1}{2}}$ and the maximum of the absolute values of the $Z_i$ is called the maximum normed residual, MNR. Important work in this area was done by Stefansky (1972), but Galpin and Hawkins (1981) have a better and more extensive set of tables (Table 12.4). If the MNR exceeds the tabulated critical values, the observation which gave rise to it is judged to be an outlier. The statistic given by C. Daniel (1960) in his work on half normal plots is equivalent to this test.

## E 12.4.1 Example

Snedecor and Cochran (1967) give an example of the use of organic phosphorus $(x_2)$ inorganic phosphorus $(x_1)$ on the yield of corn (y) on 18 Iowa soils. The data are shown below:

| Soil Sample | $Y_i$ | $X_{1i}$ | $X_{2i}$ | $\hat{Y}_i$ | $e_i$ | $r_i$ |
|---|---|---|---|---|---|---|
| 1 | 64 | 0.4 | 53 | 61.56 | 2.44 | 0.14 |
| 2 | 60 | 0.4 | 23 | 58.96 | 1.04 | 0.06 |
| 3 | 71 | 3.1 | 19 | 63.45 | 7.55 | 0.42 |
| 4 | 61 | 0.6 | 34 | 60.27 | 0.73 | 0.04 |
| 5 | 54 | 4.7 | 24 | 66.74 | -12.74 | -0.67 |
| 6 | 77 | 1.7 | 65 | 64.93 | 12.07 | 0.79 |
| 7 | 81 | 9.4 | 44 | 76.89 | 4.11 | 0.21 |
| 8 | 93 | 10.1 | 31 | 77.01 | 15.99 | 0.81 |
| 9 | 93 | 11.6 | 29 | 79.53 | 13.47 | 0.70 |
| 10 | 51 | 12.6 | 58 | 83.83 | -32.83 | -1.72 |
| 11 | 76 | 10.9 | 37 | 78.97 | -2.97 | -0.15 |
| 12 | 96 | 23.1 | 46 | 101.58 | -5.58 | -0.29 |
| 13 | 77 | 23.1 | 50 | 101.93 | -24.93 | -1.29 |
| 14 | 93 | 21.6 | 44 | 98.72 | -5.72 | -0.29 |
| 15 | 95 | 23.1 | 56 | 102.45 | -7.45 | -0.39 |
| 16 | 54 | 1.9 | 36 | 62.77 | -8.77 | -0.45 |
| 17 | 168 | 26.8 | 58 | 109.24 | 58.76 | 3.18 |
| 18 | 99 | 29.9 | 51 | 114.18 | -15.18 | -0.84 |

Snedecor and Cochran thought the residual on soil 17 was "suspiciously large." Using the above test for one outlier, we fit the equation $E(y) = \alpha + \beta_1 x_1 + \beta_2 x_2$ to the data and obtain the predicted values shown above. The residuals are also shown, as well as the studentized residuals. The value of $r_m$ is 3.18, and from Table 12.3 with q = 3 (number of parameters) significant at the .05 level, hence is declared an outlier.

### E 12.4.2 Example

The following two-way layout was given by Daniel (1960):

| Data | | | | Fitted | | | | Residuals | | | |
|------|----|----|----|----|----|----|-----|----|----|----|----|
| 35 | 32 | 40 | 37 | 33 | 34 | 39 | 38 | +2 | -2 | +1 | -1 |
| 29 | 29 | 36 | 34 | 29 | 30 | 35 | 34 | +0 | -1 | +1 | +0 |
| 25 | 29 | 20 | 30 | 23 | 24 | 29 | 28 | +2 | +5 | -9 | +2 |
| 19 | 25 | 35 | 25 | 23 | 24 | 29 | -28 | -4 | +1 | +6 | -3 |
| 22 | 20 | 29 | 29 | 22 | 23 | 28 | 27 | +0 | -3 | +1 | +2 |

The residual in the third row and column seems large.

For these data, Stefansky's maximum normed residual is $9/\sqrt{202}$ = .6332. Consulting Table 12.4 we find a 5% critical value of .590 and a 1% critical value of .640, hence the observation in the third row and third column would be judged an outlier at the 5% level but not at the 1% level.

## 12.5 MULTIPLE OUTLIERS IN THE LINEAR MODEL

In the simple linear regression case, the observation with largest residual may not be the one with the largest test statistic, $r_m$, so that judging residuals by eye begins to fail as a tool. In a two-way table, Gentleman and Wilk (1975) declare that "in the null case of no outliers, the residuals behave much like a normal sample. When one outlier is present, the direct statistical treatment of residuals provides a complete basis for data-analytic judgements, especially through judicious use of probability plots. When two outliers are present, however, the resulting residuals will often not have any noticeable statistical peculiarities." These authors devised a test statistic $Q_k = \hat{e}'\hat{e} - \tilde{e}'\tilde{e}$ where "$\tilde{e}'\tilde{e}$ is the sum of squares of revised residuals resulting from fitting the basic model to the data remaining after the omission of k data points" and $\hat{e}'\hat{e}$ is the sum of squares of residuals obtained by fitting the model to all of the data. They envisioned the computation of $Q_k$ for each of the $\binom{n}{k}$ possible data partitions, and used the largest of the $Q_k$ to identify the "k most likely outlier subset." Such a procedure can be computationally awesome. Methods of reducing the labor have been devised, but they are still formidable, since one first chooses a maximum value for k (no easy task) then proceeds.

If the $Q_k$ is not significant, the value of k is reduced by 1 and the process repeated. A statistic useful in judging $Q_k$ is $F = (n - p - k)Q_k/k(\hat{e}'\hat{e} - Q_k)$ with k and $n - p - k$ degrees of freedom. The authors suggest that a plot be made of the largest $Q_k$ values against typical values obtained as medians from 10 Monte Carlo trials.

John and Draper (1978) showed that $Q_k$ is "the sum of squares of k successive revised normalized uncorrelated residuals," and that one only need examine the subset of $Q_k$ which arose from the subset of the larger residuals. In practice they suggested examining a plot of the original residuals to see if one outlier appeared to be present. If so, that value was omitted and treated as a missing value and estimated as usual (by minimizing the error sum of squares). After estimation, the same process was repeated to see if a second outlier could be detected by examining the residuals. Continuing this process, John and Draper (1980) show how to conduct a three-stage test for up to three outliers, using simulated critical values.

A good example would take up more space than we have available, hence the reader is referred to the papers mentioned above.

## 12.6 ACCOMMODATION OF OUTLIERS

We have thus far concentrated on detecting or identifying outliers. We shall now discuss the use of robust regression techniques to get around the outliers without detecting them and without allowing them to "devastate" our analyses. Besides the use of trimmed means, Winsorized means, and medians, there is a set of techniques known as Huber M-estimation. Since the derivative of the log of the likelihood function for any $f(x_j; \theta)$ is equal to $\Sigma f'(x_j - \theta)/f(x_j - \theta)$, we set this equal to zero and have as a result that the maximum likelihood estimator of $\theta$ is $\hat{\theta}$, the solution of $\Sigma \psi(x_j - \hat{\theta}) = 0$, where $\psi(x) = -f'(x)/f(x)$. Since $f(x)$ is not known, we can choose a $\psi(x)$ which will have suitable robust estimation properties. Many forms of $\psi(x)$ have been suggested, but we shall confine ourselves to one. Andrews (1974) suggested a sine-wave function for $\psi(x)$, and showed that it could be carried out by using any iterative weighted least squares procedure in which the model for the data is $y_i = x_i'\beta + \epsilon_i$, the least squares estimate of $\beta$ is $b$, and the ith residual is $r_i = y_i - x_i'b$. To do robust regression, we start with an initial estimate of $\beta$, denoted by $b_0$, obtained as described below. From this estimate we calculate the residuals $r_j$. We let $s = \text{median}|r_j|$, and solve, by weighted least squares, the system of p equations $\Sigma w_i x_{ij}(y_i - x_i'b) = 0$. The summation runs from $i = 1$ to n where $j = 1, 2, \ldots, p$. The weights are $w_i^2 = \sin(r_i/s)/r_i$ if $|r_i| < 1.5\pi$ and zero otherwise. The solution of the equations provides a new estimate $b_1$, from which we obtain new residuals and new weights and get another estimate $b_2$. This continues until the $b_i$ converge. On the last iteration, very small or zero weights for some residuals indicate that they are outliers or nearly so. By letting $x$ be a vector of 1's, the model will do for the univariate case as well. In most cases a starting value $b_0$, obtained by solving the original equations for $\beta$, will do nicely. Andrews has a rather involved alternative when the data are "far from Gaussian."

I would recommend the use of robust regression to accompany the usual parametric regression procedure. If the answers are quite different, there is an indication of outliers or influential observations which indicate that the situation should be studied further. Andrews also suggests that the procedure may give better results after a few iterations than if it is carried to convergence.

### E 12.6.1 Example

A set of data from Brownlee (1965) on observations from a plant for oxidation of ammonia to nitric acid is given below.

| Observation number | Stack loss y | Air flow $x_1$ | Cooling water inlet temperature $x_2$ | Acid concentration $x_3$ |
|:---:|:---:|:---:|:---:|:---:|
| 1 | 42 | 80 | 27 | 89 |
| 2 | 37 | 80 | 27 | 88 |
| 3 | 37 | 75 | 25 | 90 |
| 4 | 28 | 62 | 24 | 87 |
| 5 | 18 | 62 | 22 | 87 |
| 6 | 18 | 62 | 23 | 87 |
| 7 | 19 | 62 | 24 | 93 |
| 8 | 20 | 62 | 24 | 93 |
| 9 | 15 | 58 | 23 | 87 |
| 10 | 14 | 58 | 18 | 80 |
| 11 | 14 | 58 | 18 | 89 |
| 12 | 13 | 58 | 17 | 88 |
| 13 | 11 | 58 | 18 | 82 |
| 14 | 12 | 58 | 19 | 93 |
| 15 | 8 | 50 | 18 | 89 |
| 16 | 7 | 50 | 18 | 86 |
| 17 | 8 | 50 | 19 | 72 |
| 18 | 8 | 50 | 19 | 79 |
| 19 | 9 | 50 | 20 | 80 |
| 20 | 15 | 56 | 20 | 82 |
| 21 | 15 | 70 | 20 | 91 |

Daniel and Wood (1971) did careful work on this problem. Their fit to the original data was $E(y) = -39.9 + .72x_1 + 1.30x_2 - .15x_3$. After much consideration and plotting of the data, they discarded observations 1, 3, 4, and 21 as outliers, and refitted the equation, obtaining $E(y) = -37.6 + .80x_1 + .58x_2 - .07x_3$. A robust regression yields $E(y) = -37.2 + .82x_1 + .52x_2 - .07x_3$, and deletion of the four points does not alter the coefficients. The residuals from the four fits are shown below. The size of the residuals for points 1, 3, 4, and 21 indicate, somewhat subjectively, that they are outliers.

| | | Residuals | | | |
|---|---|---|---|---|---|
| | | Least squares | | Robust fit c = 1.5 | |
| Observation number | Response | with 1,3,4,21 | without | with 1,3,4,21 | without |
| 1 | 42 | 3.24 | 6.08 | 6.11 | 6.11 |
| 2 | 37 | -1.92 | 1.15 | 1.04 | 1.04 |
| 3 | 37 | 4.56 | 6.44 | 6.31 | 6.31 |
| 4 | 28 | 5.70 | 8.18 | 8.24 | 8.24 |
| 5 | 18 | -1.71 | -0.67 | -1.24 | -1.24 |
| 6 | 18 | -3.01 | -1.25 | -0.71 | -0.71 |
| 7 | 19 | -2.39 | -0.42 | -0.33 | -0.33 |
| 8 | 20 | -1.39 | 0.58 | 0.67 | 0.67 |
| 9 | 15 | -3.14 | -1.06 | -0.97 | -0.97 |
| 10 | 14 | 1.27 | 0.35 | 0.14 | 0.14 |
| 11 | 14 | 2.64 | 0.96 | 0.79 | 0.79 |
| 12 | 13 | 2.78 | 0.47 | 0.24 | 0.24 |
| 13 | 11 | -1.43 | -2.51 | -2.71 | -2.71 |
| 14 | 12 | -0.05 | -1.34 | -1.44 | -1.44 |
| 15 | 8 | 2.36 | 1.34 | 1.33 | 1.33 |
| 16 | 7 | 0.91 | 0.14 | 0.11 | 0.11 |
| 17 | 8 | -1.52 | -0.37 | -0.42 | -0.42 |
| 18 | 8 | -0.46 | 0.10 | 0.08 | 0.08 |
| 19 | 9 | -0.60 | 0.59 | 0.63 | 0.63 |
| 20 | 15 | 1.41 | 1.93 | 1.87 | 1.87 |
| 21 | 15 | -7.24 | -8.63 | -8.91 | -8.91 |

## 12.7 MULTIVARIATE OUTLIERS

A generalization of the univariate technique has been used here. Let $\underline{S}$ be the sample covariance matrix for all the data and $\underline{S}_i$ be the covariance matrix obtained by deleting k observations from the sample. $\left(\text{The } \binom{n}{k} \text{ possible}\right.$ ways of choosing the k outliers are indexed by i.$\left.\right)$ Wilks (1963) suggested forming $R_i = |\underline{S}_i| \,/\, |S|$ and using the minimum of the $R_i$ as a test statistic and gave some critical values for one and two outliers. His test is recommended.

## 12.8 OUTLIERS IN TIME SERIES

Fox (1972) was apparently the first to take into account the correlations between successive observations in processing time series for outliers. He considered a type I outlier as one in which a gross error of observation or recording errors affects only a single observation. The type II outlier occurs when a single "innovation" is extreme and will affect that observation and the subsequent ones.

As a model for the type I outlier, Fox used a stationary pth order autoregressive process in which the qth observation has d added to it. The null hypothesis is $H_0 : \Delta = 0$, and the alternative, $H_1 : \Delta \neq 0$. Using an asymptotic expression for the elements $W^{ij}$ of the inverse of the covariance matrix, Fox obtained a likelihood ratio criterion,

$$\Delta_{q,n} = \frac{(y - \tilde{\Delta})'\tilde{W}^{-1}(y - \tilde{\Delta})}{y'\hat{W}^{-1}y} \tag{12.8.1}$$

where $\tilde{W}^{-1}$ is the estimated inverse under $H_1$, $\hat{W}^{-1}$ is the estimated inverse under $H_0$, and $\tilde{\Delta}$, the displacement in the qth observation, is a vector of zeros except for the qth component which is estimated asymptotically by $\Sigma \tilde{W}^{qj}/\tilde{W}^{qq}$. The distribution of $\Delta_{q,n}$ is not known except with W known. Using simulation, Fox compared the distribution of $\Delta_{q,n}$ with (a) the distribution obtained by replacing $\tilde{W}^{-1}$ by $\hat{W}^{-1}$ and (b) the distribution obtained by assuming that W is known instead of estimated in Eq. (12.8.1). The distributions in the three cases were very close, hence Fox recommended that we act as though the estimated W's were known in order to obtain the distribution (and critical values) of $\Delta_{q,n}$. Where the position of the outlier is not known, Fox simulates the situation and obtains a small table of critical values.

For the type II outlier, Fox again employs a likelihood ratio criterion and obtains an approximate distribution for both cases (position known and unknown). In the case where the type of outlier is unknown, he suggests seeing whether one can detect the effect of the observation on subsequent observations. If not, the type I outlier is assumed. The above approach is shown to be superior to the one in which the observations are considered to be independent, and we recommend its use. Note that it has the characteristic form of a (weighted) sum of squares omitting outliers divided by a weighted sum of squares for the total sample.

REFERENCES

Andrews, D. F. (1971). Significance tests based on residuals, Biometrika 58, 139-148.

Andrews, D. F. (1974). A robust method for multiple linear regression, Technometrics 16, 523-531.

Anscombe, F. J. (1960). Rejection of outliers, Technometrics 2, 123-147.

Barnett, V. and Lewis, T. (1978). Outliers in Statistical Data. New York: Wiley.

Beckman, R. J. and Cook, R. D. (1983). Outlier. . .s, Technometrics 25, 119-149.

Daniel, C. (1960). Locating outliers in factorial experiments, Technometrics 2, 149-156.

Fox, A. J. (1972). Outliers in time series, Journal of the Royal Statistical Society, Ser. B 34. 350-363.

Galpin, J. S. and Hawkins, D. M. (1981). Rejection of a single outlier in two- or three-way layouts, Technometrics 23, 65-70.

Gentleman, J. F. and Wilk, M. B. (1975a). Detecting outliers in a two-way table: I. Statistical behavior of residuals, Technometrics 17, 1-14.

Gentleman, J. F. and Wilk, M. B. (1975b). Detecting outliers II. Supplementing the direct analysis of residuals, Biometrics 31, 387-410.

Grubbs, F. E. (1950). Sample criteria for testing outlying observations, Annals of Mathematical Statistics 21, 27-58.

Grubbs, F. E. (1969). Procedures for detecting outlying observations in samples, Technometrics 11, 1-21.

Guttman, I. (1973). Premium and protection of several procedures for dealing with outliers when sample sizes are moderate to large, Technometrics 15, 385-404.

Guttman, I. and Smith, D. E. (1969). Investigation of rules for dealing with outliers in small samples from the normal distribution I: Estimation of the mean, Technometrics 11, 527-550.

Guttman, I. and Smith, D. E. (1971). Investigation of rules for dealing with outliers on small samples from the normal distribution II: Estimation of the variance, Technometrics 13, 101-111.

Jain, R. B. (1981). Percentage points of many-outlier detection procedures, Technometrics 23, 71-75.

John, J. A. and Draper, N. R. (1978). On testing for two outliers or one outlier in two-way tables, Technometrics 20, 69-78.

Joshi, P. C. (1972). Some slippage tests of mean for a single outlier in linear regression, Biometrika 59, 109-120.

Kruskal, W. H. (1960). Some remarks on wild observations, Technometrics 2, 1-3.

Lund, R. E. (1975). Tables for an approximate test for outliers in linear regressions, Technometrics 17, 473-476.

Rosner, B. (1975). On the detection of many outliers, Technometrics 17, 221-227.

Rosner, B. (1977). Percentage points for the RST many outlier procedure, Technometrics 19, 307-312.

Srikantan, K. S. (1961). Testing for a single outlier in a regression model, Sankhya, Ser. A 23, 251-260.

Stefansky, W. (1971). Rejecting outliers by maximum normal residual, Annals of Mathematical Statistics 42, 35-45.

Tietjen, G. L. and Moore, R. H. (1972). Some Grubbs-type statistics for the detection of several outliers, Technometrics 14, 583-597.

Wilks, S. S. (1963). Multivariate statistical outliers, Sankhya 25, 407-426.

# Appendix

1. Table 1, Cumulative Distribution Function of the
      Standard Normal Distribution

2. Table 2, Critical Values of the Chi-Square Distribution

3. Simulated Data Sets

| Set Name | Distribution | $\sqrt{\beta_1}$ | $\beta_2$ |
|---|---|---|---|
| NOR | Normal $\mu = 100$, $\sigma = 10$ | .0 | 3. |
| UNI | Uniform on Interval 0 to 10 | .0 | 1.80 |
| EXP | Negative Exponential $\mu = 5$ | 2. | 9. |
| LOG | Logistic $\mu = 100$ | .0 | 4.2 |
| WE.5 | Weibull k = .5 | 6.62 | 87.72 |
| WE2 | Weibull k = 2 | .63 | 3.25 |
| SU(1,2) | Johnson Unbounded (1,2) | -.87 | 5.57 |
| SU(0,3) | Johnson Unbounded (0,3) | .0 | 3.53 |
| SU(0,2) | Johnson Unbounded (0,2) | .0 | 4.51 |
| SB(0,2) | Johnson Bounded (0,2) | .0 | 2.63 |
| SB(0,.5) | Johnson Bounded (0,.5) | .0 | 1.63 |
| LCN(.05,3) | Contaminated Normal p = .05, $\mu = 3$ | .68 | 4.35 |
| LCN(.1,3) | Contaminated Normal p = .1, $\mu = 3$ | .80 | 4.02 |
| LCN(.2,3) | Contaminated Normal p = .2, $\mu = 3$ | .68 | 3.09 |
| SB(1,1) | Johnson Bounded (1,1) | .73 | 2.91 |
| SB(1,2) | Johnson Bounded (1,2) | .28 | 2.77 |
| SB(.533,.5) | Johnson Bounded (.533,.5) | .65 | 2.13 |

4. Real Data Sets

   BLIS  Data Set
   CHEN  Data Set
   BLAC  Data Set
   EMEA  Data Set
   BAEN  Data Set

APPENDIX

1. TABLE 1  Cumulative Distribution Function of the
Standard Normal Distribution

Areas under the standard normal curve (Areas to the left)

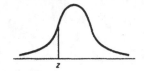

| z | 0 | 1 | 2 | 3 | 4 | 5 | 6 | 7 | 8 | 9 |
|---|---|---|---|---|---|---|---|---|---|---|
| -3.0* | .0013 | .0013 | .0013 | .0012 | .0012 | .0011 | .0011 | .0011 | .0010 | .0010 |
| -2.9 | .0019 | .0018 | .0017 | .0017 | .0016 | .0016 | .0015 | .0015 | .0014 | .0014 |
| -2.8 | .0026 | .0025 | .0024 | .0023 | .0023 | .0022 | .0021 | .0020 | .0020 | .0019 |
| -2.7 | .0035 | .0034 | .0033 | .0032 | .0031 | .0030 | .0029 | .0028 | .0027 | .0026 |
| -2.6 | .0047 | .0045 | .0044 | .0043 | .0041 | .0040 | .0039 | .0038 | .0037 | .0036 |
| -2.5 | .0062 | .0060 | .0059 | .0057 | .0055 | .0054 | .0052 | .0051 | .0049 | .0048 |
| -2.4 | .0082 | .0080 | .0078 | .0075 | .0073 | .0071 | .0069 | .0068 | .0066 | .0064 |
| -2.3 | .0107 | .0104 | .0102 | .0099 | .0096 | .0094 | .0091 | .0089 | .0087 | .0084 |
| -2.2 | .0139 | .0136 | .0132 | .0129 | .0125 | .0122 | .0119 | .0116 | .0113 | .0110 |
| -2.1 | .0179 | .0174 | .0170 | .0166 | .0162 | .0158 | .0154 | .0150 | .0146 | .0143 |
| -2.0 | .0228 | .0222 | .0217 | .0212 | .0207 | .0202 | .0197 | .0192 | .0188 | .0183 |
| -1.9 | .0287 | .0281 | .0274 | .0268 | .0262 | .0256 | .0250 | .0244 | .0239 | .0233 |
| -1.8 | .0359 | .0351 | .0344 | .0336 | .0329 | .0322 | .0314 | .0307 | .0301 | .0294 |
| -1.7 | .0446 | .0436 | .0427 | .0418 | .0409 | .0401 | .0392 | .0384 | .0375 | .0367 |
| -1.6 | .0548 | .0537 | .0526 | .0516 | .0505 | .0495 | .0485 | .0475 | .0465 | .0455 |
| -1.5 | .0668 | .0655 | .0643 | .0630 | .0618 | .0606 | .0594 | .0582 | .0571 | .0559 |
| -1.4 | .0808 | .0793 | .0778 | .0764 | .0749 | .0735 | .0721 | .0708 | .0694 | .0681 |
| -1.3 | .0968 | .0951 | .0934 | .0918 | .0901 | .0885 | .0869 | .0853 | .0838 | .0823 |
| -1.2 | .1151 | .1131 | .1112 | .1093 | .1075 | .1056 | .1038 | .1020 | .1003 | .0985 |
| -1.1 | .1357 | .1335 | .1314 | .1292 | .1271 | .1251 | .1230 | .1210 | .1190 | .1170 |
| -1.0 | .1587 | .1562 | .1539 | .1515 | .1492 | .1469 | .1446 | .1423 | .1401 | .1379 |
| -.9 | .1841 | .1814 | .1788 | .1762 | .1736 | .1711 | .1685 | .1660 | .1635 | .1611 |
| -.8 | .2119 | .2090 | .2061 | .2033 | .2005 | .1977 | .1949 | .1922 | .1894 | .1867 |
| -.7 | .2420 | .2389 | .2358 | .2327 | .2296 | .2266 | .2236 | .2206 | .2177 | .2148 |
| -.6 | .2743 | .2709 | .2676 | .2643 | .2611 | .2578 | .2546 | .2514 | .2483 | .2451 |
| -.5 | .3085 | .3050 | .3015 | .2981 | .2946 | .2912 | .2877 | .2843 | .2810 | .2776 |
| -.4 | .3446 | .3409 | .3372 | .3336 | .3300 | .3264 | .3228 | .3192 | .3516 | .3121 |
| -.3 | .3821 | .3783 | .3745 | .3707 | .3669 | .3632 | .3594 | .3557 | .3520 | .3483 |
| -.2 | .4207 | .4168 | .4129 | .4090 | .4052 | .4013 | .3974 | .3936 | .3897 | .3859 |
| -.1 | .4602 | .4562 | .4522 | .4483 | .4443 | .4404 | .4364 | .4325 | .4286 | .4247 |
| -.0 | .5000 | .4960 | .4920 | .4880 | .4840 | .4801 | .4761 | .4721 | .4681 | .4641 |

*For $z \leq -4$ the areas are 0 to four decimal places.

TABLE 1 (continued)

| z | 0 | 1 | 2 | 3 | 4 | 5 | 6 | 7 | 8 | 9 |
|---|---|---|---|---|---|---|---|---|---|---|
| .0 | .5000 | .5040 | .5080 | .5120 | .5160 | .5199 | .5239 | .5279 | .5319 | .5359 |
| .1 | .5398 | .5438 | .5478 | .5517 | .5557 | .5596 | .5636 | .5675 | .5714 | .5753 |
| .2 | .5793 | .5832 | .5871 | .5910 | .5948 | .5987 | .6026 | .6064 | .6103 | .6141 |
| .3 | .6179 | .6217 | .6255 | .6293 | .6331 | .6368 | .6406 | .6443 | .6480 | .6517 |
| .4 | .6554 | .6591 | .6628 | .6664 | .6700 | .6736 | .6772 | .6808 | .6844 | .6879 |
| .5 | .6915 | .6950 | .6985 | .7019 | .7054 | .7088 | .7123 | .7157 | .7190 | .7224 |
| .6 | .7257 | .7291 | .7324 | .7357 | .7389 | .7422 | .7454 | .7486 | .7517 | .7549 |
| .7 | .7580 | .7611 | .7642 | .7673 | .7704 | .7734 | .7764 | .7794 | .7823 | .7852 |
| .8 | .7881 | .7910 | .7939 | .7967 | .7995 | .8023 | .8051 | .8078 | .8106 | .8133 |
| .9 | .8159 | .8186 | .8212 | .8238 | .8264 | .8289 | .8315 | .8340 | .8365 | .8389 |
| 1.0 | .8413 | .8438 | .8461 | .8485 | .8508 | .8531 | .8554 | .8577 | .8599 | .8621 |
| 1.1 | .8643 | .8665 | .8686 | .8708 | .8729 | .8749 | .8770 | .8790 | .8810 | .8830 |
| 1.2 | .8849 | .8869 | .8888 | .8907 | .8925 | .8944 | .8962 | .8980 | .8997 | .9015 |
| 1.3 | .9032 | .9049 | .9066 | .9082 | .9099 | .9115 | .9131 | .9147 | .9162 | .9177 |
| 1.4 | .9192 | .9207 | .9222 | .9236 | .9251 | .9265 | .9279 | .9292 | .9306 | .9319 |
| 1.5 | .9332 | .9345 | .9357 | .9370 | .9382 | .9394 | .9406 | .9418 | .9429 | .9441 |
| 1.6 | .9452 | .9463 | .9474 | .9484 | .9495 | .9505 | .9515 | .9525 | .9535 | .9545 |
| 1.7 | .9554 | .9564 | .9573 | .9582 | .9591 | .9599 | .9608 | .9616 | .9625 | .9633 |
| 1.8 | .9641 | .9649 | .9656 | .9664 | .9671 | .9678 | .9686 | .9693 | .9699 | .9706 |
| 1.9 | .9713 | .9719 | .9726 | .9732 | .9738 | .9744 | .9750 | .9756 | .9761 | .9767 |
| 2.0 | .9772 | .9778 | .9783 | .9788 | .9793 | .9798 | .9803 | .9808 | .9812 | .9817 |
| 2.1 | .9821 | .9826 | .9830 | .9834 | .9838 | .9842 | .9846 | .9850 | .9854 | .9857 |
| 2.2 | .9861 | .9864 | .9868 | .9871 | .9875 | .9878 | .9881 | .9884 | .9887 | .9890 |
| 2.3 | .9893 | .9896 | .9898 | .9901 | .9904 | .9906 | .9909 | .9911 | .9913 | .9916 |
| 2.4 | .9918 | .9920 | .9922 | .9925 | .9927 | .9929 | .9931 | .9932 | .9934 | .9936 |
| 2.5 | .9938 | .9940 | .9941 | .9943 | .9945 | .9946 | .9948 | .9949 | .9951 | .9952 |
| 2.6 | .9953 | .9955 | .9956 | .9957 | .9959 | .9960 | .9961 | .9962 | .9963 | .9964 |
| 2.7 | .9965 | .9966 | .9967 | .9968 | .9969 | .9970 | .9971 | .9972 | .9973 | .9974 |
| 2.8 | .9974 | .9975 | .9976 | .9977 | .9977 | .9978 | .9979 | .9979 | .9980 | .9981 |
| 2.9 | .9981 | .9982 | .9982 | .9983 | .9984 | .9984 | .9985 | .9985 | .9986 | .9986 |
| 3.0[†] | .9987 | .9987 | .9987 | .9988 | .9988 | .9989 | .9989 | .9989 | .9990 | .9990 |

[†]For $z \geq 4$ the areas are 1 to four decimal places.
Adapted from Probability with Statistical Applications, second edition, by
F. Mosteller, R. E. K. Rourke, and G. B. Thomas, Jr. Reading, Mass.:
Addison-Wesley, 1970, p. 473.

2. TABLE 2 Critical Values of the Chi-Square Distribution

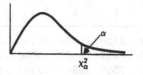

| d.f. | .995 | .99 | .975 | .95 | .90 | .10 | .05 | .025 | .01 | .005 |
|---|---|---|---|---|---|---|---|---|---|---|
| 1 | .00 | .00 | .00 | .00 | .02 | 2.71 | 3.84 | 5.02 | 6.63 | 7.88 |
| 2 | .01 | .02 | .05 | .10 | .21 | 4.61 | 5.99 | 7.38 | 9.21 | 10.60 |
| 3 | .07 | .11 | .22 | .35 | .58 | 6.25 | 7.81 | 9.35 | 11.34 | 12.84 |
| 4 | .21 | .30 | .48 | .71 | 1.06 | 7.78 | 9.49 | 11.14 | 13.28 | 14.86 |
| 5 | .41 | .55 | .83 | 1.15 | 1.61 | 9.24 | 11.07 | 12.83 | 15.09 | 16.75 |
| 6 | .68 | .87 | 1.24 | 1.64 | 2.20 | 10.64 | 12.59 | 14.45 | 16.81 | 18.55 |
| 7 | .99 | 1.24 | 1.69 | 2.17 | 2.83 | 12.02 | 14.07 | 16.01 | 18.48 | 20.28 |
| 8 | 1.34 | 1.65 | 2.18 | 2.73 | 3.49 | 13.36 | 15.51 | 17.54 | 20.09 | 21.96 |
| 9 | 1.73 | 2.09 | 2.70 | 3.33 | 4.17 | 14.68 | 16.92 | 19.02 | 21.67 | 23.59 |
| 10 | 2.16 | 2.56 | 3.25 | 3.94 | 4.87 | 15.99 | 18.31 | 20.48 | 23.21 | 25.19 |
| 11 | 2.60 | 3.05 | 3.82 | 4.57 | 5.58 | 17.28 | 19.68 | 21.92 | 24.72 | 26.76 |
| 12 | 3.07 | 3.57 | 4.40 | 5.23 | 6.30 | 18.55 | 21.03 | 23.34 | 26.22 | 28.30 |
| 13 | 3.57 | 4.11 | 5.01 | 5.89 | 7.04 | 19.81 | 22.36 | 24.74 | 27.69 | 29.82 |
| 14 | 4.07 | 4.66 | 5.63 | 6.57 | 7.79 | 21.06 | 23.68 | 26.12 | 29.14 | 31.32 |
| 15 | 4.60 | 5.23 | 6.26 | 7.26 | 8.55 | 22.31 | 25.00 | 27.49 | 30.58 | 32.80 |
| 16 | 5.14 | 5.81 | 6.91 | 7.96 | 9.31 | 23.54 | 26.30 | 28.85 | 32.00 | 34.27 |
| 17 | 5.70 | 6.41 | 7.56 | 8.67 | 10.09 | 24.77 | 27.59 | 30.19 | 33.41 | 35.72 |
| 18 | 6.26 | 7.01 | 8.23 | 9.39 | 10.86 | 25.99 | 28.87 | 31.53 | 34.81 | 37.16 |
| 19 | 6.84 | 7.63 | 8.91 | 10.12 | 11.65 | 27.20 | 30.14 | 32.85 | 36.19 | 38.58 |
| 20 | 7.43 | 8.26 | 9.59 | 10.85 | 12.44 | 28.41 | 31.41 | 34.17 | 37.57 | 40.00 |
| 21 | 8.03 | 8.90 | 10.28 | 11.59 | 13.24 | 29.62 | 32.67 | 35.48 | 38.93 | 41.40 |
| 22 | 8.64 | 9.54 | 10.98 | 12.34 | 14.04 | 30.81 | 33.92 | 36.78 | 40.29 | 42.80 |
| 23 | 9.26 | 10.20 | 11.69 | 13.09 | 14.85 | 32.01 | 35.17 | 38.08 | 41.64 | 44.18 |
| 24 | 9.89 | 10.86 | 12.40 | 13.85 | 15.66 | 33.20 | 36.42 | 39.36 | 42.98 | 45.56 |
| 25 | 10.52 | 11.52 | 13.12 | 14.61 | 16.47 | 34.38 | 37.65 | 40.65 | 44.31 | 46.93 |
| 26 | 11.16 | 12.20 | 13.84 | 15.38 | 17.29 | 35.56 | 38.89 | 41.92 | 45.64 | 48.29 |
| 27 | 11.81 | 12.88 | 14.57 | 16.15 | 18.11 | 36.74 | 40.11 | 43.19 | 46.96 | 49.65 |
| 28 | 12.46 | 13.56 | 15.31 | 16.93 | 18.94 | 37.92 | 41.34 | 44.46 | 48.28 | 50.99 |
| 29 | 13.12 | 14.26 | 16.05 | 17.71 | 19.77 | 39.09 | 42.56 | 45.72 | 49.59 | 52.34 |
| 30 | 13.79 | 14.95 | 16.76 | 18.49 | 20.60 | 40.26 | 43.77 | 46.98 | 50.89 | 53.67 |
| 50 | 27.99 | 29.71 | 32.36 | 34.76 | 37.69 | 63.17 | 67.50 | 71.42 | 76.15 | 79.49 |
| 100 | 67.33 | 70.06 | 74.22 | 77.93 | 82.36 | 118.5 | 124.3 | 129.6 | 135.8 | 140.2 |
| 500 | 422.3 | 429.4 | 439.9 | 449.1 | 459.9 | 540.9 | 553.1 | 563.9 | 576.5 | 585.2 |
| 1000 | 888.6 | 898.8 | 914.3 | 927.6 | 943.1 | 1058 | 1075 | 1090 | 1107 | 1119 |

Adapted from D. B. Owen, Handbook of Statistical Tables. Courtesy of the Atomic Energy Commission. Reading, Mass.: Addison-Wesley, 1962.

## 3. Simulated Data Sets

There are 17 simulated data sets, each of 100 observations. Throughout the book these have been used to illustrate and compare procedures.

| Set Name | Distribution | $(\sqrt{\beta_1}, \beta_2)$ |
|---|---|---|
| NOR | Normal distribution, $\mu = 100$, $\sigma = 10$ | (0, 3) |
| UNI | Uniform distribution, $f(x) = \frac{1}{10}$ for $0 < x < 10$ | (0, 1.80) |
| EXP | Negative exponentially $f(x) = \frac{1}{5} e^{-x/5}$ for $x > 0$ | (2, 9) |
| LOG | Logistic, $F(x) = \dfrac{1}{1 + e^{-\left(\frac{x-100}{10}\right)}}$ | (0, 4.2) |
| WE.5 | Weibull, $F(x) = 1 - e^{-x^{\frac{1}{2}}}$ | (6.62, 87.72) |
| WE2 | Weibull, $F(x) = 1 - e^{-x^2}$ | (.63, 3.25) |
| SU(1, 2) | Johnson Unbounded $(\gamma, \delta)$ $\sinh^{-1}(x)$ is standard normal $\quad (-\infty < x < \infty)$ | (-.87, 5.59) |

$S_I$
$\gamma=1$, $\delta=2$
$\beta_1=0.76$, $\beta_2=5.59$

| | | |
|---|---|---|
| SU(0, 3) | Johnson Unbounded (0, 3) | (0, 3.53) |
| SU(0, 2) | Johnson Unbounded (0, 2) | (0, 4.51) |

$S_I$
$\gamma=0$, $\delta=2$
$\beta_1=0$, $\beta_2=4.51$

| | | |
|---|---|---|
| SB(0, 2) | Johnson Bounded $(\gamma, \delta)$ | (0, 2.63) |

$\gamma + \delta \ln\left(\dfrac{x}{1 - x}\right)$ is a standard normal $(0 < x < 1)$

$S_{II}$
$\gamma=0$, $\delta=2$
$\beta_1=0$, $\beta_2=2.63$

| Set Name | Distribution | $(\sqrt{\beta_1}, \beta_2)$ |
|----------|--------------|-----------------------------|
| SB(0, .5) | Johnson Bounded (0, .5) | (0, 1.63) |

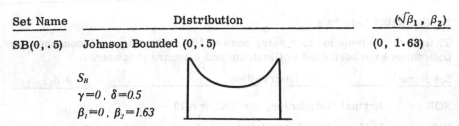

$$S_B$$
$$\gamma = 0, \ \delta = 0.5$$
$$\beta_1 = 0, \ \beta_2 = 1.63$$

The above distributions were the first ones generated. To increase the number of skewed distributions, the following were added.

Location Contaminated Normals

$$pN(\mu, 1) + (1 - p)N(0, 1)$$

The three included here are:

| p | $\mu$ | $\sqrt{\beta_1}$ | $\beta_2$ | Data Set Name |
|-----|-----|------|------|----------------|
| .05 | 3 | .68 | 4.35 | LCN (.05, 3) |
| .10 | 3 | .80 | 4.02 | LCN (.10, 3) |
| .20 | 3 | .68 | 3.09 | LCN (.20, 3) |

The last three simulated data sets are from Johnson Bounded SB($\gamma$, $\delta$) distribution where SB($\gamma$, $\delta$) is defined above. The three samples here are from:

| $\gamma$ | $\delta$ | $\sqrt{\beta_1}$ | $\beta_2$ | Data Set Name |
|------|-----|------|------|----------------|
| 1 | 1 | .73 | 2.91 | SB (1, 1) |
| 1 | 2 | .28 | 2.77 | SB (1, 2) |
| .533 | .5 | .65 | 2.13 | SB (.533, .5) |

NOR Data Set $\qquad$ Normal Distribution $\qquad$ $\mu = 100$ $\qquad$ $\sqrt{\beta_1} = 0$

$\sigma = 10$ $\qquad$ $\beta_2 = 3$

| No. | Obser- vation | No. | Obser- vation | No. | Obser- vation | No. | Obser- vation |
|---|---|---|---|---|---|---|---|
| 1 | 92.55 | 26 | 102.56 | 51 | 111.38 | 76 | 88.13 |
| 2 | 96.20 | 27 | 79.43 | 52 | 103.22 | 77 | 102.98 |
| 3 | 84.27 | 28 | 105.48 | 53 | 113.17 | 78 | 103.71 |
| 4 | 90.87 | 29 | 85.29 | 54 | 108.39 | 79 | 95.14 |
| 5 | 101.58 | 30 | 83.53 | 55 | 103.60 | 80 | 85.71 |
| 6 | 106.82 | 31 | 104.21 | 56 | 103.90 | 81 | 103.56 |
| 7 | 98.70 | 32 | 100.75 | 57 | 89.35 | 82 | 89.44 |
| 8 | 113.75 | 33 | 92.02 | 58 | 124.60 | 83 | 88.26 |
| 9 | 98.98 | 34 | 100.10 | 59 | 104.34 | 84 | 97.80 |
| 10 | 100.42 | 35 | 87.83 | 60 | 85.29 | 85 | 97.33 |
| 11 | 118.52 | 36 | 89.00 | 61 | 97.78 | 86 | 103.90 |
| 12 | 89.90 | 37 | 108.67 | 62 | 109.76 | 87 | 96.38 |
| 13 | 92.45 | 38 | 103.09 | 63 | 94.92 | 88 | 94.33 |
| 14 | 115.92 | 39 | 99.12 | 64 | 95.12 | 89 | 99.62 |
| 15 | 103.61 | 40 | 91.46 | 65 | 88.56 | 90 | 95.94 |
| 16 | 96.13 | 41 | 125.28 | 66 | 115.95 | 91 | 104.89 |
| 17 | 95.45 | 42 | 91.45 | 67 | 100.79 | 92 | 83.34 |
| 18 | 108.52 | 43 | 92.56 | 68 | 104.87 | 93 | 87.04 |
| 19 | 112.69 | 44 | 102.66 | 69 | 95.89 | 94 | 89.80 |
| 20 | 90.03 | 45 | 101.91 | 70 | 110.72 | 95 | 83.07 |
| 21 | 111.56 | 46 | 76.35 | 71 | 86.28 | 96 | 112.14 |
| 22 | 109.26 | 47 | 111.30 | 72 | 107.97 | 97 | 113.90 |
| 23 | 83.67 | 48 | 89.33 | 73 | 117.23 | 98 | 100.46 |
| 24 | 112.97 | 49 | 79.89 | 74 | 104.12 | 99 | 110.39 |
| 25 | 116.87 | 50 | 110.17 | 75 | 95.97 | 100 | 98.43 |

UNI Data Set      Uniform Distribution on Interval 0 to 10      $\sqrt{\beta_1} = 0$
                                                                 $\beta_2 = 1.80$

| No. | Obser-vation | No. | Obser-vation | No. | Obser-vation | No. | Obser-vation |
|-----|--------------|-----|--------------|-----|--------------|-----|--------------|
| 1   | 8.10         | 26  | 6.54         | 51  | 3.93         | 76  | 4.26         |
| 2   | 2.06         | 27  | 8.24         | 52  | 0.08         | 77  | 3.32         |
| 3   | 1.60         | 28  | 9.12         | 53  | 3.51         | 78  | 9.29         |
| 4   | 8.87         | 29  | 0.31         | 54  | 0.44         | 79  | 2.57         |
| 5   | 9.90         | 30  | 2.63         | 55  | 1.22         | 80  | 0.55         |
| 6   | 6.58         | 31  | 6.20         | 56  | 1.12         | 81  | 6.53         |
| 7   | 8.68         | 32  | 5.47         | 57  | 2.34         | 82  | 2.33         |
| 8   | 7.31         | 33  | 7.80         | 58  | 1.86         | 83  | 9.01         |
| 9   | 2.85         | 34  | 1.30         | 59  | 8.35         | 84  | 7.86         |
| 10  | 6.09         | 35  | 9.39         | 60  | 3.53         | 85  | 7.06         |
| 11  | 6.10         | 36  | 8.67         | 61  | 5.05         | 86  | 8.54         |
| 12  | 2.94         | 37  | 1.87         | 62  | 5.28         | 87  | 9.71         |
| 13  | 1.85         | 38  | 6.67         | 63  | 6.87         | 88  | 8.49         |
| 14  | 9.04         | 39  | 5.90         | 64  | 2.96         | 89  | 2.08         |
| 15  | 9.38         | 40  | 0.15         | 65  | 2.35         | 90  | 0.50         |
| 16  | 7.30         | 41  | 3.91         | 66  | 4.02         | 91  | 3.54         |
| 17  | 2.11         | 42  | 8.87         | 67  | 1.44         | 92  | 3.75         |
| 18  | 4.55         | 43  | 2.50         | 68  | 9.63         | 93  | 9.46         |
| 19  | 7.66         | 44  | 7.49         | 69  | 9.44         | 94  | 0.04         |
| 20  | 9.63         | 45  | 0.55         | 70  | 5.44         | 95  | 7.79         |
| 21  | 9.48         | 46  | 5.25         | 71  | 3.71         | 96  | 8.08         |
| 22  | 5.31         | 47  | 5.61         | 72  | 4.21         | 97  | 3.60         |
| 23  | 5.76         | 48  | 1.00         | 73  | 2.22         | 98  | 8.85         |
| 24  | 9.66         | 49  | 3.23         | 74  | 2.87         | 99  | 1.50         |
| 25  | 4.37         | 50  | 1.05         | 75  | 0.72         | 100 | 0.18         |

EXP Data Set          Negative Exponential with Mean = 5          $\sqrt{\beta_1} = 2$
                                                                  $\beta_2 = 9$

| No. | Obser- vation | No. | Obser- vation | No. | Obser- vation | No. | Obser- vation |
|-----|------|-----|------|-----|------|-----|------|
| 1 | 8.15 | 26 | 11.89 | 51 | 1.27 | 76 | 5.19 |
| 2 | 4.69 | 27 | 7.26 | 52 | 1.56 | 77 | 0.26 |
| 3 | 2.17 | 28 | 14.71 | 53 | 16.81 | 78 | 9.46 |
| 4 | 0.37 | 29 | 0.23 | 54 | 6.07 | 79 | 0.95 |
| 5 | 16.69 | 30 | 1.21 | 55 | 3.89 | 80 | 0.51 |
| 6 | 0.06 | 31 | 0.18 | 56 | 9.60 | 81 | 1.39 |
| 7 | 6.48 | 32 | 1.24 | 57 | 3.12 | 82 | 3.74 |
| 8 | 2.63 | 33 | 12.94 | 58 | 4.16 | 83 | 4.37 |
| 9 | 0.44 | 34 | 4.78 | 59 | 0.07 | 84 | 3.87 |
| 10 | 0.89 | 35 | 18.53 | 60 | 1.67 | 85 | 5.40 |
| 11 | 6.96 | 36 | 9.20 | 61 | 3.80 | 86 | 2.41 |
| 12 | 5.15 | 37 | 1.65 | 62 | 1.52 | 87 | 5.93 |
| 13 | 9.78 | 38 | 2.20 | 63 | 2.79 | 88 | 39.12 |
| 14 | 6.47 | 39 | 1.13 | 64 | 0.36 | 89 | 1.05 |
| 15 | 0.99 | 40 | 5.20 | 65 | 4.49 | 90 | 0.47 |
| 16 | 7.70 | 41 | 14.74 | 66 | 9.76 | 91 | 9.57 |
| 17 | 1.61 | 42 | 2.86 | 67 | 2.37 | 92 | 8.29 |
| 18 | 1.68 | 43 | 0.19 | 68 | 9.91 | 93 | 3.79 |
| 19 | 0.92 | 44 | 0.08 | 69 | 6.60 | 94 | 2.35 |
| 20 | 1.87 | 45 | 3.22 | 70 | 0.17 | 95 | 1.09 |
| 21 | 14.80 | 46 | 1.21 | 71 | 14.68 | 96 | 4.19 |
| 22 | 9.96 | 47 | 3.51 | 72 | 3.72 | 97 | 12.21 |
| 23 | 25.92 | 48 | 5.67 | 73 | 6.92 | 98 | 1.57 |
| 24 | 3.37 | 49 | 10.50 | 74 | 2.53 | 99 | 3.52 |
| 25 | 2.76 | 50 | 10.45 | 75 | 4.77 | 100 | 0.48 |

LOG Data Set                Logistic                $\mu = 100$            $\sqrt{\beta_1} = 0$
                                                                            $\beta_2 = 4.2$

| No. | Obser-vation | No. | Obser-vation | No. | Obser-vation | No. | Obser-vation |
|-----|--------------|-----|--------------|-----|--------------|-----|--------------|
| 1   | 96.91        | 26  | 86.98        | 51  | 112.50       | 76  | 98.51        |
| 2   | 109.99       | 27  | 79.23        | 52  | 109.82       | 77  | 107.91       |
| 3   | 102.97       | 28  | 110.70       | 53  | 94.66        | 78  | 132.40       |
| 4   | 118.54       | 29  | 98.58        | 54  | 107.08       | 79  | 103.32       |
| 5   | 63.35        | 30  | 76.52        | 55  | 108.22       | 80  | 116.01       |
| 6   | 94.63        | 31  | 93.44        | 56  | 81.61        | 81  | 111.18       |
| 7   | 144.28       | 32  | 89.81        | 57  | 102.90       | 82  | 65.87        |
| 8   | 104.47       | 33  | 100.62       | 58  | 85.94        | 83  | 96.30        |
| 9   | 111.81       | 34  | 108.75       | 59  | 66.35        | 84  | 83.74        |
| 10  | 78.32        | 35  | 103.91       | 60  | 97.12        | 85  | 91.97        |
| 11  | 109.91       | 36  | 87.71        | 61  | 90.09        | 86  | 94.95        |
| 12  | 98.07        | 37  | 145.33       | 62  | 111.92       | 87  | 98.95        |
| 13  | 82.45        | 38  | 121.83       | 63  | 83.89        | 88  | 98.21        |
| 14  | 114.97       | 39  | 99.52        | 64  | 77.45        | 89  | 98.71        |
| 15  | 103.08       | 40  | 116.58       | 65  | 74.29        | 90  | 108.88       |
| 16  | 78.48        | 41  | 106.05       | 66  | 102.90       | 91  | 68.44        |
| 17  | 97.45        | 42  | 92.55        | 67  | 113.41       | 92  | 118.92       |
| 18  | 107.64       | 43  | 79.07        | 68  | 104.37       | 93  | 117.01       |
| 19  | 83.73        | 44  | 111.59       | 69  | 100.46       | 94  | 89.22        |
| 20  | 116.99       | 45  | 103.18       | 70  | 104.14       | 95  | 123.39       |
| 21  | 103.82       | 46  | 105.03       | 71  | 51.90        | 96  | 85.30        |
| 22  | 131.24       | 47  | 101.19       | 72  | 105.34       | 97  | 123.58       |
| 23  | 95.86        | 48  | 102.81       | 73  | 108.94       | 98  | 113.79       |
| 24  | 111.90       | 49  | 106.17       | 74  | 103.43       | 99  | 102.86       |
| 25  | 60.57        | 50  | 112.12       | 75  | 81.17        | 100 | 88.22        |

WE.5 Data Set                   Weibull with k = .5                    $\sqrt{\beta_1}$ = 6.62
                                                                       $\beta_2$ = 87.72

| No. | Obser- vation | No. | Obser- vation | No. | Obser- vation | No. | Obser- vation |
|-----|---------------|-----|---------------|-----|---------------|-----|---------------|
| 1   | .30           | 26  | .06           | 51  | 2.26          | 76  | .39           |
| 2   | 1.72          | 27  | .01           | 52  | 1.69          | 77  | 1.36          |
| 3   | .73           | 28  | 1.86          | 53  | .21           | 78  | 10.75         |
| 4   | 4.00          | 29  | .39           | 54  | 1.23          | 79  | .76           |
| 5   | .00           | 30  | .01           | 55  | 1.41          | 80  | 3.19          |
| 6   | .21           | 31  | .17           | 56  | .02           | 81  | 1.96          |
| 7   | 19.71         | 32  | .09           | 57  | .72           | 82  | .00           |
| 8   | .89           | 33  | .52           | 58  | .05           | 83  | .28           |
| 9   | 2.10          | 34  | 1.50          | 59  | .00           | 84  | .03           |
| 10  | .01           | 35  | .82           | 60  | .31           | 85  | .14           |
| 11  | 1.71          | 36  | .07           | 61  | .10           | 86  | .22           |
| 12  | .36           | 37  | 20.64         | 62  | 2.12          | 87  | .41           |
| 13  | .03           | 38  | 5.24          | 63  | .03           | 88  | .37           |
| 14  | 2.89          | 39  | .45           | 64  | .01           | 89  | .40           |
| 15  | .74           | 40  | 3.36          | 65  | .01           | 90  | 1.52          |
| 16  | .01           | 41  | 1.08          | 66  | .72           | 91  | .00           |
| 17  | .33           | 42  | .15           | 67  | 2.47          | 92  | 4.13          |
| 18  | 1.31          | 43  | .01           | 68  | .87           | 93  | 3.49          |
| 19  | .03           | 44  | 2.05          | 69  | .51           | 94  | .09           |
| 20  | 3.48          | 45  | .75           | 70  | .85           | 95  | 5.91          |
| 21  | .81           | 46  | .95           | 71  | .00           | 96  | .04           |
| 22  | 10.03         | 47  | .57           | 72  | .99           | 97  | 6.00          |
| 23  | .26           | 48  | .71           | 73  | 1.53          | 98  | 2.57          |
| 24  | 2.12          | 49  | 1.10          | 74  | .77           | 99  | .72           |
| 25  | .00           | 50  | 2.17          | 75  | .02           | 100 | .07           |

WE2 Data Set  Weibull with k = 2  $\sqrt{\beta_1}$ = .63
$\beta_2$ = 3.25

| No. | Obser- vation | No. | Obser- vation | No. | Obser- vation | No. | Obser- vation |
|---|---|---|---|---|---|---|---|
| 1 | .74 | 26 | .49 | 51 | 1.23 | 76 | .79 |
| 2 | 1.15 | 27 | .34 | 52 | 1.14 | 77 | 1.08 |
| 3 | .92 | 28 | 1.17 | 53 | .68 | 78 | 1.81 |
| 4 | 1.41 | 29 | .79 | 54 | 1.05 | 79 | .93 |
| 5 | .16 | 30 | .30 | 55 | 1.09 | 80 | 1.34 |
| 6 | .68 | 31 | .65 | 56 | .38 | 81 | 1.18 |
| 7 | 2.11 | 32 | .56 | 57 | .92 | 82 | .18 |
| 8 | .97 | 33 | .85 | 58 | .47 | 83 | .72 |
| 9 | 1.20 | 34 | 1.11 | 59 | .18 | 84 | .42 |
| 10 | .33 | 35 | .95 | 60 | .75 | 85 | .61 |
| 11 | 1.14 | 36 | .51 | 61 | .56 | 86 | .69 |
| 12 | .78 | 37 | 2.13 | 62 | 1.21 | 87 | .80 |
| 13 | .40 | 38 | 1.51 | 63 | .43 | 88 | .78 |
| 14 | 1.30 | 39 | .82 | 64 | .32 | 89 | .79 |
| 15 | .93 | 40 | 1.35 | 65 | .27 | 90 | 1.11 |
| 16 | .33 | 41 | 1.02 | 66 | .92 | 91 | .20 |
| 17 | .76 | 42 | .62 | 67 | 1.25 | 92 | 1.43 |
| 18 | 1.07 | 43 | .34 | 68 | .97 | 93 | 1.37 |
| 19 | .42 | 44 | 1.20 | 69 | .85 | 94 | .54 |
| 20 | 1.37 | 45 | .93 | 70 | .96 | 95 | 1.56 |
| 21 | .95 | 46 | .99 | 71 | .09 | 96 | .45 |
| 22 | 1.78 | 47 | .87 | 72 | 1.00 | 97 | 1.56 |
| 23 | .71 | 48 | .92 | 93 | 1.11 | 98 | 1.27 |
| 24 | 1.21 | 49 | 1.02 | 74 | .94 | 99 | .92 |
| 25 | .14 | 50 | 1.21 | 75 | .36 | 100 | .52 |

SU(1, 2) Data Set          Johnson Unbounded (1, 2)                 $\sqrt{\beta_1} = .87$
                                                                   $\beta_2 = 5.59$

| No. | Observation | No. | Observation | No. | Observation | No. | Observation |
|-----|------------|-----|------------|-----|------------|-----|------------|
| 1   | -.41       | 26  | -.10       | 51  | -1.00      | 76  | -.47       |
| 2   | -.90       | 27  | .11        | 52  | -.89       | 77  | -.82       |
| 3   | -.63       | 28  | -.93       | 53  | -.34       | 78  | -1.88      |
| 4   | -1.25      | 29  | -.47       | 54  | -.78       | 79  | -.64       |
| 5   | .50        | 30  | .18        | 55  | -.83       | 80  | -1.15      |
| 6   | -.34       | 31  | -.30       | 56  | .05        | 81  | -.95       |
| 7   | -2.46      | 32  | -.19       | 57  | -.63       | 82  | .44        |
| 8   | -.68       | 33  | -.54       | 58  | -.07       | 83  | -.39       |
| 9   | -.97       | 34  | -.85       | 59  | .43        | 84  | -.01       |
| 10  | .13        | 35  | -.66       | 60  | -.42       | 85  | -.25       |
| 11  | -.90       | 36  | -.13       | 61  | -.20       | 86  | .35        |
| 12  | -.45       | 37  | -2.51      | 62  | -.98       | 87  | -.48       |
| 13  | .02        | 38  | -1.40      | 63  | -.02       | 88  | -.46       |
| 14  | -1.10      | 39  | -.50       | 64  | .16        | 89  | -.48       |
| 15  | -.63       | 40  | -1.17      | 65  | .24        | 90  | -.85       |
| 16  | .13        | 41  | -.74       | 66  | -.63       | 91  | .38        |
| 17  | -.43       | 42  | -.27       | 67  | -1.04      | 92  | -1.27      |
| 18  | -.81       | 43  | .11        | 68  | -.68       | 93  | -1.19      |
| 19  | -.01       | 44  | -.96       | 69  | -.54       | 94  | -.17       |
| 20  | -1.19      | 45  | -.64       | 70  | -.67       | 95  | -1.47      |
| 21  | -.66       | 46  | -.71       | 71  | .76        | 96  | -.06       |
| 22  | -1.83      | 47  | -.56       | 72  | -.72       | 97  | -1.48      |
| 23  | -.38       | 48  | -.62       | 73  | -.86       | 98  | -1.05      |
| 24  | -.98       | 49  | -.75       | 74  | -.64       | 99  | -.62       |
| 25  | .56        | 50  | -.98       | 75  | .09        | 100 | -.14       |

SU(0,3) Data Set          Johnson Unbounded (0,3)                    $\sqrt{\beta_1} = 0$
                                                                     $\beta_2 = 3.53$

| No. | Obser- vation | No. | Obser- vation | No. | Obser- vation | No. | Obser- vation |
|-----|-----|-----|-----|-----|-----|-----|-----|
| 1 | .06 | 26 | .27 | 51 | -.26 | 76 | .03 |
| 2 | -.21 | 27 | .42 | 52 | -.20 | 77 | -.16 |
| 3 | -.06 | 28 | -.22 | 53 | .11 | 78 | -.63 |
| 4 | -.38 | 29 | .03 | 54 | -.15 | 79 | -.07 |
| 5 | .70 | 30 | .47 | 55 | -.17 | 80 | -.33 |
| 6 | .11 | 31 | .14 | 56 | .37 | 81 | -.23 |
| 7 | -.83 | 32 | .21 | 57 | -.06 | 82 | .66 |
| 8 | -.09 | 33 | -.01 | 58 | .29 | 83 | .08 |
| 9 | -.24 | 34 | -.18 | 59 | .65 | 84 | .33 |
| 10 | .43 | 35 | -.08 | 60 | .06 | 85 | .17 |
| 11 | -.20 | 36 | .25 | 61 | .20 | 86 | .11 |
| 12 | .04 | 37 | -.85 | 62 | -.25 | 87 | .02 |
| 13 | .36 | 38 | -.44 | 63 | .33 | 88 | .04 |
| 14 | -.31 | 39 | .01 | 64 | .45 | 89 | .03 |
| 15 | -.06 | 40 | -.34 | 65 | .51 | 90 | -.18 |
| 16 | .43 | 41 | -.13 | 66 | -.06 | 91 | .61 |
| 17 | .05 | 42 | .15 | 67 | -.28 | 92 | -.38 |
| 18 | -.16 | 43 | .42 | 68 | -.09 | 93 | -.35 |
| 19 | .33 | 44 | -.24 | 69 | -.01 | 94 | .22 |
| 20 | -.35 | 45 | -.07 | 70 | -.09 | 95 | -.47 |
| 21 | -.08 | 46 | -.10 | 71 | .89 | 96 | .30 |
| 22 | -.61 | 47 | -.02 | 72 | -.11 | 97 | -.47 |
| 23 | .09 | 48 | -.06 | 73 | -.19 | 98 | -.28 |
| 24 | -.24 | 49 | -.13 | 74 | -.07 | 99 | -.06 |
| 25 | .75 | 50 | -.25 | 75 | .40 | 100 | .24 |

SU(0, 2)                Johnson Unbounded (0, 2)                $\sqrt{\beta_1} = 0$
$\beta_2 = 4.51$

| No. | Obser-vation | No. | Obser-vation | No. | Obser-vation | No. | Obser-vation |
|-----|------|-----|------|-----|------|-----|------|
| 1 | .10 | 26 | .41 | 51 | -.39 | 76 | .05 |
| 2 | -.31 | 27 | .65 | 52 | -.31 | 77 | -.25 |
| 3 | -.09 | 28 | -.33 | 53 | .17 | 78 | -1.01 |
| 4 | -.58 | 29 | .04 | 54 | -.22 | 79 | -.10 |
| 5 | 1.15 | 30 | .73 | 55 | -.26 | 80 | -.50 |
| 6 | .17 | 31 | .21 | 56 | .57 | 81 | -.35 |
| 7 | -1.39 | 32 | .32 | 57 | -.09 | 82 | 1.07 |
| 8 | -.14 | 33 | -.02 | 58 | .44 | 83 | .12 |
| 9 | -.37 | 34 | -.27 | 59 | 1.05 | 84 | .51 |
| 10 | .68 | 35 | -.12 | 60 | .09 | 85 | .25 |
| 11 | -.31 | 36 | .38 | 61 | .31 | 86 | .16 |
| 12 | .06 | 37 | -1.42 | 62 | -.37 | 87 | .03 |
| 13 | .55 | 38 | -.68 | 63 | .50 | 88 | .06 |
| 14 | -.47 | 39 | .01 | 64 | .70 | 89 | .04 |
| 15 | -.10 | 40 | -.52 | 65 | .80 | 90 | -.28 |
| 16 | .67 | 41 | -.19 | 66 | -.09 | 91 | .98 |
| 17 | .08 | 42 | .23 | 67 | -.42 | 92 | -.59 |
| 18 | -.24 | 43 | .65 | 68 | -.14 | 93 | -.53 |
| 19 | .51 | 44 | -.36 | 69 | -.01 | 94 | .34 |
| 20 | -.53 | 45 | -.10 | 70 | -.13 | 95 | -.73 |
| 21 | -.12 | 46 | -.16 | 71 | 1.51 | 96 | .46 |
| 22 | -.97 | 47 | -.04 | 72 | -.17 | 97 | -.74 |
| 23 | .13 | 48 | -.09 | 73 | -.28 | 98 | -.43 |
| 24 | -.37 | 49 | -.19 | 74 | -.11 | 99 | -.09 |
| 25 | 1.23 | 50 | -.38 | 75 | .62 | 100 | .37 |

SB(0, 2) Data Set          Johnson Bounded (0, 2)                $\sqrt{\beta_1} = 0$
                                                                  $\beta_2 = 2.63$

| No. | Obser-vation | No. | Obser-vation | No. | Obser-vation | No. | Obser-vation |
|-----|------|-----|------|-----|------|-----|------|
| 1 | .52 | 26 | .60 | 51 | .41 | 76 | .51 |
| 2 | .42 | 27 | .65 | 52 | .42 | 77 | .44 |
| 3 | .48 | 28 | .42 | 53 | .54 | 78 | .29 |
| 4 | .37 | 29 | .51 | 54 | .45 | 79 | .47 |
| 5 | .73 | 30 | .66 | 55 | .44 | 80 | .38 |
| 6 | .54 | 31 | .55 | 56 | .63 | 81 | .42 |
| 7 | .24 | 32 | .58 | 57 | .48 | 82 | .72 |
| 8 | .47 | 33 | .50 | 58 | .60 | 83 | .53 |
| 9 | .41 | 34 | .43 | 59 | .71 | 84 | .62 |
| 10 | .65 | 35 | .47 | 60 | .52 | 85 | .56 |
| 11 | .42 | 36 | .59 | 61 | .58 | 86 | .54 |
| 12 | .52 | 37 | .24 | 62 | .41 | 87 | .51 |
| 13 | .63 | 38 | .35 | 63 | .62 | 88 | .51 |
| 14 | .39 | 39 | .50 | 64 | .66 | 89 | .51 |
| 15 | .48 | 40 | .38 | 65 | .68 | 90 | .43 |
| 16 | .65 | 41 | .45 | 66 | .48 | 91 | .70 |
| 17 | .52 | 42 | .56 | 67 | .40 | 92 | .36 |
| 18 | .44 | 43 | .65 | 68 | .47 | 93 | .38 |
| 19 | .62 | 44 | .41 | 69 | .50 | 94 | .58 |
| 20 | .38 | 45 | .48 | 70 | .47 | 95 | .34 |
| 21 | .47 | 46 | .46 | 71 | .77 | 96 | .61 |
| 22 | .30 | 47 | .49 | 72 | .46 | 97 | .34 |
| 23 | .53 | 48 | .48 | 73 | .43 | 98 | .40 |
| 24 | .41 | 49 | .45 | 74 | .47 | 99 | .48 |
| 25 | .74 | 50 | .41 | 75 | .64 | 100 | .59 |

SB(0, .5)  Johnson Bounded (0, .5)  $\sqrt{\beta_1} = 0$
$\beta_2 = 1.63$

| No. | Obser-vation | No. | Obser-vation | No. | Obser-vation | No. | Obser-vation |
|---|---|---|---|---|---|---|---|
| 1 | .60 | 26 | .83 | 51 | .18 | 76 | .55 |
| 2 | .23 | 27 | .92 | 52 | .23 | 77 | .27 |
| 3 | .41 | 28 | .21 | 53 | .66 | 78 | .03 |
| 4 | .10 | 29 | .54 | 54 | .29 | 79 | .40 |
| 5 | .98 | 30 | .94 | 55 | .27 | 80 | .13 |
| 6 | .66 | 31 | .69 | 56 | .90 | 81 | .20 |
| 7 | .01 | 32 | .78 | 57 | .41 | 82 | .98 |
| 8 | .36 | 33 | .48 | 58 | .85 | 83 | .61 |
| 9 | .19 | 34 | .25 | 59 | .98 | 84 | .88 |
| 10 | .93 | 35 | .38 | 60 | .59 | 85 | .73 |
| 11 | .23 | 36 | .82 | 61 | .77 | 86 | .65 |
| 12 | .56 | 37 | .01 | 62 | .19 | 87 | .53 |
| 13 | .89 | 38 | .07 | 63 | .87 | 88 | .56 |
| 14 | .14 | 39 | .51 | 64 | .93 | 89 | .54 |
| 15 | .40 | 40 | .12 | 65 | .95 | 90 | .25 |
| 16 | .93 | 41 | .32 | 66 | .41 | 91 | .97 |
| 17 | .58 | 42 | .72 | 67 | .16 | 92 | .10 |
| 18 | .28 | 43 | .92 | 68 | .37 | 93 | .12 |
| 19 | .88 | 44 | .19 | 69 | .49 | 94 | .79 |
| 20 | .12 | 45 | .40 | 70 | .37 | 95 | .06 |
| 21 | .38 | 46 | .35 | 71 | .99 | 96 | .86 |
| 22 | .03 | 47 | .46 | 72 | .34 | 97 | .06 |
| 23 | .63 | 48 | .41 | 73 | .25 | 98 | .16 |
| 24 | .19 | 49 | .32 | 74 | .39 | 99 | .41 |
| 25 | .98 | 50 | .19 | 75 | .91 | 100 | .81 |

LCN(.05,3) Data Set    Contaminated Normal    $\sqrt{\beta_1} = .68$
$(p = .05, \mu = 3)$    $\beta_2 = 4.35$
$.95N(0,1) + .05N(3,1)$

| No. | Observation | No. | Observation | No. | Observation | No. | Observation |
|-----|-------------|-----|-------------|-----|-------------|-----|-------------|
| 1 | .19 | 26 | -.76 | 51 | .60 | 76 | 3.77 |
| 2 | -.19 | 27 | .33 | 52 | -.50 | 77 | -.46 |
| 3 | 1.96 | 28 | -.51 | 53 | .69 | 78 | 1.40 |
| 4 | -2.26 | 29 | -.18 | 54 | .28 | 79 | .13 |
| 5 | -.72 | 30 | 1.83 | 55 | 1.15 | 80 | -.08 |
| 6 | -.61 | 31 | .61 | 56 | -.24 | 81 | .71 |
| 7 | 1.05 | 32 | .97 | 57 | -.02 | 82 | 1.14 |
| 8 | -.19 | 33 | 1.47 | 58 | -1.31 | 83 | -1.21 |
| 9 | .16 | 34 | -.82 | 59 | 1.82 | 84 | -.32 |
| 10 | .98 | 35 | -.03 | 60 | .48 | 85 | .00 |
| 11 | -.24 | 36 | 2.41 | 61 | -1.52 | 86 | .58 |
| 12 | .26 | 37 | -.55 | 62 | .84 | 87 | -.24 |
| 13 | 2.07 | 38 | 1.17 | 63 | -.77 | 88 | -1.88 |
| 14 | 1.22 | 39 | -.49 | 64 | -1.17 | 89 | .44 |
| 15 | .09 | 40 | -.21 | 65 | -.65 | 90 | .27 |
| 16 | .41 | 41 | 2.31 | 66 | .79 | 91 | .66 |
| 17 | -.04 | 42 | .23 | 67 | -.97 | 92 | -.51 |
| 18 | -.24 | 43 | .50 | 68 | -.14 | 93 | -1.24 |
| 19 | -2.30 | 44 | .07 | 69 | .20 | 94 | -1.19 |
| 20 | .03 | 45 | .08 | 70 | .19 | 95 | .27 |
| 21 | -.38 | 46 | 1.74 | 71 | .20 | 96 | .13 |
| 22 | 1.23 | 47 | -1.02 | 72 | .50 | 97 | 1.68 |
| 23 | -.20 | 48 | -1.35 | 73 | -.13 | 98 | -.05 |
| 24 | -.07 | 49 | -1.36 | 74 | -2.25 | 99 | -.30 |
| 25 | -.38 | 50 | -.18 | 75 | 1.77 | 100 | -1.70 |

LCN(.10,3)

Contaminated Normal
(p = .10, $\mu$ = 3)
.90N(0,1) + .10N(3,1)

$\sqrt{\beta_1}$ = .8
$\beta_2$ = 4.02

| No. | Obser-vation | No. | Obser-vation | No. | Obser-vation | No. | Obser-vation |
|---|---|---|---|---|---|---|---|
| 1 | -.78 | 26 | .96 | 51 | -2.01 | 76 | .78 |
| 2 | -1.98 | 27 | -.81 | 52 | -1.12 | 77 | .24 |
| 3 | .57 | 28 | -.04 | 53 | .18 | 78 | -.40 |
| 4 | 2.10 | 29 | .88 | 54 | .31 | 79 | -1.42 |
| 5 | .36 | 30 | .07 | 55 | 3.54 | 80 | .37 |
| 6 | 2.33 | 31 | 3.78 | 56 | .00 | 81 | -.47 |
| 7 | .99 | 32 | 2.61 | 57 | 1.66 | 82 | -.09 |
| 8 | 1.23 | 33 | 3.93 | 58 | .74 | 83 | .11 |
| 9 | -1.05 | 34 | -.06 | 59 | .87 | 84 | -1.24 |
| 10 | .00 | 35 | 1.48 | 60 | -1.02 | 85 | 3.67 |
| 11 | -1.05 | 36 | -.92 | 61 | .67 | 86 | .67 |
| 12 | -.55 | 37 | .87 | 62 | 3.03 | 87 | .70 |
| 13 | -.26 | 38 | -2.60 | 63 | 3.42 | 88 | -1.43 |
| 14 | .09 | 39 | .28 | 64 | -1.95 | 89 | -1.70 |
| 15 | .08 | 40 | .26 | 65 | .83 | 90 | .33 |
| 16 | -.81 | 41 | -.13 | 66 | -.67 | 91 | .44 |
| 17 | 3.30 | 42 | -.69 | 67 | -1.71 | 92 | -1.00 |
| 18 | -1.21 | 43 | -2.13 | 68 | -.10 | 93 | .49 |
| 19 | -1.31 | 44 | -.27 | 69 | 3.54 | 94 | -.10 |
| 20 | .07 | 45 | .49 | 70 | .26 | 95 | -.10 |
| 21 | .71 | 46 | .48 | 71 | .32 | 96 | 1.84 |
| 22 | .89 | 47 | .02 | 72 | .04 | 97 | -.99 |
| 23 | -1.54 | 48 | .69 | 73 | .16 | 98 | .15 |
| 24 | 2.45 | 49 | -.25 | 74 | -.51 | 99 | 1.67 |
| 25 | -.91 | 50 | -.84 | 75 | .52 | 100 | 1.30 |

LCN(.20,3) Data Set          Contaminated Normal          $\sqrt{\beta_1}$ = .68
                                    (p = .20, $\mu$ = 3)                    $\beta_2$ = 3.09
                              .80N(0,1) + .20N(3,1)

| No. | Observation | No. | Observation | No. | Observation | No. | Observation |
|-----|-------------|-----|-------------|-----|-------------|-----|-------------|
| 1   | .09         | 26  | -.95        | 51  | 1.42        | 76  | 4.12        |
| 2   | -.24        | 27  | 2.03        | 52  | 1.17        | 77  | 3.57        |
| 3   | 1.44        | 28  | -.87        | 53  | 2.00        | 78  | .13         |
| 4   | 1.25        | 29  | -1.61       | 54  | .26         | 79  | -1.55       |
| 5   | 2.24        | 30  | 2.93        | 55  | -1.87       | 80  | -.49        |
| 6   | .16         | 31  | -1.14       | 56  | .80         | 81  | .47         |
| 7   | .05         | 32  | -1.05       | 57  | 3.45        | 82  | 1.08        |
| 8   | -.62        | 33  | -.32        | 58  | -.29        | 83  | -.59        |
| 9   | 3.44        | 34  | 1.12        | 59  | -.27        | 84  | .64         |
| 10  | -.04        | 35  | -.25        | 60  | 1.29        | 85  | 1.49        |
| 11  | -.41        | 36  | 1.12        | 61  | -.04        | 86  | 2.39        |
| 12  | 1.49        | 37  | 1.31        | 62  | .78         | 87  | -.43        |
| 13  | -.87        | 38  | -.31        | 63  | .62         | 88  | .46         |
| 14  | -2.61       | 39  | -1.17       | 64  | .79         | 87  | -.68        |
| 15  | .08         | 40  | .05         | 65  | .43         | 90  | -.08        |
| 16  | -.83        | 41  | -.02        | 66  | 2.83        | 91  | -.16        |
| 17  | .32         | 42  | 3.86        | 67  | .69         | 92  | 4.80        |
| 18  | .43         | 43  | .09         | 68  | .55         | 93  | 1.71        |
| 19  | 2.86        | 44  | -.34        | 69  | .35         | 94  | 3.59        |
| 20  | 1.40        | 45  | -1.05       | 70  | 1.78        | 95  | 2.25        |
| 21  | -.08        | 46  | -2.20       | 71  | .02         | 96  | -.71        |
| 22  | 2.03        | 47  | -.22        | 72  | 2.01        | 97  | -.53        |
| 23  | .31         | 48  | 2.49        | 73  | 3.86        | 98  | -1.82       |
| 24  | 4.58        | 49  | -1.41       | 74  | .06         | 99  | -.03        |
| 25  | .07         | 50  | .49         | 75  | -.54        | 100 | 5.04        |

SB(1,1) Data Set              Johnson Bounded (1,1)              $\sqrt{\beta_1}$ = .73
$\beta_2$ = 2.91

| No. | Observation | No. | Observation | No. | Observation | No. | Observation |
|-----|-------------|-----|-------------|-----|-------------|-----|-------------|
| 1 | .31 | 26 | .45 | 51 | .15 | 76 | .29 |
| 2 | .17 | 27 | .55 | 52 | .17 | 77 | .18 |
| 3 | .23 | 28 | .16 | 53 | .34 | 78 | .06 |
| 4 | .11 | 29 | .29 | 54 | .19 | 79 | .23 |
| 5 | .72 | 30 | .59 | 55 | .18 | 80 | .12 |
| 6 | .34 | 31 | .36 | 56 | .52 | 81 | .16 |
| 7 | .04 | 32 | .41 | 57 | .23 | 82 | .70 |
| 8 | .22 | 33 | .26 | 58 | .46 | 83 | .32 |
| 9 | .15 | 34 | .18 | 59 | .70 | 84 | .49 |
| 10 | .57 | 35 | .22 | 60 | .31 | 85 | .38 |
| 11 | .17 | 36 | .44 | 61 | .40 | 86 | .34 |
| 12 | .29 | 37 | .04 | 62 | .15 | 87 | .28 |
| 13 | .51 | 38 | .09 | 63 | .49 | 88 | .29 |
| 14 | .13 | 39 | .27 | 64 | .58 | 89 | .29 |
| 15 | .23 | 40 | .12 | 65 | .61 | 90 | .18 |
| 16 | .56 | 41 | .20 | 66 | .23 | 91 | .68 |
| 17 | .30 | 42 | .37 | 67 | .14 | 92 | .11 |
| 18 | .19 | 43 | .56 | 68 | .22 | 93 | .12 |
| 19 | .49 | 44 | .15 | 69 | .26 | 94 | .42 |
| 20 | .12 | 45 | .23 | 70 | .22 | 95 | .09 |
| 21 | .22 | 46 | .21 | 71 | .80 | 96 | .47 |
| 22 | .06 | 47 | .25 | 72 | .21 | 97 | .09 |
| 23 | .32 | 48 | .24 | 73 | .17 | 98 | .14 |
| 24 | .15 | 49 | .20 | 74 | .23 | 99 | .24 |
| 25 | .75 | 50 | .15 | 75 | .54 | 100 | .43 |

SB(1, 2) Data Date          Johnson Bounded (1, 2)              $\sqrt{\beta_1}$ = .28
                                                                $\beta_2$ = 2.77

| No. | Obser-vation | No. | Obser-vation | No. | Obser-vation | No. | Obser-vation |
|-----|------|-----|------|-----|------|-----|------|
| 1 | .35 | 26 | .39 | 51 | .41 | 76 | .49 |
| 2 | .26 | 27 | .46 | 52 | .42 | 77 | .43 |
| 3 | .29 | 28 | .46 | 53 | .33 | 78 | .44 |
| 4 | .21 | 29 | .43 | 54 | .29 | 79 | .17 |
| 5 | .36 | 30 | .31 | 55 | .27 | 80 | .57 |
| 6 | .24 | 31 | .37 | 56 | .22 | 81 | .37 |
| 7 | .46 | 32 | .39 | 57 | .15 | 82 | .37 |
| 8 | .42 | 33 | .68 | 58 | .38 | 83 | .34 |
| 9 | .21 | 34 | .61 | 59 | .55 | 84 | .25 |
| 10 | .43 | 35 | .51 | 60 | .22 | 85 | .34 |
| 11 | .42 | 36 | .37 | 61 | .25 | 86 | .28 |
| 12 | .42 | 37 | .23 | 62 | .21 | 87 | .45 |
| 13 | .26 | 38 | .27 | 63 | .32 | 88 | .39 |
| 14 | .22 | 39 | .32 | 64 | .37 | 89 | .47 |
| 15 | .29 | 40 | .33 | 65 | .53 | 90 | .45 |
| 16 | .46 | 41 | .29 | 66 | .29 | 91 | .33 |
| 17 | .39 | 42 | .34 | 67 | .30 | 92 | .44 |
| 18 | .38 | 43 | .30 | 68 | .41 | 93 | .65 |
| 19 | .29 | 44 | .24 | 69 | .32 | 94 | .43 |
| 20 | .45 | 45 | .30 | 70 | .37 | 95 | .26 |
| 21 | .37 | 46 | .42 | 71 | .22 | 96 | .50 |
| 22 | .31 | 47 | .44 | 72 | .24 | 97 | .44 |
| 23 | .31 | 48 | .33 | 73 | .24 | 98 | .44 |
| 24 | .27 | 49 | .45 | 74 | .43 | 99 | .53 |
| 25 | .38 | 50 | .33 | 75 | .26 | 100 | .68 |

SB(.535,.5) Data Set        Johnson Bounded (.535,.5)        $\sqrt{\beta_1}$ = .65
                                                             $\beta_2$ = 2.13

| No. | Observation | No. | Observation | No. | Observation | No. | Observation |
|-----|-------------|-----|-------------|-----|-------------|-----|-------------|
| 1 | .07 | 26 | .00 | 51 | .02 | 76 | .03 |
| 2 | .03 | 27 | .01 | 52 | .02 | 77 | .15 |
| 3 | .45 | 28 | .05 | 53 | .62 | 78 | .05 |
| 4 | .80 | 29 | .49 | 54 | .05 | 79 | .41 |
| 5 | .14 | 30 | .32 | 55 | .82 | 80 | .15 |
| 6 | .02 | 31 | .02 | 56 | .91 | 81 | .25 |
| 7 | .16 | 32 | .24 | 57 | .39 | 82 | .39 |
| 8 | .22 | 33 | .01 | 58 | .04 | 83 | .11 |
| 9 | .12 | 34 | .00 | 59 | .14 | 84 | .08 |
| 10 | .56 | 35 | .30 | 60 | .44 | 85 | .43 |
| 11 | .41 | 36 | .07 | 61 | .69 | 86 | .55 |
| 12 | .03 | 37 | .20 | 62 | .01 | 87 | .30 |
| 13 | .93 | 38 | .33 | 63 | .71 | 88 | .42 |
| 14 | .56 | 39 | .23 | 64 | .36 | 89 | .93 |
| 15 | .66 | 40 | .55 | 65 | .01 | 90 | .95 |
| 16 | .03 | 41 | .04 | 66 | .78 | 91 | .04 |
| 17 | .23 | 42 | .10 | 67 | .81 | 92 | .05 |
| 18 | .57 | 43 | .15 | 68 | .22 | 93 | .30 |
| 19 | .07 | 44 | .42 | 69 | .49 | 94 | .63 |
| 20 | .39 | 45 | .26 | 70 | .02 | 95 | .72 |
| 21 | .41 | 46 | .06 | 71 | .03 | 96 | .98 |
| 22 | .08 | 47 | .52 | 72 | .06 | 97 | .15 |
| 23 | .56 | 48 | .25 | 73 | .01 | 98 | .09 |
| 24 | .07 | 49 | .22 | 74 | .25 | 99 | .09 |
| 25 | .45 | 50 | .15 | 75 | .36 | 100 | .08 |

4. Real Data Sets

There are five data sets which are given here. These can be used to illustrate and compare various techniques. Some are used in the text.

BLIS Data Set    Data Set for Two Independent Normal Populations

Body weight in grams of white Leghorn chicks at 21 days; from two laboratories in a collaborative vitamin D assay (Bliss's data)

| Series A | | Series B | |
|---|---|---|---|
| Weight | Frequency | Weight | Frequency |
| 156 | 1 | 130 | 1 |
| 162 | 1 | 147 | 1 |
| 168 | 1 | 155 | 1 |
| 182 | 1 | 156 | 1 |
| 186 | 1 | 167 | 1 |
| 190 | 2 | 177 | 1 |
| 196 | 1 | 179 | 1 |
| 202 | 1 | 183 | 1 |
| 210 | 1 | 187 | 2 |
| 214 | 1 | 193 | 1 |
| 220 | 1 | 195 | 1 |
| 226 | 1 | 196 | 1 |
| 230 | 2 | 199 | 1 |
| 236 | 2 | 203 | 1 |
| 242 | 1 | 208 | 1 |
| 246 | 1 | 225 | 1 |
| 270 | 1 | 231 | 1 |
|  | 20 | 232 | 1 |
|  |  | 236 | 1 |
|  |  | 246 | 1 |
|  |  |  | 21 |

Source: Bliss, C. I. (1946). Collaborative comparison of three ratios for the chick assay of vitamin D. J. Assoc. Off. Agr. Chem. 29: 396-408, given in Bliss, C. I. (1967). Statistics in Biology, vol. 1, New York: McGraw-Hill, p. 108.

CHEN Data Set    Data Set for Lognormal Population

Lethal dose of the drug cinobufagin in 10 (mg/kg), as determined by titration to cardiac arrest in individual etherized cats (Chen et al., 1931)

| Dose | log Dose | f | Dose | log Dose | f |
|------|----------|---|------|----------|---|
| 1.26 | 0.100 | 1 | 2.34 | 0.369 | 1 |
| 1.37 | 0.137 | 1 | 2.41 | 0.382 | 1 |
| 1.55 | 0.190 | 1 | 2.56 | 0.408 | 1 |
| 1.71 | 0.233 | 1 | 2.63 | 0.420 | 2 |
| 1.77 | 0.248 | 1 | 2.67 | 0.427 | 1 |
| 1.81 | 0.258 | 1 | 2.82 | 0.450 | 2 |
| 1.89 | 0.276 | 2 | 2.84 | 0.453 | 1 |
| 1.98 | 0.297 | 1 | 2.99 | 0.476 | 1 |
| 2.03 | 0.308 | 3 | 3.65 | 0.562 | 1 |
| 2.07 | 0.316 | 1 | 3.83 | 0.583 | 1 |
|      |       |   |      |          | 25 |

Source: Chen, K. K., H. Jensen, and A. L. Chen (1931). The pharmacological action of the principles isolated from ch'an su, the dried venom of the Chinese toad. J. Pharmacol. Expt. Therap. 43, 13-50. Given in Bliss, C. I. (1967). Statistics in Biology, vol. 1, New York: McGraw-Hill, p. 114.

BLAC Data Set    Data Set for Testing for Poisson Model

Density of Eryngium Maritimum (Blackman's data)

| Number of plants per quadrat square | Frequency |
|---|---|
| 0 | 16 |
| 1 | 41 |
| 2 | 49 |
| 3 | 20 |
| 4 | 14 |
| 5 | 5 |
| 6 | 1 |
| 7 | 1 |
| 8 and over | 0 |
|  | 147 |

Source: Blackman, G. E. (1935). A study by statistical methods of the distribution of species in grassland associations. Ann. Bot. 49, 749-777. Given in K. Mather (1966). Statistical Analysis in Biology. London: Methuen, p. 37.

EMEA Data Set      Data Set for Testing for Normality

### Distribution of the Heights of Maize Plants (in decimeters)
### (Emerson and East's data)

| Height of plants in dms (class center) | Frequency |
|:---:|:---:|
| 7 | 1 |
| 8 | 3 |
| 9 | 4 |
| 10 | 12 |
| 11 | 25 |
| 12 | 49 |
| 13 | 68 |
| 14 | 95 |
| 15 | 96 |
| 16 | 78 |
| 17 | 53 |
| 18 | 26 |
| 19 | 16 |
| 20 | 3 |
| 21 | 1 |
| | 530 |

$\bar{X} = 14.5396$

$\dfrac{\Sigma(X - \bar{X})^2}{N - 1} = 4.9936$

$4.9936 - .0833$ (Sheppard's correction) $= 4.9103$

$S = \sqrt{4.9103} = 2.2159$

Source: Emerson, K. A., and E. M. East (1913). Inheritance of quantitative characters in maize, <u>Neb. Exp. Stat. Res. Bull. 2</u>. Given in K. Mather (1966). <u>Statistical Analysis in Biology</u>. London: Methuen, p. 29.

BAEN Data Set    Data Set for Testing for Double Exponential (Laplace)

Differences in flood stages for two stations on the Fox River in Wisconsin
(Bain and Engelhardt's data)

1.97, 1.96, 3.60, 3.80, 4.79, 5.66, 5.76, 5.78, 6.27, 6.30, 6.78, 7.65,
7.84, 7.99, 8.51, 9.18, 10.13, 10.24, 10.25, 10.43, 11.45, 11.48, 11.75,
11.81, 12.34, 12.78, 13.06, 13.29, 13.98, 14.18, 14.40, 16.22, 17.06

Source: Bain, L. J., and M. Engelhardt (1973). Interval estimation for the
two-parametric double exponential distribution. Technometrics 15, 875-
887, p. 885.

# Index

Printed in the United States
by Baker & Taylor Publisher Services

Printed in the United States
by Baker & Taylor Publisher Services